World Sustainability Series

Series Editor

Walter Leal Filho, European School of Sustainability Science
and Research, Research and Transfer Centre "Sustainable Development
and Climate Change Management", Hamburg University of Applied Sciences,
Hamburg, Germany

Due to its scope and nature, sustainable development is a matter which is very interdisciplinary, and draws from knowledge and inputs from the social sciences and environmental sciences on the one hand, but also from physical sciences and arts on the other. As such, there is a perceived need to foster integrative approaches, whereby the combination of inputs from various fields may contribute to a better understanding of what sustainability is, and means to people. But despite the need for and the relevance of integrative approaches towards sustainable development, there is a paucity of literature which address matters related to sustainability in an integrated way.

Notes on the quality assurance and peer review of this publication

Prior to publication, the works published in this book are initially assessed and reviewed by an in-house editor. If suitable for publication, manuscripts are sent for further review, which includes a combined effort by the editorial board and appointed subject experts, who provide independent peer-review. The feedback obtained in this way was communicated to authors, and with manuscripts checked upon return before finally accepted. The peer-reviewed nature of the books in the "World Sustainability Series" means that contributions to them have, over many years, been officially accepted for tenure and promotion purposes.

Walter Leal Filho · Laís Viera Trevisan ·
Gabriel Brito Costa · Ismar Borges de Lima
Editors

Amazon 2030—Sustainability Issues in the World's Largest Rainforest Region

Springer

Editors
Walter Leal Filho
Faculty of Life Sciences
Hamburg University of Applied Sciences
Hamburg, Germany

Laís Viera Trevisan
School of Administration
Federal University of Rio Grande do Sul (UFRGS)
Porto Alegre, Brazil

Gabriel Brito Costa
Institute of Biodiversity and Forests
Federal University of West Pará
Santarém, Brazil

Ismar Borges de Lima
Roraima State University
Boa Vista, Roraima, Brazil

ISSN 2199-7373 ISSN 2199-7381 (electronic)
World Sustainability Series
ISBN 978-3-031-81467-9 ISBN 978-3-031-81465-5 (eBook)
https://doi.org/10.1007/978-3-031-81465-5

This Springer imprint is published by the registered company Springer Nature Switzerland AG
The registered company address is: Gewerbestrasse 11, 6330 Cham, Switzerland

Contents

The Yanomami Crisis, Ethno-Ecocide Practices and the 2030 Agenda: Challenges to Territorialize the SDGs in the Amazon

Yara Resende Marangoni Martinelli and Thiago Gehre Galvão

Abstract The violence systematically perpetuated against Indigenous peoples and their lands, mark the 2030 Agenda territorialization context in Brazil and other Amazonian countries. The Indigenous Peoples rights includes public policies related to territory, health, education, and to traditional cultures. Indigenous languages and cosmologies have a profound impact on countries' ability to achieve Sustainable Development Goals, in particular SDGs 1/2/3/5/13/15. The Brazilian Indigenist policy has been historically violent against Indigenous Peoples and Local Communities and their cosmopraxes. The indigenist policy implemented between 2019 and 2022 negatively captured international community's attention, for its practices and discourses' serious impacts. This chapter looks at the relation between necropolicies, that caused thousands of deaths and deprived lives of dignity, and its connection to the 2030 Agenda territorialization in Brazil. The Yanomami Case (YC), evidenced in 2023, is presented as a study case for understanding the dynamics that connect ethno-ecocidal policies with illegal mining and invasion of in Indigenous Lands (ILs), and the weakening of inspection/regulatory bodies that protect Indigenous rights, their territories, and the environment.

Keywords Indigenous rights · Sustainable development · 2030 agenda · Amazon · Ethnocide · Ecocide

1 Introduction

In 2023 a public health emergency on Yanomami Indigenous Land (IL) was triggered and achieved international repercussions. The responsiveness of the newly elected president was remarkable. Lula da Silva signed a decree to face the health

---

Y. R. M. Martinelli (✉) · T. G. Galvão
International Relations Institute, University of Brasília, Brasília, Brazil
e-mail: yarammartinelli@gmail.com

Campus Universitário Darcy Ribeiro-Asa Norte, Brasília, DF 70910-900, Brazil

W. Leal Filho et al. (eds.), *Amazon 2030—Sustainability Issues in the World's Largest Rainforest Region*, World Sustainability Series,
https://doi.org/10.1007/978-3-031-81465-5_1

and food insecuriFty situation of Yanomami people and visited the Roraima Indigenous Health Center (Rodrigues et al. 2023), establishing the Yanomami Emergency Operations Center (COE) (Brazilian Health Ministry 2023). The international visibility of the Yanomami Case (YC) pushed Lula's administration to conduct several policies and strategies, weekly reported on the severity and scope of malnutrition, mercury contamination and malaria cases, respiratory syndromes, as well as the violence suffered, especially by women and children, with the presence of illegal invaders, mostly illegal miners (Putini 2023).

The singularity of the Yanomami Case, a multidimensional inequality issue, crossed by many forms of structural violence, raises the debate on the limits and possibilities of global agendas. More specifically, it allows understanding the latitude in which the 2030 Agenda and its 17 Sustainable Development Goals are in fact achievable. This chapter focuses on the Yanomami Case's constitutive elements that affect directly and indirectly the accomplishment of the 169 targets defined for the SDGs. In this chapter is presented the argument that a twin practice of ethnocide and ecocide deteriorates the Brazilian capabilities to sustain the 2030 Agenda commitment.

Methodologically, the research design beholds on four elements of qualitative content analysis (QCA). First, the interpretation of primary and secondary sources related to the Yanomami case and related to the SDGs; second, a situational approach to focus on and evaluate official SDGs documents from Indigenous issues perspective; and third a case-oriented study on the Yanomami's triggered crisis. This method aims to shed light on how Indigenous rights guarantees are a crucial variable to understand the political and cultural range of the 2030 Agenda and how deliberate practices of violence are weakening Brazilian possibilities to territorialize the SDGs in a singular context such as the Amazon.

The theoretical framework considers a 'pluriversal' existence as a path to overcome the imposition of Western scientific universality, considering Indigenous cosmopraxes as co-creators of the reality that constitutes the world (Inoue et al. 2020). A relational ontology perspective is the starting point for this reflection, considering the constant transformation that arises from intersubjective encounters, opening paths for thinking sustainability and development from cosmopraxes diversity (Querejazu 2022) and emancipatory possibilities in a pluriverse of epistemologies and possibilities for 'our common world' construction.

2 Understanding the Yanomami Case

In this section, the contextual and conceptual frame to understand the Yanomami Case is delineated. A brief presentation of the Yanomami people's culture and history is presented, focusing on dimensions related to the humanitarian crisis that undermined Brazilian efforts to accomplish the 2030 Agenda. It is important to highlight that this brief presentation is not intended to cover the complexity and historical-cultural diversity and cosmopraxes of the Yanomami people, but only build the context on which the interpretation and situational perspective makes sense to evaluate a crisis arising in a political moment of expectations on the political transition.

2.1 The Yanomami People and the Emergence of a Humanitarian Crisis

The Yanomami Peoples are located in the interfluve region between the *Orinoco* and Amazon rivers on the Brazil-Venezuela borders. They are considered a people of recent contact with non-Indigenous society, and constitute a cultural and linguistic group composed of at least four subgroups that speak the same family languages (*Yanomae, Yanõmami, Sanima and Ninam*) (ISA 2023). They are descendants of an Indigenous group that has remained relatively isolated since ancient times and occupied the area at the headwaters of the *Orinoco* and *Parima* rivers for around a millennium (ISA 2023). For the Yanomamis, the word '*urihi*' designates the forest and its ground, and '*yanomae thëpë urihipë*' means "the forest of human beings", or the forest that their ancestral deity '*Omama*' gave to the Yanomami People to live through generations (ISA 2023). In Yanomami cosmopraxis, '*urihi*', the forestland, is not an inert source of resources, but a living entity that creates life itself (Kopenawa 2009).

The Yanomami people have long suffered the impacts of developmentalism. In the 1970s and 1980s, the Brazilian military dictatorship development projects subjected the Yanomami to massive contact with the regional economic frontier, through roads' construction, large projects implementation, farms creation, sawmills, construction sites and mines. These contacts caused an epidemiological shock, which caused countless deaths and massive social disruption (ISA 2023). The 1970s were marked by the military government's "National Integration Plan", with the opening of the Perimetral Norte road and 'public colonisation programs' that invaded the southeast of Yanomami lands. In 1975, the Amazon resources survey project named RADAM detected the existence of mineral deposits in the region. It was published nationally as a mineral potential, which triggered a progressive movement of mining invasion and opening of clandestine mining tracks (ISA 2023).

Also at the Military Dictatorship, various diseases were fatal to Yanomami villages, brought especially by the presence of invaders and people hired to work on roads and construction sites, such as malaria, measles, sexually transmitted diseases, whooping cough and malnutrition. The report by the National Truth Commission, released in 2014, also denounced the Brazilian State responsibility for violence in Indigenous Lands. Yet, the use of the term genocide to classify the violence committed against the Yanomami people has not been certified by national and international courts, despite mentions to the 'risk of genocide' (ADPF 709, UN 2021). Historically, the only case of genocide crime recognized by Brazilian justice was called "Haximu Massacre", the murder of 16 Yanomamis by miners in 1993 (Souza 2023).

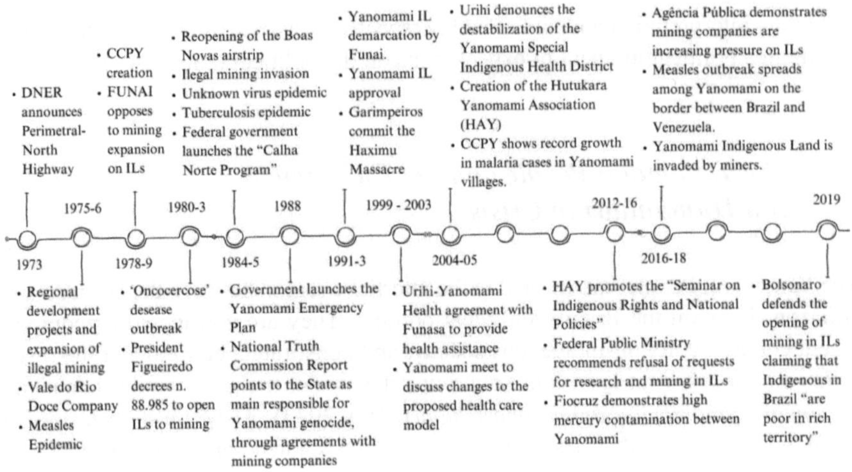

Fig. 1 Yanomami indigenous land conflict timeline. *Source* Made by the authors based on 'Conflict Map' (Fiocruz 2023)

Between the 2000 and 2010 s, the national and international political articulation of the Yanomami people was strengthened, including by the use of United Nations mechanisms to denounce historical violences and the continuity of development projects in Yanomami IL, even during the progressive governments of the Workers' Party. The xamã and activist Davi Kopenawa Yanomami and his son, Dário Vitório Kopenawa Yanomami, became great leaderships, guiding the creation of the Yanomami Park Commission (CCPY) and the Hutukara Yanomami Association (HAY) (Souza 2023).

Threats to the Yanomami People are historically linked to illegal mining, reflected in bills and government decisions. The work of then senator Romero Jucá, from Roraima state, must be highlighted in this process, for the presidency of the National Indian Foundation (FUNAI) and the proposition of the Bill 1610/1996, on the exploration of mineral resources in Indigenous Lands (Souza 2023). To address the complexity of this history of violence, and highlight some of its milestones, the authors prepared a timeline based on research carried out by Fiocruz, through the 'Conflict Map' initiative (Fiocruz 2023). The historical delimitations of the time line follow the proposal of the Fiocruz study, and in Fig. 1 it ends in 2019. This option was made by the authors, as details of the Indigenous policy aimed at the Yanomami people will be presented in the body of the text in the following sections.

In the entire historical process since the invasion, colonialist logic of development established a pattern of exploitation of the region's natural resources. The Roraima state, as the hole Amazon Forest, still lives on the margins of a supposed 'capitalist prosperity', from the rubber economic cycle to the agricultural frontiers' expansion, livestock farming and monocultures, the region became a peripheral supply zone of commodities for the world market (Cararões and Flames 2019).

2.2 *Historical Context*

The crisis that took place in the Yanomami Land is the consequence of a State-centric perspective that generates ethnocide and ecocide practices, in years of uninterrupted violence against Indigenous peoples. The symbiotic relation between state and private corporations into the world system nourishes a machinery of multiple violences against marginalised groups, for the sake of profit and developmentalism (Longhini 2021). The expropriation of Indigenous Lands began at the time of the colonial invasion and this violence continues to be practised. The national myth on the 'discovery of Brazil', is in fact a history of invasion, exploitation, genocide and erasure. A racist project based on the assumption of a 'land without owner' (Kayapó and Brito 2015) in which invaders have the right to exploit.

The Brazilian State designed a strategy to forcefully integrate these various Indigenous communities by tactical political moves, from the prohibition of their mother tongue usage and cultural practices, physical punishments and other repressive actions against Indigenous traditions (Gonçalves 2017, p. 5). Indigenous Peoples historical 'othernization' in their own traditional territories, inhabited for more than 10,000 years ago, is a mark of colonial and racist legacies. Indigenous history and struggles invisibilization and the privilege of Eurocentric historical narratives in the Brazilian education system creates the foundations of the considered "scientific" narratives on the history of Brazil (Picq 2017, p. 343/4).

The Jair Bolsonaro Presidency, between 2019 and 2022, is recognised as an administration unfavourable to Indigenous people, that stated categorically it would not demarcate 'one centimetre more' on Indigenous lands (Observatório do Agronegócio 2018). This administration reinforced violent patterns and the situation severely worsened, especially to Indigenous Peoples and other Local communities, as *Quilombolas* Peoples.[1] During this period, important public policies were interrupted, including those aimed to combat organised crime, the international trafficking and the illegal chains of extraction and commercialization of wood and minerals at the Amazon Forest (Observatório da Mineração 2022). This fuelled the long historical process suffered by the Yanomami people, in a growing process of escalating violence, which culminated in the humanitarian crisis of 2023.

2.3 *Relevance of the Yanomami Case*

First the Yanomami Case (YC) is paradigmatic in contesting the motto of the 2030 Agenda "leave no one behind" (LNOB). Once the 2030 Agenda is a state-centric global project that places the Nation-State abstraction in a concrete place of privilege, it would be almost paradoxical to make real advancements by 'leaving Indigenous

[1] Quilombolas and Indigenous peoples have had specific legislation that protects their territories and cultures for many years in Brazil, but they have been systematically ignored and disrespected by governments and rural producers.

people behind'. It could be argued that the failure of SDG implementation so far (SDG Report 2023) is path dependent to colonialism and their structural manifestations. Indigenous cosmopraxis and struggles for Indigenous rights seem to have been silenced in SDG values and principles, as Indigenous cosmopraxis have been systematically ignored in global agendas. For instance, the Free, Prior and Informed Consent,[2] that was excluded from the SDGs, even with the Indigenous struggle to insert it (WACC 2019).

Second, the YC is a direct consequence of the 'cannibal forces of capitalism' and the global imposition of a development model made at the expense of 'epistemicides' (Santos 2010). Capitalism, in that sense, deteriorates the aspirational politics of the 2030 Agenda by inducing the 'cannibalization of land and natural resources' (Fraser 2022). The last five years have been marked by an exponential expansion of illegal mining on Indigenous lands aligned with models of extraction and management of predatory natural resources (Observatório da Mineração 2022).

Third, the YC is a multidimensional crisis, permeated by multiple forms of violence, that are pushing the SDGs implementation backward. Indigenous peoples and Local Communities live under a Nation-State grounded on economic and political elites that rules on territorial spoliation and colonialist domination practices. Ethnic-racial prejudice is transversal to this relationship, especially in Brazil where Indigenous peoples belong to different races, ethnicities and cultures from the white world order (Vitalis 2015), often with different mother tongues than the 'national language', imported from the colonisers and used in the State institutional bureaucracies and the educational system.

Fourth, the YC highlights a twin concept that defines two tracks of violences against peoples and nature. 'Ethnocide' is a systematic cultural erasure practice, largely through the use of violent mechanisms, and the imposition of a hegemonic narrative validated by the State apparatus itself (Pruim 2014). This invisibilization is so profound in Brazil that only in 1991 the term 'Indigenous' was included in official censuses. Before that, the category was not even recognized by the State as part of Brazil's ethnic-racial identity. 'Ecocide' is understood as a long-term intentional human action that imposes an extensive damage and destruction of ecosystems, altering life quality and beings' survival capacity (Higgins 2015, p. 11).

Fifth, the YC demonstrates the tendency of silencing Indigenous people's voices and violating their rights. The seriousness of the Yanomami humanitarian crisis have been warned in 2020 by the Federal University of Minas Gerais (UFMG) and the Socio-Environmental Institute (ISA), that released a report pointing a "risk of genocide with the complicity of the Brazilian state" regarding illegal mining and Covid-19. Still in 2020, the Complaint of Noncompliance with Fundamental Precept (ADPF) 709, presented by the Indigenous Peoples of Brazil Association (APIB) to

[2] Free, Prior, and Informed Consent (FPIC) is a specific right granted to Indigenous Peoples recognised in the UN Declaration on the Rights of Indigenous Peoples (UNDRIP), which aligns with their universal right to self-determination. FPIC allows Indigenous Peoples to provide or withhold/withdraw consent, at any point, regarding projects impacting their territories. FPIC allows Indigenous Peoples to engage in negotiations to shape the design, implementation, monitoring, and evaluation of projects. See: https://www.fao.org/indigenous-peoples/en/.

the Brazilian Federal Supreme Court, pointed to a "ongoing genocide" against the Yanomami people (APIB 2020a). Between 2021 and 2022, the Hutukara Yanomami Association sent 37 letters denouncing the abuses and violence committed, and didn't obtain any response to the violations reported. Furthermore, the Supreme Court recommended the formulation of territorial protection and health care measures, but the request was ignored by the previous administration (HAY et al. 2022).

Sixth, the YC unveiled a shaming policy strategy, through access to international forums for denouncing the precarious situations faced. At the 20th Session of the United Nations Permanent Forum on Indigenous Affairs, in 2021, Brazilian Indigenous leaders denounced the President for the crime of genocide. Still in 2021, during the UN Human Rights Council, the UN High Commissioner for Human Rights, Michelle Bachelet, stated her concern: *"I am alarmed by the recent attacks on members of the Yanomami and Munduruku peoples by illegal miners in the Amazon. Attempts to legalise the entry of companies into Indigenous territories and limit the demarcation of Indigenous lands (…) are also a cause for serious concern"* (OHCHR 2021).

These six points demonstrate that the humanitarian crisis experienced by the Yanomami people unveil challenges and limitations on SDGs implementation in the Amazon context. Illegal invasion and occupation of the territory by illegal miners, the enticement of young people, especially women, the contamination of rivers and fishery by mercury (used in gold extraction), armed violence, murders, and the compromised food and nutritional security of the Indigenous people (ISA 2023) are clear manifestations of a move away from the 'Leave No One Behind' imperative.

The Yanomami Case encapsulates a multidimensional humanitarian crisis that is affecting the territorialization of the 2030 Agenda in Brazil. The crisis is rooted in the ongoing colonial violence against Indigenous Peoples, which is exacerbated by the capitalist system and the Brazilian state's complicity. It has a devastating impact on Yanomami people's culture, survival, social dynamics, and in the protection of their territory. It is also undermining the SDGs, which are designed to achieve a more just and sustainable world for all. In the two next sections is presented the discussion on the systematic exclusion of Indigenous Peoples from the conception of development and how the deliberate practice of silencing their voices is preventing the SDGs from being territorialized in a way that is truly inclusive and sustainable.

3 Brazilian Indigenist Policy Between 2019 and 2022

This section focuses on how a policy unfavourable to Indigenous peoples was designed in the Brazilian state, especially implemented during the 2019–2022 administration, and how this impacted the incorporation of the 2030 Agenda in the Amazon. A comprehensive overview of the challenges faced by Indigenous peoples in Brazil between 2019 and 2022 is presented, underlining a range of issues, including the lack of Indigenous lands demarcation, legislative and administrative measures restricting Indigenous rights, environmental degradation, violence, and inadequate healthcare.

In this part it also emphasised the impact of illegal mining activities on Indigenous Peoples and Local Communities, and the consequences of government inaction. The focus is aimed on the compatibilities to achieve SDGs targets by addressing these challenges to protect the rights, well-being, and livelihoods of Indigenous peoples in Brazil.

The policy against Indigenous Peoples and Local Communities was identified as an attempt to reinsert a 'integralist policy'[3] to the Brazilian state (Resende et al. 2023). This policy aims to 'integrate' Indigenous peoples, in a colonialist perspective, to an occidentalized and capitalist social organisation model and culture, by the implementation of ethno-ecocidal practices. Among them, Indigenous cultures depreciation, lack of basic health and food security policies, leadership coaptation, illegal deforestation, land grabbing, illegal burning and Indigenous Lands invasions that generated territorial conflicts (Martinelli 2022). It is possible to detect, especially in the former administration actions and inactions, "the exercise of necropolitics, a decision of death" (Krenak 2020, p. 5).

Other elements of that necropolitics are related to the non-demarcation of any Indigenous Land from 2019 to 2022; the rigging of environmental protection and Indigenous rights bodies; active action in favour of the execution of extractive/mining activities in Indigenous Lands; the active promotion of violent and prejudiced speeches; and the failure to meet basic food needs for populations made vulnerable by the Covid-19 crisis. This led to an incalculable number of deaths and losses that could have been avoided (CIMI 2022).

The pandemic crisis has heightened the impacts of government' action and inaction regarding the lack of basic health care, especially for Indigenous peoples, even with sufficient public facilities capable of providing it. According to the Brazilian Association of Collective Health, the mortality rate from Covid-19 among the Indigenous population was 150% higher than the Brazilian average in 2020 (Fellows et al. 2020).

The governmental management during the pandemic, in pursuit of profits and the unrestricted maintenance of economic activities, under the declaration that "some will die, and it is inevitable" was "an unreasonable statement [...] a trivialization of life, but it is also a trivialization of the power of the word. Someone who says this is pronouncing a condemnation" (Krenak 2020, p. 7). These disposable lives, those that serve as a bargaining chip for the maintenance of profit, are those already vulnerable and marginalised by social, racial, ethnic and cultural contexts. The Brazilian indigenist policy could be highlighted in several provisional measures and projects proposed, presented below in a mapping of the most significant acts throughout the 2019–2022 administration.

---

[3] The 'Integralist Indigenist Policy' was an ethnocidal policy, that aimed to integrate Indigenous peoples into a colonial perception of what would be 'national union', in the sense of exterminating 'other' cultures and creating a 'national identity' based on coloniality and Eurocentric paradigms. Integralism marked the Brazilian Indigenist Policy from the creation of the Brazilian State until the Federal Constitution of 1988, the first Constitution to recognize the Indigenous Peoples Rights to practice and perpetuate their culture and to territory.

These various measures have given rise to the alarming threat of genocide against Indigenous populations in Brazil. The UN Special Adviser for the Prevention of Genocide Office, Alice Waimiru Nderitu, asserted Brazil's urgent need to "protect communities at risk and ensure accountability for the crimes committed" (UN 2021). That set the stage for understanding the latitude of the challenge in accomplishing global agendas—like the SDGs—within the Amazon and its great sociobiodiversity.

4 Challenges to Territorialize the SDGs in the Amazon

According to the 2022 Spotlight Report, inequality in Brazil continued to increase, especially in relation to socio-environmental wellbeing and the rights of Indigenous Peoples and Local Communities. The remarkable setbacks in SDGs 1, 2, 5 and 8, and the impoverishment of population affected mostly women, black and Indigenous peoples, especially in the North and Northeast states of Brazil. The setbacks in SDGs 6, 7, 12 and 13 are manifested in the lack of access to basic sanitation and energy for nearly half the population, and the climate emergency that triggers natural disasters (CSWG 2030A 2022). The Amazon region was deeply affected, not only by the expansion of organised crime, as by the increase in deforestation, fires and illegal mining, by an anti-environmental internal policy and diplomacy, and by the neglect of the region during the pandemic (Ramos 2021) (Table 1).

The 2030 Agenda, as an international commitment, gains relevance and power in terms of national and international struggles in such political crisis's context, as the Brazilian during the pandemics and between 2019 and 2022 administration regarding Indigenous issues (Resende et al. 2023). Considering these interconnected challenges, it's important to highlight that the SDGs approach critical issues for Indigenous Peoples. Some of these are mapped by the authors in Table 2.

Even addressing critical issues for Indigenous Peoples and Local Communities, and although seeking to promote Human and Environmental Rights, it's important to underscore that the 2030 Agenda represents the imposition of a social, political and economic life model. This reflects in the implementation of projects with major social and environmental impacts. Strategies to measure progress towards 'sustainability' are often misaligned with Indigenous lives—indicators are unable to encompass Indigenous realities, or to measure wellbeing, harmony and 'sustainability' within Indigenous cosmopraxis (Yap and Wekene 2019). The decision to exclude Free, Prior and Informed Consent (FPIC) from the text of the Agenda, and all of the possible systemic impacts of this decision, makes the debate on its legitimacy and potential for fair implementation a triggering point.

Table 1 Indigenist acts proposed by the Brazilian administration from 2019 to 2022

Provisional measure 870/2019 and its two associated decrees[4]	Altered the National Indian Foundation (FUNAI), transferring the ILs' demarcation responsibility to the Ministry of Agriculture
Provisional measure no. 866/19[5] and law no. 13,844/2019[6]	Another attempt to transfer the ILs demarcation to the Ministry of Agriculture, despite the congress previous restriction to the proposed measures. The federal superior court unanimously endorsed a precautionary measure for suspension of the provisional measure (STF, 2019)
Provisional measure 910/2019 and its two reports[7]	Provided 'Land Regularization' for occupations of lands located in Union areas (as Indigenous lands). It lost its effectiveness within the congress deadline for voting (CONGRESSO NACIONAL, 2020). The measure would amend law no. 11,952/2009, and would hinder Indigenous peoples access to their territories
Bill 191/2020[8]	Aimed to regulate the Federal Constitution' Article 176 (Paragraph 1) and Article 231 (Paragraph 3), establishing specific criteria for mineral resources and hydrocarbons research and mining; and water resources use for electricity generation; on Indigenous lands, including demarcated lands of peoples in voluntary isolation. The bill also amended Law No. 6,001/1973, attesting the legal permission to develop economic activities in ILs, such as agriculture, livestock, extractivism and tourism
Normative Instruction (IN) no. 9 of FUNAI[9]	Promoted administrative changes that allowed INCRA requests, through self-declaratory registration, and subsequent certification for squatters, land grabbers and allotters of Indigenous Lands. It only vetoed the Declaration of Recognition of Limits (DRL) for lands approved/registered
Presidential vetoes to bill 1142/2020[10]	Presidential vetoes to the proposed bill 1142/2020 that recognized Indigenous peoples and Local Communities as "groups of extreme vulnerability" during the covid-19 pandemic, and that determined emergency actions directed at these groups

(continued)

[4] Available at: https://www.in.gov.br/materia/-/asset_publisher/Kujrw0TZC2Mb/content/id/57510830.

[5] Available at: http://www.planalto.gov.br/ccivil_03/_Ato2019-2022/2019/Mpv/mpv886.htm#art1.

[6] Available at: http://www.planalto.gov.br/ccivil_03/_Ato2019-2022/2019/Lei/L13844.htm.

[7] Available at: https://www.planalto.gov.br/ccivil_03/_ato2019-2022/2019/mpv/mpv910.htm.

[8] Available at: https://www.camara.leg.br/proposicoesWeb/prop_mostrarintegra%3Bjsessionid=node0h4mwgn2lxwc2uo6y5fama9541664590.node0?codteor=1855498&filename=PL+191/2020.

[9] Available at: https://www.gov.br/funai/pt-br/arquivos/conteudo/dpt/pdf/instrucao-normativa-09.pdf.

[10] Available at: https://www.in.gov.br/en/web/dou/-/despachos-do-presidente-da-republica-265632873.

Table 1 (continued)

Bill 490/07[11]	Submitted the ILs demarcation to the legislative power, through bill of law, removing the responsibility of FUNAI, and created a requirement to prove the possession and occupation of the territory claimed by Indigenous peoples before the enactment of the constitution (instituting the thesis of the 'Temporal framework'[12])
PEC 215/00[13]	Intended to amend the constitutional text by transferring to the National Congress the competence to approve ILs demarcations; to make new ILs demarcations unfeasible, including through the enshrinement of the 'Temporal framework' thesis; and to end the indigenous exclusive usufruct right to their lands, allowing the agricultural, mineral and forestry exploitation of the demarcated lands

Source Prepared by the authors based on Brazilian 2019–2022 indigenist policy

Table 2 Direct mention to indigenous peoples into the 2030 agenda

Goals	
2.3	By 2030, double the agricultural productivity and incomes of small-scale food producers, in particular women, **Indigenous peoples**, family farmers, pastoralists and fishers, including through secure and equal access to land, other productive resources and inputs, knowledge, financial services, markets and opportunities for value addition and non-farm employment
4.5	By 2030, eliminate gender disparities in education and ensure equal access to all levels of education and vocational training for the vulnerable, including persons with disabilities, **Indigenous peoples** and children in vulnerable situations
6.b	Support and strengthen the participation of **local communities** in improving water and sanitation management
13.b	Promote mechanisms for raising capacity for effective climate change related planning and management in least developed countries and small island developing States, including focusing on women, youth and **local and marginalized communities**
5.c	Enhance global support for efforts to combat poaching and trafficking of protected species, including by increasing the capacity of **local communities** to pursue sustainable livelihood opportunities

Source Created by authors based on the United Nations Sustainable Development Summit 2015—"Transforming our world: the 2030 Agenda for Sustainable Development"

[11] Available at: https://www.camara.leg.br/proposicoesWeb/prop_mostrarintegra?codteor=444088&filename=PL+490/2007.

[12] The 'Temporal Framework' (*Marco Temporal,* in Portuguese) is a constitutional thesis that dates back to the Federal Supreme Court judgement on the Petition 3.388-4/RR/2009, the *Raposa Serra do Sol* Indigenous Land case. The former STF minister, Carlos Ayres, conceived the 'Temporal Framework' thesis, which created a fixed conditioning date (5th of October 1988, promulgation of the Federal Constitution) for recognition of the Indigenous fundamental right to land demarcation. The Framework proposes that Indigenous peoples would only have this right if they occupied the land on the Constitution promulgation day, denying the Brazilian Constitution and the Declaration on Indigenous Peoples Rights, and ignoring the expulsion and persecution processes promoted by the State itself against Indigenous Peoples.

[13] Available at: https://imagem.camara.gov.br/Imagem/d/pdf/DCD19ABR2000.pdf#page=69.

4.1 Understanding the Main Challenges to SDG Territorialization in Amazon

The non-demarcation of any Indigenous Lands between 2019 and 2022, added to the legislative proposals and administrative acts presented to restrict Indigenous people's rights, and the authorization for ILs occupation and exploitation (see Table 1), the Yanomami Case has an aggravating factor: the decreed launch of the Artisanal Mining Development Support Programme (Pró-Mape), that boosted mining and land grabs across Indigenous Lands (CSWG 2030A 2022). These impacts are evident in deforestation rates, which grew 138% in Indigenous Lands between 2019 and 2021, in comparison to the previous three years, added to the cut by BRL 35 million in the budget for the Environment Ministry in 2022 (CSWG 2030A 2022). This evidences an institutional drive for deforestation, or at least the State's flexibility to these practices.

The total paralysis of territorial demarcation procedures in Brazil has also led to security weakening in ILs, to health, sanitary and educational precariousness for Indigenous peoples. In 2021, 118 conflict cases concerning territorial rights were registered. The violence escalation against Indigenous peoples has been alarmingly noted—305 cases of invasions, illegal exploitation of natural resources and various damages to heritage, which affected at least 226 Indigenous lands throughout the country (CIMI 2022, p. 53). It is noteworthy the complaints made by the Yanomami Ayrca and Kumirayoma Associations, through letters sent to the Brazilian Army denouncing the violent invasion of villages that lead to deaths and torture of the Yanomami People, especially woman and children (CIMI 2022).

In 2021, 34 lack of assistance cases were recorded 34, many linked to food shortages and hunger (CIMI 2022). In health issues, there were 107 lack of assistance cases, many linked to the Covid-19 pandemic. Data from the Mortality Information System showed that 847 Indigenous persons died in 2021 due to coronavirus infection (CIMI 2022). The federal government denied to Indigenous people in urban contexts and those living outside demarcated ILs the right to the priority vaccination plan, but was lately forced by the Federal Supreme Court to insert them.

The invasions of Indigenous territories by miners during the pandemic, encouraged by the omission in the supervision and protection of ILs, were among the major vectors of Covid-19 contamination. The case was registered in precautionary measures of the Inter-American Commission on Human Rights (IACHR), in favour of the Yanomami IL, in Roraima and Amazonas, the Munduruku IL, in Pará, and the Guajajara and Awá peoples of the Araribóia IL, in Maranhão (CSWG 2030A 2022).

Among the territorial conflicts that marked the 2019–2022 administration, the overlapping of certifications in Indigenous Lands was responsible for the official federal government certification of 415 farms in 239,000 ha within ILs (CIMI 2022, p. 74). The overlapping of mining requirements, which manifests the strengthening of prospectors and mining companies over Indigenous territories, was recorded in several states, especially in the Amazon region. More than 500 overlapping requirements were registered in Yanomami Indigenous Land, adding up to an area of 3.28

million ha. Within the Yanomami Case, the launch of the "Anomaly Charts" by the Geological Survey of Brazil, that pointed out places with greater chances for precious metals mining precisely in the region inhabited, has boosted this expansion (CIMI 2022, p. 74).

FUNAI went through a series of transformations during Bolsonaro's administration. Marcelo Xavier, appointed in 2019 by the President to chair the Foundation, opened criminal investigations against officials, Indigenous leaders and Indigenous rights activists; hindered initiatives to protect Indigenous Lands; and adopted policies that drove invasion (HRW 2022), promoting censorship, persecution, disciplinary administrative proceedings (INESC 2021). The number of political appointees in the FUNAI staff grew by 20.20% in Direction and Superior Advisory positions (INESC 2021).

With Instruction No. 9 issued by FUNAI in 2020, the government weakened the protection of Indigenous Lands. The Instruction allowed land registration to individuals claiming ownership in Indigenous areas awaiting demarcation (HRW 2022). FUNAI also issued official notes stating that its staff should not act in the protection of non-demarcated Lands and stopped intergovernmental and civil society partnerships to protect ILs. This empowered criminal groups involved in environmental destruction: illegal logging, mining, poaching and land grabbing of Indigenous Lands increased by 137% in 2020, compared to 2018 (CIMI 2021) and the deforested areas in ILs was 138% higher than in the previous three years in the Amazon (ISA 2021).

The emblematic escalation of violence against the Yanomami people was systematically denounced, within the invasion from thousands of illegal miners, territory devastation, attacks and violence scale up (CIMI 2022). The attacks against the *Palimiú* community were filmed, including with the use of tear gas bombs (which is controlled and authorised only by the Brazilian Army) (CIMI 2022). There were also recurrent reports of sexual violence in the Yanomami Land, mainly by illegal miners, including cases of food and basic health items being offered in exchange for sex; grooming of women; and use of alcohol to make women and children more vulnerable to violence (CIMI 2022).

The National Institute for Space Research (INPE) recorded an eightfold increase in illegal mining activities on Indigenous Lands in the northern region between 2016 and 2022. The invasions were detected in the Yanomami Indigenous Land, in Roraima state, and in six ILs in Pará state (INPE 2023). Luciana Gatti, INPE's Greenhouse Gas Laboratory coordinator, asserted that a project of economic exploitation was implemented in the Amazon during the 2019–2022 government, manifested in large contracts of illegal deforestation for soybeans and corn planting, and mining, through the amendment of public policies and laws, and weakening of environmental inspection bodies (Stabile and Casemiro 2023).

According to the Yanomami Indigenous Land's Illegal Mining Monitoring System, from October 2018 to December 2022, the area impacted by mining grew by 300%, reaching 5053.82 hectares of devastated area, directly affecting almost 60% of the Yanomami population (HAY et al. 2022). This project has had violent and irreversible impacts, which are clearly manifested in the humanitarian crisis to which the Yanomami people were subjected. From direct impacts on the environment, especially on the soil and rivers, and on the local fauna and flora, to the health, survival, culture perpetuation and dignity of the populations that inhabit the invaded territories (Stabile and Casemiro 2023).

The Federal Public Ministry published a Report (MPF 2023a, b) that analyses the Yanomami Case, highlighting the preventable deaths among children under five years old. According to the report, it almost doubled in the last 8 years, with peaks in 2020 and 2022. The numbers on childhood mortality from preventable causes from 2019 to 2022 was recorded in 570 (Sumaúma 2023), but the numbers may be higher than those recorded by the Indigenous Health Special Secretariat (SESAI), once some FUNAI base poles located in areas dominated by mining remained closed throughout 2022 (Sumauma 2023). In addition, the massive presence of illegal miners in the Yanomami Indigenous Land was identified as one of the main vectors of malaria, covid-19 and other diseases. The lack of federal government support has forced the Yanomamis to resort to illegal mining camps to communicate with the FUNAI to request urgent health care (CIMI 2022).

The 2022 Spotlight Report highlighted targets and setbacks on the Indigenous issues, especially regarding those Goals with specific targets directed to Indigenous populations, but also those that directly/ indirectly affected them during the year of 2021. The setbacks that mostly impacted Indigenous Peoples could be organised in that Table 3.

Final Considerations

The Yanomami Case is a warning that the 2030 Agenda cannot be achieved without addressing the root causes of colonial violence and systemic racism. Violence perpetrated against IPLCs is directly related to the multiple SDGs targets and the 2030 Agenda compromise. Indigenous peoples and Local Communities are critical agents in implementing a sustainable existence that 'leaves no one behind', and the pluriverse of cosmopraxes (Querejazu 2022) constitute essential science and knowledge to deal with environmental, climatic and socio-biodiverse global dimensions.

Although the Yanomami Land situation now seems to be 'under control', in June 2023, Hutukara Association (HAY) recorded illegal miners' movement in the *Apiaú* and *Couto Magalhaẽs* rivers. Yanomami People reported that, after successive operations, the presence of miners near *Papiú* decreased significantly, but that invaders groups attempted to return to the old camps. This situation was also reported in the *Parafuri*, *Xitei* and *Homoxi* regions, where miners continue to work near the villages, being supplied with food and fuel with regular helicopter flights (HAY et al. 2022).

The 2019–2022 power shift in Brazil to a far-right government (Van Djik and Theun 2023) ideologically averse to Indigenous People Rights (APIB 2020a, b), have had a disastrous impact on the territorialization of the 2030 Agenda in the country, driven by a distressing and multifaceted crisis for Indigenous communities. The prevailing policies, led by the presidential administration, have contributed to a host of pressing issues, from the lack of ILs demarcation and legislative constraints on Indigenous rights to environmental degradation, violence, and inadequate healthcare.

Table 3 SDG target setbacks related to indigenous people

Target 1.4 setback	Suspension of the Agrarian Reform and an 1,100% increase in the number of murders in rural conflicts in 2021. Economic model favouring agribusiness impacts Indigenous peoples and Quilombolas
Target 1.5 setback	Budgetary cuts in programmes for the prevention and containment of natural disasters A total of 41,273,561 people were affected by floods in 2021, the damage totalling more than BRL $60 billion16, disproportionately impacting black and Indigenous people, who predominantly live in the most unstable regions, a direct result of environmental racism
Target 2.5 setback	Official data regarding the number of protected plant and animal genetic resources for food and agriculture has not been updated since 2020 State's decision to deregulate instruments for monitoring and combating organised crime in Indigenous and traditional Lands These decisions have increased hunger amongst quilombola, riverine and forest peoples, human rights violation and the destruction of plant and animal species
Target 4.5 setback	Racial exclusion increased, amongst the 6 + million people not in school in 2020, black and Indigenous people outnumber white people three to one
Target 5.2 setback	Brazilian women were subjected to at least 4.3 million physical assaults (6.3%) in 2020, eight every minute, while roughly 13 million faced verbal abuse (18.6%); violence against black and Indigenous women also increased. Of 56,098 reports of sexual violence (an increase of 3.7%), 18,681 were against children, 73.7% were against already vulnerable populations, and 60.6% of victims were under the age of 14 The sexual exploitation of Indigenous women and girls has increased at an alarming rate
Target 6.4 setback	The loss of treated water in distribution increased to 40.1% in 2020 There were 304 conflicts over water in 2021, impacting 56,135 families in Brazil. Water dumped without adequate cleaning, from submarine outfalls and mining residuals, continues to contaminate water bodies and populations, increasing levels of phosphorus, nitrogen, mercury and other heavy metals Mining activity increased incidences of violence against Indigenous populations, especially girls and women
Target 11.7 setback	Increased violence made urban environments particularly hostile to women, LGBTQIP + and black people, and urban Indigenous peoples
Target 16.7 setback	There is only one representative from the 256 Indigenous nations in the country
Target 16.8 setback	Federal Government's negligence in terms of deforestation; climate crises denialism and associated environmental racism; measures taken against the demarcation of Indigenous and quilombola Lands

Source Created by authors based on Spotlight Report 2022

The expansion of illegal mining activities, encroaching deeper into Indigenous Lands, has not only ravaged the environment but has also posed grave threats to the health and security of Indigenous peoples. Inadequate government response to these challenges has exacerbated the humanitarian crisis within these communities. The horror seen at the Yanomami Case demonstrates the setback in relation to guaranteeing Indigenous rights and the Sustainable Development Goals.

Recent new-elected government actions seem to indicate a transition in Brazilian Indigenist Policy, such as the creation of the Indigenous Peoples Ministry, the appointment of an Indigenous woman to preside FUNAI, and the recreation of a series of crucial national programs and projects for Indigenous peoples in Brazil.

Although, there are still many challenges to be tackled. It is crucial, to guarantee truly sustainable development, the consideration of the Free, Prior and Informed Consent (FPIC) principle. Any initiative that impacts directly or indirectly an Indigenous Land needs to be built together with the local communities in a truly participatory way. This Consent is a premise for any project to be sustainable and expanded upon to ensure a more secure and just future for Indigenous Peoples and Local Communities. The plight of Indigenous peoples in Brazil underscores the need for continued advocacy, both domestically and internationally, to protect their rights, lands, and way of life. It is a reminder that the struggle for Indigenous rights is far from over and requires the collective efforts of governments, civil society, and global communities to secure a better future for all Indigenous peoples.

If the Brazilian government, and the world, seeks to achieve a sustainable development agenda, they must consider the inclusion of Indigenous peoples and their knowledge in decision-making. Public policies that guarantee rights to life, territory, and traditional cultures practised in its diversity have a profound impact on countries' ability to achieve Sustainable Development Goals objectives. SDGs 1, 2, 3, 5, 13 and 15 are crucial for Indigenous struggles, nationally and internationally.

References

Associação dos Povos Indígenas do Brasil (APIB) (2020a) Arguição de Descumprimento de Preceito Fundamental (ADPF) 709. Associação dos Povos Indígenas do Brasil, Brasília. Available at https://acervo.socioambiental.org/sites/default/files/documents/g3s00122.pdf

Associação dos Povos Indígenas do Brasil (APIB) (2020b) Nota Técnica: a Instrução Normativa da Funai nº 09/2020 e a gestão de interesses em torno da posse de terras públicas. Associação dos Povos Indígenas do Brasil, Brasília. Available at https://apiboficial.org/2020/04/28/nota-tecnica-a-instrucao-normativa-da-funai-no-092020-e-a-gestao-de-interesses-em-torno-da-posse-de-terras-publicas/#:~:text=A%20Instru%C3%A7%C3%A3o%20Normativa%20n.%2009%2F2020%20transforma%20a%20Funai,posseiros%2C%20grileiros%20e%20loteadores%20de%20Terras%20Ind%C3%ADgenas%20%28TIs%29

Baniwa G (2006) O Índio Brasileiro: o que você precisa saber sobre os povos indígenas no Brasil de hoje. Ministério da Educação, Brasília; LACED/Museu Nacional

Brasil (1988) Constituição da República Federativa do Brasil. Brasília, DF: Senado Federal: Centro Gráfico

Brazilian Health Ministry Centro de Operações de Emergência em Saúde—Yanomami. COE Yanomami. Available at https://www.gov.br/saude/pt-br/composicao/svsa/resposta-a-emergencias/coes/coe-yanomami

Casarões G, Flemes D (2019) Brazil first, climate last: Bolsonaro's foreign policy. Giga Focus—Latin America, no 05, p 14

Civil Society Working Group for the 2030 Agenda (CSWG 2030A) (2022) 2030 agenda for sustainable development spotlight report 2022 Brazil synthesis. Available at https://brasilnaagenda2030.files.wordpress.com/2022/07/en_sr_2022_final_web-1.pdf

Conselho Indigenista Missionário (CIMI) (2022) Relatório Violência Contra os Povos Indígenas no Brasil—Dados de 2021. Conselho Missionário Indigenista—Conferência Nacional dos Bispos do Brasil, Brasília. ISSN 1984–7645. Available at https://cimi.org.br/wp-content/uploads/2022/08/relatorio-violencia-povos-indigenas-2021-cimi.pdf

Djik V, Theun, (2023) Conferência "Ideologias e discursos de extrema direita." Universidade de Brasília, Laboratório de Estudos Críticos do Discurso

Ecocide Act (2012) Ecocide as the 5th international crime against peace. Available at https://ecocidelaw.com/wp-content/uploads/2012/06/Earth-is-Our-Business-Appendix-II.pdf

Faustini D, Fanon F (2017) A branquitude e a racialização: aportes introdutórios a uma agenda de pesquisas. In: Muller, Tânia MP (ed) vol 18, no 2

Fellows M et al (2020) Não são números, são vidas! A ameaça da covid-19 aos povos indígenas da Amazônia brasileira. Relatório do Instituto de Pesquisa Ambiental da Amazônia (IPAM) e da Coordenação das Organizações Indígenas da Amazônia Brasileira (COIAB). Available at https://ipam.org.br/wp-content/uploads/2020/06/NT-covid-indi%CC%81genas-amazo%CC%82nia.pdf

Fiocruz (2023) Mapa de Conflitos—O garimpo ilegal e o genocídio yanomami. Mapa de Conflitos de injustiça ambiental e saúde no Brasil. Available at https://mapadeconflitos.ensp.fiocruz.br/conflito/rr-invasao-de-posseiros-e-garimpeiros-em-terra-yanomami/

Fraser N (2022) Cannibal capitalism. Editora Verso

Gonçalves B (2017) Parecer Psicossocial da Violência contra os Povos Indígenas Brasileiros: o Caso Reformatório Krenak. Revista Psicologia, Ciência e Profissão, pp 01–10

Gordilho H, Ravazzano F (2017) Ecocídio e o Tribunal Penal Internacional. Justiça Do Direito. 31(3):688–704

Higgins P (2015) Eradicating ecocide, 2nd edn. Shepheard-Walwyn publishers, Londres

Human Rights Watch (HRW) (2022) Relatório Brasil—Eventos de 2021. Report 2022/ Country, New York. Available at https://www.hrw.org/pt/world-report/2022/country-chapters/380707#48851c

Hutukara Associação Yanomami (HAY) (2022) Cicatrizes na Floresta II: Evolução do garimpo ilegal na Terra Indígena Yanomami em 2021. Hutukara Associação Yanomami/Associação Wanasseduume Ye'Kwana, Boa Vista. Available at https://acervo.socioambiental.org/acervo/documentos/cicatrizes-na-floresta-evolucao-do-garimpo-ilegal-na-ti-yanomami-em-2020

Hutukara Associação Yanomami (HAY) (2022) Associação Wanasseduume Ye'kwana, Instituto Socioambiental (ISA) Yanomami sob ataque: garimpo ilegal na Terra Indígena Yanomami e propostas para combatê-lo. Editora HAY, Boa Vista

Hutukara Associação Yanomami (HAY) et al (2023) Relatório Yamakɨ nɨ ohotaɨ xoa. Hutukara Associação Yanomami, Associação Wanasseduume Ye'kwana, Urihi Associação Yanomami. Available at https://org.br/wp-content/uploads/2023/08/Yamaki-ni-ohotahi-Xoa-Nos-ainda-estamos-sofrendo-yad00620.pdf

INESC (2021) Fundação anti-indígena: um retrato da FUNAI durante o governo Bolsonaro. Report by INESC—Instituto de Estudos Socioeconômicos; INA—Indigenistas Associados. Available at https://inesc.org.br/wp-content/uploads/2022/07/Fundacao-anti-indigena_Inesc_INA.pdf

Instituto Socioambiental (ISA) (2023) Povo Yanomami. Available at https://pib.socioambiental.org/pt/Povo:Yanomami

Inoue CY, Mello RA, Urt JN (2020) Abordagem dos muitos mundos aplicada ao estudo da política ambiental global no Antropoceno: vozes indígenas na Amazônia. Monções Rev Relações Int Da UFGD 9(18):435–460. https://doi.org/10.30612/rmufgd.v9i18.13640

Kayapó E, Brito T (2015) A pluralidade étnico-cultural indígena no Brasil: o que a escola tem a ver com isso? Mneme—Revista De Humanidades 15(35):38–68

Kopenawa YD (2009) Depoimento recolhido, traduzido da língua Yanomami e editado por Bruce Albert. In: Albert B, William M (ed) Urihi A: A terra-floresta Yanomami. Instituto Socioambiental (ISA). Available at https://xapuri.info/urihi-a-a-terra-floresta-yanomami/

Krenak A (2020) O amanhã não está à venda. [S. l.]: Companhia das Letras. ISBN 978-85-5451-732-8

Longhini G (2021) Da cor da Terra: etnocídio e resistência indígena. Revista Tecnologia & Cultura—Rio de Janeiro—Edição especial

Martinelli Y (2022) Etno-ecocídio: uma análise decolonial do Estado e da política indigenista brasileira no governo Bolsonaro. Relações Internacionais. Brasília: Universidade de Brasília, Monografia

Ministério Público Federal (MPF) (2023a) Inquérito Civil nº 1.32.000.000084/2023–17

Ministério Público Federal (MPF) (2023b) Nota Pública—Território Indígena Yanomami. Procuradoria Geral da República (PGR). PGR-00022463/2023. Available at https://www.mpf.mp.br/pgr/documentos/pgr-00022463-2023-ti-yanomami-1.pdf

Moreira A (2020) Valor econômico. Macron alega "ecocídio" para barrar acordo da União Europeia para Mercosul. Available at https://valor.globo.com/brasil/

Munanga K (2004) O negro na sociedade brasileira: resistência, participação e contribuição. Fundação Cultural Palmares, Brasília

National Aeronautics and Space Administration (NASA) (2020) A new tool for tracking Amazon fires. Available: https://earthobservatory.nasa.gov/images/147075/a-new-tool-for-tra ckingamazon-fires

Observatório da Mineração (2022) Dinamite pura: como a política mineral do governo Bolsonaro armou uma bomba climática e anti-indígena. Relatório do Observatório da Mineração & Sinal de Fumaça. Available at https://observatoriodamineracao.com.br/dinamite-pura-relatorio-pro duzido-por-observatorio-da-mineracao-e-sinal-de-fumaca-traz-analise-do-legado-explosivo-da-politica-mineral-do-governo-bolsonaro/

Observatório do Agronegócio no Brasil (2018) Nem um centímetro a mais para terras indígenas, diz Bolsonaro. De Olho nos Ruralistas. Available at https://deolhonosruralistas.com.br/2018/02/08/nem-um-centimetro-mais-para-terras-indigenas-diz-bolsonaro/

Office of the United Nations High Commissioner for Human Rights (OHCHR) (2021) Human rights council opens forty-eighth regular session, hearing the high commissioner's global human rights update and her separate updates on Venezuela, Afghanistan, Nicaragua and Sri Lanka. Available at https://www.ohchr.org/en/press-releases/2021/09/human-rights-council-opens-forty-eig hth-regular-session-hearing-high

Picq M (2017) Visões indígenas desafiando o global: mulheres Kichwa pluralizando a soberania. Monções: Rev Relações Int Da UFGD, 6 (11), (Tradução)

Povos Indígenas do Brasil (PIB) do Instituto Socioambiental (ISA) (2018) Demarcações. PIB Socioambiental. Available at https://pib.socioambiental.org/pt/Demarca%c3%a7%c3%b5es

Povos Indígenas do Brasil (PIB) do Instituto Socioambiental (ISA) (2018) Povo Yanomami. PIB Socioambiental. Available at https://pib.socioambiental.org/pt/Povo:Yanomami

Pruim S (2014) Ethnocide and indigenous peoples: article 8 of the declaration on the rights of indigenous peoples. 269 Adelaide Law Rev 17; 35(2). Available at https://law.adelaide.edu.au/ua/media/573/ch4-alr-35-2-pruim.pdf

Putini J (2023) Ministério da Saúde decreta emergência de saúde pública para combater desassistência de indígenas Yanomami. Available at https://g1.globo.com/saude/noticia/2023/01/20/ministerio-da-saude-decreta-emergencia-de-saude-publica-para-combater-desassistencia-de-indigenas-yanomami.ghtml

Querejazu A (2022) Cosmopraxis: Relational methods for a Pluriversal IR. Rev Int Stud 48(5):875–890. https://doi.org/10.1017/S0260210521000450

Ramos A (2021) Amazônia sob Bolsonaro. Aisthesis, Santiago, (70): 287–310. https://doi.org/10.7764/aisth.70.13

Resende V et al (2023) Contra o etnoecocídio: da violência política da extrema direita à resistência coletiva dos povos indígenas no Brasil. Fórum Lingüístico 20(3):9342–9357. https://doi.org/10.5007/1984-8412.2023.e93113
Rodrigues C et al (2023) Lula visita Casa de Saúde Yanomami em Roraima e diz que situação de indígenas é desumana. Available at https://g1.globo.com/rr/roraima/noticia/2023/01/21/lula-chega-a-roraima-e-deve-anunciar-acoes-contra-crise-na-saude-e-desnutricao-de-criancas-yanomami.ghtml
SDG Report (2023) The sustainable development goals report 2023: Special edition. Towards a Rescue Plan for People and Planet
Souza, Oswaldo (2023) O que você precisa saber para entender a crise na Terra Indígena Yanomami. Instituto Socioambiental (ISA). Available at https://www.socioambiental.org/noticias-socioambientais/o-que-voce-precisa-saber-para-entender-crise-na-terra-indigena-yanomami
Stabile A, Casemiro P (2023) Garimpo aumentou 787% em terras indígenas entre 2016 e 2022, aponta INPE. G1. Globo, 11 de fevereiro de 2023. Available at https://g1.globo.com/meio-ambiente/noticia/2023/02/11/garimpo-aumenta-787percent-em-terras-indigenas-entre-2016-e-2022-aponta-inpe-infografico.ghtml
Sumaúma (2023) Não estamos conseguindo contar os corpos. Available at https://sumauma.com/nao-estamos-conseguindo-contar-os-corpos/
Thomé A, Nunes N, Thomé R (2020) A degradação ambiental na Amazônia brasileira e os desafios para a inclusão do crime de ecocídio no Estatuto de Roma. Revista Brasileira de Desenvolvimento Territorial Sustentável GUAJU, Matinhos, vol 6, no 2
Tupinambá, Casé Angatu Xukuru (2019) Nós não somos donos da terra, nós somos a terra. Entrevista especial com Casé Angatu Xukuru Tupinambá, concedida a Ricardo Machado. In: Revista instituto humanitas unisinos
United Nations (UN) (2008) Declaração das Nações Unidas sobre os Direitos dos Povos Indígenas. Organização das Nações Unidas. Rio de Janeiro. Available at https://www.un.org/esa/socdev/unpfii/documents/DRIPS_pt.pdf
United Nations (UN) (2021) Human rights council holds separate interactive dialogues on prevention of genocide and on violence against women. United Nations Geneva. Available at https://www.ungeneva.org/en/news-media/meeting-summary/2021/06/morning-human-rights-council-holds-separate-interactive
Viana T (2020) Ecocídio e o tribunal penal internacional: A crise ambiental na Amazônia brasileira reacende o debate. Revista O Cosmopolítico, 7(2)
Vitalis R (2015) White world order. The Birth of American International Relations. Cornell University Press, Black Power Politics
WACC (2019) What do the SDGs mean for the world's indigenous peoples? World Association for Christian Communication (WACC). Media Development. Available at https://waccglobal.org/what-do-the-sdgs-mean-for-the-worlds-indigenous-peoples/#:~:text=Indigenous%20Peoples%E2%80%99%20visions%20of%20development%20were%20not%20included,affirms%20Indigenous%20Peoples%20right%20to%20self-determined%20development.%206
Yap M, Watene K (2019) The Sustainable Development Goals (SDGs) and indigenous peoples: Another missed opportunity? J Hum Dev Capab 20(4):451–467

The History Behind the Land Use Change in Rondônia State: Deforestation Culture for the Exploitation of the Land

Verônica Garcia Donoso and Christa Reicher

Abstract Amazonian deforestations have received international attention and political pressure over recent years due to their environmental impact on local and global climates. The concern is real: from 1985 to 2021, the Brazilian Amazon lost 44.1 million hectares of forest. Deforestation is particularly intense in the state of Rondônia in northern Brazil, as land is cleared for cattle ranching, agriculture and urban areas. This process of deforestation in Rondônia is especially linked to the arrival of migrant settlers from the late 1960s and their interest in making the land profitable. This book chapter examines the history of land-use change in Rondônia as a cause of the process of vegetation loss. Using a mixed-methods approach, it highlights the specific aspects of deforestation linked to the historical background of the colonisation period with the creation of urban settlements. It aims to show that deforestation is a cultural legacy of the occupation of the territory; therefore, the need for ecological education is necessary to ensure greater changes and the protection of the forest.

Keywords Rondônia · Amazon · History · Land-use change · Slash-and-burn agriculture

V. G. Donoso (✉)
Architecture and Urbanism Course, Federal University of Santa Maria, Cachoeira Do Sul Campus. Cachoeira Do Sul-RS, Santa Maria, Brazil
e-mail: veronica.donosco@ufsm.br

Alexander Von Humboldt Foundation, Berlin, Germany

C. Reicher
Chair of Urban Design and Institute for Urban Design and European Urbanism and UNESCO Chair in Cultural Heritage and Urbanism, RWTH Aachen University, Aachen, Germany

W. Leal Filho et al. (eds.), *Amazon 2030—Sustainability Issues in the World's Largest Rainforest Region*, World Sustainability Series,
https://doi.org/10.1007/978-3-031-81465-5_2

1 Introduction

The aim of this book chapter is to analyse the deforestation process in the state of Rondônia, Brazilian Amazon, in relation to the historical background of the occupation of the territory in the last 100 years, especially with the arrival of migrant settlers from the late 1960s and the expansion of agriculture, urban areas and cattle ranching.

The Amazon is the world's largest tropical forest, with an area of 944 million hectares, divided between nine countries. Brazil shares the largest area of the biome, with an area of 420 million hectares. Equivalent to 49.50% of the Brazilian territory. This area is shared by nine states in northern Brazil: Acre, Amapá, Amazonas, Maranhão, Mato Grosso, Pará, Roraima, Tocantins and Rondônia.

Deforestation is one of the greatest challenges facing the Amazon rainforest, with the main drivers of deforestation being anthropogenic activities, such as agricultural and cattle ranching expansion, coupled with illegal logging with slash-burn activities (Raisg 2020).

In Brazil, the processes of deforestation and land use change are directly linked (Gomes and Bastos 2018; Alves et al. 2021). This issue has gained international attention and political pressure in the last decade due to the ecological importance of the Amazon for the local and global climate and the need to respond to the Sustainable Development Goals (SDGs) and the New Urban Agenda (NUA) to combat climate change (Alves et al. 2021; Gatti et al. 2021; Berenguer et al. 2021; Zhang et at. 2021; Esquivel-Muelbert et al. 2018; Lawrence and Vandecar 2015; United Nations 2015).

Discussing land use change in the Amazon is essential and is linked to a broad conceptual framework that combines ongoing debates on sustainability and urban growth (Ndubisi 2002; Mokkonen et al. 2018; Macedo et al. 2018; Silva and Vergara-Perucich 2021).

In Brazil, unsustainable development is just one of the major concerns considering climate change; the discussion is also about deforestation, the growth of urbanisation and plantations over natural vegetation areas and rising inequality (Donoso and Queiroga 2023; Macedo et al. 2018).

Rondônia is a state in northern Brazil that exemplifies these challenges. Its history is linked with the crucial role of infrastructure and cattle ranching expansion as drivers of deforestation. In recent decades, Rondônia has been the destination of many Brazilian migrants through massive government colonisation projects. Originally almost entirely rainforest, the state of Rondônia has experienced a major loss of natural vegetation due to the expansion of agriculture and cattle ranching. In 1985, natural forest vegetation covered 87.10% of the state. In 2021, the natural forest vegetation was 57.58%. On the other hand, pastureland, which covered 8.21% of the state in 1985, increased to 36.34% in 2021.

In addition, the urban areas of the State of Rondônia are quite new, some of them being less than 60 years old, representing 0.21% of the state's territory in 2021. Although they were built in the middle of the Amazon and in the XX century, their planning does not follow the recent discussions of sustainability.

This book chapter therefore presents a Brazilian region that raises important global and local issues for sustainable urban planning in the context of growing global environmental and ecological awareness. It highlights the specific aspects of deforestation in this Brazilian state, exploring the historical background of the colonisation period, which increased urban settlements but above all drove the expansion of agriculture and cattle ranching in the region.

2 Methods

The research gathers a mixed method approach to articulate quantitative and qualitative data. The methods include preliminary research, bibliographic research and analysis, mapping and morphological analysis using the Mapbiomas method (2021, 2023).

In terms of mapping methodology, a geographic information system is used to locate and map land use changes, with the support of QGIS for spatial modelling and analysis. The material to carry out the operation is the land use assessment dataset provided by the Mapbiomas project, which is based on satellite imagery and covers land use changes between 1985 and 2022 (Mapbiomas 2023). It is available in the Geographic Coordinate System with a spatial resolution of 30 or 10 m and covers the entire Brazilian territory. Other data support our mapping methodology, such as data from the Brazilian Institute of Geography and Statistics (IBGE).

3 History

3.1 Occupation of the Amazon Territory

The first records of the occupation of the Amazon region show the presence of settlements approximately 14,000 years ago. This period of indigenous population before the arrival of Europeans is divided into three historical phases: palaeoindigenous, archaic and late prehistoric. In the first phase, approximately 11,200 years ago, the population was small and nomadic. Traces of these civilisations have been found, for example, in Monte Alegre, Pará. In the second, archaic phase, between 8,000 and 3,000 years ago, indigenous populations settled along the Amazon River. The third phase begins between 3,000 and 1,000 years ago, with the development of more complex societies in flooded areas, such as the Marajoaras on the island of Marajó (Neves 2022).

When Europeans arrived in the Amazon in the sixteenth century, they found a forest inhabited by many culturally diverse indigenous tribes. The archaeologist Eduardo Neves (2022) estimates that there were approximately 5 million indigenous people in the Amazon before the Europeans arrived. During the period of European invasion, this indigenous population was drastically reduced, mainly due to the introduction of unknown diseases and the use of firearms. The anthropologist Lévi Strauss (1981) estimated that he came across more than 50,000 during his expedition to explore the Amazon through the Machado and Pimenta rivers in the 1950s.

The present territory of Rondônia was one of the last areas to be colonised by Europeans in Brazil. It was a territory of conflicts by the treaty of conquest limits of colonies from Spain and Portugal until the Treaty of Madrid.

3.2 A Brief History of Rondônia

Rondônia is an Amazonian state with a very special history. The region was developed as a result of an international agreement between Brazil and Bolivia (Tratado de Petrópolis, 17 November 1903), which allowed the construction of the Madeira-Mamoré railway at the beginning of the twentieth century. Some cities, such as the capital Porto Velho in 1908, were founded thanks to this infrastructure (Morelato and Cardoso 2020; Ferreira 2017).

The construction of the Madeira-Mamoré railway between 1907 and 1912 and the network of telegraph lines between 1907 and 1915, which connected the region and shortened the distances for the transport of rubber production, were responsible for the first large influx of immigrants from different parts of the world to the region, especially to work on laying the tracks.

Previously, the territory of Rondônia belonged to other Brazilian states (Amazonas and Mato Grosso). During the New State of Getúlio Vargas (1937–1945), another attempt was made to occupy the Amazon region with the "March to the West" project, which created seven new federal territories, including the federal territory of Guaporé in 1943 (the future state of Rondônia). As part of this project, the BR-364 highway was inaugurated, linking the territory of Guaporé to the state of Mato Grosso.

In 1956, the name of the territory was changed to Rondônia, in honour of Marshal Cândido Rondon, who led expeditions to the region to bring the telegraph line to the region. Rondon was also known for his defence of the causes of the indigenous peoples.

Population and economic growth accelerated after the Second World War as a result of migration and mineral discoveries.

The transformation of Rondônia's landscape is the result of economic interests, infrastructure projects and political ideals. Rondônia's history has been marked by economic cycles, including rubber, minerals and agriculture, but above all by the colonisation process, as will be shown in the next section.

3.3 *Encouraging Migration to Northern Brazil at the End of the 1960s*

On 31 March 1964, a coup d'état took place in Brazil. Amidst the curtailment of freedoms and rights, there were policies and incentives to develop the northern region of the country, such as with the development of highways and new settlements. This is because the military government considered the Amazon to be empty or unproductive, as was the case for Rondônia.

As we have seen in the previous section, this was not the first governmental approach to the development and occupation of the Amazon, but it was undoubtedly one of the most efficient for the advancement of the agricultural frontier.

The BR-364 highway was the main axis for this colonisation process in Rondônia. After the highway was built, urbanisation along its axis increased, partly as a result of the land grants made by the Institute of Colonisation and Agrarian Reform (INCRA). This process created several municipalities, such as Pimenta Bueno, Ariquemes, Cacoal, Ji-Paraná, Ouro Preto do Oeste and Jarú.

New waves of migrants and their families arrived in the Amazon region for work as a result of INCRA's distribution of land and for other reasons, such as work in the rubber plantations, the reactivation of the Madeira-Mamoré railway and the mining of minerals such as cassiterite and gold from the 1960s onwards (Franco 2019; Dall'igna 1996; Coy 1988).

By attracting migrant settlers with an interest in occupying the territory, the government's strategy disregarded the existence of indigenous, riverine and other populations already established in the region. The military slogan was clear: 'land without people for people without land'. It brought large waves of rural workers to the Amazon as a strategy for occupying the territory and to meet demands for land reform.

Those who benefited from the distribution of land included both large landowners/entrepreneurs and individual workers. In the 1970s, the development programmes carried out in Rondônia were the Integrated Colonisation Plans (PICs), Directed Settlement Programmes (PADs) Settlement Projects (PAs), the Polamazônica Programme and Polonoroeste. To distribute the land, trees had to be felled to demarcate the plots, and roads had to be opened for the arriving settler families.

In the 1970s, INCRA was responsible for the demarcation of seven territorial units into PICs along the BR-364 highway for the settlement of arriving migrants.

The colonisation projects were both official (carried out by the government) and private (carried out by private companies registered with INCRA) (Cunha and Moser 2010).

The occupation of Rondônia was a slow process on the part of INCRA and a fast one on the part of the settlers. The INCRA was responsible for providing the land, and the local government was responsible for providing the infrastructure, which was not the case. As a result, settlers opened trails and roads to facilitate the entry and disposal of small agricultural products, increasing deforestation.

Plant species with commercial value, such as mahogany, were exploited by timber companies; plant species with no commercial value were burned. After commercialising the wood, agricultural activities were developed after clearing the remaining forest by slashing and burning.

The deforestation process was the settlers' way of securing ownership of the land, and the strategy was perhaps also recognised by the official government as beneficial to the development of the territory. It was said that INCRA's documentation of land ownership stipulated that 50 percent of the land received had to be deforested to guarantee ownership. However, according to Fiori et al. (2013), who analysed Ordinance No. 839/INCRA of 13 June 1988, this information was not contained in the form of a law but was unofficial information that circulated among settlers.[1] The land that had not been deforested was considered to be uncultivated and at risk of being invaded. Therefore, there was an informal social pact that it was necessary to cut down the forest to protect the land from invasion by third parties.

INCRA was one of the organisations most responsible for the agricultural colonisation of Rondônia. However, it did not have the capacity to deal with the number of people arriving in the area, and the planning got out of hand. There were no settlement areas prepared to distribute land to everyone, nor were there enough programmes, such as PICs and PADs, to cope with the migratory movement that arrived in the 1970s and 1980s. In the absence of INCRA, migrants moved to vacant land and waited for INCRA to expand its actions to regularise the new land they took possession of (Oliveira 2010). A few examples of the history of the region will be presented in the next section.

4 Land-Use Change Analysis, Population Growth and Deforestation in Rondônia State

Impressive land-use changes related to forest loss and an increase in pasture areas occur in the central area of Rondônia state, following the axis of BR-364 and the INCRA's division of land for settlers (Fig. 1).

Here are some features of land use change and population growth in the following municipalities: Ariquemes, Jaru, Ji-Paraná, Cacoal, Ouro Preto do Oeste, Presidente Médici, Ministro Andreazza, Castanheiras and Pimenta Bueno (Fig. 2). Those cities were selected because of their relationship with the historical process of land use

[1] According to the authors, the legislation requires that 50 percent of the plot be maintained as a legal reserve for rural property in the legal Amazon, in accordance with the rules of the Forest Code in force at the time, but does not explain what happens to the other half of the plot. However, the authors find conflict and room for interpretation in Articles 3 and 10, the first of which deals with the use of the property for cattle ranching. Article 3 states that when the property is used for livestock farming, the area actually cultivated is recognised as the area represented by artificial pasture, which implies the removal of forest cover for artificial pasture.

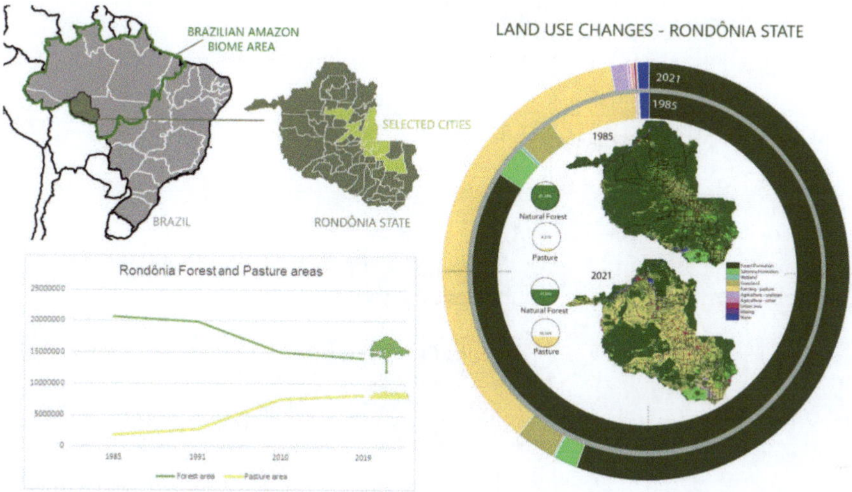

Fig. 1 Land use changes in Rondônia state. *Source* Elaborated by the author. Data sources: Mapbiomas (2021), IBGE (2022, 2019)

change in the State of Rondônia, especially the BR-364 highway, the colonisation projects carried out by INCRA, and the deforestation rates resulting from the expansion of cattle ranching (Fig. 3).

Ji-Paraná and Cacoal have a similar peculiarity: both have indigenous territories in their areas. The indigenous territories, together with the conservation units, play an important role in the protection of the forest areas in the Amazon region, as shown in Fig. 4.

The data source for this book chapter session is the data collected by the satellite imagery of Mapbiomas (2021, 2023) and the IBGE data.

The Mapbiomas data allowed the analysis of the land use change process from 1985 to 2022. The IBGE data allowed the confirmation of historical facts and the collection of figures such as population and cattle data.

Ariquemes

The Ariquemes territory was the home of the Arikêmes indigenous group, now extinct due to the violent process of occupation of the territory by explorers, miners and settlers (Meira 2017).

From the 1900s, due to the rubber exploration process and the construction of telegraph lines, the region was also occupied by the first small migratory group. In 1958, when cassiterite was discovered in the region, more migrants arrived. The construction of BR-364 also increased the number of settlers coming to the region.

INCRA's first settlement projects in the region took place between 1972 and 1975 (settlements of Buraqueiro and Marechal Dutra). One year later, the new town of Ariquemes was planned, a little further away from the Jamari River area, where the original urban nucleon was born (Vila Velha). In 1977, Ariquemes was declared a city.

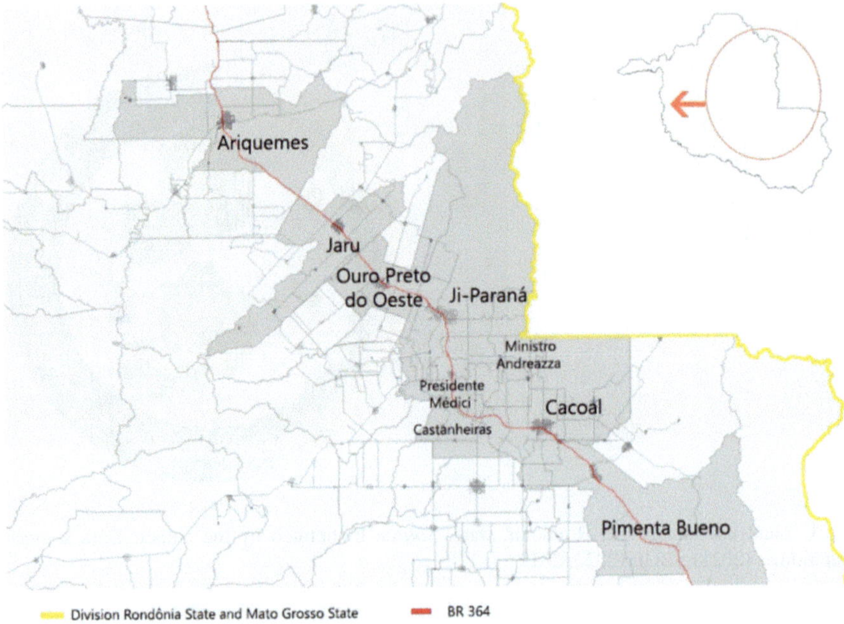

Fig. 2 The selected municipalities. *Source* Elaborated by the author. Data sources: IBGE (2022; 2019)

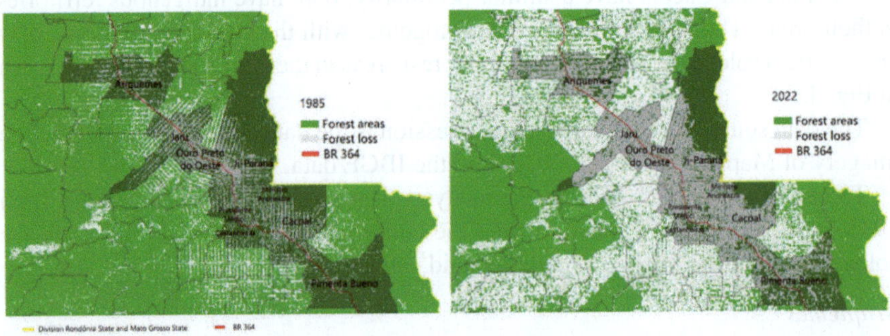

Fig. 3 The selected municipalities and the forest loss from 1985 to 2022. *Source* Elaborated by the author. Data sources: IBGE (2022, 2019) and Mapbiomas (2023)

From that period, the population of Ariquemes almost dobled: In 1980, Ariquemes had a population of 53,365 inhabitants. In 2022, the estimated population was 96,833 inhabitants.

In 1985, the forest areas occupied 80.91% of the municipal territory (358,093 ha), which decreased to 31.92% (141,267 ha) in 2022. Even if the mining areas (legal and illegal) have represented a percentage of the history of the occupation of the territory, the main cause of the loss of vegetation since the 1970s was the growth of

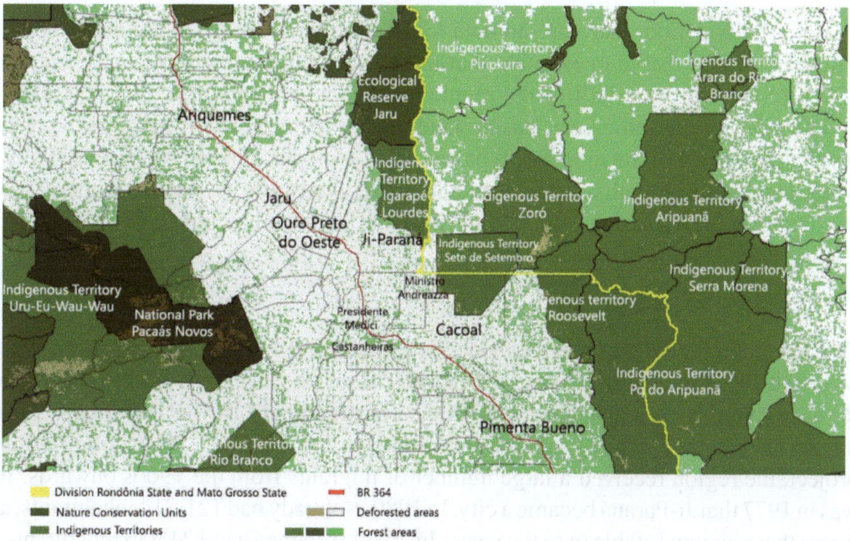

Fig. 4 The role of indigenous territories and conservation units in forest protection. *Source* Elaborated by the author. Data sources: IBGE (2022, 2019) and Mapbiomas (2023)

cattle pastures. In 1985, it occupied 16.60% (73,481 ha) of the territory, increasing to 64.25% (284,365 ha) of the territory in 2022.

Jaru

Until 1981, Jaru belonged to Ariquemes, but the history of the occupation of the region is linked to the Jaru indigenous group, the expansion of the telegraph line and rubber explorers in the 1910s, and the creation of the PIC Padre Adolpho Rohl settlement by INCRA in 1975.

In 1996, the population was 47,981 inhabitants, which increased to 50,591 in 2022 (IBGE 2022). Even if the increase in the population has not been significant, this is not the case for the increase in livestock, which in 2021 had 534,672 cattle heads.

In 1985, the pasture areas occupied 33.01% (97,187 ha) of the territory, which increased to 83.85% (246,873 ha). On the other hand, the forest area decreased drastically. While in 1985, it covered 65.54% (192,949 ha) of the territory, in 2022, it covered only 14.93%.

Ouro Preto Do Oeste

The municipality of Ouro Preto do Oeste was named after the river Ouro Preto, where the PIC Ouro Preto was created by INCRA in 1970. At that time, it was under the jurisdiction of Ji-Paraná, and in 1981, it became a city. Although it was planned for only 2,000 families, migration to the region was intense until the 1990s. It has an interesting growth curve: while the population decreased from 52,095 in 1996 to 35,044 in 2022, the cattle population remained stable, from 351,553 in 2004 to 361,116 in 2021.

Before this period, there are no data to measure the increase in the cattle population, but it is possible to estimate that there was intense growth by analysing the land use: in 1985, only half of the municipal area was covered by forest (51.05%, 100,566 ha). In the same year, pastureland already represented 46.66% of the area (91,907 ha). Most of the increase in pasture land took place in the 1980s and 1990s, when population growth was also intense. Afterwards, the pasture area increased continuously until it reached 86.57% (170,530 ha) of the municipal territory in 2022, leaving only 11,89% of forest area.

Ji-Paraná

The Ji-Paraná region was the territory of the Jarus and Urupás indigenous groups. Marchall Rondon arrived in the area in 1909 along the Ji-Paraná River (Machado River) to build a telegraph station. With the construction of the BR-364 highway, a private settlement company called Calama and the INCRA Ouro Preto colonisation project, the region received a large number of migrants from the 1960s onwards. It was in 1977 that Ji-Paraná became a city. In 1980, it already had 121,714 inhabitants, a figure that remained stable over the years. In 2022, Ji-Paraná had 124,333 inhabitants.

The municipality of Ji-Paraná is also home to the Igarapé Lourdes Indigenous Territory, land of the Arara and Gavião ethnic groups, in the division with the state of Mato Grosso. This indigenous territory was demarcated in 1976–1977 (Felzke 2018). Thanks to the demarcation of the indigenous territory, the forest cover in the municipality still remains in good percentages. In 1985, 82.21% (566,962 ha) of the territory was covered by forest. In 2022, it was 65.71% (453,188 ha). The loss of forest is related to the areas outside the indigenous territory, and the change in land use has been for pasture, with an increase, especially from the 1980s to 2003, and has remained mostly stable since then.

Presidente Médici

With a population of 19,327 (2022), the region was first occupied by rubber explorers in 1915 and by settlers from the 1960s, initially by irregular migrants who occupied the areas. Due to the rapid occupation of the area from the 1970s, INCRA created a sector as an extension of the PIC of Ouro Preto. The city was created in 1981, separated from Ji-Paraná. The city's name was chosen by the settlers in honour of the military president, Emílio Médici, who ruled during the most repressive years of Brazil's military dictatorship.

The settlers quickly took over the territory for pastureland. In 1985, only 61.45% (108,043 ha) of the municipality was covered by forest. In 2021, it was only 16.47% (28,962 ha). The deforestation process was especially important until 2005, when the increase in pasture areas in 2022 was responsible for 81.42% (143,148 ha) of the municipal area.

Ministro Andreazza

The municipality was part of INCRA's PIC Ji-Paraná and belonged to Cacoal until it became a city in 1992. With a declining population and growing cattle herd, it is a small town. In 2022, it had 6,466 inhabitants and 133,658 head of cattle (2021).

This small population was responsible for a large change in land use: in 1985, 63.49% (50,693 ha) of the municipality was covered by forest. In 2022, only 17.64% (14,088 ha) of the original forest remains. On the other hand, pastures already cover over 80% (64,664 ha) of the territory.

Cacoal

The origin of Cacoal is linked to the BR-364 highway and the occupation of the Machado riverbank due to the presence of native cacao trees. Most of the settlers arrived in 1965, and the PIC Ji-Paraná distributed land in Cacoal in 1972. By 1980, the population had reached 67,030. The region was also explored for coffee plantations, especially in the 1980s and 1990s, but meat production quickly took over a larger percentage of the economic activities of the territory. In 2022, the population was 86,895 inhabitants.

Due to the Indigenous Territory Sete de Setembro, which has part of its area in the town of Cacoal, the remaining forest area (37.73%) in 2022 is mainly located in the northern part of the municipality, together with the indigenous territory. In 1985, the forest covered 71.26% of the municipal territory (270,285 ha). Deforestation is linked to an increase in pastureland, which occupies more than 60% of the municipality in 2022.

Castanheiras

Castanheiras is a small town whose origins are linked to the Rolim de Moura colonisation project. It was a rural urban nucleus (NUAR), named for the presence of Castanheira trees (*Bertholletia excelsa*) in the area.

With a population of 3,233 inhabitants (2022) and 113,913 cattle (2021), its land use history is also related to the increase in pasture land and the decrease in forests. This process was particularly rapid until 2005. In 2022, forest covers 29.32% (26,176 ha) of the territory, while pastures cover 66.22% (59,128 ha).

Pimenta Bueno

The telegraph line was responsible for the birth of the small town of Pimenta Bueno. Prior to the colonisation project organised by INCRA, the region was sparsely populated, and the economy was organised around the exploitation of rubber and mining. With the arrival of the BR-364 highway in the 1960s and the INCRA project in 1969, several migrants arrived to settle in the area. In 1977, it was defined as a city, and in 1980, the population was 29,993 inhabitants. The population increased until the end of the 1990s and then began to decline. In 2022, the population was 34,998 inhabitants. The cattle population was 318,608 head in 2021.

In 2018, a conservation unit was created in the municipality, the Estate Park Abaitará, with 152.25 ha. In the same year, local legislation was responsible for the cancellation of the creation of the park. In 2021, it was reinstated. This is an example of the process of Protected Areas Downgrading, Downsising and Degazettement (PADDD) when legal changes are made to reduce or eliminate legal protection.

Pimenta Bueno has been facing a reduction in its forest areas since 1985 but still gathers a good percentage of forest areas in the territory (58.98%, or 368,089 ha, in

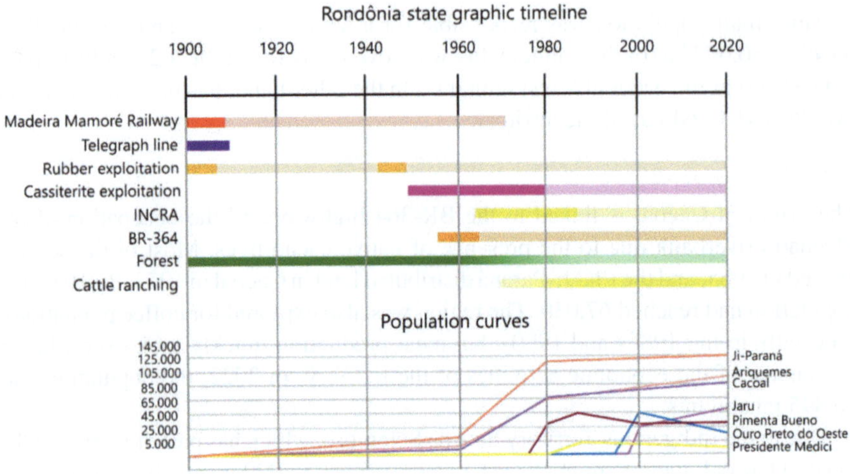

Fig. 5 History of Rondônia and the selected cities. *Source* Elaborated by the author. Data sources: IBGE (2022, 2019)

2022). The process of deforestation is the same as in the other municipalities studied, with a change of land use to pasture made by a slash-and-burn process.

The compiled data show that land use change and the history of the 1960s colonisation projects in the state of Rondônia are linked. A graphical summary of the data is presented in Fig. 5.

5 Conclusion

This chapter has examined the intense deforestation process that has taken place in Rondônia, linked to the clearing of forest areas by slash-and-burn due to the arrival of settlers from colonisation projects. The opening of the BR-364 highway, together with the colonisation projects developed by INCRA in the 1960s, were key factors in the rapid occupation of the territory and deforestation.

Using historical and land use change data for the municipalities of Ariquemes, Jaru, Ji-Paraná, Cacoal, Ouro Preto do Oeste, Presidente Médici, Ministro Andreazza, Castanheiras and Pimenta Bueno as case studies, the research was able to link the deforestation process to a cultural legacy of deforestation.

Some of the cities analysed have already reached critical scenarios, and there is an interesting relationship: the smaller the city is, the greater the deforestation process is, only if there are no protected areas in the territory. For example, Ouro Preto do Oeste had only 11.89% of its territory covered by forest. Jaru, only 14.93%; Presidente Médici, 16.47%; Ministro Andreazza, 17.64%.

On the other hand, the medium cities of the selected area (Ji-Paraná, Ariquemes and Cacoal) have more than 31% of forest areas in the territory.

The cities with the largest area of forest in the region analysed are Ji-Paraná and Pimenta Bueno. This is due to the presence of an indigenous territory in the former and a state park in the latter. The case of Pimenta Bueno deserves special attention, as there are ongoing local disputes over the demarcation of the protected area for the expansion of agriculture and cattle ranching.

The results point to the need to further investigate local environmental legislation in smaller cities to understand this correlation with the size of the city and the rate of deforestation. There is also a clear need to strengthen local actions of ecological education to support the necessary change in the cultural mindset of exploitation of the territory. On a global scale, it is necessary to change the attitude of consumption of animal products to support the protection of the forest.

Acknowledgements The authors gratefully acknowledge the support of the Alexander von Humboldt Foundation.

Author Contributions Conceptualisation, V.G.D.; analysis, V.G.D.; writing, V.G.D.; funding acquisition, V.G.D.; supervision, C.R.

Funding This research had the support of the Alexander von Humboldt Foundation (Experienced Researcher Fellowship Grant for the first author).

References

Alves MTR, Piontekowski VJ, Buscardo E, Pedlowski MA, Sano EE, Matricardi EAT (2021) Effects of settlement designs on deforestation and fragmentation in the Brazilian amazon. Land Use Policy 109:105710. https://doi.org/10.1016/j.landusepol.2021.105710

Berenguer, E et al (2021) Tracking the impacts of El Niño drought and fire in human-modified Amazonian forests. PNAS

Coy M (1988) Desenvolvimento regional na periferia amazônica: organização do espaço, conflitos de interesses e programas de planejamento dentro de uma região de "fronteira": o caso de Rondônia. In: Aubertin Catherine (ed.), Becker B. (pref.). Fronteiras. Brasilia (BRA). Universidade de Brasília, ORSTOM, Paris, pp 167–194. ISBN 85–230–0239–1

Cunha ET da, Moser LM (2010) Os projetos de colonização em Rondônia. Dossiê "A construção do Sujeito Contemporâneo: Perspectiva para a Amazônia—parte II". Revista Labirinto—Ano X, n.14, dez

Dall'igna LF (1996) E o garimpo de cassiterita em Rondônia. Revista Da CPRM 1:56–61

Donoso VG, Queiroga EF (2023) Social landscape, peripheral inclusion and un-practice: concepts for understanding social housing daily life in open spaces. Sustainability 15:12672. https://doi.org/10.3390/su151712672

Esquivel-Muelbert A et al (2018) Compositional response of amazon forests to climate change. Glob Chang Biol. https://doi.org/10.1111/gcb.14413

Felzke L (2018) A trama dos documentos em torno da demarcação da Terra Indígena Igarapé Lourdes. Rev Bras Linguística Antropológica 10(20):287–300. https://doi.org/10.26512/rbla.v10i2.20979

Ferreira IA (2017) Povos indígenas de rondônia: rumo aos 518 anos de resistência. Teoria e Debate, ed. 167, 18 dez. 2017. https://teoriaedebate.org.br/2017/12/18/povos-indigenas-rondonia-rumo-518-anos-de-resistencia/. Accessed on 11 Oct 2023

Fiori MF, Fiori LE, Nenevé M (2013) Colonização agrícola de Rondônia e (não) obrigatoriedade de desmatamento como garantia de posse sobre a propriedade rural. Novos Cadernos NAEA, 16(1), Suplemento:9–22. ISSN 1516–6481

Franco GY (2019) A construção do "Brasil Potência": a propaganda de estímulo a migração para o Norte do Brasil—um estudo a partir do caso de Rondônia (1968–1981). História Unisinos, 23(1):96–109. https://www.redalyc.org/journal/5798/579865455015/html/. Access in Mar 2023

Gatti LV, Basso LS, Miller JB et al (2021) Amazonia as a carbon source linked to deforestation and climate change. Nature 595:388–393. https://doi.org/10.1038/s41586-021-03629-6

Gomes FB, Bastos A (2018) Geomorfologia e sua relação com a configuração espacial do município de Machadinho D'Oeste no Estado de Rondônia—sudoeste da Amazônia. Ver Geogr (Recie) 35:242–253

IBGE (2019) Brazilian Institute of geography and statistics. Pesquisa da Pecuária Municipal. https://www.ibge.gov.br/estatisticas/economicas/agricultura-e-pecuaria/9107-producao-da-pecuariamunicipal.html?=&t=series-historicas. Accessed 23 Aug 2023

IBGE (2022) Brazilian Institute of geography and statistics. population census and projection. https://cidades.ibge.gov.br/brasil/ro/pesquisa/53/49645?ano=2050. Accessed 23 Aug 2023

Lawrence D, Vandecar K (2015) Effects of tropical deforestation on climate and agriculture. Nature Clim Change 5:27–36. https://doi.org/10.1038/nclimate2430

Macedo S, Queiroga EF, Campos ACA, Galender F, Custodio V (2018) Os sistemas de espaços livres e a constituição da esfera pública contemporânea no Brasil, 1 edn. Editora da Universidade de São Paulo, São Paulo

Mapbiomas (2021) Brazil annual land use and land cover mapping project. Collection 6.0. http://www.mapbiomas.org. Accessed at Nov 2021

Mapbiomas (2023) Brazil annual land use and land cover mapping project. Collection 8.0. http://www.mapbiomas.org. Accessed at Sept 2023

Meira GL de (2017) O desconhecimento da história dos indígenas Arikêmes pela população do município de Ariquemes. Graduation course dissertation in History. Federal Unviersity of Fronteira Sul, Erechim, RS

Monkkonen P, Comandon A, Escamilla JAM, Guerra E (2018) Urban sprawl and the growing geographic scale of segregation in Mexico, 1990–2010. Habitat Int 73:89–95

Morelato AAHN, Cardoso ACD (2020) Madeira mamoré railway company: O legado da company-town à cidade de Porto Velho, Rondônia. In: XII SIIU—Seminario Internacional de Investigação em Urbanismo. São Paulo and Lisboa

Ndubisi F (2002) Ecological planning. Johns Hopkins University Press

Neves E (2022) Sob os tempos do equinócio. Oito mil anos de história na Amazônia central. Edusp, São Paulo

Oliveira JL de (2010) Rondônia: geopolítica e estrutura fundiária. Grafiel, Porto Velho

Raisg (2020) Amazonia under pressure. www.amazoniasocioambiental.org. Accessed at Dec 2023

Silva C, Vergara-Perucich F (2021) Determinants of urban sprawl in Latin America: evidence from Santiago de Chile. SN Soc Sci 1:202. https://doi.org/10.1007/s43545-021-00197-4

Strauss L (1981) Tristes trópicos. Lisboa/São Paulo, Ed. 70/Martins Fontes

United Nations (2015) United Nations—sustainable development actions. www.un.org/sustainabledevelopment/cities/. Accessed in 07 Nov 2021

Zhang H et al (2021) Forest canopy changes in the southern Amazon during the 2019 fire season based on passive microwave and optical satellite observations. Remote Sens. https://doi.org/10.3390/rs13122238

Risk Assessment Through the Consumption of Mercury-Contaminated Fish in Riverine Communities, Brazilian Amazon

Sabrina do Carmo Vieira Pereira, Moniky Rayanne Silva dos Santos, Johnnasson de Medeiros Soares, Marcela Poenna de Sousa Farias, Edenise Garcia, and Heloísa do Nascimento de Moura Meneses

Abstract Fish can contain high levels of mercury, which can harm health when consumed in large quantities. This work aimed to assess mercury exposure and fish consumption and calculate the health risks associated with consumption of mercury-contaminated fish in the riverine communities of Baixo Amazonas area, in the Brazilian Amazon. *Materials and methods*: A health risk assessment was carried out according to the methodology of the World Health Organization. Weight and fish consumption data were obtained in 2022 through a structured questionnaire covering social, demographic, and behavioral variables, fish consumption and symptoms associated to Hg exposure; and blood mercury was measured. Different scenarios of Hg contamination were used. *Results*: 108 participants were selected, of whom 89% (n = 96) were exposed and frequently consumed fish. Fish consumption indicated that the daily intake of methylmercury exceeded the recommendation of 0.23 μg MeHg/kg for women of childbearing age and 0.45 μg MeHg/kg for adult men. In these two groups, the risk ratio was higher than 1.0; in women of childbearing age (18–49 years), the ratio was 1.8, and in adult men the ratio was 1.2.

S. do Carmo Vieira Pereira (✉) · J. de Medeiros Soares · H. do N. de Moura Meneses
Programa de Pós-graduação em Ciências da Saúde (PPGCSA), Universidade Federal do Oeste do Pará, Santarém 68040-470, Brazil
e-mail: sabrina.larisa@gmail.com

H. do N. de Moura Meneses
e-mail: heloisa.meneses@ufopa.edu.br

M. R. S. dos Santos · M. P. de Sousa Farias · H. do N. de Moura Meneses
Laboratório de Epidemiologia Molecular (LEpiMol), Instituto de Saúde Coletiva (ISCO), Universidade Federal do Oeste do Pará, Santarém 68040-470, Brazil

E. Garcia
Instituto de Conservação Ambiental, The Nature Conservancy Brasil, São Paulo 01311-936, Brazil
e-mail: egarcia@tnc.org

W. Leal Filho et al. (eds.), *Amazon 2030—Sustainability Issues in the World's Largest Rainforest Region*, World Sustainability Series,
https://doi.org/10.1007/978-3-031-81465-5_3

Conclusions: These results indicate that fish consumption although essential for the food and nutritional security of the riverine community, can cause significant health risks due to the consumption of mercury-contaminated fish.

Keywords Mercury · Fish · Risk assessment · Riverine community

1 Introduction

Mercury (Hg) is a neurotoxic element that affects not only the central nervous system but also other organs such as the kidneys, liver, and cardiovascular system, and can cause symptoms such as dizziness, headaches, numbness in the hands and feet, fatigue, and neuromotor changes (Crespo-Lopez et al. 2023; Nogara et al. 2019). Other health conditions have already been related to mercury exposure, such as hypertension, cardiovascular diseases and diabetes, with the Methylmercury (MeHg) being the Hg form the most associated with adverse health effects, especially in populations that frequently consume fish and are environmentally exposed to this metal (Farkhondeh et al. 2020; Fillion et al. 2006; Tsai et al. 2019).

Mercury exposure has been increasingly studied since the Minanata disease episode in Japan that highlighted this metal's adverse effects from acute exposure through consumption of contaminated fish (WHO 1990). Therefore, the food security of populations that have a diet based on fish is compromised when it is contaminated with MeHg and when access to a variety of foods is restricted. However, this consumption habit is still relevant from a nutritional, cultural and economic perspective (Gama et al. 2022).

In Brazil, the primary source of exposure to Hg is through the dietary route, via consumption of fish. In this context, riverine populations, which rely on fish as the basis of their diet and main source of protein, are those most at risk of mercury exposure. Therefore, the prevalence of mercury exposure tends to be higher in these communities, as observed by Meneses et al. (2022), in which 90% of riverine participants in the study were exposed.

Other risk groups are women of childbearing age, pregnant women and fetuses, as Hg can cross the placental barrier and affect fetal neurodevelopment (WHO 1990). The Minamata episode, when women with few or no clinical signs gave birth to children with severe neurological disorders indicates the risks to the fetus associated with high maternal exposure to mercury (Crammer 2020).

However, even at low Hg exposure levels, negative effects can be observed, such as impairment to neurobehavioral development. This fact can be partially explained by the passage of mercury through the placenta, as mercury levels are higher in the umbilical cord compared to maternal blood, with the fetus being more vulnerable to toxic effects such as genotoxicity and oxidative stress as its system is still developing (Chen and Dong 2022; Crespo-Lopez et al. 2009; Saavedra et al. 2022). In this context, in the study by Reuben et al. (2020), children between 5- and 12-year-old living near artisanal and small-scale gold mining had exposure to mercury and impaired cognitive development.

Due to the processes of biomagnification and bioaccumulation carnivorous and older fish tend to have the highest Hg levels (Burger and Gochfeld 2007; FAO/WHO 2003; Nogara et al. 2019; Wang and Wang 2019). Therefore, measures to assess or reduce exposure to Hg need to consider the frequency and quantity of consumption and the dietary habits of the fish consumed. It should be noted that, despite the risks associated with the consumption of fish, the exclusion of this item from the diet of riverine populations is not recommended, as it could worsen the situation of food insecurity in these communities. On the other hand, eating fish during pregnancy has been associated with positive outcomes too, such as healthy neurodevelopment. There are also evidences that certain nutrients in fish could mitigate toxic effects of Hg in children, but further studies are needed (Saavedra et al. 2022; Snetselaar et al. 2020).

This paper aims to assess mercury exposure and fish consumption and estimate the health risks associated with consuming mercury-contaminated fish in the riverine communities of lower Amazonas area, in the Brazilian Amazon.

2 Materials and Methods

The study was developed in communities of lower Amazonas area, on the Amazon riverbanks. This area is located near the municipality of Santarém in Pará, Brazil. Data was collected between November and December 2022 by mobilizing volunteers through the local Basic Health Unit.

A questionnaire with sociodemographic variables was used, and the weight of adult participants over 18 years of age was measured, excluding pregnant women. A portable Omron® bioimpedance scale with a maximum capacity of 150 kg and an accuracy of 0.1 kg was used to measure weight. In addition, a food frequency questionnaire (FFQ) was used to collect data on fish consumption, including information on types of fish, frequency, and amount in grams. The frequency of fish consumption was classified as daily (every day), frequent (1–6 times per week), and infrequent (less than once a week). The study was submitted to the Ethics and Research Committee, and an approval was obtained under number 4.764.646.

Blood mercury levels were measured by atomic absorption spectrometry using the DMA-80 Direct Mercury Analyzer (Milestone), with the accuracy of the results determined by the calibration curve with certified reference materials (Seronorm™ Trace Elements Whole Blood). The assay's detection limit was 0.2 μg/L, and samples were analyzed in duplicate. Regarding exposure levels, the World Health Organization (WHO) (1990) limit of up to 10.0 μg/L of total Hg in blood was taken into account, with individuals with Hg levels above this recommendation considered exposed.

The health risk assessment was performed by estimating the relationship between the amount of methylmercury ingested and the reference dose, following the methodology and reference limit proposed by WHO (2008). To calculate the daily mercury intake, it was assumed that the mercury present in fish was 100% of the mercury absorbed in the human gastrointestinal tract. The amount of mercury ingested was estimated using the equation:

$$IM = FI * MC/KBW$$

The IM refers to the intake of methylmercury per kilogram of body weight per day (g of methylmercury per kg of body weight per day); FI refers to the quantity of fish consumed per day (g/day); MC refers to the concentration of mercury in the fish consumed (g/g); and KBW refers to the kilogram of body weight.

The FAO/WHO limits were 0.23 μg Hg/kg bw/day for women of childbearing age (18–49 years) and 0.45 μg Hg/kg bw/day for men. For the mercury concentration in fish, the average values found by Basta et al. (2023) for the municipality of Santarém were 0.14 μg/g for average Hg, 0.35 μg/g for carnivorous fish and 0.02 μg/g for non-carnivorous fish. The value of 0.19 μg/g observed in the study carried out in the same study area (UFOPA 2023) was also adopted.

The result of the equation is evaluated as <1 or ≥1. If the risk ratio is less than 1, the risk of disease is considered low, and the risk of exposure is below the reference levels. However, if the risk ratio is equal to or greater than 1, there is a disease risk due to mercury exposure. Using Excel software, data analysis was performed using descriptive statistics (mean, median, standard deviation, minimum, and maximum) of Hg levels according to fish consumption and sociodemographic profile.

3 Results

A total of 108 individuals participated in the study; 63% (n = 68) were female and 37% (n = 40), male; 39,8% (n = 43) were between 41 and 59 years and 33.3% (n = 36) were over 60 years. Women of childbearing age (18–49 years) accounted for 55.8% (n = 38) of the women. Concerning race, 75.9% (n = 82) identified themselves as brown and 20.4% (n = 22) as white.

Looking at the mercury levels and the sociodemographic profile of the studied population (Table 1), the highest mercury averages were found in the older population, with an average of 30.67 μg/L; this age group also presented the highest level found among the participants, with 150.65 μg/L of total mercury in the blood. Despite being less numerous, men had the highest average Hg, 35.22 μg/L, with 98% (n = 39) exposed. In addition, 89% (n = 96) of the participants were considered exposed according to the WHO values, and 76% (n = 29) of the women of childbearing age were exposed.

Regarding fish consumption, 69.4% (n = 75) reported frequent consumption (at least once a week), with average mercury levels of 23.23 μg/L; and 30.6% (n = 33)

Table 1 Sociodemographic profile and blood Hg levels of participants

	N	%	Blood total Hg levels (μg/L)				
			Mean	SD	Median	Min	Max
Age (n = 108)							
18–20 years	3	2.8	12.42	5.69	11.92	7.00	18.34
21 a 40 years	26	24.1	16.68	9.75	13.99	3.28	41.21
41 a 59 years	43	39.8	28.42	14.95	24.95	9.97	75.94
Over 60 years	36	33.3	30.67	24.72	27.13	6.81	150.65
Gender (n = 108)							
Female	68	63.0	20.42	12.08	19.28	3.28	75.94
Male	40	37.0	35.22	23.63	33.05	9.98	150.65
Race (n = 108)							
White	22	20.4	27.05	12.75	25.24	7.15	54.26
Brown	82	75.9	25.57	20.36	22.02	3.28	150.65
Black	4	3.7	26.36	3.49	26.44	22.11	30.44
Instruction (n = 108)							
Higher education	4	3.7	7.42	3.65	7.19	3.28	12.00
Basic education	102	94.4	26.41	18.71	22.86	6.78	150.65
Did not attend school	2	1.9	37.06	5.31	37.06	33.31	40.82
Fish consumption (n = 108)							
Daily	33	30.6	31.97	25.73	24.37	9.48	150.65
Frequent	75	69.4	23.23	13.82	22.23	3.28	75.48

reported daily consumption, with higher average Hg levels, 31.97 μg/L. In addition, average daily consumption (n = 108) was 169 g, with men having the highest daily average (204 g), while women of childbearing age had an average of 150 g.

The fish most commonly reported to have been consumed in the month prior to sampling were acari, curimatã, pescada, surubim, and tambaqui, two of which (pescada e surubim) are considered carnivorous species. The highest mean blood mercury levels were found in those who reported eating piranha (33.49 μg/L), traíra (31.07 μg/L), pirarara (29.21 μg/L), matrinchã (28.57 μg/L) and pacu (28.50 μg/L) (Table 2). Regarding the eating habits of the fishes consumed, about 37% of the participants reported consuming carnivorous fishes, while 63% of the participants reported predominantly consumption of non-carnivorous fishes.

The risk estimate considering the value of 0.19 μg/g for fish observed in the study area (UFOPA 2023) resulted in values greater than 1 for women of childbearing age and men, with ratios of 1.8 and 1.2, respectively (Table 3), which exceeded the reference limits recommended by FAO/WHO. Based on the lower reference values for total mercury in fish for the municipality of Santarém (0.14 μg/g) by Basta et al. (2023), the analysis of risk estimates still showed that the daily intake of methylmercury exceeded the reference limits for the group of women of childbearing

Table 2 Mercury levels according to the type of fish consumed

Fish	Eating habit	N	Blood total Hg levels (μg/L)				
			Mean	SD	Median	Min	Max
Acari	Non-carnivorous	97	26.08	19.62	22.11	3.28	150.65
Curimatã	Non-carnivorous	93	27.20	19.59	23.69	3.28	150.65
Pescada	Carnivorous	82	27.07	14.60	23.91	6.81	75.94
Surubim	Carnivorous	82	28.20	20.44	24.13	6.27	150.65
Tambaqui	Non-carnivorous	77	25.60	12.51	23.69	7.00	74.69
Tucunaré	Carnivorous	75	25.92	13.68	24.93	6.81	75.94
Pirapitinga	Non-carnivorous	53	27.86	12.60	27.61	8.11	75.94
Jaraqui	Non-carnivorous	53	26.48	11.51	25.92	9.61	54.26
Pacu	Non-carnivorous	52	28.50	21.05	24.27	6.81	150.65
Pirarucu	Carnivorous	48	26.76	22.91	21.61	6.27	150.65
Traíra	Carnivorous	47	31.07	21.57	28.24	9.98	150.65
Dourada	Carnivorous	43	27.20	11.17	27.81	8.11	54.26
Sardinha	Non-carnivorous	41	26.27	11.48	27.61	6.81	54.26
Aracu	Non-carnivorous	39	25.21	15.44	20.36	6.78	75.48
Branquinha	Non-carnivorous	38	25.83	11.11	27.97	6.81	48.41
Matrinchã	Non-carnivorous	33	28.57	15.08	27.46	8.11	75.94
Mapará	Non-carnivorous	33	27.49	10.95	30.44	6.27	44.20
Piaba	Non-carnivorous	31	27.86	11.42	28.77	8.11	48.41
Piranha	Carnivorous	25	33.49	10.65	34.59	9.98	54.26
Pirarara	Non-carnivorous	25	29.21	15.12	32.05	6.78	75.94
Charuto	Non-carnivorous	23	26.30	11.16	29.8	6.81	48.41
Caratinga	Non-carnivorous	19	24.10	10.91	28.51	9.98	40.19
Filhote	Carnivorous	19	25.55	11.78	24.37	8.11	44.20

age, that is, the risk ratio estimates had values greater than 1 (1.3, more precisely); for men the risk rate was 0.9. However, when the consumption of non-carnivorous fish was considered, the risks were lower for women of childbearing age and adult men. If the estimate was based on a value of 0.35 μg/g for carnivorous fish, the risk ratio increased to 3.0 for women of childbearing age and 2.2 for adult men. In summary, women of childbearing age in this community were at high risk of becoming ill from ingesting mercury-contaminated fish in all scenarios, except under the consumption of non-carnivorous fish.

Table 3 Risk assessment for women of childbearing age and adult males

Population group	Fish comsuption (g)	Hg ingested (μg/g)	Risk ratio			
			Mean Hg of 0.14 μg/g	Mean Hg of 0.19 μg/g	Mean Hg of carnivorous fish (0.35 μg/g)	Mean Hg of non-carnivorous fish (0.02 μg/g)
Women of childbearing age	150	0.3	1.3	1.8	3.0	0.2
Adult men	204	0.4	0.9	1.2	2.2	0.1

4 Discussion

Fish consumption indicates that the daily intake of methylmercury in the riverine communities of lower Amazonas area can exceed the recommendation of 0.23 μg MeHg/kg for women of childbearing age and 0.45 μg MeHg/kg for adult men. In these two groups, the risk ratio was higher than 1.0, even when considering the presence of MeHg in fish below the maximum limit established by the WHO, representing a health risk specially for women of childbearing age.

Fish is an essential source of nutrients that is readily available to this population. This food has several benefits for human health, including antioxidant, anti-inflammatory, cardioprotective, and neuroprotective properties, mainly attributed to polyunsaturated fatty acids (PUFAs), in addition to its role in preventing energy-protein malnutrition (Chen et al. 2022). Despite these benefits, methylmercury can reduce the protection attributed to fish consumption (Mozaffarian and Rimm 2006).

In this study, 69.4% of the riverine dwellers consumed different types of fish regularly, at least once a week, corroborating Passos et al. (2008), in which 92.5% of the participants consumed at least one meal with fish. However, Vasconcellos et al. (2021) considered regular consumption three times a week. They showed that 96% of the families in their survey had consumption within this classification.

The average daily fish consumption of the groups studied differed according to gender, with adult men consuming a higher average (204 g) than women of child-bearing age (150 g). Similar estimates were found in other studies, such as that by Vasconcellos et al. (2021), in which adult men consumed 216.75 g and women of childbearing age 168.58 g. However, in Passos et al. (2008), women consumed an average of 124 g and men 189.7 g, with a total average consumption lower than in this study. In addition, according to Passos et al., riverine residents used to eat on average a daily fish meal of around 141 g, which is lower than that found in this study.

Despite the amount of fish consumed per meal in this study, this intake is influenced by the availability of different species, which, depending on the proportion of species with a carnivorous diet, have higher or lower concentration of mercury that affect blood levels (Passos et al. 2008; Silva et al. 2020). The consumption of fish contaminated with MeHg is directly related to the presence of Hg in the blood, which indicates recent exposure. Meneses et al. (2022) assessed Hg levels in the blood of urban and riverine populations, including the community of Tapará, localized in lower Amazonas area. They showed higher levels of mercury in individuals who consumed fish daily.

In the Amazon region, according to a technical note published by Fiocruz (Basta et al. 2023), 21.3% of the fish assessed had mercury levels above the established safe limits (≥ 0.5 μg/g), with piscivorous fish having the highest concentrations (Lino et al. 2019). Thus, the type of fish consumed can influence the daily methylmercury intake. Silva and Lima (2020) observed that the highest levels of mercury were found in carnivorous/piscivorous fish, and the consumption of these species by the population of the Amazon Triple Frontier (Brazil, Peru and Colombia) represented 50% of all consumption, so the daily mercury intake was above the recommendation. However, this scenario was not observed in the present study, with results showing a higher intake of non-carnivorous fish.

This study estimates show that the daily methylmercury intake in general exceeded the FAO/WHO reference limits. Vasconcellos et al. (2021) also found risk ratio estimates above 1.0 in the Munduruku indigenous population, with women of childbearing age having an intake three times higher than the safe intake. In that study, 60% of the fish most consumed by families were piscivores.

High risks of adverse health effects were also observed in Hacon et al. (2020), with potential mercury intake dose higher than the acceptable reference dose of 0.23 μg/kg/day, in one of the scenarios studied, with risks for adults and women of reproductive age 26 and 20 times higher than the reference daily dose, respectively.

The estimated daily mercury intake varied between 0.11 and 1.48 μg/kg/day, with an average of 0.47 μg/kg/day for consuming fish containing 0.19 μg/kg of mercury. Passos et al. (2008) found a range of 0 to 11.8 μg/kg/day and an average of 0.92 μg/kg/day, and differences by gender, with women showing lower intake rates than men.

Fish is important to promote food and nutritional security in traditional communities. However, mercury contamination puts the food security of these communities into question, even under consumption within the concentration limits established by Anvisa of 0.5–1 μg/g (Ministério da Saúde 1998), which, alone, is insufficient to provide recommendations. Residents living in remote communities face a situation of food insecurity due to difficult access to food diversity, health services, and food safety, given that food can often be contaminated. Fish, being the greatest source of protein, promotes food safety and when contaminated or above the maximum recommended mercury limits questions arise about the safety of this food (Crespo-Lopez et al. 2021).

5 Conclusions

The risk assessment shows that exposure to mercury through fish consumption can put the riverine population studied at risk, particularly women of childbearing age. In addition, the quantity and frequency of fish consumption are directly related to the amount of Hg ingested. Although the levels of this metal in the fish consumed by the studied community were not quantified, the two references used were from collections made simultaneously with this study in the vicinity. Therefore, monitoring mercury levels in blood or hair is necessary to determine the level of exposure and possible related symptoms, especially in women of childbearing age and pregnant women, who are characterized as higher risk groups. Moreover, health education activities could be carried out to guide fish consumption regarding the quantity, frequency, and type of fish consumed, in order to reduce possible health risks, especially in risk groups.

Risk assessment is influenced by body mass, amount ingested and amount of MeHg present in the food. Therefore, there are some limitations inherent to this calculation, as it does not consider inter-individual variability related, for example, to genetic polymorphisms that may offer greater protection or toxicity (Andreoli and Sprovieri 2017). Furthermore, the pattern of food consumption itself influences the metabolism of this metal in the body, with greater protective factors, such as consumption of antioxidant foods (Chapman and Chan 2000). Thus, there is no consensus on the real health effects of long-term exposure in traditional communities, some studies already link mercury to cardiovascular events, changes in blood pressure and glycemia, in addition to neurological damage, but advances are needed in relation to effects in these communities (Cho 2017; Farkhondeh et al. 2020; Genchi et al. 2017; Passos and Mergler 2008).

References

Andreoli V, Sprovieri F (2017) Genetic aspects of susceptibility to mercury toxicity: An overview. Int J Environ Res Public Health 14(1):93. https://doi.org/10.3390/ijerph14010093

Basta PC, Vasconcellos ACS, Hallwass G, Yocota D, Pinto DOR, Aguiar DS, Campos C, Costa MO (2023) Análise regional dos níveis de mercúrio em peixes consumidos pela população da Amazônia brasileira. Iepé—Instituto de Pesquisa e Formação Indígena Fundação Oswaldo Cruz. https://informe.ensp.fiocruz.br/assets/anexos/2441a041be660fb7575f8fe0bf6f8f34.PDF

Burger J, Gochfeld M (2007) Risk to consumers from mercury in Pacific cod (Gadus macrocephalus) from the Aleutians: fish age and size effects. Environ Res 105(2):276–284. https://doi.org/10.1016/j.envres.2007.05.004

Chapman L, Chan HM (2000) The influence of nutrition on methyl mercury intoxication. Environ Health Perspect 108 (Suppl1):29–56. https://doi.org/10.1289/ehp.00108s129

Chen B, Dong S (2022) Mercury contamination in fish and its effects on the health of pregnant women and their fetuses, and guidance for fish consumption-a narrative review. Int J Environ Res Public Health 19(23):15929. https://doi.org/10.3390/ijerph192315929

Chen J, Jayachandran M, Bai W, Xu B (2022) A critical review on the health benefits of fish consumption and its bioactive constituents. Food Chem 1(369):130874. https://doi.org/10.1016/j.foodchem.2021.130874

Cho YM (2017) Fish consumption, mercury exposure, and the risk of cholesterol profiles: findings from the Korea national health and nutrition examination survey 2010–2011. Environ Health Toxicol 32:e2017014. https://doi.org/10.5620/eht.e2017014

Cranmer JS (2020) Special issue on fish, mercury and health. Neurotoxicology 79:209–210. https://doi.org/10.1016/j.neuro.2020.04.001

Crespo-López ME, Macêdo GL, Pereira SI, Arrifano GP, Picanço-Diniz DL, Nascimento JL, Herculano AM (2009) Mercury and human genotoxicity: critical considerations and possible molecular mechanisms. Pharmacol Res 60(4):212–220. https://doi.org/10.1016/j.phrs.2009.02.011

Crespo-Lopez ME, Augusto-Oliveira M, Lopes-Araújo A, Santos-Sacramento L, Yuki Takeda P, Macchi BM, Nascimento JLM, Maia CSF, Lima RR, Arrifano GP (2021) Mercury: What can we learn from the amazon? Environ Int 146:106223. https://doi.org/10.1016/j.envint.2020.106223

Crespo-Lopez ME, Arrifano GP, Augusto-Oliveira M, Macchi BM, Lima RR, Nascimento JLM, Souza CBA (2023) Mercury in the amazon: The danger of a single story. Ecotoxicol Environ Saf 256:114895. https://doi.org/10.1016/j.ecoenv.2023.114895

FAO/WHO (2003) Evaluation of certain food additives and contaminants: Sixty-first report of the joint FAO/WHO expert committee on food additives. In: Proceedings of the Joint FAO/WHO Expert Committee on Food Additives (JECFA), Rome, Italy, pp 10–19

Farkhondeh T, Afshari R, Mehrpour O, Samarghandian S (2020) Mercury and atherosclerosis: Cell biology, pathophysiology, and epidemiological studies. Biol Trace Elem Res 196(1):27–36. https://doi.org/10.1007/s12011-019-01899-w

Fillion M, Mergler D, Sousa Passos CJ, Larribe F, Lemire M, Guimarães JR (2006) A preliminary study of mercury exposure and blood pressure in the Brazilian amazon. Environ Health 10(5):29. https://doi.org/10.1186/1476-069X-5-29

Gama ASM, Corona LP, Tavares BM, Secoli SR (2022) Padrões de consumo alimentar nas comunidades ribeirinhas da região do médio rio Solimões—Amazonas—Brasil. Cien Saude Colet 27(7):2609–2620. https://doi.org/10.1590/1413-81232022277.20362021

Genchi G, Sinicropi MS, Carocci A, Lauria G, Catalano A (2017) Mercury exposure and heart diseases. Int J Environ Res Public Health 14(1):74. https://doi.org/10.3390/ijerph14010074

Hacon SdS, Oliveira-da-Costa M, Gama CdS, Ferreira R, Basta PC, Schramm A, Yokota D (2020) Mercury exposure through fish consumption in traditional communities in the Brazilian Northern amazon. Int J Environ Res Public Health 17(15):5269. https://doi.org/10.3390/ijerph17155269

Lino AS, Kasper D, Guida YS, Thomaz JR, Malm O (2019) Total and methyl mercury distribution in water, sediment, plankton and fish along the Tapajós river basin in the Brazilian amazon. Chemosphere 235:690–700. https://doi.org/10.1016/j.chemosphere.2019.06.212

Meneses HdNdM, Oliveira-da-Costa M, Basta PC, Morais CG, Pereira RJB, de Souza SMS, Hacon SdS (2022) Mercury contamination: A growing threat to riverine and urban communities in the Brazilian amazon. Int J Environ Res Public Health 19:2816. https://doi.org/10.3390/ijerph19052816

Ministério da Saúde (1998) Portaria n° 685, de 27 de agosto de 1998. In: N° 685, editada pela secretaria de vigilância sanitária. Diário Oficial da União, Brasil

Mozaffarian D, Rimm EB (2006) Fish intake, contaminants and human health: evaluating the risks and the benefits. JAMA 18;296(15):1885–1899. https://doi.org/10.1001/jama.296.15.1885

Nogara PA, Oliveira CS, Schmitz GL, Piquini PC, Farina M, Aschner M, Rocha JBT (2019) Methylmercury's chemistry: From the environment to the mammalian brain. Biochim Biophys Acta Gen Subj 1863(12):129284. https://doi.org/10.1016/j.bbagen.2019.01.006

Passos CJS, Mergler D (2008) Exposição humana ao mercúrio e efeitos adversos à saúde na Amazônia: uma revisão. Cad Saúde Pública [internet] 24:s503–s520. https://doi.org/10.1590/S0102-311X2008001600004

Passos CJS, Da Silva D, Lemire M, Fillion M, Guimarães JRD, Lucotte M, Mergler D (2008) Daily mercury intake in fish-eating populations in the Brazilian amazon. J Expo Sci Environ Epidemiol 18:76–87. https://doi.org/10.1038/sj.jes.7500599

Reuben A, Frischtak H, Berky A, Ortiz EJ, Morales AM, Hsu-Kim H, Pendergast LL, Pan WK (2020) Elevated hair mercury levels are associated with neurodevelopmental deficits in children living near artisanal and small-scale gold mining in Peru. GeoHealth 4:e2019GH000222. https://doi.org/10.1029/2019GH000222

Saavedra S, Fernández-Recamales Á, Sayago A, Cervera-Barajas A, González-Domínguez R, Gonzalez-Sanz JD (2022) Impact of dietary mercury intake during pregnancy on the health of neonates and children: a systematic review. Nutr Rev 80(2):317–328. https://doi.org/10.1093/nutrit/nuab029

Silva SF, Lima MO (2020) Mercury in the fish market in the amazon triple frontier and health risk assessment. Chemosphere 248:125989. https://doi.org/10.1016/j.chemosphere.2020.125989

Silva SF, Pereira JPG, Oliveira DC, Lima MO (2020) Methylmercury in predatory and non-predatory fish species marketed in the amazon triple frontier. Bull Environ Contam Toxicol 104:733–737. https://doi.org/10.1007/s00128-020-02862-5

Snetselaar L, Bailey R, Sabaté J, Van Horn L, Schneeman B, Spahn J, Kim JH, Bahnfleth C, Butera G, Terry N, Obbagy J (2020) Seafood consumption during pregnancy and lactation and neurocognitive development in the child: A systematic review [Internet]. USDA Nutrition Evidence Systematic Review, Alexandria (VA). https://doi.org/10.52570/NESR.DGAC2020.SR0502

Tsai TL, Kuo CC, Pan WH, Wu TN, Lin P, Wang SL (2019) Type 2 diabetes occurrence and mercury exposure—From the National nutrition and health survey in Taiwan. Environ Int 126:260–267. https://doi.org/10.1016/j.envint.2019.02.038

UFOPA (2023) Nível de mercúrio no pirarucu do PAE Tapará está dentro do limite permitido pela legislação vigente para consumo. https://www.ufopa.edu.br/iced/comunica/informes/nivel-de-mercurio-no-pirarucu-do-pae-tapara-esta-dentro-do-limite-permitido-pela-legislacao-vigente-para-consumo/

Vasconcellos ACSD, Hallwass G, Bezerra JG, Aciole ANS, Meneses HNDM, Lima MDO, Jesus IMD, Hacon SDS, Basta PC (2021) Health risk assessment of mercury exposure from fish consumption in munduruku indigenous communities in the Brazilian Amazon. Int J Environ Res Public Health 18(15):7940. https://doi.org/10.3390/ijerph18157940

Wang X, Wang WX (2019) The three 'B' of fish mercury in China: Bioaccumulation, biodynamics and biotransformation. Environ Pollut 250:216–232. https://doi.org/10.1016/j.envpol.2019.04.034

World Health Organization (2008) Guidance for identifying populations at risk from mercury exposure. Mercury Publications, Geneva, Switzerland

World Health Organization (1990) Environmental health criteria 101. In: Methylmercury. World Health Organization, Geneva, Switzerland

Struvite Recovery from Domestic and Agro-Industrial Wastewater: Agronomic Potential Fertilizer for the Circular Economy in Amazon Cities

Rose Caldas de Souza Meira, Dayse Drielly Souza Santana Vieira, Celeste Queiroz Rossi, Danyelle Barbosa Panassollo, and Joelma Lourenço Pereira Mendes

Abstract Phosphorus (P) is an essential macronutrients for plant growth. Brazil depends on the annual import of large quantities of fertilizers for agricultural purposes. Domestic and agro-industrial wastewater are promising secondary P sources for nutrient recovery from struvite synthesis. In this study, domestic and agro-industrial wastewater were evaluated as anthropogenic sources of P and nitrogen (N) for struvite precipitation under optimized conditions established in previous work by the group. The syntheses were performed with phosphorus concentrations of 300 mg/L, molar ratio of 1:1:2 (Mg: P: N) and a pH 10. Characterization was conducted using X-ray diffraction (XRD) and scanning electron microscopy (SEM/EDS). Preliminary agronomic tests using soils typical of the Amazon region showed promising results. Struvite is a high-quality, slow-release fertilizer suitable for agricultural use. The use of secondary P sources in struvite synthesis is a sustainable nutrient recycling alternative for agricultural applications in Amazonian soils.

Keywords Wastewater · Struvite · Smart fertilizer · Circular economy

1 Introduction

Wastewater is no longer considered waste as it can be utilized to recover nutrients and produce potential fertilizers. In recent decades, the demand for fertilizers for agricultural production has increased due to the growing global population. Currently,

---

R. C. de Souza Meira (✉) · D. D. S. S. Vieira · C. Q. Rossi · D. B. Panassollo · J. L. P. Mendes
Universidade Federal do Oeste do Pará—UFOPA, Santarém, Brazil
e-mail: rose.meira@ufopa.edu.br

W. Leal Filho et al. (eds.), *Amazon 2030—Sustainability Issues in the World's Largest Rainforest Region*, World Sustainability Series,
https://doi.org/10.1007/978-3-031-81465-5_4

the recovery of phosphorus from industrial wastewater has greater environmental and economic benefits (Zhang et al. 2023; Lyon et al. 2020).

Phosphorus is one of the essential macronutrients for plant growth. It (P) is involved in the majority of mechanisms necessary for plant development, such as photosynthesis, respiration, composition of nucleic acids, regulation of enzymes and different molecular, biochemical and physiological aspects (Hawkesford et al. 2022; Cong et al. 2020; López-Arredondo et al. 2014; Lambers 2022). Phosphorus deficiency significantly limits production and is irreplaceable for agricultural yields. However, reserves are projected to be depleted within a century, making it both a limited resource and a critical commodity (Cordell et al. 2009; Hu et al. 2023). Research into secondary sources of P is essential to ensure the sustainable use of this finite resource for future generations.

Brazil must import large quantities of fertilizer for agricultural use each year, especially since soils Amazonia, in general, soils in the region are poor in phosphorus, nonetheless this element is of great importance in forest dynamics (Reichert et al. 2022; Cunha et al. 2022).

In contrast, Brazil produces approximately 26 million cubic meters of wastewater daily, which could serve as a potential source of P (Meira et al. 2020). Coverage rates for basic sanitation in the country reveal significant deficits, underscoring the environmental and social challenges linked to this issue. In the cities of Amazonia, the deficits are the largest in the country. In this context, the evaluation of alternative ways to supply phosphorus (P) to plants becomes necessary, especially for crops that promote the development of small and medium-sized farmers, such as passion fruit.

Partially treated or untreated wastewater containing excess phosphorus and ammonium, when discharged into surface waters, can cause eutrophication and disrupt the balance of aquatic ecosystem (El-Maghrabi et al. 2022). A viable method to mitigate this issue is the utilization of wastewater for nutrient recovery (Lin et al. 2024).

Domestic and agro-industrial wastewater represent promising secondary P sources for nutrient recovery through struvite synthesis (Meira et al. 2020). According to Puyol et al. (2017), the use of wastewater for nutrient extraction and the creation of value-added products is essential for the transition of the global economy from a linear to a circular model.

In this context, struvite is an ammonium magnesium hexahydrate phosphate ($NH_4MgPO_4.6H_2O$), contains equimolar proportions of ammonium (NH_4^+), magnesium (Mg^{2+}), and phosphate (PO_4^{3-}) ions. Its formation depends on physicochemical conditions of the reaction, such as ion concentration, pH, temperature and it can be synthesized from different secondary sources of P.

Kathi et al. (2023) argue that the recycled struvite from domestic and agro-industrial wastewater offers multiple benefits, including the reduction of environmental pollution, nutrient recovery, the development of technologies to protect the environment, and generation of potential fertilizers for the circular economy in Amazonian cities. The valorization of waste into by-products, through the application of circular principle, is a central and essential factor in promoting sustainable economic development. Therefore, adopting the concept of a circular economy is critical to addressing global challenges.

2 Methodology

2.1 *Synthesis of Struvite from Anthropogenic Phosphorus Sources*

The effluent used as a source of phosphorus was collected from two anthropogenic sources. Sample 1—domestic wastewater, sample 2—agro-industrial wastewater, both samples were collected in systems located in the city of Santarém, Pará. The synthesis was performed using these samples and compared to sample 3, a standard reference sample (Meira et al. 2020).

The synthesis process used in samples 1 and 2 followed the methodology used in sample 3 (standard sample) described by Meira et al. (2020), as shown in the conditions described:

- A molar ratio of 1: 1: 2 (Mg: P: N), 300 mg/L of Phosphorous and a pH 10, at room temperature (RT);
- Solutions of ammonium chloride (NH_4Cl), magnesium hexahydrate chloride ($MgCl_2.6H_2O$) (supplied by Exodo Scientific), and domestic and agro-industrial wastewater were used as anthropogenic sources of Phosphorus;
- Sodium hydroxide (NaOH) (supplied by Sigmae Aldrich) was used for pH adjustment.

The solution was agitated in a two-liter reactor for five minutes under rapid mixing and then allowed to stand at room temperature (RT 25 °C) for an additional 20 minutes. At the end of the precipitation reaction, the solids were collected on a qualitative filter (150 mm) and then dried at 40 °C for 24 hours. The formation of struvite was confirmed by Powder X-ray diffraction PXRD (Fig. 1).

2.2 *Characterization of Anthropogenic Sources of Phosphorus*

The phosphorus sources were characterized to determine the contents of ammonium, magnesium and potassium. The analysis of ammonia, nitrogen and phosphorus was conducted using with a multiparameter photometer for nutrient analysis (Hanna Instruments Brazil).

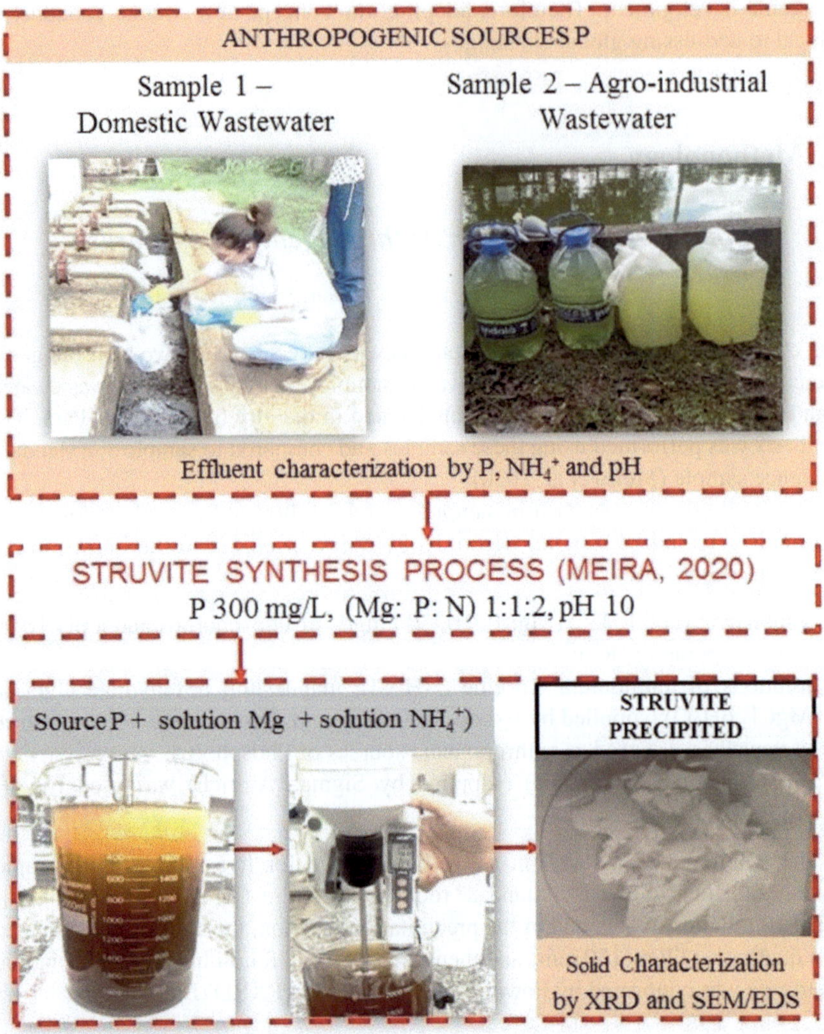

Fig. 1 Struvite synthesis process obtained from anthropogenic phosphorus sources

2.3 *Characterization of the Precipitated Material*

2.3.1 Powder X-ray Diffraction

The PXRD analysis was performed in a divergent beam diffractometer (Empyrean, PANalytical) equipped with a θ-θ goniometer, a Cu-sealed X-ray tube ($K\alpha = 1.5418$ Å), a kβ iron metal filter and a $PIXel^{3D}$ 2×2 area detector with an active length of 3.3473° 2θ (255 active channels). The instrument conditions were: (i) scan from 10 to 100°(2θ); (ii) 40 kV and 40 mA; (iii) 0.01° 2θ step size at 20 s per step; (iv) 1/2° divergent slit and 1° anti-scattering; Instrument resolution was evaluated using the LaB6 NIST/SRM 660b standard. PANalytical's X'Pert HighScore software version 3.0 was used in the identification of crystalline phases.

2.3.2 Scanning Electron Microscopy and Microchemical Analysis

SEM analysis was performed using Laboratory Equipment BRAND: OXFORD INSTRUMENTS and MODEL: X-MAX-80. The operating conditions were as follows: Detection area of $80 mm^2$ active area, Count rate of 500,000 cps, Resolution of 127 eV for Mn Ka and 48 eV for carbon, Resolution guaranteed according to ISO15632:2012, Acceleration voltage of 15 kV, Working distance of 15 mm and an element analysis counting time of 30 s.

2.4 Agronomic Experiment

The agronomic experiment was conducted out in the greenhouse of the Juruti University campus of the Federal University of Western Pará. The soil used to establish the experiment was a yellow Oxisol, with medium texture, collected at a depth of 0–20 cm. The chemical characterization of the soil used in the experiment was carried out according to the methodology of Embrapa (2017).

The cultivar of passion fruit used was BRS Gigante Amarelo, with the application of 5 treatments (100% struvite, 50% struvite, 100% NPK, 50% NPK and control), with 6 replications each. The experiment was implanted in DIC and the plants were evaluated during 45 days, measuring: plant height, number of leaves, stem diameter, number of lateral shoots, gas exchange (A, Gs and E), dry and fresh mass of root and shoot, root volume and level of available phosphorus in the soil.

The data obtained were subjected to analysis of variance and the means were compared by Tukey test at 5% level of significance, using SISVAR software. For the variables plant height, stem diameter, number of leaves, fresh mass of the aerial part, dry mass of the aerial part, fresh root mass, dry root mass, root volume, number of shoots and soil phosphorus content, 6 replications per treatment were used ($n = 6$). For the variables photosynthetic rate, stomatal conductance and transpiration, 4 replications per treatment were used ($n = 4$) (Fig. 2).

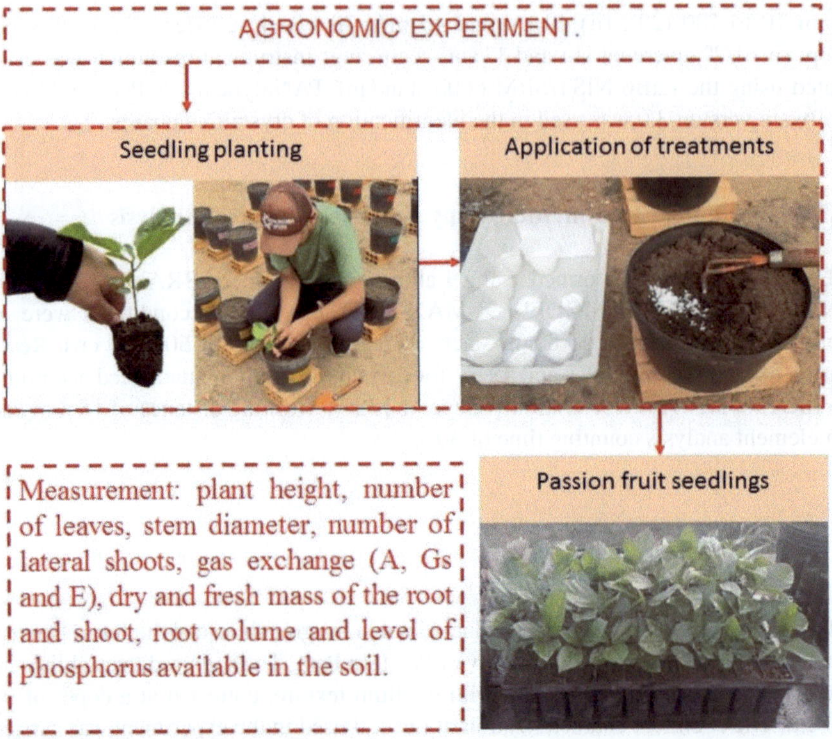

Fig. 2 Agronomic experiment process

3 Results and Discussion

3.1 Characterization of Synthesis Products by PXRD

The PXRD analysis results revealed the successful synthesis of struvite under condition S1, replicated in S2. This condition was compared with the optimal condition (standard sample S3) as shown in Fig. 3. The 23–26° 2θ CuKα region is highlighted to emphasize the subtle difference between the samples, where conditions S1 and S2 coincide with condition S3 and with the d111 values from the ICDD PDF, indicating only the struvite phase.

3.2 Synthesis Products Characterization by SEM–EDS

The SEM results presented in the SEM images show a typical morphology of struvite, with characteristic X-shaped crystals with dendritic growth, showing in all conditions

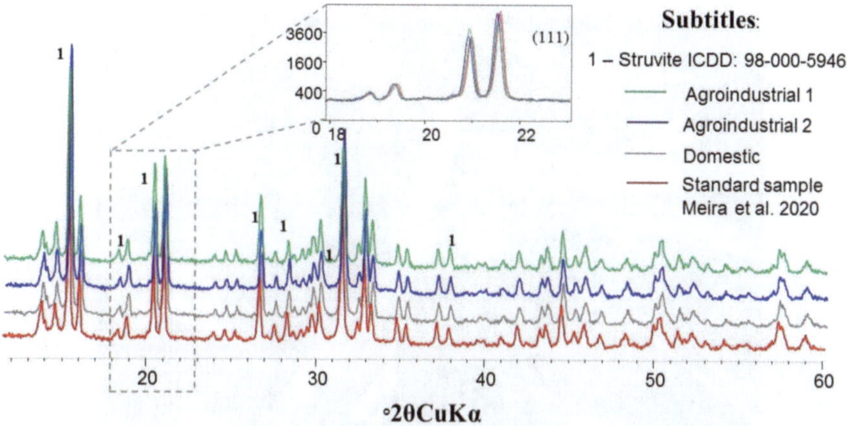

Fig. 3 Comparison of XRD patterns obtained under samples S1 and S2 with the standard sample S3 (Meira et al. 2020). The highlighted region is between 18–23° 2θCuKα

S1 (Fig. 4) and S2 (Fig. 5), similar to the standard sample S3 (Meira et al. 2020) shown in Fig. 6.

Fig. 4 Morphology struvite obtained under condition S1

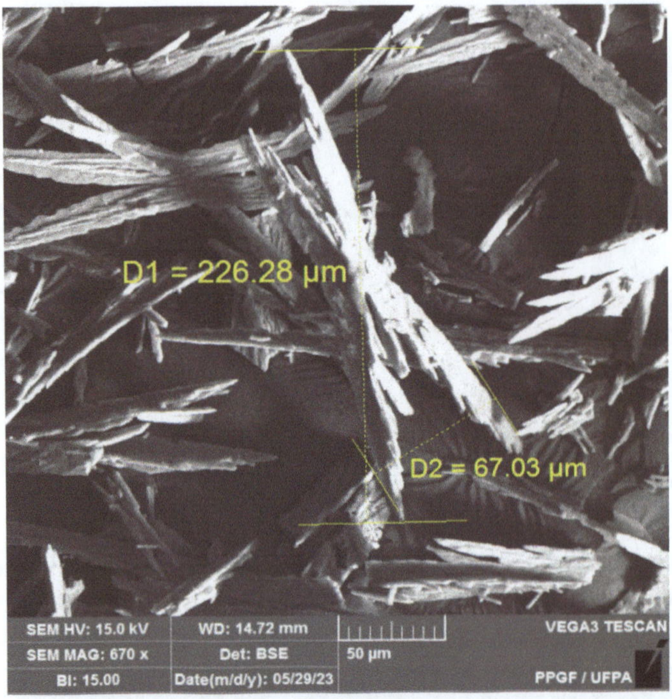

Fig. 5 Morphology struvite obtained under condition S2

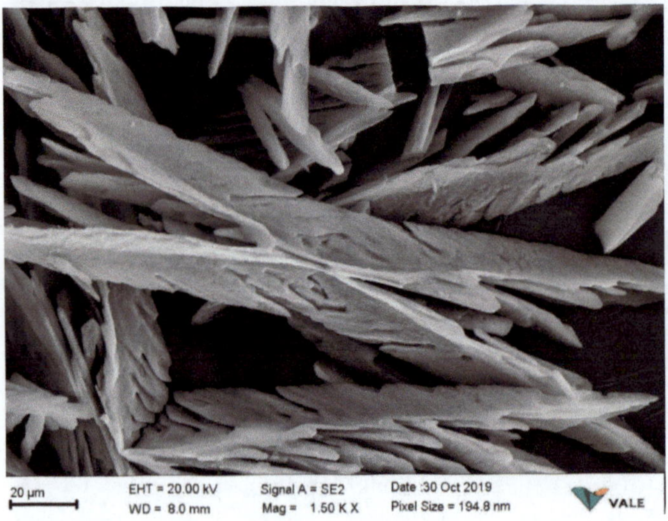

Fig. 6 Morphology struvite obtained under condition standard S3 by Meira et al. (2020)

3.3 Results of Agronomic Tests

This is the first agronomic test conducted out with this material. It was observed that treatments T1 and T2, which received 100% and 50% of the P_2O_5 dose with struvite, obtained better averages in the variables number of leaves, stem diameter, and fresh and dry mass of the aerial part (Table 1). However, they had a greater number of lateral shoots, lower values for height, the presence of leaf chlorosis and higher levels of available phosphorus in the soil, which may have caused an excess of this nutrient (P) for the plants (Figs. 7 and 8). In general, because phosphorus is an essential macronutrient for plant development, there are few studies on its excess in soils and/or plants (Jones Jr 1998).

Previous studies show that soils have phosphorus in their constitution however, they contain low availability for plants (Jones Jr 1998; Mardamootoo et al. 2021). This fact occurs due to the dynamics of this nutrient in the soil, as it remains adsorbed on the particles, or is lost through leaching (Pierzynski et al. 2000, 2005).

According to Mardamootoo et al. (2021), the effect of excess phosphorus is rarely observed in plants, as the mineral is strongly adsorbed in the soil. However, excess phosphate fertilizers can cause micronutrient deficiencies, such as zinc and iron. Other studies report that excess phosphorus causes indirect damage to the plant, especially in soils with low sorption capacity, influencing the metabolism and behavior of other nutrients (Mengel and Kirkby 2001; Fageria 2009). In view of this, additional tests with lower amounts of struvite extracted from the agro-industrial wastewater are needed to evaluate the type of phosphorus available to the plants, as well as the appropriate dosage for passion fruit and/or other crops of agronomic interest.

Table 1 Plant height, number of leaves, stem diameter and number of shoots at 45 days after transplanting passion fruit seedlings.

Treatments	N° leaves				Diameter				Height				N° shoots			
T1	40,83	±	5,71	a	6,92	±	0,47	a	70,67	±	12,71	b	8,17	±	2,64	a
T2	31,67	±	5,20	b	6,44	±	0,54	ab	77,00	±	12,03	ab	5,83	±	1,17	a
T3	18,33	±	1,63	c	5,38	±	0,26	c	93,28	±	9,30	a	1,50	±	1,05	b
T4	19,83	±	2,64	c	5,78	±	0,76	bc	92,17	±	8,66	a	1,17	±	0,41	b
T5	17,17	±	1,47	c	5,26	±	0,38	c	93,83	±	9,26	a	0,83	±	0,75	b

Note In Table 1, means followed by different letters in the columns are statistically different by the Tukey test, at the 5% probability level. The treatments are: T1 = 100% struvite; T2 = 50% struvite; T3 = 100% NPK; T4 = 50% NPK; and T5 = pure soil

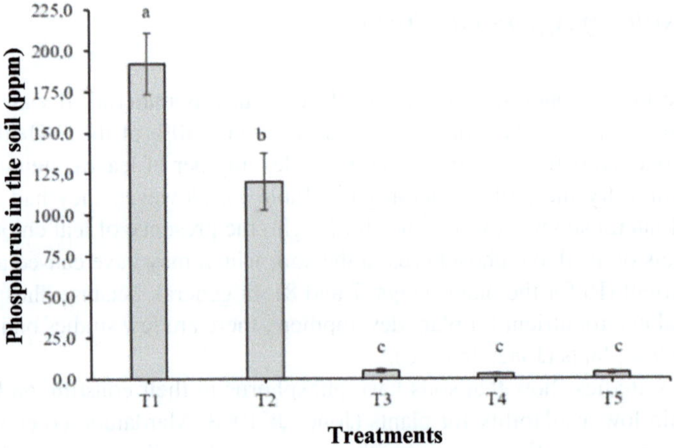

Fig. 7 Levels of phosphorus **a** available in the soil after 45 days of the experiment; Plants at the end of experiment (**b**)

Fig. 8 Development of passion fruit plants 45 days after transplanting, with the different treatments applied

4 Conclusions

The crystal-chemical analyses of the synthesis product showed the efficiency in the formation of struvite from domestic and agro-industrial wastewater. The XRD results indicate that struvite was the only phase formed under the synthesis condition, confirmed by the SEM/EDS analyses, which were similar to the struvite standard of Meira et al. (2020).

Struvite has the potential to become a source of nutrients for agronomic purposes, particularly for phosphorus recycling, showing results similar to commercial fertilizers in most of the evaluated variables.

Secondary P is a sustainable option in the context of agriculture in Amazonian soils as an alternative for recycling waste from struvite synthesis. However, more studies are needed with different struvite doses on the same material, in addition to testing other species to determine the actual form in which the P from struvite will be available to promote plan development and, ultimately, contribute to agricultural production in Amazonian cities.

Acknowledgements The authors acknowledge to FAPESPA (Fundação Amazônia de Amparo a Estudos e Pesquisas) and CNPq (Conselho Nacional de Desenvolvimento Científico e Tecnológico) for their financial support (Edital FAPESPA/CNPq Nº 002/2021, SICONV Nº 794276/2013). The authors acknowledge the LABNANO-AMAZON/UFPA network for its support through SEM facilities used in this work. The authors would also like to thank the MICROANALYSIS LABORATORY—IG/UFPA for qualified support.

References

Cong W-F, Suriyagoda LDB, Lambers H (2020) Tightening the phosphorus cycle through phosphorus-efficient crop genotypes. Trends Plant Sci 25:967–975

Cordell D, Drangert JO, White S (2009) The story of phosphorus: Global food security and food for thought. Glob Environ Chang 19:292–305. https://doi.org/10.1016/j.gloenvcha.2008.10.009

Cordell D, Rosemarin A, Schröder JJ, Smit AL (2011) Towards global phosphorus security: A systems framework for phosphorus recovery and reuse options. Chemosphere 84:747–758. https://doi.org/10.1016/j.chemosphere.2011.02.032

Cunha HFV, Andersen KM, Lugli LF (2022) Direct evidence for phosphorus limitation on Amazon forest productivity. Nature 608:558–562. https://doi.org/10.1038/s41586-022-05085-2

El-Maghrabi N, Fawzy M, Mahmoud AE (2022) Efficient removal of phosphate from wastewater by a novel phyto-graphene composite derived from palm byproducts. ACS Omega 7(49):45386–45402. https://doi.org/10.1021/acsomega.2c05985

Fageria NK (2009) The use of nutrients in crop plants. CRC Press, New York USA

Hawkesford MJ, Cakmak I, Coskun D, De Kok LJ, Lambers H (2022) Functions of macronutrients. In: Rengel Z, Cakmak I, White PJ (eds) Marschner's mineral nutrition of plants, 4th edn. Elsevier, London. In press

Hu S, Ma S, Li C, Shuyan Yu, Li C, Yang W (2023) Selective recovery of phosphorus from the leachate of incinerated sewage sludge ash using the Zr-modified acid-leaching residue as adsorbent for two-fold resource utilization. Chem Eng J, 477:147174. ISSN 1385-8947. https://doi.org/10.1016/j.cej.2023.147174

Jones JB Jr (1998) Phosphorus toxicity in tomato plants: when and how does it occur. Commun Soil Sci Plant Anal 29(11–14):1779–1784. https://doi.org/10.1080/00103629809370068

Kathi S, Singh S, Yadav R, Singh AN, Mahmoud AED (2023) Wastewater and sludge valorization: a novel approach for treatment and resource recovery to achieve circular economy concept. Front Chem Eng 5:1129783. https://doi.org/10.3389/fceng.2023.1129783

Lambers H (2022) Phosphorus acquisition and utilization in plants. Annu Rev Plant Biol 73:17–42. https://doi.org/10.1146/annurev-arplant-102720-125738

Lin H, Wang Y, Dong Y (2024) A review of methods, influencing factors and mechanisms for phosphorus recovery from sewage and sludge from municipal wastewater treatment plants. J Environ Chem Eng, 12(1):111657. ISSN 2213-3437. https://doi.org/10.1016/j.jece.2023.111657

López-Arredondo DL, Leyva-González MA, González-Morales SI, López-Bucio J, Herrera-Estrella L (2014) Phosphate nutrition: improving low-phosphate tolerance in crops. Annu Rev Plant Biol 65:95–123

Lyon C, Cordell D, Jacobs B, Martin-Ortega J, Marshall R, Camargo-Valero MA, Sherry E (2020) Five pillars for stakeholder analyses in sustainability transformations: The global case of phosphorus. Environ Sci & Policy, 107:80–89. ISSN1462-9011, https://doi.org/10.1016/j.envsci.2020.02.019

Mardamootoo T, du Preez CC, Barnard JH (2021) Phosphorus management issues for crop production: A review. Afr J Agric Res 17(7):939–952. https://doi.org/10.5897/AJAR2020.15205

Meira RCS, Paz SPA, Corrêa JAM (2020). XRD-Rietveld analysis as a tool for monitoring struvite analog precipitation from wastewater: P, Mg, N and K recovery for fertilizer production. J Mater Res Technol 9(6):15202–15213. https://doi.org/10.1016/j.jmrt.2020.10.082

Mengel K, Kirkby (2001) Principles of plant nutrition, 5th edn. Kluwer Academic Publishers, Dordrecht, The Netherlands

Pierzynski GM, Sims JT, Vance FG (2000) Soil phosphorus and environmental quality. In: Soils and environmental quality, 2nd edn. CRC Press, Boca Raton, Florida, USA

Pierzynski GM, McDowell RW, Sims, JT (2005) Chemistry, cycling, and potential movement of inorganic phosphorus in soil. In: Sims JT, Sharpley AN (eds) Phosphorus: agriculture and the environment. American Society of Agronomy, Madison, Wisconsin, USA

Puyol D, Batstone DJ, Hülsen T, Astals S, Peces M, Krömer JO (2017) Resource recovery from wastewater by biological technologies: Opportunities, challenges, and prospects. Front Microbiol 7:2106. https://doi.org/10.3389/fmicb.2016.02106

Reichert T, Rammig A, Fuchslueger L, Lugli LF, Quesada CA, Fleischer K (2022) Plant phosphorus-use and -acquisition strategies in Amazonia. New Phytol. https://doi.org/10.1111/nph.17985

Teixeira WG (eds) (2017) Manual de Métodos de Análise de Solo. 3 ed. Embrapa, Brasília, p 573. ISBN 9788570357717. https://www.embrapa.br/busca-de-publicacoes/-/publicacao/990374/manual-de-metodos-de-analise-de-solo

Zhang Y, Qin J, Chen Z, Chen Y, Zheng X, Guo L, Wang X (2023) Efficient removal and recovery of phosphorus from industrial wastewater in the form of vivianite. Environ Res, 228:115848. ISSN 0013–9351. https://doi.org/10.1016/j.envres.2023.115848

Environmental Policy, Climate Emergency and Inequalities in the Amazon (2015–2023)

Roberto Goulart Menezes, Ana Tereza Reis da Silva, and Carlos Potiara Castro

Abstract The current period is one of unprecedented historical inflection, in which the destiny of the Amazon is irreversibly intertwined with the destiny of the planet. One of the challenges in this context is the need to overcome the colonial imaginary of the Amazon. It is against this background that studies and an important research agenda have emerged to analyze the dynamics between development, inequalities, and environmental protection with a focus on the Legal Amazon. Therefore, the objective of this chapter is to analyze some aspects that permeate the debate on climate emergencies, global governance and the Amazon, with an emphasis on the Sustainable Development Goals (SDGs) 10 and 13. We analyze environmental protection policies for the region in the period between 2015 and 2023 along three axes: developmentalism in the Amazon and its correlation with ecological modernization, situating the disputes between sustainability and neo-extractivist projects; the process of discontinuity and rupture of the country's environmental protection policy under the Temer and Bolsonaro governments; and the resumption of the environmental agenda by the Lula da Silva government, within the scope of domestic politics and international relations, indicating the advances already achieved and the challenges of 2030 Agenda for the Amazon biome.

Keywords Amazon · Environmental governance · Inequalities · Agenda 2030 · Climate emergency

R. G. Menezes · A. T. R. da Silva
University of Brasília (UnB), Brasília, Brazil
e-mail: rgmenezes@unb.br

A. T. R. da Silva
e-mail: tapajuara@gmail.com

C. P. Castro (✉)
Federal University of Para (UFPA), Belem, PA 66075-110, Brazil
e-mail: poticastro@ufpa.br

W. Leal Filho et al. (eds.), *Amazon 2030—Sustainability Issues in the World's Largest Rainforest Region*, World Sustainability Series,
https://doi.org/10.1007/978-3-031-81465-5_5

1 Introduction

The Amazon biome is considered strategic in the global debate on climate governance (Wit and Mourato 2022). In this sense, one of the greatest challenges faced by countries in the region is to ensure the protection of nature combined with the promotion of social justice and equity (Hecht 2014). This concern was central to the historic speech that President Lula da Silva gave at the opening of the United Nations (UN) General Assembly in September 2023. According to him, "acting against climate change implies thinking about tomorrow and facing historical inequalities". The president also reaffirmed the leading role and sovereignty that the Pan-Amazonian countries must have in the construction of a global environmental policy that includes the region: "The whole world has always talked about the Amazon. Now, the Amazon is speaking for itself" (Lula da Silva 2023).

This phrase evokes, on the one hand, the strategic importance of the region in confronting climate collapse and the growing interest that the biome has aroused around the world, especially in protecting its forests and the people who live there. On the other hand, the phrase can be interpreted as a gesture of recognition of the agency capacity of the forest and its people, indicating the need for an effective role of the region in the construction and formulation of a new civilizing path—a path that respects the life improvement projects of local communities and considers nature as a subject of law (Gudynas 2019). This involves a reconciliation between environmental protection and social justice, which proposes the idea of full citizenship within the forest—*forestry*, as formulated by the Alliance of Forest Peoples (Cunha and Almeida 2000)—and development whilst keeping the forest intact, as proposed by geographer Bertha Becker (2006)

The *new climate regime* (Latour 2020a, b) to which we are subjected also implies that we are going through a period of unprecedented historical inflection, in which the destiny of the Amazon is irreversibly intertwined with the destiny of the planet. One of the challenges is to overcome the colonial imaginary of the Amazon that remains in force, stigmatizing its people either as an obstacle to progress or as an obstacle to ill-fated preservationist projects that aim to safeguard the Amazon without regard to the people who live there (Reis da Silva 2015). Resembling the modern myth of untouched nature (Diegues 2008), this view disregards how the biome is the result of a historical interaction between its people and the landscape, of ancestral practices of use and management of the territory that have enabled the expansion of agrobiodiversity whilst preserving the forest (Adams et al. 2006).

It is in this context that studies and an important research agenda have emerged to analyze the dynamics between development, inequalities, and environmental protection with regard to the Legal Amazon.[1] Therefore, the objective of this chapter is to analyze some aspects that permeate the debate on climate emergencies, global

---

[1] The Legal Amazon was established by Law 1,806 of 1953. Today it corresponds to 59% of the Brazilian territory—an area of around 5 million square kilometers. It encompasses nine states and 772 municipalities distributed as follows: Acre (22 municipalities), Amapá (16), Amazonas (62), Mato Grosso (141), Pará (144), Rondônia (52), Roraima (15), Tocantins (139) and part of Maranhão

governance, and the Amazon, with an emphasis on Sustainable Development Goals (SDGs) 10 and 13. We examine environmental protection policies for the region in the period between 2015 and 2023. To this end, the first section contextualizes developmentalism in the Amazon and its correlation with ecological modernization, outlining the disputes between sustainability and neo-extractivist projects, between the market economy and life prospects of traditional peoples and communities (Porto et al. 2013; North and Grinspun 2016; Menezes and Kraychete 2022). The second part analyzes the process of discontinuity and rupture of the country's environmental protection policy under the Temer and Bolsonaro governments. Finally, the third section looks at the resumption of the environmental agenda by the Lula da Silva government, within the scope of internal politics and international relations, pointing out the already achieved advances and the challenges of the Agenda 2030 for the Amazon biome.

The reflections developed here place the Amazon at the center of a global debate and aims to build another, more sustainable and fair civilizing path (Garret et al. 2021). This starting point demands a rejection of the colonial imaginary and the economic model imposed on the region since colonization and predicated on the plunder and the intensive and disorderly extractivism that have historically transformed the Amazon into a *zone of sacrifice* (Oliveira and Hecht 2019).

2 Environmental Governance in the Context of Ecological Modernization: Between Sustainability and Neo-extractivism

The search for an economic model that is sensitive to the biophysical limits of ecosystems has constantly come up against a mainstream vision of sustainable development based on the notion of ecological modernization, a middle way of sorts or conciliatory route between economic growth and sustainability (Romeiro 2012). Critics of ecological modernization have identified at least three limits of this perspective: the unbalanced relationship between economy and sustainability, in which the first overshadows the second; the predominance of a utilitarian conception of the environment; and the absence of social concerns regarding poverty, inequality, and environmental injustice. Moreover, they warn that the focus on technological innovations to mitigate environmental problems tends to exclude a deeper debate about the need for structural changes in current economic models (Altvater 2001; Acselrad 2004; Alier 2011). In 2008, the idea of ecological modernization gained new momentum when the United Nations Environment Program (UNEP) adopted the notion of *green economy* as a developmentalist ideal, based on clean technologies, aimed at promoting the "[...] improvement of human well-being and social equality" as well as a significant reduction "of environmental risks and ecological scarcity" (PNUMA 2011:

(181). It is home to 56% of the country's indigenous population and presents great biodiversity, including the tropical forest, 37% of the Cerrado biome and 40% of the Pantanal.

9). Despite the emphasis on social dimensions and ecological sustainability, which may be interpreted as an effort to reassess the historical predominance of economic aspects, nothing has changed with regard to the blind trust in clean technologies to correct environmental problems. What is more, by proposing that clean technologies can increase the efficiency of natural resource use (eco-efficiency) and, in the long term, decouple economic growth from its material and energy base, the green economy concept actually reinforces solutions based on scientific and technological development.

In some respects, the previously noted limits of ecological modernization appear the same in the context of the green economy. After all, if the problems caused by the capitalist logic of production and consumption can be indefinitely corrected by technological advances, it follows that nature does not impose restrictions on economic growth or, if it does, these would be merely circumstantial. Furthermore, according to Almeida (2013), we must consider the commercial interests at stake and the distribution of power between countries in the Global South and the Global North. Instead of supporting the diffusion of more sustainable economic practices, the clean technology market may deepen global inequalities. Additionally, there is another aspect that needs to be taken into account: that, by restricting solutions to the field of production and consumption of technologies, the green economy reduces the environmental issue to a market issue, superimposing mercantile interests on the collective interests of populational well-being, quality of life, and social equity (Tienhaara 2014).

In summary, the concept of green economy strengthens the conviction that "future developments in our ability to control nature will allow us to deal with any new problem and any undesirable side effects that arise from its implementation" (Lacey 2008, p. 163). Thus, the limits of ecosystems will never be seen as a "restriction to the possibilities of expanding the economy". They can, at most, "create relative and temporary obstacles that will eventually be overcome by innovations" (Veiga 2010: 122).

In the Amazon, this reasoning underpinned the developmentalist project implemented in the region by the military dictatorship, whose motto to "integrate so as to not surrender" created the image of an empty, uninhabited, and barbaric territory, a green *inferno* that needed to be dominated, tamed, explored, and modernized so it could contribute to the nation's progress (Milanez 2019). In the last 20 years, in particular, there has been a rapid expansion of economic frontiers in the region, including an increase in megaprojects (hydroelectric plants, ports, and railways, among others), the spread of large-scale monocultures, extensive livestock farming, mining, and forestry.

This advance of aggressive economic ventures, characterized by a joint participation of the state and private capital, has been identified by specialized literature as neo-extractivism. According to Svampa (2019: 18), transnational neo-extractivism in the twenty-first century is characterized by a "new dynamic of capital accumulation marked by increased pressure on natural assets, lands and territories". According to Svampa, the neo-extractivist logic is not restricted to countries led by governments known to be conservative or neoliberal but is also present in progressive governments in the first decade of the twenty-first century in Latin America (Coates and Sandroni 2023; Diniz 2013).

The Belo Monte hydroelectric power plant (HPP) is an emblematic example of transnational neo-extractivism. Conceived during the military regime (1964–1985), the project was carried out in 2011 under full democratic rule and in the context of a progressive government. The construction of Belo Monte was only possible thanks to the participation of the state, which acted to "circumvent" environmental laws and conditions, which, if applied, would make the project unfeasible. As interested parties, government bodies acted to make rules more flexible, ignored human rights violations, and turned a blind eye to a variety of environmental crimes. The environmental and social cost of Belo Monte involved an unprecedented violation of the human rights of the people of Volta Grande do Xingu, as well as rights of nature (Brum 2021). Both humans and animals were displaced, buried in the dam's construction site, or dammed along with the Xingu River. Furthermore, people and social movements in the region have denounced effects that had not been foreseen in environmental impact studies, associated with the HPP's full operation stage, such as the disappearance of species of fish and flora (Camara et al. 2005; Campelo and Neri 2013). These effects may exacerbate over time, irreversibly disrupting the biology of the river, the surrounding forest, and the lives of local residents. A similar disaster is predicted for the Tapajós basin, in western Pará. If approved, the São Luiz do Tapajós HPP will cover an area of 722.25 km^2, which includes protected areas and territories occupied by riverside communities and the Munduruku indigenous people (Barbosa 2023).

These examples demonstrate that neo-extractivism is directly associated with the phenomenon of expulsions caused by neoliberal policies since the 1980s (Sassen 2016, 2018). According to Sassen, these "predatory formations" constitute a feature of contemporary capitalism and occur through the "combination of elites and systemic capabilities", in which the financial markets enhance the concentration of wealth, whilst promoting expulsions and inequalities (2016: 22).

Still, according to Sassen, "systemic capabilities are a variable combination of technical, market and financial innovations, endorsed by the government, that exclude people", whether it be through the implementation of austerity policies, the suppression and violation of social rights, the acquisition of large tracts of land by corporations for large-scale monoculture, extensive livestock farming, logging or infrastructure megaprojects (Sassen 2016: 23). Although Sassen focuses on the transformations of contemporary developed capitalism, the analytical keys formulated by the author favor an examination of contemporary international reality, in particular new spaces for extracting profits and new forms of expulsion in the Global South, as is the case of the Amazon region.

Despite the lobbying and political force behind the developmentalist model, which has strong support from conservative and ultra-conservative sectors of the Brazilian Congress, the hegemony of this colonial and predatory economic project has been contested due to a growing ecological interest in the Amazon. Especially since the UN Conference on Environment and Development (ECO 92), the Amazon has gained visibility, as has the struggle of indigenous peoples in defense of their territories and ways of life (Posey 1983).

It follows from this that the advancement of neo-extractivist practices (such as the rapid expansion of the commodity frontier, the implementation of new mining and energy megaprojects, among others) does not occur without conflict. On the contrary, It has given rise to new processes of mobilization and resistance from traditional peoples and communities, especially those whose rights have been violated and their ways of life threatened by the advancement of economic development in their territories. What these people seek, in addition to the protection of their rights, is a nation-building project that includes them. It is, in the words of many community leaders, to "never again have a Brazil that excludes us" (Krenak 2019); nor a development model without civil participation or that disregards ancestral sustainability practices, specific knowledge, and the worldviews of traditional peoples.

Traditional peoples and communities therefore stand for a sustainability that goes beyond mere intention, environmentally correct discourses, and the "green" makeup of capital. As political agents who have taken a leading role in the face of climate challenges, these people aspire to a radical change in structures and dismantling of the hegemonic developmentalist model, which includes considering and taking seriously the development alternatives that are inspired by their ways of life and well-being, as societies not centered on the market economy (Castro et al. 2020).

The way in which these people have formulated an alternative perspective on sustainability, whether through their historical and ancestral practices, or through their worldviews and intellectual productions, profoundly challenges the conventional conceptions of sustainability, the latter still being dominant in the official discourse of world leaders, decision makers, international organizations, and science.

Regarding the conceptual aspect of sustainable development in the Brazilian context, Zhouri (2014: 131) emphasized that, "with a surprising catalytic capacity, the growing prestige of the notion of sustainable development was accompanied by a process of depoliticization of debates and concealment of conflicts, opening space for the paradigm of ecological modernization and its operational logic of adequacy within the scope of environmental licensing". In other words, such hollowing out of the definition of sustainable development ends up serving the purposes of the hegemonic development concept with its predatory logic.

Beyond any semantic disputes, and far from evoking a real break with developmentalism and the technical-scientific progress that is at the basis of the ecological crisis, the notion of sustainable development has always represented a wager on modernization as the choice path to "efficient economic growth sustained in the long term" (Romeiro 2012: 69). In recent years, this debate has gained new contours with the emergence of an "alternative" approach to clean technologies, the so-called bioeconomy. Although disputed, the formulations around this concept indicate that it is yet another effort to reconcile economic growth with environmental protection. An effort that, it is worth remembering, initially emerged in developed countries and has gradually gained prominence in the private sector and government projects in Brazil. This is the case with the Action Plan on Science, Technology, and Innovation in Bioeconomy (PACTI-Bioeconomia), which emerged in 2018 under the leadership of the Strategic Studies Management Center (CGEE), within the scope of Brazil's Ministry of Science, Technology, and Innovation (CGEE 2017).

Branco (2021: 8) remembers that, although the Action Plan is in line with the SDGs and the goals of the Paris Agreement, it does not make "[…] any mention of the Amazon or traditional peoples and their knowledge of its ecosystems. On the other hand, the raw materials most addressed by the Plan as a counterpart to the use of fossil derivatives are sugar cane, soy, corn, eucalyptus, sunflower, cotton, among other raw materials linked to agribusiness." The author states that Brazil's insertion in this topic has taken place in a subordinate way, with practices that, despite a discourse of sustainability and bioeconomy, maintain the same modus operandi of exploiting natural resources, based on the production or extraction of commodities. "Thus, it is possible to relate Brazil's insertion in the bioeconomy to the concept of 'commodity consensus' […], that is, a wager on an economy based on agribusiness and mineral extraction, which results in many environmental conflicts due to its destructive potential to the environment and social development, resulting in widening inequalities" (Branco 2021: 11).

This fact reveals that bioeconomy, as a practice and concept in dispute, is at risk of being turned into a fad and sold as a magical solution. Moreover, it may also serve as a cover for economic projects with predatory and neo-extractive intentions.

3 The Bolsonaro Government: Rupture in Environmental Policy

Between 2019 and 2022, Brazil experienced not only discontinuity in the country's environmental policy, but a rupture in the institutional and legal framework for the protection of nature (ASCEMA National 2020; Menezes and Barbosa 2021). The Bolsonaro government's predatory environmental policy dismissed any and all legislation that represented an obstacle to economic interests related to the exploitation of public lands for the expansion of agribusiness, mining, and logging. To this end, experienced public servants committed to the strict application of legislation were removed from their posts, while military and police officials came to hold key positions. The dismantling of the National Environmental Council (CONAMA) was another step in this process of rupture. Established in 1982, it had 96 counselors until April 2019. However, Decree 9,806/2019 reduced it to 23 members, nearly eliminating civil society representation. As a result, CONAMA, whose role it is to formulate the government's National Environmental Policy, became dominated by the economic sectors supportive of Bolsonaro's policies and that openly and radically defend the repeal of almost all Brazilian environmental legislation enacted over the last four decades and which had contributed to Brazil's reputation at international forums.

The objective of this decree was to eliminate the spaces for civil participation, social control, and dialogue between society and the authorities provided for in the 1988 Constitution and expanded since the country's redemocratization process. On another front, in September 2021, the then Minister of Economy, Paulo Guedes, presented a package with 14 measures to "loosen environmental protection", including the adoption of automatic environmental licensing, the relaxation of pesticide regulations and withdrawal of the Atlantic Rainforest's protected status (Miola 2021).

In the face of an unprecedented paralysis in the demarcations of Indigenous Lands, the advance of illegal mining, record numbers of deforestation and fires, declared government collusion with all sorts of environmental crimes, Brazil left the 2021 United Nations Climate Change Conference (COP26) discredited and unable to convince any of its interlocutors regarding the target plan to deal with the climate emergency. At COP 26, Brazil was directly questioned and asked to resume the Action Plan for the Prevention and Control of Deforestation in the Legal Amazon (PPCDAm), created in 2004. It was the excellent results of this Plan between 2004 and 2012 that had led to the creation of the Amazon Fund by Norway and Germany. At present, Brazil is responsible for 3.2% of global greenhouse gas emissions, so the supposed commitments made by that government were nothing more than narratives that sought to disguise its anti-environmental policy. Furthermore, the European Union had decided that there would be no Mercosur-EU Agreement while Bolsonaro was in office (Menezes and Sabatovicz 2019; Valor 2023).

It should be noted that, even during the military dictatorship and its repression of social movements, censorship and political persecution, the struggle to defend the environment and the ways of life of traditional peoples had not ceased. The Amazon was the scene of several of these struggles, as it was one of the biomes most affected by the dictatorship's developmentalist project. Infrastructure megaprojects (like the Tucuruí power plant and the Trans-Amazonian Highway), the distribution of land to large companies, incentives to mining, and the expropriation of traditional peoples' ancestral lands (indigenous people, *quilombolas*, river-side communities, and fishing communities, among others) comprised the strategies for exploiting the region with a view to achieving a so-called "economic miracle". During this period Brazil recorded high rates of economic growth, while violence in the countryside, hunger, poverty, and social inequality grew dramatically. However, despite the existence of more specific institutional mechanisms in the environmental area, many conflicts have remained and, in some cases, even worsened.

The events during Bolsonaro's tenure did not constitute a debate about which environmental policy to follow, but rather represented an attempt to deconstruct the entire legal framework that imposes environmental limits on economic practices. The government not only reinforced a perverse bureaucratic logic similar to the practices of the military regime, but also sought to break with the institutional framework of environmental protection under the guise of reducing bureaucracy to assist the country's development. In practice, this meant opening the door to unlimited exploitation of natural resources. Part of this push was to reverse the demarcations of indigenous lands already carried out in the country, to block the approval of new demarcations, and to authorize mineral and forestry exploitation and other economic activities harmful to the environment within indigenous lands and protected areas. Not by chance, it was precisely in the Amazon region, where 180 indigenous peoples live (around 80% of all indigenous people in Brazil), that the Bolsonaro government concentrated most of its predatory policies (Guajajara 2020; Hochstetler 2021; Medeiros 2023; Malheiro 2023). His government defended the spurious thesis of the time frame: only indigenous peoples who inhabited their lands on the date of promulgation of the 1988 Constitution would be recognized as having a right to claim territories—a policy widely supported by agribusiness in the country. In September 2023, Brazil's Federal Supreme Court declared the time frame proposal unconstitutional.

The territories of traditional peoples, as we know, are the last barriers of resistance to the advance of agribusiness, neo-extractive projects (mining and logging), and infrastructure megaprojects that threaten the maintenance of ecosystems. The presence of these peoples on their lands is what keeps the forest standing and mobilizes strategic environmental assets for policies to protect nature and natural assets (Reis da Silva 2015, 2019). Therefore, even though it has not been officially declared, burning the forest became a prime aspect of Bolsonaro's predatory environmental policy—not only because burning dramatically removes vegetation cover, favoring land grabbing and the implementation of prohibited economic activities in these areas, but also because the destruction of forests makes the life of the populations that live in them unfeasible, diminishing their ability to resist the advance of predatory economic interests (Isa 2023; UOL 2023).

In other words, given that the ways of life of traditional peoples are directly associated with the future of the forest and that their well-being depends on the well-being of nature, undermining the physical and symbolic integrity and the areas where they live became a constitutive part of this policy of destruction. The latter is, in essence, a perverse version of neo-extractivism, involving expulsions that not only annihilate nature and natural resources, but also subject entire populations to extreme situations of physical violence, illness and material poverty—as is the case of the Munduruku and Yanomami indigenous peoples, whose rights and human dignity were directly violated. As was widely reported by the national and international press, the situation worsened drastically during the Covid-19 pandemic.

It is in these terms that we see the protagonism and historical agency of the traditional peoples and communities of Brazil in an adverse context, marked by democratic setbacks. Through national and international networks and alliances, they resist a concept of development that seeks to domesticate, pillage, and commodify their homelands. In doing so, they resist a nefarious project that wishes to transform them from landowners into unskilled workers in agribusiness or on the outskirts of cities. These peoples would pursue another civilizing path, based on their own ways of life and relational ontology, in which nature figures as a family member, not a commodity. Thus they put into practice economies and ecological knowledge that may "sustain the sky" (Kopenawa 2015) and "postpone the end of world" (Krenak 2019).

4 The Lula Government: Challenges of Sustainable Development in the Amazon

The adoption of measures for environmental protection has been a hallmark of Lula da Silva's third term as president. Among the actions implemented in 2023 to reverse the predatory environmental policy implemented between 2016 and 2022, the following stand out: the resumption of the Plan to Combat Deforestation in the Amazon (PPCDAm) (Esteves 2023); the reestablishment of the Amazon Fund; the reconfiguration of CONAMA to ensure more social control and civil society participation; the resumption of environmental protection resolutions, as well as the annulment of measures taken by the Bolsonaro government that had aimed to make environmental regulations more flexible and to allow for the advancement of illegal mining in indigenous lands and protected areas.

From an institutional point of view, one of Lula's innovations was the creation of the Ministry of Indigenous Peoples. The indigenous leader Sônia Guajajara was invited to head this ministry, making her the first indigenous Minister of State in Brazilian history. Additionally, Lula has ordered the creation of secretariats, departments, directorates, and coordinating and advisory organs in 14 other ministries to be dedicated exclusively to the environmental agenda. Simultaneously, environmental control and inspection bodies are being reconfigured and reequipped.

One example of how this set of measures represents an important change of direction stands out. The resumption of inspections by organs like the Chico Mendes Institute for Biodiversity Conservation (ICMBIO) and the Brazilian Institute of the Environment and Natural Resources (IBAMA) in the fight against environmental crimes has already brought a 54% decrease in illegal deforestation in the Amazon and an 80% decrease in illegal mining in the Yanomami Reserve, compared to the same period in 2022 (Censipam 2023).

The renewed focus on the environmental agenda indicates that this is becoming one of the priorities of Brazil's international activities. In fact, environmental diplomacy is the area in which the Lula government has made the most progress in its first year of tenure (Spring 2022). During a meeting with US President Biden in February 2023, Lula obtained a contribution to the Amazon Fund, in addition to contributions from the European Union and the United Kingdom. Furthermore, the Amazon Summit in August 2023 in Belém, Brazil, and the choice for Brazil to host COP 30 in 2025 demonstrate how the environmental area has contributed to removing Brazil from international isolation. These developments represent a complete reversal in relation to Jair Bolsonaro's government.

It is no coincidence that Lula's inauguration in January 2023 and the appointment of Marina Silva as Minister of Environment and Climate Change was received as a boost in Brazil and around the world by all who fight for social and environmental justice.

However, despite the importance of these changes and the unequivocal message that Brazil has sent to the world by emphasizing its recommitment to the environmental agenda, in addition to other positive points, deforestation rates in the Legal Amazon still remain high.

Furthermore, the launch of the new Growth Acceleration Plan (PAC 2023) and the National Energy Transition Plan (PLANTE) of the Ministry of Mines and Energy are likely to increase pressure on the Amazon biome. It is worth noting, in particular, that almost half of the new PAC involves projects interrupted by previous governments. The proposed oil exploration at the mouth of the Amazon River is a contradiction in the Lula government's proclaimed environmental agenda. This demonstrates that despite the change of direction and the Lula government's strong environmental concern, the economic agenda remains largely developmentalist and hegemonic, revealing the contradictions and challenges to building an economic project for the country guided by sustainability and social justice.

4.1 Agenda 2030 and the Amazon: Climate Emergency and Inequalities

The Agenda 2030 was adopted by the UN in 2015 and consists of 17 Sustainable Development Goals and 169 targets. The fundamental principle is transversality in achieving this action plan. In this work, we have focused on SDGs 10 (Reducing Inequalities) and 13 (Action against global climate change). Eight years after its launch, Brazil and the vast majority of nations are far from reaching the 169 goals.

According to the 2023 Social Progress Index (IPS) in the Brazilian Amazon, produced by Imazon, poverty and inequality persist in the nine states of the region. According to the report, the creation of the IPS results from "the understanding that development measures based solely on economic indicators are insufficient, as economic growth without social progress can result in environmental degradation, increased inequality, exclusion and social conflicts" (Imazon 2023: 7). The 772 Amazonian municipalities were classified into five groups (tiers) based on social and environmental indicators aggregated into three dimensions: Basic Human Needs, Foundations for Well-being, and Opportunities. The following twelve components were also added: Nutrition and Basic Medical Care; Water and Sanitation; Housing; Personal Security; Access to Basic Knowledge; Access to Information and Communication; Health and Well-being; Quality of the Environment; Individual Rights; Individual Freedoms and Choice; Social Inclusion; and Access to Higher Education" (2023: 6).

In 2023, the Amazon's IPS was 54.32 compared to 67.94 in Brazil as a whole. In the Basic Human Needs dimension, the ratio was 64.83 to 76.30; Fundamentals for Well-being was 57.83 to 69.14, and Opportunities, the lowest of all, 40.31 to 58.38. The study has revealed new dimensions of a paradoxical reality in the country: the Amazon region, so rich in natural and energy resources, is also one of the poorest in

Brazil. Inequality in access to basic services captured by the 12 IPS indicators reveals that the population in the Amazon continues to be deprived of fundamental social rights. The conclusion of the report is that "among the 15 municipalities with the worst IPS scores, some are strongly associated with deforestation, forest degradation and social conflicts".

This reality was addressed at the 2023 Amazon Summit in Belém. The final document produced by the Summit, the Belém Declaration, contains 113 objectives and principles, related to issues such as climate change, social protection, and combating social inequalities. On climate emergency, the text makes reference to the perspective of common action in international forums to defend the interests of countries in the Amazon region. This includes, in particular, the collection of resources for climate financing from developed countries in the amount of US$ 100 billion/year to be directed to the adaptation and mitigation needs of countries in the Global South. Furthermore, the document encourages cooperation between scientific and university institutions in the Amazon to support the understanding of the "interrelations between climate change and forest ecosystems", which should include, in particular, the region's peoples and traditional communities. With reference to the fight against inequality, this issue appears throughout the Belém Declaration, but the fact that it appears in the second line of the document indicates the importance awarded to it (Amazon Cooperation Treaty 2023).

5 Conclusion

In this chapter, we have analyzed some aspects of the debate on the intersections between climate emergency, Brazilian environmental policy, and inequalities in the Amazon, for the years 2015–2023. During this period, the environmental protection policy in Brazil underwent processes of significant discontinuity (Temer government, 2016–2018) and rupture (Bolsonaro government, 2019–2022). In our view, the way the latter subverted institutional devices in place to protect nature, curtailed civil society's participation in consultative and deliberative bodies, violated the rights of traditional peoples and communities, and encouraged the destruction of biomes constituted a rupture rather than a mere dismantling of environmental protection (Castro 2022; Castro and Castro 2022). The Amazon was at the center of this predatory policy: illegal deforestation broke records, the invasion of miners and loggers in indigenous territories increased violence, and public-land grabbing became routine. With the aim of making the legal framework inoperative, the Bolsonaro government prohibited, in practice, inspections in areas where environmental crimes occurred. The number of fines applied fell sharply and their payment decreased by 93% (Hopwood et al. 2005; IPEA 2011; Kröger 2020a, b).

The reports published annually by Mapbiomas and Imazon have recorded a trail of destruction in this short period of time. Along with the destruction of the forest, the indicators have pointed to an increase in poverty and inequality in the region. The Social Progress Index is below the national average, even though the area possesses immense natural wealth, minerals, and energy potential. Subjected to the neo-extractivist logic, a colonial imaginary (Loureiro 2022), and expulsions, the Amazon remains permanently threatened.

The climate emergency, nature conservation, and recognition of the rights of traditional peoples and communities to their territories and ways of life were repeatedly ignored by that government. The fact is that Bolsonaro's ultra-neoliberal economic agenda was politically supported by agribusiness, big banks, and a large part of the industrial and commercial sector. The only privately expressed fear on the part of agribusiness was related to possible sanctions or even a boycott of Brazilian meat or soy exports. This was the case of the Mercosur-European Union Agreement, signed in July 2019, which was paralyzed due to the Bolsonaro government's predatory environmental policy.

To clarify, we understand that the dismantling of a policy can be reverted after a change of government, while a rupture shatters achievements that had taken years to consolidate and which cannot be easily repaired in the short term. This notwithstanding, the reconstruction process has been underway in Brazil since President Lula took office in January 2023.

At the international level, Bolsonaro tried to cover up the destruction of the Amazon Forest with the establishment of the National Council for the Legal Amazon (February 2019). Presided over by then vice-president Hamilton Mourão, this Council was made up of several cabinet ministers, but did not count with the participation of the Amazonian state governors or the region's civil society. The Council's objective was to convince foreign investors and Brazil's main commercial partners that the federal government was committed to environmental conservation, despite evidence to the contrary. However, the reality is that, during the Bolsonaro government, the Council had no credibility and merely served as a smokescreen for further setbacks.

Nevertheless, in the face of Bolsonaro and his destructive policies, traditional peoples and communities have resisted and maintained the long-term counter-colonial struggle in defense of their bodily integrity, territories, and rights. As Quijano (2009) reminds us, the protagonism and agency of these peoples do not shape the utopia of a world that is to come, but rather their projects of happiness in both the present and the past. Precisely because they are radically divergent from the interests of capital, their ways of life not only inform us about alternative civilizational paths, they also sustain the sky and postpone the end of the world.

In this context, it is important to remember the warning issued by the traditional peoples and communities, indicating that there is no magic solution to develop the region in a sustainable manner without popular participation. Just as there is no future country without the democratization of society and the inclusion of groups historically excluded from power, there can be no viable future for the Amazon without the political and intellectual protagonism of those who inhabit it. There is no possible future without their ancestral ecological knowledge and low impact local economies to formulate development alternatives for the Amazon region, which reconcile environmental protection, social justice, citizenry in the forest, and the standing forest.

Acknowledgements We thank the Research Support of the National Council for Scientific and Technological Development—CNPq, Brazil (Process 404006/2021-3 and 312103/2021-2). Menezes thanks the financial support of National Council for Scientific and Technological Development (CNPq) Fellow Researcher programme for supporting this research and the Federal District Research Foundation (FAPDF—Process nº 291/2023).

References

Acselrad H (2004) Justiça Ambiental: ação coletiva e estratégias argumentativas. In: Acselrad H et al (eds) Justiça ambiental e cidadania. Relume Dumará, Rio de Janeiro, pp 23–40

Adams C et al (2006) Sociedades caboclas amazônicas: modernidade e invisibilidade. Annablume, São Paulo

Almeida M (2013) Caipora e outros conflitos ontológicos. Revista de Antropologia da UFSCAR 5(1):7–28

Alier JM (2011) O ecologismo dos pobres. Contexto, São Paulo

Altvater E (2001) O preço da riqueza: pilhagem ambiental e a nova (des)ordem mundial. Unesp, São Paulo

Amazon Cooperation Treaty (2023) Belém declaration. https://www.gov.br/planalto/pt-br/assuntos/cop28/belem-declaration

ASCEMA National (2020) Chronology of an announced disaster: actions of the Bolsonaro government to dismantle environmental policies in Brazil. Associação Nacional dos Servidores de Meio Ambiente

Barbosa AL (2023) Ancestralidade, encantaria e resistência nas comunidades de Belo Monte de Vitória do Xingu e Belo Monte do Pontal na Volta Grande do Xingu. Programa de Pós-Graduação Povos e Comunidades Tradicionais, Universidade de Brasília, Dissertação de Mestrado

Becker B (2006) Amazônia: geopolítica na virada do III Milênio. Garamond, Rio de Janeiro

Branco L (2021) Bioeconomia: um caminho para o futuro ou um padrão de subordinação? OPEB. https://opeb.org/2021/06/28/bioeconomia-um-caminho-para-o-futuro-ou-um-padrao-de-subordinacao-renovado/

Brum E (2021) Banzeiro òkotó: uma viagem à Amazônia centro do mundo. Companhia das Letras, São Paulo

Camara G et al (2005) Amazonian deforestation models. Science 307(5712):1043–1044

Campelo T, Neri M (2013) (eds) Programa Bolsa Família: uma década de inclusão e cidadania. IPEA, Brasília. http://www.mds.gov.br/webarquivos/publicacao/bolsa_familia/Livros/Bolsa10anos.pdf

Castro C (2022) A ciência e a construção do campo de estudos amazônicos. CRV, Curitiba

Castro E, Castro C (2022) Desmatamento na Amazônia, desregulação socioambiental e financeirização do mercado de terras e de commodities. Novos Cadernos NAEA 25(1):11–36

CGEE (2017) Relatório Anual. CGEE/MCTI, Brasília. https://www.cgee.org.br/documents/10182/2994811/Rel_Anual_2017.pdf

Castro F, Lopes GR, Brondizio ES (2020) The Brazilian Amazon in Times of COVID-19: from crisis to transformation? Ambiente Sociedade 23:1–11. https://doi.org/10.1590/1809-4422asoc20200123vu2020L3ID

Censipam (2023) Resultados do PRODES 2022/23 Bioma Amazônia. Ministério do Meio Ambiente e Mudança Climática. https://www.gov.br/mma/pt-br/taxa-de-desmatamento-na-amazonia-cai-22-3-em-2023/coletiva-de-imprensa-prodes-22-23.pdf

Coates R, Sandroni L (2023) Protected truths: neoextractivism, conservation, and the rise of posttruth politics in Brazil. Ann Am Assoc Geogr 113(9):2048–2067. https://doi.org/10.1080/24694452.2023.2209627

Cunha MC, Almeida MWB (2000) Indigenous people, traditional people, and conservation in the Amazon. Daedalus 129(2):315–338
Diegues AC (2008) O mito da natureza intocada. Hucitec, São Paulo
Diniz E (2013) Desenvolvimento e Estado Desenvolvimentista: tensões e desafios da construção de um novo modelo para o Brasil do século XXI. Revista De Sociologia e Política 21(47):9–20. https://doi.org/10.1590/S0104-44782013000300002
Esteves B (2023) A descupinização: Como Lula vai combater o desmatamento. Revista Piauí:36–43
Garret RD et al (2021) Forests and sustainable development in the Brazilian Amazon: history, trends and future prospects. Annu Rev Environ Resour 46:625–652
Guajajara S (2020) Governo Bolsonaro: o retrato da barbárie contra os povos indígenas e a vida. In: Conflitos no Campo Brasil 2019. Goiânia, CPT Nacional, pp 182–89
Gudynas E (2019) Direitos da natureza: ética biocêntrica e políticas ambientais. Elefante, São Paulo
Hecht S (2014) The new Amazon geographies: insurgent citizenship, Amazon Nation and the politics of environmentalims. In: Vodjunec J, Schmink M (eds) Amazonian geographies. Routledge, London, pp 203–233
Hochstetler K (2021) O meio ambiente no governo Bolsonaro. In: Avritzer L et al (eds) Governo Bolsonaro. Autêntica, Belo Horizonte, pp 271–285
Hopwood B, Mellor M, O'brien G (2005) Sustainable development: mapping different approaches. Sustain Dev 13(1):38–52. https://doi.org/10.1002/sd.244
IMAZON (2023) Índice de Progresso Social na Amazônia Brasileira: IPS Amazônia 2023. https://amazonia2030.org.br/indice-de-progresso-social-na-amazonia-brasileira-ips-amazonia-2023/
IPEA (2011) Desigualdades: nos estados da região norte e no Brasil por regiões (comparativo). IPEA, Brasília. https://www.ipea.gov.br/code2011/chamada2011/pdf/area4/area4-artigo31.pdf
Isa (2023) Diálogos amazônicos: contribuições para o debate sobre sustentabilidade e inclusão. São Paulo
Kopenawa D (2015) A queda do céu: palavras de um xamã Yanomami. Companhia das Letras, São Paulo
Krenak A (2019) Ideias para adiar o fim do mundo. Companhia das Letras, São Paulo
Kröger M (2020a) Field research notes on Amazon deforestation during the Bolsonaro era. Globalizations 17:1080–1083. https://doi.org/10.1080/14747731.2020.1763063
Kröger M (2020b) Deforestation, cattle capitalism and neodevelopmentalism in the Chico Mendes extractive reserve, Brazil. J Peasant Stud 47:464–482. https://doi.org/10.1080/03066150.2019.1604510
Lacey H (2008) Valores e atividade científica 1. Editora, Sào Paulo, p 34
Latour B (2020a) Onde aterrar? como se orientar politicamente no Antropoceno. Bazar do Tempo, Rio de Janeiro
Latour B (2020b) Diante de Gaia: oito conferências sobre a natureza no Antropoceno. UBU, São Paulo
Loureiro V (2022) Amazônia: colônia do Brasil. Valer Editora, Manaus
Lula da Silva LI (2023) Discurso do presidente Luiz Inácio Lula da Silva na abertura da 78ª Assembleia da ONU. https://www.gov.br/planalto/pt-br/acompanhe-o-planalto/discursos-e-pronunciamentos/2023/discurso-do-presidente-luiz-inacio-lula-da-silva-na-abertura-da-78a-assembleia-da-onu
Malheiro B (2023) Geografias do Bolsonarismo: Entre a expansão das commodities do negacionismo e da fé evangélica no Brasil. Amazônia Latitude Press, Tallahassee, Manaus
Medeiros M (2023) Os ricos, os pobres. Desigualdade no Brasil. Cia. Das Letras, São Paulo
Menezes RG, Barbosa R (2021) Environmental governance under Bolsonaro: dismantling institutions, curtailing participation, delegitimising opposition. Z Vgl Polit Wiss 15(2):229–247. https://doi.org/10.1007/s12286-021-00491-8
Menezes R, Kraychete E (2022) (eds) Desigualdade Global e Desenvolvimento, Caderno CRH 35, Dossiê

Menezes RG, Sabatovicz S (2019) Agora vai? As negociações do Acordo Mercosul-União Européia no contexto da nova onda de liberalização radical da economia brasileira. In: Esposito Neto T, Sartori H (Org) União Europeia: visões brasileiras. Editora Appris, Curitiba, pp 195–205

Milanez F (2019) Colonialism, extractivism and re-existence in the Brazilian Amazon. In: Chertkovskaya E, Paulsson A, Barca S (eds) Towards a political economy of degrowth. Rowman and Littlefield, London

Miola I (2021) Despachante da destruição ambiental. Jornal Nexo. https://www.nexojornal.com.br/ensaio/2021/Despachantes-da-destruição-ambiental

North L, Grinspun R (2016) Neo-extractivism and the new Latin American developmentalism: the missing piece of rural transformation. Third World Q 37:1483–1504. https://doi.org/10.1080/01436597.2016.1159508

Oliveira GLT, Hecht SB (2019) Sacred groves, sacrifice zones and soy production: globalization, intensification of neo-nature in South America. Soy, globalization and environmental politics in South America. Routledge, London, London, pp 1–35

PNUMA (2011) Caminhos para o desenvolvimento sustentável e a erradicação da pobreza: síntese para tomadores de decisão. UN, New York

Porto MF, Pacheco T, Leroy JP (eds) (2013) Injustiça ambiental e saúde no Brasil: o mapa de conflitos. Editora Fiocruz, Rio de Janeiro

Posey D (1983) Indigenous ecological knowledge and development of the Amazon. In: Moran E (ed) The Dilemma of Amazonian development. Routledge, London

Quijano A (2009) Outro horizonte de sentido histórico. América Latina Em Movimento 441:2–5

Reis da Silva AT (2015) A conservação da Biodiversidade entre os saberes da tradição e a ciência. Estudos Avançados 29(83):233–259

Reis da Silva AT (2019) Áreas protegidas, populações tradicionais da Amazônia e novos arranjos conservacionistas. Revista Brasileira De Ciências Sociais 34(99):1–22

Romeiro A (2012) Desenvolvimento sustentável: uma perspectiva econômico-ecológica. Estudos Avançados 26(74):65–92

Sassen S (2018) Predatory logics: going well beyond inequality. In: Langman L, Smith D (eds) Twenty-first century inequality & capitalism. Brill, Leiden, pp 64–85

Sassen S (2016) Expulsões. brutalidade e complexidade na economia global. Paz e Terra, São Paulo

Spring J (2022) Brazil's green new deal: Lula promises environmental policy overhaul. https://www.reuters.com/world/americas/brazils-green-new-deal-lula-promises-environmental-policy-overhaul-2022-10-27/

Svampa M (2019) As fronteiras do neoextrativismo na América Latina. Elefante, São Paulo

Tienhaara K (2014) Varieties of green capitalism: economy and environment in the wake of the global financial crisis. Environ Polit 23(2):187–204

UOL (2023) Garimpo ilegal põe em risco ao menos 13 mil indígenas mundurukus e kayapós. https://noticias.uol.com.br/politica/ultimas-noticias/2023/01/26/garimpo-ilegal-poe-em-risco-20-mil-indigenas-de-povos-munduruku-e-kaiapo.htm

Valor (2023) Macron diz que é "totalmente contra"o acordo Mercosul-UE. https://valor.globo.com/mundo/noticia/2023/12/02/macron-diz-que-totalmente-contra-o-acordo-mercosul-ue.ghtml?li_source=LI&li_medium=news-page-widget

Wit F, Mourato J (2022) Governing the diverse forest: polycentric climate governance in the Amazon. World Dev 157. https://doi.org/10.1016/j.worlddev.2022.105955

Zhouri A (2014) Mapeando desigualdades ambientais. mineração e desregulação ambiental. In: Zhouri A, Valencio N (eds) Formas de matar, de morrer e de resistir. Belo Horizonte, UFMG, pp 111–141

Roberto Goulart Menezes is an Associate Professor at the Institute of International Relations at the University of Brasília (UnB), Professor of the Graduate Program in International Relations (IREL/UnB). He has also been Visiting Professor at Johns Hopkins University (2018–2019). Coordinator of the Center for Latin American Studies at IREL/UnB. Fellow researcher of the National Council for Scientific and Technological Development (CNPq).

Ana Tereza Reis da Silva is a riverside-cabocla-indigenous woman, originally from the floodplain region of Santarém-PA. Professor at the Faculty of Education at the University of Brasília in the Postgraduate Programs of Education (PPGE/UnB), Master's Degree Course in Sustainability of Traditional Peoples and Territories (MESPT/UnB). She coordinates the Education, Knowledge and Decoloniality Research Group (Gpdes/UnB/CNPq). Topics of interest: Decolonial pedagogy, Intercultural Education, Anti-racist Education, Ethnic-racial Relations, Socio-environmental Education, Decolonial thought and epistemologies of the South, Dialogue of knowledge, Epistemic and Cognitive Justice.

Carlos Potiara Castro is an Adjunct Professor at the Faculty of Social Sciences at the Federal University of Para (UFPA) in the Postgraduate Programs of Sociology and Anthropology (PPGSA/UFPA). He worked at the Secretariat of Biodiversity, Forests and Animal Rights of the Ministry of Environment and Climate Change, with international cooperation. He completed a Post-Doctorate at the Institute of International Relations at the University of Brasilia (IREL/UnB) and is the Executive secretary of the National Association of Postgraduate Studies and Research in Environment and Society (ANPPAS). He can be reached through the e-mail address poticastro@ufpa.br.

Mercury Exposure in Riverside Tapajós Communities in the Brazilian Amazon

Raissa Vasconcelos Rego, Ana Sofia Sousa de Holanda, Johnnasson de Medeiros Soares, Marcela Poenna de Sousa Farias, Edenise Garcia, and Heloísa do Nascimento de Moura Meneses

Abstract Mercury (Hg) is considered one of the most dangerous toxic metals for the environment and human health. The present study aimed to investigate mercury exposure in riverside communities of the Lower and Middle Tapajós River. This is an observational, cross-sectional, descriptive, and quantitative study, including 126 individuals from eight communities, carried out between 2021 and 2022. A questionnaire was administered, and hair samples were collected to measure total Hg levels. Results showed that Hg levels exceeded the World Health Organization (WHO) recommended exposure limit of 2 μg/g in all the communities, and in 95% of the participants. The highest levels of Hg were observed in two individuals from the São Luís do Tapajós community (18.60 μg/g). In relation to the symptoms reported, "muscle and joint pain" was mentioned by participants from all the communities and "decreased motor coordination" in four of the communities visited. These results indicate that participants from communities in the study region are exposed to Hg and the levels observed are sufficient to trigger the reported symptoms. Results from this study can influence public policies aimed at curbing illegal activities that increase Hg contamination of aquatic ecosystems and aquatic biota, including key species for the food security of riverine populations in the Amazon region.

R. V. Rego (✉) · A. S. S. de Holanda · J. de Medeiros Soares · H. do Nascimento de Moura Meneses
Laboratory of Molecular Epidemiology, Postgraduate Program in Health Sciences, Federal University of West Para, Santarém 68040-470, Brazil
e-mail: raissastm20@gmail.com

H. do Nascimento de Moura Meneses
e-mail: heloisa.meneses@ufopa.edu.br

M. P. de Sousa Farias
Laboratory of Molecular Epidemiology, Pharmacy Graduate, Federal University of West Para, Santarém 68040-470, Brazil

E. Garcia
The Nature Conservancy, Brasilia, Brazil
e-mail: egarcia@tnc.org

W. Leal Filho et al. (eds.), *Amazon 2030—Sustainability Issues in the World's Largest Rainforest Region*, World Sustainability Series,
https://doi.org/10.1007/978-3-031-81465-5_6

Keywords Environmental health · Public health · Environmental toxicology · Mercury exposure · Tapajós river · Amazon

1 Introduction

Mercury (Hg) is considered one of the most dangerous metals for the environment and human health due to its high toxicity and mobility in different ecosystems. It is also considered a global pollutant, which once released into the air can travel long distances and impact places far from its source of emission (NRC 2000).

Hg is released into the environment through natural and/or anthropogenic sources, undergoes complex chemical transformations and circulates between the atmosphere and terrestrial and aquatic environments (Wright and Welbourn 2002). There are three chemical forms of Hg: (1) elemental or metallic (Hg0); (2) inorganic (Hg+ and Hg2+); and (3) organic, with methylmercury (MeHg) being the most common form of organic Hg found in the food chain (ATSDR 1999).

Hg can be transformed in the environment (NRC 2000; Farina et al. 2013) through oxidoreduction and methylation processes. When transformed into methylmercury (MeHg), it poses a particular risk to humans and other animals, as it is considered toxic and can cause irreversible damage to the central nervous system (Clarkson et al. 2007; Abelsohn et al. 2002).

In the aquatic environment, Hg accumulates in sediments in lakes and rivers, where it changes to the methylmercury form, and enters the food chain, absorbed by phytoplankton and ingested by zooplankton, and ends up contaminating fish (WHO 2018; Silva et al. 2007). Fish contamination depends on where they are in the food chain, their size and age. Herbivorous have lower levels of contamination; this group includes tambaqui, jaraqui, pirapitinga and pacu, among the most fished Amazonian species. Carnivorous, on the contrary, tend to show high concentrations of MeHg, as they occupy the top of the aquatic food chain; in the Amazon Basin, this group includes sea bream, piranha, tucunaré and hake. Scientific studies show that 75% to 95% of the total amount of Hg accumulated in fish tissues is MeHg (Castilhos and Bidone 2000).

Consumption of fish is the most frequent component in the diet of riverine populations, and consequently the main route of ingestion of mercury (Gonçalvez and Gonçalvez 2004; Sá et al. 2006). Basta et al. (2023) have shown that more than a fifth of the fish sold in urban centers in the Brazilian northern region, which reach the tables of families in the Amazon region, have mercury levels above the safe limits established by the Food and Agriculture Organization and the World Health Organization (FAO/WHO) and the Brazilian Health Surveillance Agency (ANVISA). A study by Junior et al. (2018) showed increasing levels of Hg in the Tapajós riverine population, with average values of 7.25 μg in 2013 and 10.80 μg in 2014, both above the limit recommended by the WHO (6.0 μg), indicating increased chronic exposure to MeHg in these populations. These studies show that mercury contamination can affect millions of people who consume fish from the rivers that make up the Amazon basin, even if these consumers are hundreds of kilometers away from areas directly impacted by mining operations, a main source of Hg.

Mercury contamination has worrying effects on the human population, including in the Amazon region. A study conducted by Gibb and O'Leary (2014) documented the serious impacts of mercury exposure in riverine communities in the Brazilian Amazon. The results revealed that the consumption of mercury-contaminated fish led to a high prevalence of health problems, such as neurological disorders, delays in child development, impaired renal and cardiovascular function, among others. In addition, indigenous populations, who depend on the Amazon's natural resources for subsistence, including fish, are particularly at risk due to the ingestion of contaminated fish (Santos et al. 2002).

The artisanal gold mining activities carried out in the Amazon region, and particularly in the Tapajós River basin, since the early 1970s, are responsible for the release of Hg into the environment and are considered to be the main cause of Hg accumulation in fish and riverine communities (Nevado et al. 2010). However, in addition to mining, activities related to land use change, such as deforestation and fires, and the construction of hydroelectric dams, contribute to Hg contamination of the region's aquatic and terrestrial ecosystems (Fearnside 2003).

Slash-and-burn activities used to convert forests into pastures increase the remobilization of Hg accumulated in the soil (Lacerda and Malm 2008) and intensify the erosion process, accelerating the transport of Hg to aquatic environments (Fostier et al. 2000), where it can be transformed into MeHg, which has the capacity to bioaccumulate in organisms and biomagnify along the trophic chain, thus contaminating aquatic biota and humans.

Exposure to Hg raises concerns about adverse effects on the central and peripheral nervous systems, extending to the lungs, kidneys and gastrointestinal tract. Measuring Hg in biological samples such as hair, blood, nails and urine allows exposure to be quantified and associated with health effects (WHO 2018). The main symptoms of Hg exposure include tremors; insomnia; memory loss; headaches, as well as cognitive and motor dysfunction, which are easily confused with other diseases. Consequently, the effects of Hg on human health are often overlooked. This is particularly worrying in the Tapajós region, where mining and deforestation are widespread throughout the river basin, and where fish consumption is the main source of protein for riverine populations. This combination of factors puts these populations at risk of showing symptoms of the toxic effects of Hg exposure. In this context, the aim of this study was to investigate exposure to Hg in riverine communities in the Lower and Middle Tapajós River to provide scientific evidence that can support future public health actions aimed at mitigating the effects of this exposure on human health. This study can also influence public policies aimed at curbing illegal activities that increase Hg contamination of aquatic ecosystems and aquatic biota, including key species for the food security of Amazonian riverine populations.

2 Materials and Methods

A cross-sectional, descriptive, and quantitative observational study was carried out, including 126 individuals from eight riverine communities in the Lower and Middle Tapajós River, in Pará State, in the Brazilian Amazon. The studied communities were: Parauá, Suruacá (Santarém municipality—Lower Tapajós); Apacê, Cassuepá, Cametá (Aveiro municipality—Middle Tapajós), Nazaré, São Luís do Tapajós and Barreiras (Itaituba municipality—Middle Tapajós) (Fig. 1). The study included men and women, both smokers and non-smokers, aged between 18 and 80. Hair was used as a biomarker of Hg exposure. The research was conducted in accordance with Resolution 466 of 2012 of the National Health Council and the research protocol was submitted to and approved by the CEP UEPA (CAAE: 46,276,921.1.0000.5168). The research was carried out at the Laboratory of Molecular Epidemiology of the University of Western Pará State (LEpiMol/UFOPA).

2.1 Data Collection

Data were collected between September 2021 and March 2022. Participants were informed of the aims and procedures of the study and voluntarily signed the Free and Informed Consent Form. A questionnaire was administered to the participants covering age, sex, fish consumption, education level and involvement in mining activities, besides socio-economic and health conditions issues.

2.2 Samples Collection

A hair sample was collected from all participants to measure Hg. Sampling was performed according to WHO guidelines (2018). On average, 250 to 500 mg of hair strands were collected from the occipital region near the scalp. This was done using blunt laser scissors, sterilized with ethyl alcohol. For participants with long hair, the strands were collected about 10 cm closer to the roots using a white adhesive label to mark the strands with a pen marking to identify the roots.

The samples were packed in transparent plastic bags with airtight seals, identifying the donor's name, date of birth and community. They were stored at room temperature, under dry and clean conditions until they arrived at the laboratory, where the samples were prepared for subsequent analysis. It is worth noting that the use of hair as a bioindicator is recommended to verify the chronicity of mercury exposure (Serra et al. 2021a, 2021b).

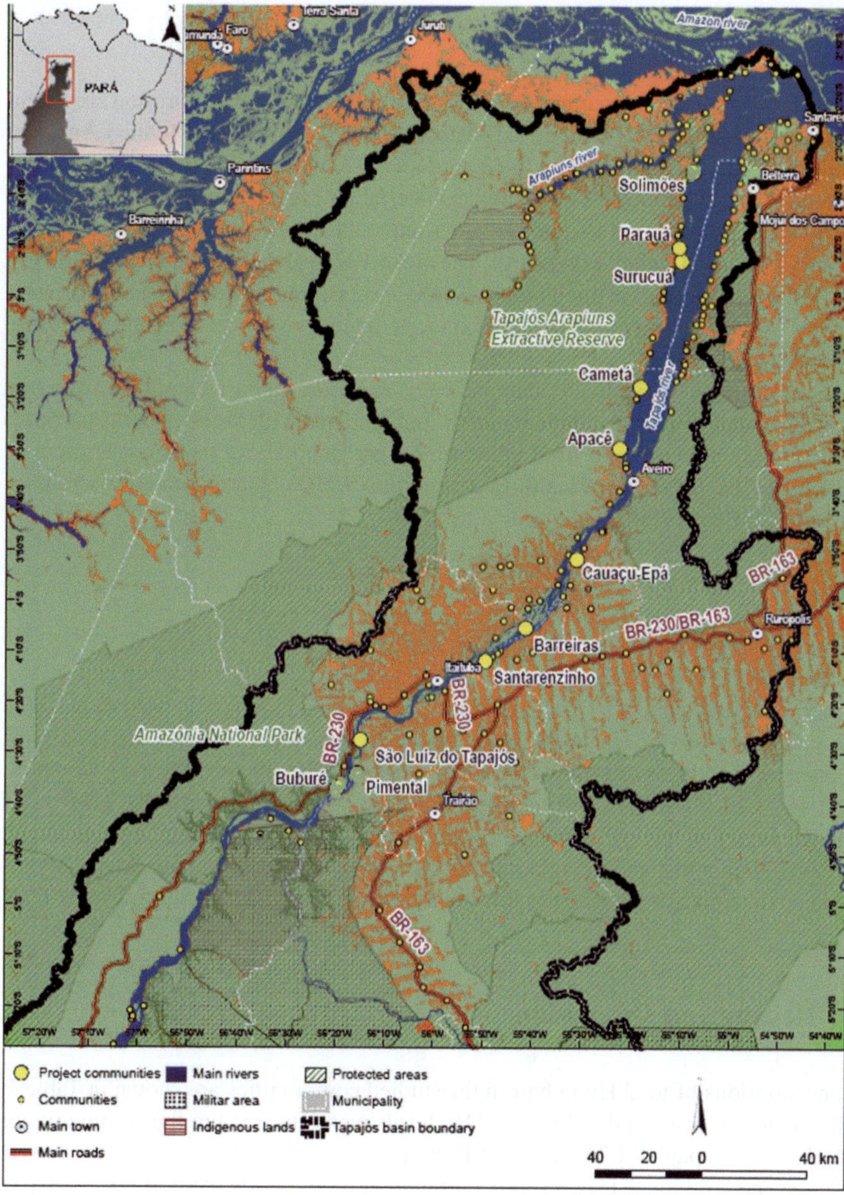

Fig. 1 Map showing the study area in the Lower and Middle Tapajós River Basin, in Pará State, Brazil, and riverine communities[1] participating in the research

[1] According to Franco et al. (2015), the expression. "riverine communities", or "comunidades ribeirinhas", in Portuguese, refers to people who live on the banks of rivers. It is worth noting that due to their tough location, they have difficulty accessing the written, television and radio media.

2.3 Quantification of Hg Levels

Total Hg levels were determined in 126 hair samples by atomic absorption spectrometry. The measurements were carried out using the DMA-80 Direct Mercury Analyzer (Milestone), in accordance with the manufacturer's recommendations. Each sample was measured in duplicate The reference material for human hair (NIES CRM No.13 Human Hair; National Institute for Environmental Studies, Japan Environmental Agency) was used for quality control. The mean T-Hg value determined was 4.32 μg/mg, falling within the certified range of 4.22–4.62 μg/mg. A coefficient of variation of 0.8% was obtained from five repeated measurements when analyzing the standard solution (50 ng Hg/mL).

The presence of Hg in the hair indicates exposure in recent months, which is directly related to the consumption of fish contaminated with MeHg. For the purposes of this study, an exposure limit value of 2 μg/g of total Hg in hair was used, as established by the World Health Organisation (WHO) (International Program on Chemical Safety 1989).

2.4 Data Analysis

Descriptive data analysis (mean, median, standard deviation) was used to examine the concentrations of Hg levels in the biomarker in relation to the community of residence. In addition, these data were examined considering the following variables: age, sex, education, and fish consumption. The relationship between THg levels and frequency of fish consumption or sex was analyzed through the non-parametric Kruskal–Wallis test and the relationship between THg levels and age was analyzed by Pearson correlation using the R Studio 4.3.0 software (significance level = 5%).

3 Results and Discussions

Concentrations of total Hg in hair in the studied communities are shown in Table 1. The highest average and individual THg levels were observed in the community of São Luís do Tapajós (8.74 μg/g and 18.60 μg/g, respectively), whereas the lowest average and individual THg levels occurred in the community of Barreiras (3.90 μg/g and 1.06 μg/g, respectively). In five communities, the minimum Hg level observed exceeded WHO recommended limits in hair, and in all communities maximum THg levels were 4.12 to 9.30 times higher the WHO limit.

Overall, these results are comparable to the findings of other studies in the Amazon region. Junior et al. (2018), who studied riverine communities along the Tapajós River

---

In addition, the river plays an important role in building contacts, an obstacle and an environmental bridge, which can create and restrict interactions (Cohen-Carneiro et al. 2009; Brasil 2012).

Table 1 Descriptors of THg levels in eight communities in the lower and middle Tapajós River, in Pará State, Brazil

Communities	n	%	Total hair mercury levels (μg/g)				
			Mean	Standard deviation	Median	Minimum	Maximum
Apacê	23	18.3	6.22	2.61	6.43	2.32	10.67
Barreiras	23	18.3	3.90	2.44	2.89	1.06	9.44
Cametá	22	17.5	8.28	4.14	8.42	2.17	16.92
Cassuepá	13	10.3	6.66	2.64	6.57	2.39	10.15
Nazaré	1	0.8	–	–	–	–	5.18
Parauá	17	13.4	4.90	2.90	4.30	1.51	13.79
São Luís do Tapajós	11	8.7	8.74	5.17	7.12	3.27	18.60
Suruacá	16	12.7	4.98	1.78	4.98	2.14	8.24

for two years (2013 and 2014), observed average THg levels ranging from 7.25 ± 5.61 μg/g and 10.80 ± 10.54 μg/g, respectively. Basta et al. (2021), in a study involving indigenous populations, including children and adults, in the region of Itaituba (Middle Tapajós), found THg levels in hair ranging from 1.42 to 23.9 μg/g, with an average of 6.6 μg/g for the entire sample. In indigenous people over 12 years old, the average was 8.4 (±4.2) μg/g and the median was 7.4 μg/g.

Pinheiro et al. (2000), working with communities along the Tapajós River, also revealed evidence of Hg exposure in hair samples, with concentrations going up to 25.3 ± 12.5. Santos et al. (2003), in a study involving indigenous communities along the Madeira River, another tributary to the Amazon River, observed average THg levels of 8.37 μg/g, with a range between 0.52 and 83.89 μg/g. Farripas (2010), while monitoring the prevalence of Hg in the communities of São Luís do Tapajós and Barreiras, from 1996 to 2008, noted an increase of 68.3% in Hg levels in the São Luís do Tapajós community. However, in the Barreiras community, a 32.3% reduction was observed.

The communities on the banks of the Tapajós River have a different dynamic, due to their location and profile, which is closely related to socioeconomic and health conditions, and very different from that of urban populations (Serra et al. 2021a, b).

Among the participants, 126 (95.2%) had Hg concentrations above the World Health Organization (WHO) limit, and were stratified as an exposed group, whereas six individuals (4.8%) represented the low exposed group. Mean THg level in the low exposed group was 1.92 μg/g, whereas in the exposed group it was 6.31 μg/g, with a maximum level of 18.6 μg/g, as shown in Table 2. The difference between low exposed and exposed groups was significant ($\chi^2 = 17.01, p = 0.0001$). The Hg exposure prevalence observed in this study (95.2%) is convergent with a previous study in the Lower Tapajós-Lower Amazon area that found a Hg exposure prevalence of 90% in participants from riverine communities (Meneses et al. 2022).

Table 2 Total Hg levels in hair of riverine dwellers (n = 126) from the Lower and Middle Tapajós River in the west of the state of Pará (Brazil), measured in 2021 and 2022, divided by participants with levels above (exposed) and below (low exposed) the limits recommended by the WHO (2.0 μg/g)

Exposure groups (N = 126)	Total Hg level in hair (μg/g)				
	Mean	Max	Min	Standard deviation	Percentage of the sampled individuals (%)
Low exposed (N = 6)	1.5	1.9	1.1	3.6	4.8
Exposed (N = 120)	6.3	18.6	2.1	3.5	95.2
Total overall (N = 126)	6.08	18.6	1.06	7.1	100

Epidemiological characteristics of the subjects of the present study in relation to the exposure to THg are presented in Table 3. Regarding age, 2.5% (n = 3) were between 18 and 21 years old; 49% (n = 56), between 21 and 40 years old; 40% (n = 47), between 41 and 60 years old; and 10 (8.5%), 60 years or older; eight individuals did not declare their age. Mean THg level was the highest in the oldest group (8.37 μg/g), whereas in the youngest group the mean THg level was almost four times lower (2.15 μg/g). For the whole group, there was a significant correlation between THg levels and age ($r^2 = 0.21, p = 0{,}03$). This correlation was higher among men, when sexes were considered separately ($r^2 = 0.34, p = 0.006$). The high prevalence of Hg exposure in the oldest group (60 +), and the correlation between Hg concentration and age can be explained by the phenomenon of Hg bioaccumulation (Lacerda and Malm 2008).

Men represented 56% of the participants (n = 71) and showed mean THg level (6.50 μg/g) 15% higher than the average estimated for women (5.54 μg/g; n = 55), but this difference was not statistically significant ($\chi^2 = 2.09, p = 0.15$). The proportion of men in the exposed and low exposed groups was similar. Among women, most belonged to the low exposed group. In the study of Basta et al. (2021), also in the Tapajós region, 66.7% of male indigenous riverine participants had Hg levels above the exposure limit, corroborating our results. Regarding the epidemiological characteristics related to gender, our study showed that the average levels of mercury above the recommended limit were found in the male group, due to higher fish consumption than women, corroborating other scholars (Serra et al. 2021a, b; Castilhos et al. 2015; Da Silva-Junior et al. 2018).

Most (94%) of the participants who declared their level of education (n = 114) had basic education (primary or secondary degree, complete or not), seven declared having high degree and two had not received any formal education; 12 individuals did not mention their education level. Similarly, Hacon et al. (2014), when evaluating Hg levels in the Madeira River region, also assessed the education situation of exposed riverine residents and identified that the majority had less than 10 years of schooling. In fact, in remote populations with difficult access to higher levels of schooling, education levels are commonly low. In the present study, although the two individuals without formal education showed relatively high Hg levels, and the average THg in

Table 3 Epidemiological characteristics in relation to exposure to THg in riverine communities of the Lower and Middle Tapajós River located in West Pará (Brazil), measured in 2021 and 2022

Epidemiological results	Exposed		Low exposed		Total Hg level in hair (mg/g)				
	n	%	n	%	Mean	Median	SD	Min	Max
Age range (n = 120)									
18–20 (n = 3)	1	33.3	2	66.7	2.1	1.5	1.5	1.1	3.9
21–40 (n = 59)	56	94.9	3	5.1	5.9	5.0	3.6	1.3	18.6
41–60 (n = 47)	46	97.9	1	2.1	5.9	5.0	3.6	1.3	18.6
> 60 (n = 11)	11	100	0	0	8.0	6.7	5.3	2.5	16.9
Gender (n = 125)									
Female (n = 55)	52	94.5	3	5.5	5.5	4.9	3.1	1.1	14.4
Male (n = 70)	67	95.7	3	4.5	6.5	5.8	3.7	1.3	18.6
Schooling (n = 114)									
Illiterate (n = 2)	2	100	–	–	10.3	10.3	7.3	5.1	15.5
Basic (n = 15)	14	93.3	1	6.7	4.8	4.9	1.9	1.5	8.1
Incomplete basic (n = 34)	33	97.1	1	2.9	7.6	7.9	3.7	1.5	16.9
High school (n = 34)	33	97.1	1	2.9	5.2	4.9	2.8	1.1	13.8
Incomplete high school (n = 22)	20	90.9	2	9.1	6.5	5.1	4.5	1.5	18.6
Graduate (n = 2)	2	100	–	–	5.0	5.0	0.2	4.8	5.1
Undergraduate (n = 5)	4	80	1	20	4.4	3.3	3.3	9.4	1.25
Fish consumption (n = 114)									
High consumption (n = 106)	103	97.2	3	2.8	6.1	5.7	3.4	1.3	18.6
Low consumption (n = 8)	06	75.0	2	25.0	4.1	4.2	2.3	7.9	1.1

the basic education group was higher than in the high education one, there was no significant evidence that exposure to Hg was associated to education level.

In relation to fish consumption frequency, most of the participants reported daily or frequent consumption, as expected. This is one of the main intrinsic characteristics of riverine communities exposed to Hg, since they live in remote and difficult-to-access locations, thus lacking diversity in their diet (Hacon et al. 2008).

The group with daily consumption habits showed the highest mean THg levels (7.21 μg/g), with 34 out 53 individuals with levels above WHO limit. In the group with frequent fish consumption, the average THg concentration was 5.24 μg/g, and only 15 out 50 participants had levels above WHO recommendation.

When consumption data was stratified and classified as “low consumption” and “high consumption”, taking into account the exposed and low exposed groups, it was possible to note that even people with low fish consumption were exposed to THg (Fig. 2). This can be explained by a variability in the amount of fish consumed (Barbieri and Gardon 2009), as well as by other factors that can influence exposure to Hg, including involvement in mining activities (Vianna et al. 2022). In this study, nine participants reported working in artisanal gold mining, and all of them belonged to the exposure group.

In the low exposure group, on the contrary, even the participants with high consumption of fish showed Hg levels within the recommended limits (Fig. 3).

It was noted that median Hg levels tended to be higher among participants with higher fish consumption in both the exposed and low exposed groups.

A descriptive analysis of the symptoms reported by the research participants in the questionnaires was carried out. Participants were asked about the occurrence of symptoms and the frequency with which they occurred. Among the self-reported symptoms (Fig. 4), “anxiety” (in a group with an average THg of 6.4 μg/g), “headache” (in a group with an average THg of 6.16 μg/g) and “muscle and joint pain” (in a group with an average THg of 6.57 μg/g) were the most frequently reported by the research participants. These three symptoms were observed in groups where all participants had THg levels above the limit recommended by the WHO.

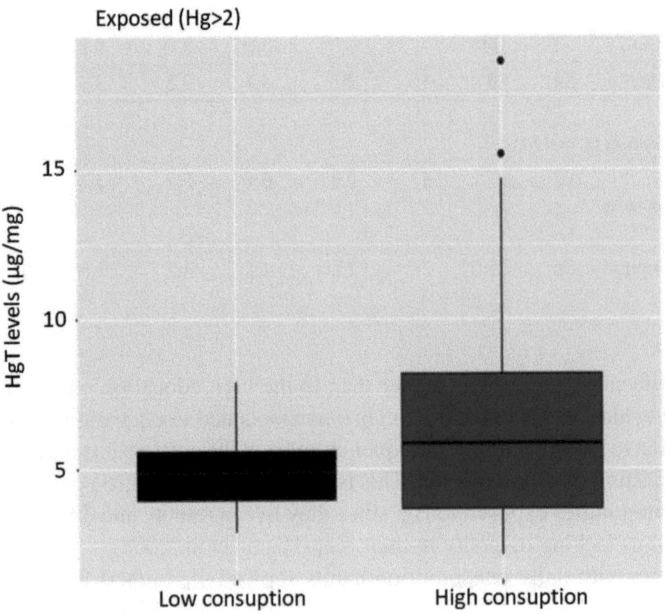

Fig. 2 Boxplots showing THg levels in the “exposed group”, according to their level of fish consumption (low or high), in riverine communities from the Lower and Middle Tapajós River, measured in 2021 and 2022

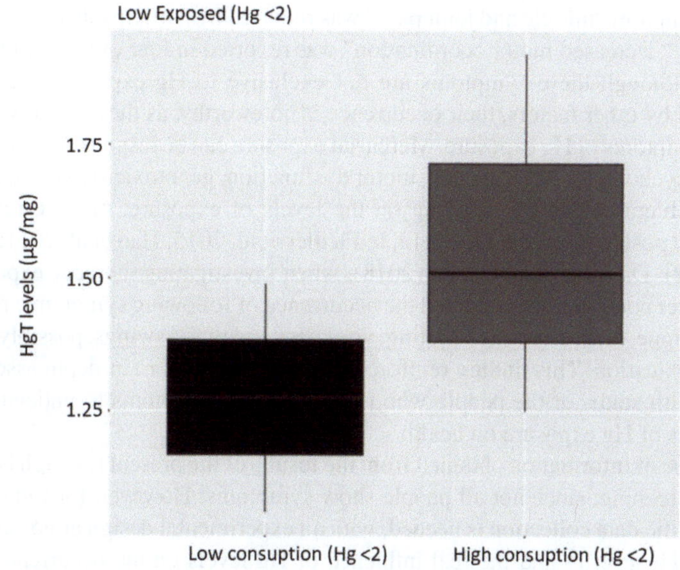

Fig. 3 Boxplots showing THg levels in the low exposed group, according to the level of fish consumption (low or high) in riverine communities of the lower and middle Tapajós River, measured in 2021 and 2022

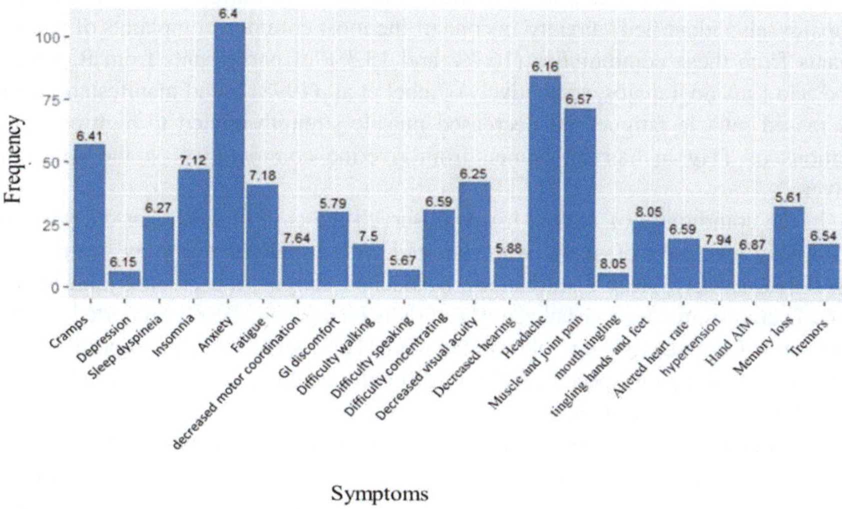

Fig. 4 Distribution of self-reported symptoms among the study participants. Numbers above bars indicate average THg levels (in µg/g) in the group of participants who reported the symptom

The symptom "muscle and joint pain" was reported by participants in all communities, and "decreased motor coordination" was reported in four of the communities visited. Although these symptoms are not exclusive to Hg exposure and can be influenced by other factors, their occurrence is noteworthy, as they are very characteristic symptoms of Hg exposure. Mercurial exposure can be responsible for causing delayed psychomotor development, motor dysfunction, genotoxicity, as well as other health problems, which, depending on the levels of exposure, may be related to mercury exposure, (Serra et al. 2021a, b; Fiedler et al. 2015; Han et al. 2011; Ratner et al. 2002). Da Silva Junior et al. (2018), when investigating mercury exposure in Xingu River communities observed the occurrence of following symptoms: memory deficit, fatigue, difficulty concentrating, weakness and mood swings, possibly related to Hg intoxication. This finding reinforces the need for a more in-depth assessment of the health status of the people who reported these symptoms to understand the true effects of Hg exposure on health.

The type of information obtained from the results of the present research is important for screening, since not all people show symptoms. However, for future steps, more specific data collection is needed, with an experimental design aimed at understanding the severity and the real influence of Hg levels on the occurrence of the symptoms reported. It should be noted that the World Health Organization (WHO) in 2003 estimated a limit of 14 μg/g of hair to generate toxic effects in exposed humans (Serra et al. 2021a, b; Grandjean et al. 2005).

Khoury et al. (2013), in a study of the communities of Barreiras and São Luís do Tapajós, also identified "anxiety" as one of the most common complaints of participants from these communities (16.7% and 13.3% of participants from Barreiras and São Luís do Tapajós, respectively). Lebel et al. (1998) found manifestations of increased muscle fatigue and decreased muscle strength related to high concentrations of THg in hair in women from riverine communities on the Amazon River.

In the community of Apacê, it was observed the case of a 48-year-old woman with THg level of 8.50 μg/g, who reported feeling various everyday symptoms: insomnia, anxiety, sleep disturbances, headaches, fatigue, muscle and joint pains, muscle cramps, reduced visual acuity and reduced hearing; she also reported hypertension. In Cametá, a 41-year-old woman with THg level of 2.17 μg/g reported daily muscle and joint pain, speech problems, difficulty walking, decreased visual acuity and abnormal eye movements.

In general, manifestations of Hg exposure are associated with neurological symptoms, according to studies such as Maruyama et al. (2012), where sensory disturbances of the distal extremities were the most common, followed by imbalance, hearing impairment, and ataxia. Auger et al. (2005) found that an increase in capillary Hg concentration was associated with tremor in adults under 40 years of age. Dolbec et al. (2000), studying populations along the banks of the Tapajós River, found mild motor impairment manifestations associated with low Hg levels. A research monitoring Hg exposure calls for control studies to investigate the direct relationship between Hg levels and the observed clinical manifestations (Sá et al. 2006).

In this study, no consensus could be reached on the level of Hg required to induce symptoms. Although most people who reported the occurrence of symptoms had Hg levels in their hair above the WHO recommended limit (2 μg/g), it was also possible to observe the occurrence of symptoms in participants with levels below 2 μg/g. This finding underscores the importance of broadening the discussion about reference limits for Hg in humans.

4 Conclusions

In general, the results presented here indicate that the participants from the Lower and Middle Tapajós communities are exposed to Hg and that the levels observed are sufficient to induce the exposure symptoms they reported. Despite the need for more specific studies to assess the extent and severity of the symptoms reported, including the use of appropriate neurological tests, the results of this study can help to demonstrate that health problems and environmental degradation go hand in hand in the Amazon region. Therefore, the prevention and mitigation of both problems should be part of a systemic approach, rather than adopting separate perspectives.

In this context, two main approaches should be considered. The first is related to minimizing exposure to Hg through dietary education. This can be done by considering the dietary peculiarities of the communities studied. Information on type of fish suitable for consumption, considering size, age, eating habits (carnivorous or not) etc. should be combined with guidelines for the consumption of fruits from the region with a high antioxidant content that could neutralize the effects of high mercury exposure. To this end, the involvement of decision-makers from different sectors closest to the population, such as education and primary health care, is essential to ensure not only the flow of information to communities, but also care for people who are exposed to methylmercury.

The second approach to mitigating Hg exposure in riverine communities in the Tapajós River basin and the Amazon region is to tackle the root causes of Hg contamination. To do this, it is necessary to implement effective environmental public policies combined with command-and-control. It is essential to control widespread and degrading activities such as deforestation, burning and artisanal mining, which contribute to the mobilization of Hg in soils, contamination of aquatic systems or alterations in the Hg biogeochemistry, in addition to all the other social, biodiversity and ecosystem impacts they cause.

Finally, epidemiological studies like this contribute to society, especially with regard to the biomonitoring of populations exposed to methylmercury, providing data that can reinforce health surveillance actions through the powers granted to local health authorities. Hence the importance of being encouraged and improved with the aim of understanding the genetic profiles of the exposed population.

Acknowledgements This study is part of the "Águas do Tapajós" (Waters of Tapajós) Project, a multidisciplinary initiative with multiple local partners, coordinated by The Nature Conservancy

(TNC), which aims to combine academic and traditional knowledge in support of the conservation of the Tapajós and Lower Amazon rivers and their biodiversity, contributing for the well-being of riverine communities.

References

Abelsohn A, Gibson BL, Sanborn MD, Weir E (2002) Identifying and managing adverse environmental health effects: persistent organic pollutants. CMAJ 166(12):1549–1554

Agency for Toxic Substances and Disease Registry (ATSDR) (1999) Toxicological profile for mercury, 617p

Auger N, Kofman O, Kosatsky T, Armstrong B (2005) Low-level methylmercury exposure as a risk factor for neurologic abnormalities in adults 26(2):149–157. https://doi.org/10.1016/j.neuro.2004.12.005

Barbieri FL, Gardon J (2009) Hair mercury levels in Amazonian populations: spatial distribution and trends. Int J Health Geograph 8(1):71

Basta PC, Viana PVS, Vasconcellos ACS, Périssé ARS, Hofer CB, Paiva NS et al (2021) Mercury exposure in Munduruku Indigenous Communities from Brazilian Amazon: Methodological background and an overview of the principal results. Int J Environ Res Public Health 18(17):9222

Basta PC, Vasconcellos ACS, Yokota D, Pinto DD, Aguiar DS, Campos C, Costa MO (2023). Análise regional dos níveis de mercúrio em peixes consumidos pela população da amazônia brasileira: um alerta em saúde pública e uma ameaça à segurança alimentar. FIOCRUZ/ENSP, WWF Brasil, Greenpeace, Instituto Socioambiental, Instituto de Pesquisa e Formação Indígena, Rio de Janeiro, 10p

Brasil, Ministry of Health, Secretariat of Health Care (2012) Department of Primary Care. Step-by-step actions of the Primary Care Department. Ministry of Health, Brasília

Castilhos C, Bidone D (2000) Hg Biomagnification in the ichthyofauna of the Tapajós River Region, Amazonia, Brazil. Bull Environ Contam Toxicol

Castilhos Z, Rodrigues-Filho S, Cesar R, Rodrigues AP, Villas-Bôas R, de Jesus I, Lima M, Faial K, Miranda A, Brabo E, Beinhoff C, Santos E (2015). Human exposure and risk assessment associated with mercury contamination in artisanal gold mining areas in the Brazilian Amazon. Environ Sci Pollut Res Int 22(15):11255–11264

Clarkson TW, Vyas JB, Ballatori N (2007) Mechanisms of mercury disposition in the body. Am J Ind Med (Jornal Americano de Medicina) Ind 50(10):757–764 (CMAJ 2002;166(12):1549–1554)

Cohen-Carneiro F, Souza-Santos R, Pontes DG, Salino AV, Rebelo MAB (2009) Provision and utilization of dental services in Amazonas State, Brazil: a case study in a riverine population in Coari Municipality. Rep Public Health 25(8):1827–1838

Da Silva-Junior FMR, Oleinski RM, Azevedo AES, Monroe KCMC, Dos Santos M, Da Silveira TB, De Oliveira AMN, Soares MCF, Pereira TDS (2018) Vulnerability associated with "symptoms similar to those of mercury poisoning" in communities from Xingu River, Amazon basin. Environ Geochem Health 40(3):1145–1154

Da Silva DS, Lucotte M, Roulet M, Poirier H, Mergler D, Crossa M (2007) Mercúrio nos peixes do Rio Tapajós, Amazônia Brasileira. InterfacEHS—Revista de Gestão Integrada em Saúde do Trabalho e Meio Ambiente 1(1) (Art 6)

Dolbec J, Mergler D, Sousa Passos CJ, Sousa de Morais S, Lebel J (2000) Methylmercury exposure affects motor performance of a riverine population of the Tapajós river, Brazilian Amazon. Int Arch Occup Environ Health 73(3):195–203. https://doi.org/10.1007/s004200050027. PMID: 10787135

Farina M, Ávila DS, Da Rocha JTR, Aschner M (2013) Metals, oxidative stress and neurodegeneration: a focus on iron, manganese and Mercury. Neurochem Int 62(5):575–594

Farripas SSM (2010) Epidemiological aspects of mercury exposure in the Tapajós region from 1994 to 2008 [dissertation]. Federal University of Pará, Belém, p 2010

Fearnside PM (2003) The Amazon rainforest in global change. INPA, 134p

Fiedler N, Rohitrattana J, Siriwong W, Suttiwan P, Ohman Strickland P, Ryan PB et al (2015) Neurobehavioral effects of exposure to organophosphates and pyrethroid pesticides among Thai children. NeuroToxicology 48:90–99

Fostier AH, Forti MC, Guimarães JRD, Melfi AJ, Boulet R, Santo CME, Krug FJ (2000) Mercury fluxes in a natural forested Amazoniancatchment (Serra do Navio, Amapa State, Brazil). Sci Total Environ 260(1–3):201–211

Franco EC, Santo CE, Arakawa AM, Xavier A, França ML, Oliveira AN, Machado MAMP, Bastos RS, Bastos JRM, Caldana ML (2015) Health promotion of the riverside population of the Amazon region: experience report. Rev CEFAC 17(5):1521–1530. https://doi.org/10.1590/1982-0216201517518714

Gibb H, O'Leary KG (2014) Mercury exposure and health impacts among individuals in the artisanal and small-scale gold mining community: a comprehensive review. Environ Health Perspect 122(7):667–672. https://doi.org/10.1289/ehp.1307864. https://www.ncbi.nlm.nih.gov/pmc/articles/PMC4080518/

Gonçalves A, Gonçalves NN (2004) Exposição humana ao mercúrio na Amazônia brasileira: uma perspectiva histórica. Rev Panam Salud Publica 16(6):415–419

Grandjean P, Budtz-Jørgensen E, Jørgensen PJ, Weihe P (2005) Umbilical cord mercury concentration as biomarker of prenatal exposure to methylmercury. Environ Health Perspect 113:905–908

Hacon S, Barrocas PRG, Vasconcellos ACS, Barcellos C, Wasserman JC, Campos RC, Ribeiro C, Azevedo-Carloni F (2008) An overview of mercury contamination research in the Amazon basin with an emphasis on Brazil. Cad Saúde Pública 24(7):1479–1492

Hacon SS, Dórea JG, Fonseca MDF, Oliveira BA, Mourão DS, Ruiz CMV, Gonçalves RA, Mariani CF, Bastos WR (2014) The influence of lifestyle changes and mercury exposure on riverine populations of the Madeira River (Amazon Basin) near a hydroelectric project. Int J Environ Res Public Health 11(3):2437–2455. https://doi.org/10.3390/ijerph110302437

Han DY, Hoelzle JB, Dennis BC, Hoffmann MA (2011) Brief review of cognitive assessment in neurotoxicology. Neurol Clin 29(3):581–90 (Health effects: persistent organic pollutants. Identifying and managing adverse environmental)

International Program on Chemical Safety (1989) Environmental health criteria 86: mercury—environmental aspects. WHO, Geneva

Junior JMFC, Silva CIM, Lima AAS, Rodrigues Júnior D, Silveira LCL, Souza GS, Pinheiro MCN (2018) Mercury levels in hair and fish consumption in riverside communities in the Brazilian Amazon, Tapajós region. Ciênc saúde coletiva 23(3):805–812

Khoury EDT, Souza GS, Silveira LCL, Costa CA, Araújo AA, Pinheiro MCN (2013) Neurological manifestations in riverside residents in areas exposed to mercury in the Brazilian Amazon. Cadernos de Saúde Pública 29(11):2307–2318

Lacerda LD, Malm O (2008) Mercury contamination in aquatic ecosystems: an analysis of critical areas. Estudos avançados 22(63)

Lebel J, Mergler D, Branches F, Lucotte M, Amorim M, Larribe F, Dolbec J (1998) Neurotoxic effects of low-level methylmercury contamination in the Amazonian basin. Environ Res 79(1):20–32

Maruyama K, Yorifuji T, Tsuda T, Sekikawa T, Nakadaira H, Saito H (2012) Methyl mercury exposure at Niigata, Japan: results of neurological examinations of 103 adults. J Biomed Biotechnol:1–7

Meneses HNM, Costa MO, Basta PC, Morais CG, Pereira RJB, Souza SMS, Hacon SS (2022) Mercury contamination: a growing threat to riverine and urban communities in the Brazilian Amazon. Int J Environ Res Public Health 19:2816

National Research Council NRC (2000) Toxicological effects of methylmercury. National Academy Press, 344p

Nevado JJB, Rodríguez Martín-Doimeadios RC, Guzmán Bernardo FJ, Morenom J, Am H, Do Nascimento JL, Crespo-lópez ME (2010) Mercury in the Tapajós river basin, Brazilian amazon: a review. Environ Int 36:593–608

Pinheiro MCN, Guimarães GA, Nakanishi J, Oikawa T, Vieira JL, Quaresma M, Cardoso B, Amoras W (2000) Evaluation of mercury contamination through the analysis of total Hg content in hair samples from riverside communities in the Tapajós region, Pará, Brazil. Rev Soc Bras Med Trop 33(2):181–184

Ratner MH, Feldman RG, White RF (2002) Neurobehavioral toxicology. In: Ramachandran VS (eds) Encyclopedia of the human brain, vol 3. Elsevier Science, New York, pp 423–39.27

Sá AL, Herculano AM, Pinheiro MC, Silveira LCL, Do nascimento JLM, Crespo-López ME (2006) Exposição humana ao mercúrio na região oeste do estado do Pará. Revista Paraense de Medicina 20(1)

Santos ECO, Câmara VM, Brabo ES, Loureiro ECB, Jesus IM, Fayal K, Sagica F (2003) Evaluation of mercury exposure levels among Pakaanóva indigenous people, Amazon, Brazil. Cad Saude Publica 19(1):199–206

Santos ECO, Jesus IM, Câmara VM, Brabo EDS, Loureiro ECB, Mascarenhas A, Weirich J, Luiz JR, Clearya D (2002) Mercury exposure in munduruku Indians from the community of Sai Cinza, State of ParaH, Brazil. Environ Res Sect A 90:98–103

Serra CV, Figueiredo WDS, Bernardi JVE (2021a) Dinâmica de exposição ao mercúrio em população do alto rio Madeira, Rondônia—Brasil 17(3). ISSN 1980-0827

Serra CV, Figueiredo WDS, Bernardi JVE (2021b) Evolução temporal da exposição humana ao mercúrio na Amazônia: uma abordagem cienciométrica 17(3). ISSN 1980-0827

Vianna AS, Câmara VM, Barbosa MCM, Santos ASE, Asmus CIRF, Luiz RR et al (2022) Mercury exposure and anemia in children and adolescents from six communities in the Brazilian Amazon. Ciência Saúde Coletiva 27(5):1859–1871. https://doi.org/10.1590/1413-81232022275.08842021

WHO—World Health Organization (2018) Exposure to mercury: a major public health concern, second edition. Preventing disease through healthy environments

Wright DA, Welbourn P (2002) Toxicologia ambiental. Cambridge University Press, Cambridge. http://doi.org/10.1017/CBO9780511805998

Companion Modeling for Bringing Closer Fishers and Policymakers for Sustainable Fisheries Management on the Lower Amazon Floodplain

Neriane Nascimento da Hora, Nathan Almeida da Silva, Marie-Paule Bonnet, Gustavo Melo, Fabíola Dourado, João Gabriel Paes de Sousa Guedes, Nathalia Gonçalves Meyer, Catarine Marques Rodrigues Gaspar, Joine Cariele Evangelista do Vale, Kevin Chapuis, Pierre Bommel, and Christophe Le Page

Abstract Since the late twentieth century, social mobilization for the sustainable management of fisheries on the lower Amazon floodplain has been driven by community-based fishing agreements. To support this process, we developed *Pesca-Viva,* a serious game designed through participatory modeling. The *Lago Grande do Curuai* (LGC) region was selected as the study area due to its fishing significance and the pressures on its fishing stocks. Fieldwork began in 2021 with the formation of a focus group comprising fishing leaders, representatives from the LGC region, and environmental agencies. To date, we have conducted ten game co-construction workshops with the focus group and nine workshops with local fishers in their communities. A key outcome of this collaborative process is the creation and maintenance of a

N. N. da Hora (✉)
Sociedade Para Pesquisa e Proteção do Meio Ambiente-Sapopema, Av. Rui Barbosa, Santarém, Pará 3040, Brazil
e-mail: neriane.dahora@ird.fr

N. A. da Silva · G. Melo · F. Dourado · J. G. P. de Sousa Guedes · N. G. Meyer · C. M. R. Gaspar
Universidade Federal do Rio de Janeiro, Rio de Janeiro, Brazil

M.-P. Bonnet · K. Chapuis
Espace-DEV, Institut de Recherche Pour le Développement, Montpellier, France

J. C. E. do Vale
Instituto Chico Mendes de Conservação da Biodiversidade, ICMBio, Santarém-PA, Brazil

P. Bommel · C. Le Page
UMR SENS, Centre de Coopération Internationale en Recherche Agronomique Pour le Développement, Montpellier, France

N. N. da Hora
Institut de recherche pour le développement - IRD, Brasília, Brazil

M.-P. Bonnet
Centro de Desenvolvimento Sustentável, Universidade de Brasília, Brasília, Brazil

W. Leal Filho et al. (eds.), *Amazon 2030—Sustainability Issues in the World's Largest Rainforest Region*, World Sustainability Series,
https://doi.org/10.1007/978-3-031-81465-5_7

space for dialogue and interaction between grassroots organizations and government institutions, fostering sustainable fisheries management in the region.

Keywords Serious game · Participatory research · Fisheries co-management

1 Introduction

In addition to its vast rainforest, the Amazon region is distinguished by its extensive freshwater ecosystems (Castello and Macedo 2016). Among them, the floodplains—locate along the banks of white-water rivers and subject to annual flooding for approximately six months—stand out for their high fishing potential and the agriculturally productive soils. These areas are inhabited by human populations organized into communities that rely on diversified livelihood strategies, including agriculture, fishing, animal husbandry, and other activities (McGrath et al. 1993). Among these activities, fishing is particularly significant, serving as major source of income and the primary source of animal protein for floodplain inhabitants (Almeida 2004; Isaac and Almeida 2011). Fishing is the region is predominantly small-scale, conducted using small vessels, commonly referred to as canoes, which may be motorized or not. Most fishing occurs in floodplain lakes, which surplus production directed to local markets (Almeida et al. 2001; Isaac et al. 2008).

In the second half of the twentieth century, a new form of fishing emerged in the region, carried out by vessels known as *geleiras*. These vessels, which primarily operate in river channels, but have increasingly expanded into floodplain lakes, possess significant fishing power, catch capacity and storage capacity. Their production is largely aimed at supplying external markets (Almeida et al. 2001; McGrath et al. 2004). This expansion has led to growing tensions between small-scale and large-scale fisheries, particularly concerning access to and use of floodplain lakes. The conflicts stem primarily from two key issues: (a) local small-scale fisheries consider floodplain lakes as communal properties and perceive the presence of external fisheries as a direct threat, and (b) small—scale fishers argue that large-scale operations deplete fish stocks in floodplain lakes, endangering their livelihoods (Furtado et al. 1998; Hartmann 2001; Silva and Ferreira 2018).

Fisheries agreements, introduced as community-led initiatives in the 1990s, were established to reduce pressure on fish stocks in floodplain lakes and mitigate social tensions surrounding fishing activities. These agreements function as governance instruments for managing fisheries resources within shared territories (Castro 2000). Initially operating without legal backing, they were later formally recognized within the framework of environmental legislation (McGrath et al. 2004).

The implementation of legal instruments for governing natural resources represents an effort by the State to reduce tensions among various social actors (D'Almeida 2006). In Brazil, the evolution of these instruments initially followed a centralized management approach. However, in response to socio-environmental movements advocating for user participation in resource governance, environmental policies shifted toward a co-management model of natural resources, promoting shared responsibility for natural resource management (Seixas et al. 2011). Within this framework, fishing agreements were formally integrated into the federal government's fishery resource management.

By the late 2000s, influenced in part by the large-scale fishing industry, the responsibility for managing fisheries resources was decentralized to the federal states. Pará, the second-largest in the Brazilian Amazon, the State Agency for Environment and Sustainability (SEMAS) and the Institute for the Development of Forests and Biodiversity of the State of Pará (IDEFLOR) assumed the role of regulating fisheries management. However, for nearly a decade, the State provided no formal guidance on the regulation of fishing agreements. It was only in 2021 that a state decree was issued, outlining procedures for users to formalize their fishing agreements (Lima et al. 2023).

Fisheries co-management remains a complex challenge. Beyond the intricacies of negotiations among resource users, significant communication barriers between communities and government agencies often result in delays in reaching agreements. To facilitate dialogue on fisheries co-management in the Lower Amazon, we employed participatory research using the companion modeling methodology (Etienne 2014) in the *Lago Grande do Curuai* (LGC) region. This article provides a concise overview of the research process and highlights key outcomes from the collaborative meetings between fishermen and environmental agencies, focusing on discussions surrounding fisheries sustainability.

2 The *Lago Grande do Curuai* Region

The *Lago Grande do Curuai* region consists of uplands and floodplains, located on the right bank of the Amazon River, in the lower Amazon. It encompasses communities primarily within the municipality of *Santarém*, as well as, to the west and north, in the municipalities of *Juruti* and *Óbidos*. Fishing is the main economic activity, and along with agriculture, serve as the primary source of livelihood for local families. In 2005, the region was designated as an "Agroextractivist Settlement Project" (PAE)—commonly referred to as *Lago Grande* PAE—, by the National Institute for Colonization and Agrarian Reform (INCRA), comprising more than 128 communities within its territory (Folhes 2016) (Fig. 1).

In an effort to mitigate the impact of commercial fishing in the LGC region, local fishers initially established a fishing agreement, which was formalized by the Brazilian Institute for the Environment and Renewable Natural Resources (IBAMA) in 1999, the agency responsible for fishing regulation until 2011. This agreement was later revised, resulting in the publication of a new legal instrument by the Ministry of the Environment in 2004. Despite the efforts of local fishers and their fisheries councils to update their agreement since 2019, the revision of LGC region fishing agreement was only recently published in July 2023 by SEMAS. Today, social mobilization among fishers in the *Lago Grande do Curuai* region is centered on ensuring the

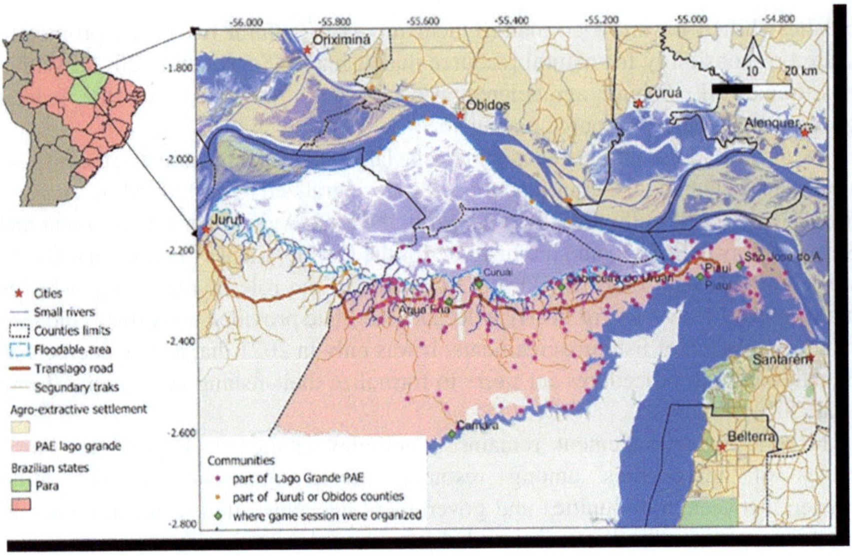

Fig. 1 *Lago Grande do Curuai* region. *Source* Hora et al. (2023)

effective implementation of their fishing agreement, with their main demand being stricter environmental enforcement by SEMAS.

3 The Strengths of the Companion Modeling Methodology

Companion modeling falls into the category of "non-textual" scientific investigation methods, which aim to reduce asymmetries between academic and non-academic participants (Beebeejaun et al. 2014 apud Melo et al. 2022). In other words, it is a methodology whose objective is to facilitate interaction and communication between different local social actors, contributing to the agglomeration of different knowledge in favor of the promotion of collectively shared reflections and strategies to tackle the problems that affect the territory in question, strengthening the organizational processes that, in turn, are necessary in the struggle for the survival and permanence of populations that live there (Ducrot et al. 2008).

The movement towards bridging interdisciplinary scientific knowledge and local traditional knowledge reveals an inherent horizontal character within participatory modeling. This approach, diverging from the traditional model where scientific research often supersedes local knowledge, facilitates the reevaluation of the community's role in dialogical processes. This, in turn, fosters the construction of a relationship of trust between the local population and researchers. However, as Melo et al. (2022) warn, it is only possible to build trust in a medium and long-term process, in which the parties involved interact with each other and engage in activities in an inclusive perspective.

The companion modeling methodology employs a range of tools including co-constructions of maps and videos to share diverse viewpoints and perspectives on situational problems. Subsequently, participants collaboratively create a model that represents this reality, exploring possibilities to engage with and transform social problems into tangible outcomes. Here, the term 'model' is understood as a representation of the participants' reality and can take the form of a board game assisted or not by a computer and/or an entirely 'open' simulation model (as opposed to a black box). The goal is not the mere creation of the model but rather the process of co-creation, modification, play, and observation of the model with stakeholders. The aim is to learn and think collaboratively about the real situation, fostering the creation of new relationships and knowledge among participants (Bousquet et al. 2014).

The co-creation of role-playing games is widely used in companion modelling. However, the use of role-playing games can be limited for scenario prediction, as the game sessions generally only represent a few years, limiting the analysis of the evolution of certain variables. As a result, the research team may also choose to computerize the board game, making it possible to generate robust scenarios in real time and simulate more years, thereby broadening the scope for reflection and proposals on the management and use of natural resources (Bommel et al. 2016).

A crucial aspect of the participatory process in serious games is the pedagogical training perspective, emphasizing a sustained commitment to fostering ongoing dialogue among various social actors involved. This commitment is essential because proposing a methodology capable of designing scenarios and reflecting social dynamics must align with the specific social transformations occurring in the territory. Supporting this collaborative construction are additional methods such as workshops, fishing calendars, and group dynamics. It is noteworthy that the feedback from these collaborative efforts is integral to the commitment guiding companion modeling. As articulated by Becu et al. (2008), it is during this phase that all social actors engage with and appropriate the knowledge co-constructed among researchers, local populations, and public policy agents.

4 *PescaViva*: The Co-construction of the Game as a Space to Strengthen the Co-management of Fishing Resources in Lago Grande Region

4.1 PescaViva *Co-construction: How Companion Modeling Process Started*

Before embarking on the process of co-constructing the PescaViva game, we worked with a group of farmers and fishermen from four communities in the LGC region. The aim of the research carried out as part of the Clim-FABIAM project (Climate Changes and Floodplain Lake Biodiversity in the Amazon Basin: How to Cope and Help the Ecological and Economic Sustainability) (2012–2015) was to gain a better understanding of the socio-environmental dynamics linked to the main local economic activities in the region and the social mobilization processes put in place in the face of the various anthropogenic pressures, as well as the strategies for coping with climate change.

In this context, we developed the *VarzeaViva* game (Bommel et al. 2016). Through subsequent debates following the game sessions, participants underscored the significance of fishing in their daily lives, and for some, as a source of income. They also highlighted the considerable pressure on local fishing stocks, particularly in the lakes, resulting in a series of socio-environmental conflicts and threats to their way of life.

These results prompted us to take a closer look at the issue of fishing and its sustainability by considering the growing environmental and socio-economic pressures in the region. Thanks to several grants, we have been able to propose the development of a new companion modeling process to address these issues.

4.2 Mobilization of Stakeholders: A Process at Two Spatial/ Administrative Levels

Considering the organization of small-scale fishing in the region described above, we decided to develop the mobilization process at two spatial/administrative scales. At a "regional scale", we mobilized the representatives of the fishermen's unions of the three municipalities (*Santarém, Jurutí*, and *Óbidos*), and the community association in charge of managing the *Lago Grande* territory. As fishing is often associated with agricultural activities, we also involved the farmers' unions of the three municipalities. The mobilization process began in November 2019 when we met representatives of the *Santarém* fishermen's union but was interrupted by the COVID-19 pandemic.

It was not until November 2021 that we were able to resume field activities. With the help of *Sociedade para Pesquisa e Proteção do Meio Ambiente* (SAPOPEMA) an NGO which has a good partnership with fishers and farmers' local organizations, we organized a workshop that brought together representatives from all these bodies, as well as from *Movimento dos Pescadores e Pescadoras do Baixo Amazonas* (MOPEBAM), an important collective made of representatives of fisheries unions of the lower Amazon; and the fisheries councils. At the workshop, we presented a very preliminary version of the game that we had developed as researchers. With this game, we wanted to give a slightly more concrete idea of what a serious game consists of and could lead to discussion. This workshop was essential for consolidating the objectives of the companion modeling process that we were going to undertake—they asked us to center the game upon a community-based fisheries agreement—and establishing a timetable for the activities that we would develop with the group (hereafter "focus group").

The focus group finally was composed of representants of the following grassroots organizations: the fishing unions of the three municipalities (*Juruti*/Z-42, *Santarém*/Z-20 and *Óbidos*/Z-19); the regional coordination of union Z-20 for the sub-regions of Lago Grande; MOPEBAM; the Sustainable Development Councils (or also known as Fishing Councils) of the *Arapixuna* and *Lago Grande* regions; *Federação das Associações de Moradores e Comunidades do Assentamento Agroextrativista da Gleba Lago Grande* (Feagle) and the Unions of Rural Workers of *Juruti* and *Santarém* and representatives of SAPOPEMA. The focus group participants committed to meeting monthly in *Santarém* and, when possible, in *Juruti* and *Óbidos*, Pará. From February to December 2022, we organized ten half-day meetings with the focus group which partly coincided with the process of revising the fisheries agreements.

In the fourth workshop, participants suggested that municipal and state environmental agencies should join the activities. They argued the issues related to fisheries management must be discussed with managers and must have their support. The research team conducted virtual meetings followed by a workshop, between June and July/2022, with representatives of environmental agencies from *Juruti*, *Santarém*, and *Óbidos* municipalities and of the Pará state to demonstrate the game in co-construction and benefit from that to invite them to join the focus group, to which they consented. Thus, the managers integrated the focus group in the eighth workshop, in August/2022.

In the end, we had a heterogeneous focus group composed of users (fishers and farmers, who also fish), representatives of NGOs, and managers. Because of its composition, this group will be able to tackle the issue of fishing on a regional scale and build the *PescaViva* game that is sufficiently generic for us to be able to deploy it in different communities in the region.

In parallel with the activities of the focus group, we organized nine workshops in various communities in the LGC region that represent the "smallest spatial scale" in the fishing organization in the region. The aim was to provide validation/criticism of the proposed game and to hear from fishermen about their activities and perceptions of fishing. Feedback on the game would be then discussed with the Focus Group to assess how it could be incorporated into the game.

4.3 PescaViva *Co-construction Process: How the Workshops Were Conducted and A Brief View of the Game Built Together with the Stakeholders*

Throughout the process, the beginning of the workshop was devoted to a summary of the progress of the negotiation of the agreements in the region and the key events. This was followed by a reminder of the activities carried out during the previous workshop, the decisions, and the next steps of the process. The various activities carried out during each workshop were aimed at developing the *PescaViva* role-playing game. So, several elements of the game were discussed: the game board, the representation of fish dynamics, the fishing rules, the costs associated with fishing and the income from fishing.

Between two meetings, our team would then summarize the information gathered and suggest a way of incorporating it into the game for the next meeting. At each workshop, a 'short' game session was organized to test the new features. In the last part of the workshop, a discussion was conducted to get the perceptions and suggestions of the participants about the game, and the workshop itself and to reflect on the issues coming out when playing the game.

Shortly, the *PescaViva* game is composed of two game boards, each representing the same spatial scale, but a distinct season: winter and summer. The gameboard has five fishing spots and four communities (Fig. 2). There are four species in the fish stocks: scale basic, flatfish basic, scale high-value, and flatfish high-value, which can be caught using three different gears, that differ in catch capacity (small, medium, and high) and are materialized by three different sizes of cups. The stocks are also composed of predators: caymans, dolphins, and carnivorous fish, that consume the fish in the nets and can damage the gears. Materially, fishes and predators are represented by colored pearls put inside five jars, one for each fishing spot, which is positioned below the gameboard, as shown in Fig. 3. Inside the jars, the colored pearls are mixed with dried seeds, representing the water.

To fish, players must insert a cup inside one of the jars (fishing spots) and try to catch as much as they can, without looking at the jar (Fig. 3). The catch is poured into a sieve, which helps separate colored pearls (fish and/or predator) from dried seeds (water). Besides fishing, players can sell the fish at the market, where the owner is played by one of the facilitators. The market also sells gear, food, and goods.

One year of the game is played in three rounds. The first round is played on the winter board, the second on the summer board, and the third also on the summer board. However, in this last round high-value species are protected by the closed period and players can have their catches inspected and pay fines. At the end of the first year, a facilitator plays the role of a *geleira*, who do not comply with the closed period and catch as much as they can. This whole sequence takes around forty minutes and it is supported by an agent-based model built into the Gama platform (Chapuis et al. 2022). By that, the decisions of players of the first year are registered and inserted into *PescaViva* model running in Gama, and then these decisions are repeated for four years. After, we show the results of fishing activities over the five

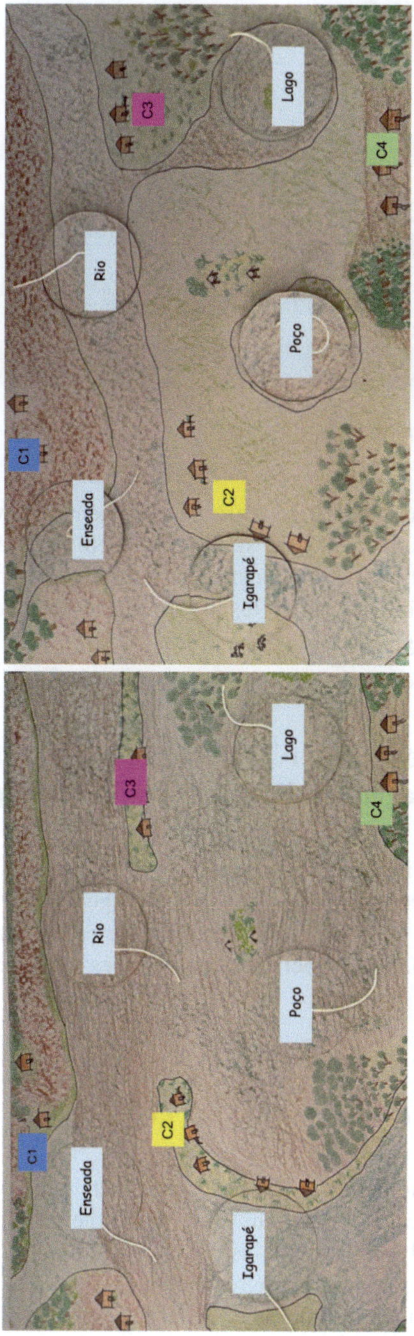

Fig. 2 On the left: winter board; right: summer board (C1: community 01; C2: community 02; C3: community 03 and C4: community 04)

Fig. 3 Fishing action on *PescaViva* game

years to players and they can decide to set a fishing agreement. Following, participants come back to the gameboard to play the sixth year, where they can apply the rules created by them. Assisted by the computer, decisions are repeated for four years, summing 10 years of playing the game. Results are shared with the participants and then we start the debriefing session.

5 Bringing Managers Closer Together to Discuss Issues of Fisheries Management in *PescaViva* Workshops

While playing the PescaViva game, participants in the workshops raised numerous issues concerning the sustainability of fisheries management in the LGC region. These concerns primarily revolved around the absence of the state in fulfilling its role and the lack of communication between communities and government agencies, as discussed below.

5.1 Large-Scale Fisheries' Impact on Fish Stocks

The focus group raised the prejudice caused by outsiders large-scale fisheries that invade the floodplain lakes of the region, even during the fishing ban period, and whose practices negatively impacts the fish stocks, resulting in fish scarcity. For them, the biggest challenge to fishing agreements' success is to ban the entrance of outsider *geleiras* into their fishing area. From their perspective, it's useless to comply internally with fishing accords if outsiders are not regulated. For the sustainability of fisheries in the LGC, they claim that enforcement of fishing regulations by environmental agencies, especially in the closed period, is essential.

The impact of large-scale fisheries on fish stocks is represented in *PescaViva* game by someone playing the *geleira* role. Initially, this role was played by a member of the teams' mediation. However, in the latest version of *PescaViva* game *geleira* is played by some of the participants. In some cases, it was played by a SEMAS technician. The characteristics of *geleiras* as the boat capacity, initial cash, number of trips by season, and others, were set in the game according to participants' knowledge. When playing the game, participants noticed the great amount of fish caught by '*geleiras*', as well as the high amount of money made by them.

Initially, during the first workshop with managers using the *PescaViva* game, and in the absence of fishers, they expressed the view that the game could serve as a valuable tool to raise awareness among fishermen about fisheries sustainability. In subsequent workshops, managers expanded their perspective on issues related to fisheries management. The game methodology enabled them to empathize and understand the challenges from the fisherman's point of view (Carvalho and Melo 2017). This shift in perspective led them to recognize that the decline in fishing stocks in the region is not solely due to a lack of awareness among some non-compliant fishers but is primarily attributed to the impact caused by large fisheries boats.

5.2 *Low Enforcement Level by Environmental Agencies*

The LGC region falls within the jurisdiction of three municipalities, and accordingly, each municipal environmental agency is responsible for monitoring its designated area. However, participants emphasize that the number of SEMAS fishing patrols conducted by environmental agencies is notably insufficient in the LGC. This observation aligns with findings from various studies, such as those by Faria Junior et al. (2018), which indicate that more concerted efforts and resources are required from the government to ensure effective fishing enforcement in the lower Amazon.

Participants argued that the insufficient level of enforcement tends to be justified by a lack of financial, infrastructure, and human resources, primarily attributed to a perceived lack of political willingness on the part of the government to allocate sufficient resources for fishing enforcement. According to their perspective, the artisanal fishing sector is considered one of the lowest priorities within Brazil's environmental policy. In the '90s and 2000s, the federal government made efforts to engage in fisheries management, exemplified by initiatives such as the community environmental agents program (IBAMA Normative Instruction nº 66/2005). However, after federated states assumed this responsibility, little progress has been observed in this regard.

Participants emphasized that environmental agencies have been delegating the enforcement task to local fishermen. Indeed, certain fishing communities are conducting patrols in the lakes within their community territory. However, communities grapple with elevated costs to sustain this effort and face life-threatening risks, as they may encounter threats from those who violate fishing agreements (Hora 2020).

To represent the enforcement task of environmental agencies in the game, we introduced the role of an environmental officer, played by one of the facilitators. The environmental officer is active only during the fishing ban period and is unable to inspect all fishing activities in the game. Out of the four players engaged in fishing, only one will be subject to inspection. Consequently, some players may exploit the opportunity of not being caught to sell prohibited species of fish.

In debriefing sessions, besides the common reason given by the managers for the low level of enforcement: environmental agencies lack sufficient financial, human, or infrastructural resources to conduct regular patrols across the entire region; they assigned that the limited patrols carried out can be frustrating, as offenders are often warned and can escape. Officials emphasized the importance of gaining community support for patrols, particularly by informing SEMAS of any illegal activities within their communities. Partnerships between environmental agencies and fishing organizations, such as the one implemented in *Óbidos*, were cited as a solution to address the issue of low-level enforcement. Additionally, environmental officials noted that having the three municipal environmental agencies in the workshop has facilitated negotiations toward a formal collaboration for sharing monitoring areas.

5.3 Lack of Knowledge About Fisheries by Environmental Agencies

The state's limited involvement in fisheries management is evident not only in the absence of controls but also in the lack of knowledge of the territories, fishing practices, total production, and the significance of fishing as a way of life, as observed by the participants of the focus group. Technicians from SEMAS, as indicated in recommended literature for the latest public tender, embody various types of knowledge in theory. This includes institutional knowledge of the environmental secretariats for approving and validating fishing agreements, technical knowledge of environmental legislation and limited literature specializing in ecology (Diário Oficial nº 35.215 2022). This lack of real-life knowledge about fishing creates a barrier for collaboration between fishers and people who possesses technical kwnoledge (Dedual et al. 2013). Besides, practical experience or firsthand knowledge of the lived reality, specifically, the practice of fishing (in our case) is something prioritized in participatory modeling (Etienne 2014).

The traditional ecological knowledge of fishers forms the foundation for constructing fishing agreements (Castro and McGrath 2001). Successful fisheries management necessitates the integration of both fishers' traditional knowledge and scientific/technical expertise (McGrath and Castello 2015). For that, environmental officers compreehension about local fisheries dynamics, the territory, and its specificities contributes to overcome misunderstandings, contributing to fisheries effective co-management.

For non-fisher participants, such as SEMAS technicians, engaging in the game also serves as a means of learning about fisheries in the lower Amazon. This learning experience enhances their communication with fishermen, fostering a better understanding of the challenges and dynamics involved in fisheries management. Although *PescaViva* game is a simplified representation of artisanal fisheries in the Amazon floodplain, and this simplification encompasses the ecological, economic, and social dimensions of the system, players nevertheless have to manage different fishing seasons and fishing spots according to the environmental dynamics of the floodplain, as well as economic aspects, such as buying different types of gear, selling fish, buying food and paying fishing costs. Each round, they have to balance their earnings and costs, which can influence where to fish and what gear to use, and so on. In the course of the game, players also have the opportunity to set fishing agreements. Through participation in *PescaViva* and engaging in post-game debriefings, SEMAS agents underscored the valuable insights gained into fishing practices and dynamics. They also underlined how the discussions provided them with an opportunity to share their own experiences as SEMAS agents.

The inclusion and integration of knowledge between different social actors, people who know the daily practice experienced in the territory, people with technical knowledge about the region or activity carried out in the region, people with institutional knowledge, people who are researchers about the region, people who know about participatory modeling and participatory modeling students is one of the principles of the companion modelling methodology. The confrontation of this diverse knowledge and viewpoints is a key moment, fostering a significant amount of learning for all participants. This process encompasses a broad understanding of various perspectives on the territory, the political and economic situation, as well as the distinct challenges and forms of resistance observed among each of these knowledge agents (Etienne 2014).

These key moments serve as opportunities for different social actors to collaboratively construct new knowledge about the territory, encompassing its myriad details. The participatory modeling facilitated by *PescaViva*'s key moments aids in shaping the understanding of state agents, who represent the State itself. During these moments, fishermen present the challenges encountered in fishing, innovative solutions devised to address real problems, verbal agreements between fishermen, techniques for handling fishing instruments, the significance of time in conducting fishing activities, the dilemmas fishermen face, and the intricacies of their trade and way of life, which is particularly complex. These key moments are essential for the State to gain awareness of the fishermen's reality and explore new, less bureaucratic, and more effective approaches to fulfill its supportive role in creating, implementing, and monitoring fishing agreements in the region. Beyond the technical aspects, this knowledge exchange between fishermen and State agents is crucial for fostering a deeper understanding within the State of another socio-ecological system—a distinct form of social organization and activity that considers fishing and fishing agreements not only in their practical and commercial dimensions but also in terms of life, culture, and ecology. In essence, it aids in the State's realization that fishing and fishing agreements are far more than organizational structures for fishing production; they represent a comprehensive way of organizing life.

By delving into the tangible reality of the daily lives of individual and collective participants in the process, considering their various dimensions and interactions—embracing real-life scenarios and experiences, and interpreting these lives and experiences as perceived and contemplated by the people with whom we engage (Brandão and Borges 2007)—this methodology enables SEMAS employees, for instance, to glean insights from fishermen during the game's workshops. The methodology aims to be pedagogical and dialogical, serving as instruments for shared learning. It inherently possesses an educational and, therefore, politically formative nature (Brandão and Borges 2007), clearly demonstrating its character as a form of popular education.

5.4 *Lack of Communication Between Users and Environmental Agencies*

During the process conducted with the fishermen, they emphasized their difficulty in communicating with the environmental agency, which complexifies and slows down the negotiation and publication of fishing agreements. This failed communication between fishers and managers is also reported by Dedual et al. (2013). LGC fishermen have shared that they are utilizing an NGO to serve as an intermediary in their negotiation with SEMAS, as can be noticed by the testimony of one the participants below:

> A Sapopema ouvia nós, levava para a Semas, contava a nossa realidade e aí eles falavam para a Sapopema e a Sapopema trazia para nós (Participant 04, 7th edition of PescaViva workshop).
>
> *Sapopema* [the NGO] listened to us, took our demands to Semas, told them our reality and then they told *Sapopema* and *Sapopema* brought it to us (Participant 04, 7th edition of PescaViva workshop, our traduction).

With municipal environmental agencies, communication is considered a bit better. For example, municipal agents and fishermen representatives made a collective effort in the last years to conduct enforcement actions in *Juruti* and *Óbidos*, as reported by participants. However, fishermen claim that the state environmental agency seems to be unapproachable to them.

By involving environmental managers in the *PescaViva* workshops, a dialogue was initiated between fishermen and environmental managers. The game served as an interface, bridging the gap in understanding between these two stakeholder groups—one with traditional knowledge grounded in everyday reality, the other with technical knowledge. The playful element also played a role in alleviating the formal and rigid nature of discussions, fostering connections among participants. Engaging in the game as teammates, for instance, allowed an agent from an environmental agency to not only discuss strategies developed in the game and real-life scenarios but also foster a more amicable bond with their teammate. This strengthened bond became apparent in the mutual receptiveness to participants' viewpoints during the debriefing sessions. Furthermore, the interaction between fishermen and managers was sustained for a minimum of four meetings, providing more opportunities for communication, knowledge-sharing, and mutual understanding to develop.

6 Conclusion

The process of co-construction of *PescaViva* game was not solely about building a game to facilitate the dialogue among fishermen for the support of establishing and implementing fishing agreements. Throughout the workshops, participants perceived that these workshops could serve a broader purpose. They saw it was a space to bring together environmental agents, allowing them to discuss fisheries management issues in a less formal and more comfortable manner. The formality often applied in interactions with state agents can create tension and barriers, making it challenging for fishermen to communicate their concerns and uncertainties about whether their demands are being heard. In this way, the construction of the game prompted a broader reflection on the dynamics of the territory, leading participants to identify the need to incorporate new contributors. They become active contributors to the research, influencing the objectives and guiding the direction of the process. The core of the participatory process revolves around the developments and proposals emerging from the participants. The research team was open to discussing all aspects of this participatory process with the fishermen, resulting in changes to many of the initially set objectives and expectations.

By involving SEMAS to workshops, the fishermen highlighted the crucial role of the state in managing fisheries management in the floodplains of the lower Amazon. During the co-construction of the PescaViva game, they seized the opportunity to emphasize the need for the State to fulfill its responsibilities, particularly in enforcing environmental legislation, to ensure the sustainability of fishing. The fishermen underscored that without the State's active involvement, it would be challenging to engage in meaningful discussions and implement effective fishing agreements.

The lighthearted atmosphere fostered by the game and other group dynamics implemented in the workshops encouraged participants to freely share not only their knowledge and perceptions about life but also their emotions and expectations. This facilitated additional connections among the participants, enabling them to view each other not just as representatives of organizations but as individuals. Sharing subjective aspects is particularly crucial in engaging people within their organizations and groups, encouraging a more active, collaborative, and open approach to dialogue about both the practical and formal aspects of the decision-making process.

For effective results in participatory researches, we claim that the involvement of the participants must be considered in all the aspects of the research, such as, who should be involved in the co-construction of the game. Without the involvement of SEMAS in *PescaViva*, as suggested by the fishermen, positive impacts related to facilitating communication between grassroots organizations and governmental sphere would not be achieved.

Acknowledgements This research is conducted within the framework of three research projects, the BONDS project (Biodiversa-Belmont Forum) with funding from the ANR, the SABERES project funded by the BNP Paribas Foundation, and the PSIOP project from the UFRJ. The first author thanks the INCT ODISSEIA (UnB/IRD) for the funding of his post-doctoral fellowship. We also acknowledge the organizations involved in this research: Colônia de Pescadores Z-20, Colônia de Pescadores Z-19, Colônia de Pescadores Z-42, MOPEBAM, FEAGLE, Conselho de Pesca do Arapixuna, Conselho de Pesca do Lago Grande, STTR-STM, STTR-Juruti, Sapopema, Guardiões do Bem Viver, SEMA-STM, SEMA-Juruti, SEMA-Óbidos and SEMAS-PA.

References

Almeida OT, McGrath DG, Ruffino M (2001) The commercial fisheries of the lower Amazon: an economic analysis. Fish Manage Ecol 8:253–269. https://doi.org/10.1046/j.1365-2400.2001.00234.x

Almeida OT (2004) Fisheries management in the Brazilian Amazon. PhD in Socio-environmental science, Imperial College, London, England, p 188

Bécu N, Neef A, Schreinemachers P, Sangkapitux C (2008) Participatory computer simulation to support collective decision-ma-king: potential and limits of stakeholder involvement. Land Use Policy 25:498–509. https://doi.org/10.1016/j.landusepol.2007.11.002

Bommel P, Bonnet MP, Coudel E, Haentjens E, Nunes C, Melo G, Nasuti S (2016) Livelihoods of local communities in an Amazonian floodplain coping with global changes. iEMSs Proceedings. https://scholarsarchive.byu.edu/iemssconference/2016/Stream-D/72/

Bousquet F, Étienne M, d'Aquino P (2014) Introduction. In: Étienne M (ed) Companion modelling. Springer, pp 1–12

Brandão CR, Borges MC (2007) A pesquisa participante: um momento da educação popular. Rev Edu Pop 6(1):51–62. https://doi.org/10.14393/REP-2007-19988

Carvalho MDS, Melo MFAQ (2017) O eu em cena: O jogo no psicodrama e os jogos eletrônicos. Rev Bras Psicodrama 25(1):94–100. https://doi.org/10.15329/2318-0498.20170011

Castello L, Macedo (2016) Large-scale degradation of Amazonian freshwater ecosystems. Glob Change Biol 22:990–1007. https://doi.org/10.1111/gcb.13173

Castro F (2000) Fishing accords: the political ecology of fishing intensification in the Amazon. PhD in Public and Environmental Affairs, Indiana University, Indiana, United States, p 308

Chapuis K, Bonnet MP, da Hora N, Evangelista-Vale CJ, Bancel N, Melo G, Le Page C (2022) Support local empowerment using various modeling approaches and model purposes: a practical and theoretical point of view. In: Czupryna M, Kamiński B (eds) Advances in social simulation. Springer Proceedings in Complexity. Springer, Cham. https://doi.org/10.1007/978-3-030-92843-8_7

D'Almeida BG (2006) Os acordos de pesca na Amazônia: Uma perspectiva diferenciada de gestão das águas. In: XV Encontro Preparatório do Conselho Nacional de Pesquisa e Pós-graduação em Direito—Recife, CONPEDI, Recife, 1–15

Dedual M, Pla OS, Arlinghau R, Clarke A, Ferter K, Geertz HP, Gerdeaux D, Hames F, Kennelly SJ (2013) Communication between scientists, fishery managers and recreational fishers: lessons learned from a comparative analysis of international case studies. Fish Manag Ecol 20:234–246. https://doi.org/10.1111/fme.12001

Diário Oficial do Governo do Estado do Pará nº 35.215 (2022) Concurso público para provimento de vagas em cargos de níveis médio e superior concurso público C-218:7–27 https://www.ioepa.com.br/pages/2022/2022.12.12.DOE.pdf

Ducrot R, Jacobi PR, Barban V, Clavel L, Carmargo ME, Carvalho YMC, Franca TJF, Sendacz S, Gunther WMR (2008) Elaboração multidisciplinar e participativa de jogos de papéis: uma experiência de modelagem de acompanhamento em torno da gestão dos mananciais da Região Metropolitana de São Paulo. Ambiente Sociedade 11(2):355–372. https://doi.org/10.1590/S1414-753X2008000200010

Etienne M (eds) (2014) Companion modelling: a participatory approach to support sustainable development. QUAE éditions and Springer Editor

Faria CH Jr, Souza JS, Pereira TM (2018) Diagnóstico da ação do IBAMA na explotação dos recursos pesqueiros no Baixo Amazonas (PA). Revista Ibero-Americana De Ciências Ambientais 9:273–282. https://doi.org/10.6008/CBPC2179-6858.2018.006.0026

Folhes RT (2016) O Lago Grande do Curuai: história fundiária, usos da terra e relações de poder numa área de transição várzea-terra firme da amazônia. PhD in Environmental Sciences, Paris 3 Sorbonne Nouvellem, Paris, France, 300 p

Furtado LG, Nascimento IH, Alencar EF, Moreira ES (1998) Lago Grande de Monte Alegre: relato sobre gente, ambiente e conflitos sociais no Médio Amazonas. Boletim do Museu Paraense Emílio Goeldi 14(1):21–95. http://repositorio.museu-goeldi.br/handle/123456789/823

Hartmann WD (2001) Conflitos de pesca em águas interiores da Amazônia e tentativas para sua solução. In: Diegues AC, Moreira ACC (eds) Espaços e recursos naturais de uso comum. NUPAUB-USP

Hora NN (2020) Sistemas de co-manejo pesqueiro com enfoque no pirarucu (Arapaima spp.) em comunidades de várzea do Baixo Amazonas: análise institucional e custos de transação. 242 f. Tese (Doutorado em Sociedade, Natureza e Desenvolvimento) Programa de Pós-graduação em Sociedade, Natureza e Desenvolvimento, Universidade Federal do Oeste do Pará, Santarém. https://repositorio.ufopa.edu.br/jspui/handle/123456789/351

Hora NN, Chapuis K, Vale JCE, Melo G, Bonnet MP, Bommel P, Le-Page C, Nasuti S (2023) Harnessing the complexity of socio-ecosystems to design games as "viable metaphors: lessons from a case study in the lower Amazon floodplain fisheries. In: Becu N (ed) Simulation and gaming for social and environmental transitions: proceedings of the 54th conference of the international simulation and gaming association, pp 399–410. https://shs.hal.science/halshs-04209935

Isaac VJ, Da Silva CO, Ruffino ML (2008) The artisanal fishery fleet of the lower Amazon Santarém, PA, Brazil. Fish Manag Ecol 15:179–187. https://doi.org/10.1111/j.1365-2400.2008.00599.x

Isaac VJ, Almeida M (2011) El consumo de pescado en la Amazonia Brasileña. Food and Agriculture Organization (FAO), COPESCAL

Lima MW, Praxedes TN, Santos VA, Alves MAMS, Bastos RZ, Henriques FM, Palheta MKS, Silva IGV (2023) Os desafios e as perspectivas dos acordos de pesca após a regulação do decreto estadual 1.686/2021 na Amazônia Oriental. In: Congresso Brasileiro de Engenharia de Pesca—XXII CONBEP

McGrath DG, Castro F, Futemma BDA, Calabria J (1993) Fisheries and the evolution of resource management on the lower Amazon floodplain. Hum Ecol 21:167–195. https://doi.org/10.1007/BF00889358

McGrath DG, Cardoso A, Sá EP (2004) Community fisheries and co-management on the lower Amazon floodplain of Brazil. In: Proceedings of the second international symposium on the management of large rivers. Phnom Penh, FAO, pp 207–221

McGrath DG, Castello L (2015) Integrating fishers' ecological knowledge and the ecosystem based management of tropical inland fisheries: an Amazon case study. In Fischer J, Jorgensen J, Josupeit H, Kalikoski D, Lucas CM (eds) Fishers' knowledge and the ecosystem approaches to fisheries: applications, experiences and lessons in Latin America. FAO Fisheries and Aquaculture and Technical Paper, pp 127–148

Melo G, Bonnet MP, Bommel P, Hora NN, Vale JCE, Chapuis K, Nasuti S, Coudel E, Le-Page C (2022) A modelagem de acompanhamento: perspectivas metodológicas para integrar conhecimentos científicos e saberes locais na Amazônia. In Nasuti S, Bursztyn M, Saito C, Bonnet M (eds) Dinâmicas socioambientais no Brasil: atores, processos e políticas. Editora IABS, pp 205–225

Seixas CS, Kalikoski DC, Almudi T, Batista VS, Costa AL, Diogo HL, Ferreira BP, Futemma CRT, Moura RL, Ruffino ML, Salles R, Thé APG (2011) Gestão compartilhada do uso de recursos pesqueiros no Brasil: elementos para um programa nacional. Ambiente Sociedade 14:23–44. https://doi.org/10.1590/S1414-753X2011000100003

Silva RE, Ferreira RR (2018) Construção de Acordos de Pesca e Políticas Públicas para gestão de recursos pesqueiros na Região de Santarém, Pará (1990–2004). O Social em Questão 41:327–354. https://www.redalyc.org/articulo.oa?id=552264297015

Tensions Between Territorial Dynamics and the Other Ways of Coexisting, Inhabiting and Reproducing Life in the Eastern Amazon: From the Collective Experience of Women *Babaçu* Coconut Breakers

María de los Ángeles Arias Guevara

> *the earth is shouting at us, the words it uses to speak to us are no longer soft, It has become necessary for it to shout at us so that we listen to it*
> *Patricia Nogueira*

Abstract The objective of the text is to understand how collective experiences of women in the use of the goods provided by the Amazon forest express other ways of inhabiting, living together and re-existing. It explores from the collective experience of women *babaçu* coconut breakers the tensions between territorial dynamics and the other ontologies of living together, of inhabiting body-territories and of care, which configure ways of feeling and thinking in opposition to the economic-reifying rationality of the modern development project on that space.

Keywords Socioeconomic dynamics · Territory · Inhabiting · Other ontologies · Socio-environmental conflicts · Women *Babaçu* coconut breakers

1 Introduction

Since the last decades of the twentieth century, we have witnessed increasing processes of commodification of nature at a global level, more expressive in those areas rich in goods that capital has historically converted into resources necessary

M. de Á. Arias Guevara (✉)
International University of Valencia, València, Spain
e-mail: ariasguevaramariadelosangeles@gmail.com
URL: http://lattes.cnpq.br/4055328507202177

W. Leal Filho et al. (eds.), *Amazon 2030—Sustainability Issues in the World's Largest Rainforest Region*, World Sustainability Series,
https://doi.org/10.1007/978-3-031-81465-5_8

for its expanded reproduction. Such is the case of the Eastern Amazon, a socially appropriated, expropriated and reterritorialized territory, crossed by multiple socioeconomic dynamics that stress the culturally constructed modes of social reproduction of life, its material and symbolic dimensions.

Extensive farming practices, agribusiness, mining, logging, large hydroelectric and road infrastructure projects have devastating consequences for the whole biodiversity, including traditional, indigenous and peasant peoples; they cause forced displacement, destroy rural livelihoods, and pollute water sources, soil and air.

A territory such as the one comprising southeastern Pará is strategic in world geopolitics, a breeding ground for the reconcentration of land according to global interests with the consent of the State and local elites. It is possible to observe that the model projected on the space accelerates deterritorialization, which as a predatory act of colonial character penetrates bodies and subjectivities, disrupts identities and destroys traditional ways and sources of life by imposing an ontology that breaks with historically constructed ways of relating to nature. These processes are constantly found in women's narratives, revealing their questioning and resistance to the socioeconomic dynamics that curtail their rights to territory and strain their modes of social reproduction of life.

The capital's logic of territorialization activates variables of a Eurocentric rationality in which the human and the non-human are considered objects of domination, lifeworlds are sacrificed, and nature is objectified by awarding it a chrematistic value. Perhaps, the concept that best qualifies the above described is the notion of "maldevelopment" taken up to explain the anomalies of 'Development' (Svampa and Viale 2014), seen as failure of developmentalist programs, insofar as these proposals widen the gaps of social and gender inequality, exacerbate eco-territorial conflicts and further stress the pressures on nature. Authors such as Hébette (2004) and Martins (1993, 2016) qualify the model of capitalism implemented in the Amazon as a dilacerator of socio-environmental life.

Although there are numerous studies on the modernization-development of the Eastern Amazon as a privileged geopolitical space for the demands of the global market, there are insufficient studies that explore with the necessary specificity the interdisciplinary theoretical tools of Political Ecology to unravel the cultural assumptions of ontologization of the multiple cleavages and hierarchies, explaining the conflicts and antagonisms in economic activities that put in tension in space and time the conditions that make possible the reproduction of the web of life by functionalizing them according to the dynamics of accumulation.

In this landscape created by neo-extractivism and agricultural monoculture is the scenario in which daily experience takes place, in which bodies are controlled, and where the simulator model of sustainability makes sense, where we identify micro-scenarios and micro-landscapes in which women and other actors express visions and strategies of relationships, interdependence and eco-dependence that allow us to understand their ontological and symbolic implications; making visible other practices of knowing and doing, other agents of knowledge production that "have elaborated what could be called an alternative political ecology based on notions of sustainability, autonomy, diversity and alternative economies that do not conform to the dominant discourse of development" (Escobar 2005, p. 22).

Studies conducted by Cunha and Almeida (2001) on peoples of the Amazon Forest showed the appropriate and sustainable ways in which these peoples relate to nature. The experiences constructed in their struggles and daily practices are critical for understanding other ontologies and cosmovisions, since the value attributed to nature differs from that attributed by utilitarian economic rationality, and its meaning involves the need for care, as it implies the goods necessary for the reproduction of life.

The epistemic aspects on which we rely are to be found in those that, based on decolonial thought, criticize the multiple separations (splits) of modernity, in the critique of the discourse on development, etc.; sustaining the understanding of capital-nature relations and the dispositives engaged for territorial control and of the corporealities that inhabit the diversity of spaces, among them, women's bodies.

In the field of Latin American studies, authors such as Quijano (2000), Mignolo (2010), Lander (2000), Escobar (2005), Walsh (2007) and Lugones (2008), among others, provide us with the theoretical arsenal to reflect on the coloniality of power, knowledge, nature and gender that the modern colonial project presents for the region. This proposal is being expanded by other authors such as Escobar (2010), Leff (2006), Toledo (2003), among others, who, from the richness provided by Political Ecology as an interdisciplinary area of study, broaden and deepen the critical analysis on the coloniality of nature and the crisis of civilization generated by the hegemonic model presented by the Western experience; a rationality that turned nature into a subordinated and controlled object. Political Ecology proposes the search for an environmental rationality based on the recognition of difference.

From the theoretical point of view, it is also necessary to dialogue with certain aspects of critical Ecofeminism, Third Worldism and the so-called indigenous and peasant ecofeminisms of Latin America respectively, given that the marginalization of women and the destruction of biodiversity originate in the patriarchal model, whose anthropocentric profile is not only directed at dominating nature, it also colonizes women's bodies. These approaches provide the analytical categories for a deeper understanding of the context from women's embodied experiences.

The text is the result of the research that under the theme: "Ruralities, Power and Resistance in Southeastern Pará. From the lived experience narrated by women" I conducted between June 2015 and February 2017 at the Federal University of the South and Southeast of Pará, together with the Programa de Pós Graduação em Dinâmicas Territoriais e Sociedade na Amazônia and its continuing ties that allowed a reflection, enabling the understanding of the ontologies of inhabiting the bodies-territories and the ontological tensions of meanings and of languages of valuation, configured on the territory, nature and the environment from the diversity of socio-cultural practices developed by the women with whom I interacted.

The methodology applied seeks to express the possibilities offered by a positioned view that looks out for the breadth of translations by a "situated knowledge", considering the situations, positionalities and circumstances, in which the pluriverse practices and relationships that reproduce life are produced and re-signified. We tried to work with the narratives of lived experiences, conflicts, ontological tensions and the languages of valuation that are built on them, making visible and incorporating the perspective that women from traditional, indigenous and peasant peoples have on them; to weave together individual with collective narratives and look through the windows they open to the fabric that relates the apparently invisible with the economic dynamics of the territory.

We focus our reflection on marginal aspects, such as the ontological relationality that underlies women's daily lives, other epistemes, expressions of inter-being, captured through narratives that re-signify and reposition them from a political point of view, expressing lives built by bodies in resistance.

This is how we can question why neo-developmentalist strategies do not consider the Amazonian experience in its diversity of expressiveness of traditional peoples, indigenous peoples, whose ways of existing and relating with nature, based on an ethic of care, guarantee the sustainability of life. Or why economic thought does not dignify those forms that value solidarity more than profit; why studies on the experiences of Amazonian women remain on the discursive margins of sociological thought and other disciplines, concentrating more on the great events and structures of the present, without giving visibility to those landscapes that weave power relations and hide other subordinations.

In order to reflect on practices and experiences that express other ways of inhabiting the body-territories, with care and in eco-dependence, on the construction of embodied subjectivities, we need a positioning and a research practice capable of producing knowledge "with" and not "about", epistemic perspectives generated in the margins that sustain and involve us, because also as a researcher, our subjectivity enters into the game of knowledge and power relations; that is to say, we are also chosen "by that what acquired meaning for us and also signified us, subjectivized us, made us subjects" (Corazza 2002: 124).

Experience as a process of signification, as a practice of giving meaning, both symbolically and narratively (Brah 2011), is taken here as a theoretical analytical substrate that allows the apprehension of the historical process lived and narrated, informing on the positionality of the subjects, furthermore it is also a methodological resource for approaching the field; likewise, the category "inhabit" taken from environmental thought allows understanding the spatialized notion of the social; but especially the inhabit as "oikos, (…), place of rootedness, of the everyday, space where we love, hate, work, relate, create bonds and affections, where we make and are life" (Nogueira and Bernal 2014: 2).

The qualitative methodologies such as narrative interviews and participant observation in the life spaces allowed the apprehension of interactions, shared practices and meanings, cognitive processes, always from the perspective of the body as a constructor of languages, as a weaver of contexts and meanings that these practices articulate, through the voices that allow their enunciation. In this sense, we mobilize the concept of languages of valuation provided by Martínez Alier (2004), since conflicts and ontological tensions are expressed through diverse languages of valuation, ecological, cultural, subsistence and economic values, always contained in the discourses of the actors, as the expressions and meanings given by the losses suffered.

2 Lived Experience of Women's Bodies that Inhabit and Are Inhabited by the Landscapes of the Eastern Amazon

The work of production and reproduction of life of traditional populations in the Amazon Forest depends on the knowledge and management of the biome, a traditional knowledge, mainly held by women as farmers, midwives, healers, nuns or others. However, the socioeconomic dynamics that are implemented in the name of development are expropriating land and territory, causing tensions and multiple socio-environmental conflicts, understood as "those generated by mega-mining, the agrarian question, water, forest goods that involve resistance of the social actors against the process of destruction of their territories and their ways of life. These are conflicts that incorporate perceptions, meanings and values about development, democracy, and relations with nature" (Guevara and Moreira 2020: 2–3).

These conflicts are also called eco-territorial (Svampa 2019) as they are related to the logics and rationalities with which the territory and its goods are perceived, or with the values given to nature; always mediated by power relations. Political Ecology defines them as ecological-distributive conflicts from which social and political movements for Environmental Justice are generated (Martínez Alier 2004, 2008). Movements framed by the same author as "environmentalism of the poor" (Martínez Alier 2005), which we can exemplify with the movements present in the context of study, such as the "*Atingidos por barragens*", the "*Movimento Anti mineração*", or the "*Movimento Interestadual das Quebradeiras de Coco Babaçu*", that express through diverse languages ecological, cultural, subsistence values as the expressions and meanings given by the losses suffered; for example, when people displaced by the Tucurui dam are heard saying "perdimos o lugar, a terra, o agua, os peixes" (Casica de los indios Gaviôes 2016).

The experience of women interrelates forms of domination intersected by class, gender, ethnicity, skin color, age, among others, and by the violence of global patriarchal power in territories where mining and livestock expansion supplanted traditional economic activities, displacing women from those spaces that generate income, such as the extractivist activity of medicinal plants, *babaçu* coconut, fruits or food production, defeating their place in the reproduction of life.

The anthropocentric profile that characterizes the modernization of the Eastern Amazon, in the same way that it colonizes nature, colonizes women's bodies. An issue with sufficient analysis from the Third World Ecofeminism by emphasizing that the marginalization of women and the destruction of biodiversity are originated from the patriarchal model (Mies and Shiva 2013). For Vandana Shiva:

> the main threat to nature and people comes from the centralization and monopolization of power and control, that drive to generate one-dimensional structures and I have termed "monocultures of the mind". Such cultures threaten diversity like a disease and create coercive structures modelling this biologically and culturally diverse world of ours on privileged categories and concepts of class, race and gender of a single kind (Shiva 2007: 53).

Therefore, the need is recognized to look at environmental issues always from power relations that intersect gender relations with other variables in the social fabric of socio-environmental conflicts in order to deconstruct the dualistic hierarchies with which modernity placed women and nature as the "other" to be dominated and even commodified.

For the purposes of this paper, our attention was focused on women whose financial autonomy and social reproduction of life depends on the maintenance of the Amazon biome: the women organized in the *"Movimiento das Mulheres Quebradeiras de Coco Babaçu"*. In order to understand their experience we comprehend the territory they inhabit as a social and symbolic construction that, following Ventura Fernandez (2022), can be defined as:

> the inhabited space, the concrete place of human experience, of relationships, of daily life; it is the place for the construction of knowledge and the physical and cultural continuity of a group. The territory is the place of knowledge and flavors, the place of cultural meaning; it is the space of belonging (p.116)

It is worth noting that the symbolic dimension of the territory results from the incorporation of the actors' representations of their spaces, customs, values and way of life, which depend on their factually available resources and their competence and willingness to use them (Bourdieu 1989). These representations operate in the territory as the actors concerned refer to spaces and extents that are not purely physical, such as, for example, the space of the immemorial times of the indigenous peoples, the mythical and sacred spaces in the forests, in the rivers and on the land that are protected by the peasants, since they have practical meanings in collective life, either to obtain a prosperous harvest, or to obtain products for their rituals of healing and protection. The value attributed to nature differs from the utilitarian economic rationality focused on profit, its meaning lies in the inseparable relationship between human beings and nature, which involves needs and care, implying the goods necessary for their life and, therefore, the need to be preserved.

This other relationship with nature is experienced by women of the *Movimiento Interestadual de las Quebradoras de Coco Babaçu*, whose symbolic territory encompasses spaces located in the valley of the Tocantins River. An area of *babuçuais* known as the "Ecological Region of the *Babaçuais*" which covers an area of approximately 18.5 million hectares belonging to the states of Maranhão, Pará, Piauí, Tocantins, Goiás and Mato Grosso (Almeida 2005). In the state of Pará it covers about 290,000 ha, mainly in the municipalities of San Domingo do Uruguaia, Palestina, Brejo Grande and San Juan do Uruguaia (Interstate Movement of Quebradeiras de Coco Babaçu 2013).

The income that emerges from extractivism is increasingly threatened by the hiring of the *babuçuais* to companies that extract the whole coconut to manufacture the charcoal that feeds the industrial park of the city of Marabá and other industries located in the state of Maranhão. The so-called *Ferro gusas* demand the charcoal to melt the iron in their furnaces. Taking into account the limits imposed by environmental legislation to continue the devastation of the Amazon forests, the companies negotiate with large landowners to buy the coconut, turning it into a strategic raw material to feed their steel mills; an issue that strains the existence of a traditional culture that organized specific ways of relating to nature, also affecting the social, cultural and physical reproduction of women who generations after generations have guaranteed the sustainability of life by cracking the *babaçu* coconut.

The *babaçu* coconut breakers by their profiles are women whose income does not exceed the minimum wage, more than 90% black, under-schooled, belonging to several generations, whose bodies bear signs of domestic patriarchal violence and that caused by the patriarchy that dominates livestock expansion, agribusiness and mineral exploitation, depriving them of the source of family sustenance: the *babaçu* coconuts (Guevara 2018). For these women, access to *babuçuais* is to situate themselves beyond the domestic economy, it is to defend a balanced way of being and existing in a landscape of which they consider themselves part. Their historical struggle led them to organize, first in rural unions and then in the social movement known as "*Movimento Interestadual das Quebradeiras de Coco Babaçu*" (MIQCB) created in 1995, after a long process of problematization of their realities, of organization and coordination of existing groups, it constitutes the political expression of the organization of women who carry out this activity in the states of Maranhão, Piauí, Tocantins and Pará. This organization demands public policies related to their gender condition and position as rural women and the raciality of their bodies, among them, agrarian reform and the sustainability of the *babuçuais* as symbolic territories, the right to free access, that is, the democratization in the access and use as a common good that nature gave them and that assures their and their families survivorship. The Interstate Movement of Quebradeiras de Coco Babaçu connects social networks such as associations, clubs, commissions, women's groups and cooperatives, constituting the representation of the *quebradeiras* as political subjects. Its struggle mainly aims at the practical realization of the "*Babaçu Livre*" Law project:

> "Queremos fazer valer essa lei, os donos da terra estão envenenando as palmeiras, colocando cerca elétrica para impedir que entremos no território, estão contratando o coco inteiro para carvão, não querem doar o coco para o povo , quando denunciamos somos perseguidas, ameaçam de morte nossas companheiras, os cocos estão cada vez mais longe e a prefeitura oferece pouco apoio (Coconut breaker, 27 years old, São Domingos do Araguaia).

Violence against women encounters the omission, the silence of State institutions, though this violence is also a matter of State (Segato 2011), it is unable to enforce the approved legislation, it is indifferent to a demand for support, to the threats and deaths of women. As another interviewee says "*Nossa luta é pelo babaçu livre, é uma questão de todas nós, queremos as mulheres e as palmeiras vivas*" (Coconut breaker, 60 years old, São Domingos do Araguaia).

This discourse expresses the conditions of vulnerability in which women and palm trees find themselves. Among women and between women and palm trees of the *babaçu* coconut there is an interdependence and eco-dependence, jeopardized by the socio-environmental conflicts that define the expansion of capital in southeastern Pará. According to Pérez Orozco (2015) it is a conflict of logics: of the logic of accumulation and the logic of sustainability of life. And that second logic is the one that nourishes the experiences of the women who break the *babaçu* coconut. The *babaçuais* can be considered symbolically constructed territories, based on social practices that form particular contexts of relationship between women and the ecosystem that sustains them and gives meaning to their way of life, from which their experience is narrated.

Coconut cracking is the main income-generating productive activity, transmitted from generation to generation, that while it is narrated is being resignified, bringing up scenes of violence experienced in the recent past. From their place of enunciation, the *quebradeiras* go back to what they call "*coco preso* time", a native expression which recalls significant episodes related to the time when they were deprived of the *coco babaçu*, because the landowners prohibited the *babuçuais* to enter the land. They began to tear down the palm trees, demand half-time jobs and even exchange harvested almonds for food from the landowners. "*O babaçu preso causou nos muito dor e sofrimento, não foi só o trabalho a meia, foram espancamentos, foram mortes*" (Coconut breaker, 65 years old, São Domingos do Araguaia).

It narrates a moment in which violence took the women's bodies as an object of domination. Death threats, murders, beatings and rapes led them first to unionize and then to organize as a movement in order to claim the rights to exist for themselves and their palm trees. A path that constitutes them as political subjects, that gives them a voice. One of the interviewees said:

> Eu não tinha voz, não podia falar em público porque não era alfabetizada, sou quebradeira de coco e também sou negra (...), o movimento fez a gente falar, nós mulheres estamos nos redescobrindo, estamos ganhando confiança em nossa própria força; nossa participação no movimento é a nossa maior conquista, eu acho (Coconut breaker, 57 years old, São Domingos do Araguaia).

From daily practices, their symbolic world is recreated, which involves the harvesting and breaking of coconuts, making charcoal, producing oil, among others. Coconut is cracked with a small axe called *machado,* sitting in a circle (a kind of horizontal relationship, face to face) with legs open, a ritual accompanied by singing. Singing is a ritual that accompanies the collective work; it has multiple meanings: "*Quando saímos de casa deixamos para trás a tristeza, cantamos para não chorar.*" (Quebradeira de coco, San Juan do Uruguaia). Metaphorical expression of human suffering, an emotion that comes from pain, from domestic violence, that looks for support networks in the collective. These songs also express re-existence, dreams, desires, emotions; affections and demands, for example the *Xote das quebradeiras de coco* whose content is directed to the preservation of the palm trees, to the demand of the fulfillment of the law "*babaçu livre*", to the palm trees as a provider of food source, of life reproduction. The *xote* is a tool of feeling-thinking within the group in daily life and outwardly, it is a cry of the territory embodied in the coconut breakers that clamors for the preservation of the palm trees and for their own lives.

This analysis disrupts the traditional concepts aligned with instrumental reason that separate culture and nature and considers nature as an object to be dominated and plundered. Thus we search for the meanings that the coconut breakers give to their palm trees to find another ontological matrix in the relationships and representations that these women have with and about the palm trees. Relationships of affectivity, mother tree, which gives and reproduces life; the space of the palm trees as their symbolic territory; common good, the palm tree as an expression of spirituality:

Xote das Quebradeiras de Coco.

Ei! Não derruba esta palmeira.
Ei! Não devora os palmerais.
Tu já sabes que não pode derrubar,
precisamos preservar as riquezas naturais
Santa Maria é a nossa companheira
Grande força verdadeira que proteje esta nação
Que fortalece a nossa luta pouco a pouco
E a mulher que quebra o coco pede a sua proteção

(Canto e encanto nos babaçuais 2014: 8)

It is a work chant that accompanies them in their daily lives, languages of valuation crossing all their work and feelings, the ways of naming, for example, the rotary fund created in the cooperative is called "*I am babaçu*", which expresses a symbiotic identity with the palm trees.

Through the content of the narrated and experienced by these women, I appreciated from the suffering, pain and joy, other ways of inhabiting and being inhabited by the territory; which connects us with community feminism: ways of inhabiting the body-territories (Cabnal 2010), ways in which the coconut breakers express their existence in relation, a kind of choreography or a geopoetic inhabiting, an inhabiting in dialogue, in construction of senses in the words of Nogueira and Bernal (2014). An inhabiting that inhabits us in the sense of the *Buen Vivir* of the Andean peoples.

3 Conclusions

The study showed the potential that traditional peoples have for the ecological sustainability of the Amazon, specifically those who socially reproduce their lives through other economies supported by symbolic traditions such as those established between coconut breakers and palm trees, constituting a model of alternative sustainability in territories of difference (Escobar 2010).

The experience of the female *babaçu* coconut breakers is related to the struggle for the life of the palm trees, which by symbiosis, as they understand it, is a struggle for their own lives, linked to a symbolic territory of which they are part and which they defend; it is a struggle for their survival as extractivists, for the right to be and exist as women, a right that is also transferred to their palm trees, considered in their rites as a mother that gives life.

Bodies inhabited by the landscapes that make up the *babuçãis*, with ontological and symbolic implications expressed in relationships of interdependence and eco-dependence such as those established with these ecosystems.

The socio-productive, cultural and symbolic practices developed by these women connect them to the common use of the goods provided by the Amazon biome and, therefore, they are interested in its sustainability. Their daily practices have linked them to care and sustainability; a sustainability strained by the economic dynamics of capital accumulation on a global scale that, by opening new frontiers, sacrifices worlds of life. Expression of a reifying ontology, which concedes chrematistic value from the subsoil to the wind, passing through land and water, leaving as externalities the so-called territories of sacrifice.

References

Almeida AWB de (2005) Guerra ecológica nos babaçuais: o processo de devastação dos palmeirais, a elevação do preço de commodities e o aquecimento do mercado de terras na Amazônia. MIQCB/Balaios Typ., San Luis

Bourdieu P (1989) O poder simbólico. Bertrand Brasil, Rio de Janeiro, p 1989

Brah A (2011) Cartografías de la diáspora. Identidades en cuestión. Traficantes de Sueños, Madrid

Cabnal L (2010) Feminismos diversos: el feminismo comunitario. ACSUR-Las Segovias. España

Canto e Encanto nos Babuçuais (2014) As encantadeiras: quebradeiras de coco babaçu que cantam e encantam. As Encantadeiras (Orgs) AMTR, MIQCB, ASSEMA, NCADRUFPA, Apoio

Corazza S (2002) Labirintos da pesquisa, diante dos ferrolhos. Caminhos investigativos: novos olhares na pesquisa em educação. DP&A, Rio de janeiro, pp 104–131. https://www.academia.edu/34853764/

Cunha M, Almeida WM (2001) Populações indígenas, povos tradicionais e preservação na Amazônia. In: Capobianco JPR (Org) Biodiversidade na Amazônia Brasileira. Avaliação e Ações Prioritárias para a Conservação, Uso Sustentável e Repartição de Benefícios. Instituto Socioambiental e Estação Liberdade, São Paulo, pp 184–193

Escobar A (2005) Más allá del tercer mundo. Globalización y diferencia. Instituto Colombiano de Antropología e Historia, Bogotá

Escobar A (2010) Territorios de diferencia: lugar, movimientos, vida, redes. Envión editores, Bogotá

Hébette J (2004) Cruzando fronteira. 30 anos de estudo do campesinato na Amazônia, V III. Editora Universitária, UFPA, Belém

Guevara MLAA, Moreira E (2020) Conflictos socioambientales, la lucha por la tierra y el territorio en el sudeste del Pará. In: Ambiente & Sociedade São Paulo 23:1–25 https://doi.org/10.1590/1809-4422asoc20180252r1vu2020L6AO

Guevara MLAA (2018) Experiencias agroecológicas amazónicas protagonizadas por mujeres en el sudeste de Pará, Brasil. In: Zuluaga P, Cataroga-Vargas G, Siliprandi E (Coords) Agroecologia en femenino: reflexiones a partir de nuestras experiencias. SOCLA, La paz, pp 107–122

Lander E (2000) La colonialidad del saber: eurocentrismo y ciencias sociales. Perspectivas latinoamericanas. CLACSO, Buenos Aires

Leff E (2006) Racionalidade Ambiental: a Reapropriação Social da Natureza. Civilização Brasileira, Rio de Janeiro

Lugones M (2008) Colonialidad y Género. In: Tabula Rasa, Bogotá (9):73–101

Martins J de S (1993) A Chegada do Estranho. Editora Hucitec, São Paulo

Martins J de S (2016) Fronteira a degradação do outro nos confins do humano. Contexto, São Paulo
Martinez-Alier J (2008) Conflictos ecológicos y justicia ambiental. En Papeles (103):11–27
Martinez-Alier J (2004) Los Conflictos Ecológico-Distributivos y los Indicadores de Sustentabilidad. Revista Iberoamericana De Economía Ecológica RJ 1:21–30
Martinez-Alier J (2005) El Ecologismo de los Pobres. Conflictos Ambientales y Lenguajes de Valoración. Barcelona, Icaria
Mignolo W (2010) Desobediencia epistémica: Retórica de la modernidad, lógica de la colonialidad y gramática de la decolonialidad. Ediciones del Signo, Buenos Aires
MIQCB (2013) Projeto Nova Cartografia Social da Amazônia Série: Movimentos sociais, identidade coletiva e conflitos. FASCÍCULO 5. Quebradeiras de coco babaçu do Pará. MIQCB, São Luís
Nogueira P, Bernal D (2014) Geografías del habitar: un habitar geopoético en la era planetaria. Geograficidade 4(2)
Pérez Orozco A (2015). La sostenibilidad de la vida en el centro ¿y eso que significa? In: Mora L, Cabello de A, Escribano J (Coords) La ecología del trabajo: el trabajo que sostiene la vida. Fundación Dialnet, La Rioja, pp 71–100
Quijano A (2000) Colonialidad del Poder, Eurocentrismo y America Latina. In: Colonialidad del Saber, Eurocentrismo y Ciencias Sociales. CLACSO-UNESCO, Buenos Aires, pp 201–246
Shiva V, Mies M (2013) Ecofeminismo Teoría, Crítica y perspectives. Icaria, Barcelona. http://www.icariaeditorial.com/pdf_libros/ecofeminismo.pdf
Shiva V (2007) Las nuevas guerras de la globalización; semillas, agua y formas de vida. España: Editorial Popular
Segato R (2011) Femi-geno-cidio como crimen en el fuero internacional de los Derechos Humanos: el derecho a nombrar el sufrimiento en el derecho. In: Fregoso R, Bejarano C (eds) Feminicidio en América Latina. UNAM-CIIECH;Red de Investigadoras por la Vida y la Libertad de las Mujeres, México, DF
Svampa M (2019) Las fronteras del neoextractivismo en América Latina. Conflictos socioambientales, giro ecoterritorial y nuevas dependencias. Editorial Universitaria/Universidad de Guadalajara. México
Svampa M, Viale E (2014) Maldesarrollo. La Argentina del extractivismo y el despojo. Katz, Buenos Aires. https://repositorio.uca.edu.ar/handle/123456789/5646
Walsh C (2007) Son posibles unas ciencias sociales/culturales otras. Reflexiones en torno a las epistemologías decoloniales. In: Nómadas (n. 26. Abril):102–107
Ventura FL (2022) O papel do Estado na violência nos territorio. In: CPT. Conflitos no campo. Centro de Documentação Dom Tomás Balduino–CPT. Goiania-GO, Brasil, pp 116–125

Tourism and Development in Amapá State, Brazil

Manuela Santana Görtz and Gutemberg de Vilhena Silva

Abstract This chapter presents aspects of an ongoing Ph.D. research regarding tourism public policies in the state of Amapá, Brazil. It aims to propound tourism as a viable activity for the promotion of sustainable development in the Brazilian Amazon, highlighting planning, local articulation, and international cooperation as important tools in this process.

Keywords Tourism · Public policies · Development · Amazon · Amapá

1 Introduction

By analyzing the issue of development from the perspective of public policies and the role of the State in promoting social well-being, we realize that the conflict between economic profitability, social needs and environmental sustainability remains current. This situation highlights the need for a joint effort by national and international organizations, governments, and civil society in search of solutions to urgent problems such as poverty, hunger, unemployment, and other basic needs, considering the sustainable approach and contributing for universal access to a decent life, including future generations.

M. S. Görtz (✉)
Researcher of the Research Group: Territorial Policies and Development (POTEDES) and the Observatory of Nationalities (ON), Public Policy at the State University of Ceará (UECE), Av. Dr. Silas Munguba, 1700—Itaperi, Fortaleza-CE 60714-903, Brazil
e-mail: manuela.gortz@aluno.uece.br

G. de V. Silva
Department of Social and Human Science, Researcher of Observatory of Borders of the Guiana Shield (OBFRON) and Leader of the Research Group: Territorial Policies and Development (POTEDES), Federal University of Amapa (UNIFAP), Rod. Juscelino Kubitschek, Jardim Marco Zero Macapá, AP KM-02, Brazil
e-mail: gutemberg@unifap.br
URL: http://www2.unifap.br/obfron

W. Leal Filho et al. (eds.), *Amazon 2030—Sustainability Issues in the World's Largest Rainforest Region*, World Sustainability Series,
https://doi.org/10.1007/978-3-031-81465-5_9

It is important to emphasize that the Amazon Region encompasses multiple realities in its vast territory, and understanding its specificities at the micro level, without disregarding the macro context, contributes to the formulation of more assertive solutions to its challenges. Amapá is the Brazilian state with the highest percentage of preserved natural ecosystems (Brito and Drumond 2022), extremely rich in biodiversity and aquifer reserves, such characteristics, combined with its transboundary location and international connections as part of the Pan-Amazon, as well as its socioeconomic challenges, touristic activities, mainly its sustainable aspects, point out as a potential alternative for its development. However, activities such as mining and soybean monoculture continue to receive strong incentives from the local government and private initiative based on the discourse of the need for economic growth.

It is in this context, and as part of one ongoing Ph. D. project, that this chapter intends to approach the matter of how tourism public policies could contribute to the sustainable regional development in the state of Amapá, Brazil. The first topic will present a small overview of the state, its socioeconomic challenges, and some historical context of the relation between Amapá and its neighbor, French Guiana. The second one, will focus on possibilities of tourism and international cooperation as means for the promotion of sustainable development. The las topic, final considerations, will resume the discussion, adding some aspects related to the contributions of decolonial studies to this matter.

2 Amapá: Overview, Potential and Challenges for Development

Amapá is a Brazilian state with a population of around 734 thousand inhabitants, located in the northern region of Brazil (IBGE 2022). Its relative isolation stands out, as it does not have a land connection with any other Brazilian state, characterizing a peripheral condition in relation to the national territory, that causes logistical and structural challenges, which reflect on the state´s socioeconomic aspects.

Analyzing data from the latest census from the Brazilian Institute of Geography and Statistics (IBGE) in 2022, the state's struggles in socioeconomic terms become evident. Amapá continues to occupy the last positions in Brazilian indicators such as the Human Development Index (IDH), as well as in indicators relating to education such as the Basic Education Development Index. In economic terms, while industry and agribusiness still contribute very low percentages to the formation of the Gross Domestic Product (GDP), the services sector, strongly driven by the state apparatus, drives the majority of the Amapá economy, more than 80% in 2015 (Rauber and Ferreira 2020). Given this scenario, the impact of large-scale enterprises such as mining companies and cultivation of eucalyptus and soybean monocultures on these indicators and on regional dynamics acquires great relevance.

At the same time, nature and cultural richness make up a striking feature of this state. The biggest tropical forest and the biggest river of the world are some of the highlights of Amapá's landscapes, while indigenous, quilombos and riverside cultural traditions compose a little known but very interesting and unique aspect of this society. However, the preservation of these areas and traditions still lacks the support of public policies to ensure the population's quality of life as an inherent part of this process.

This panorama, combined with the reality that a large part of the state's territory is protected by indigenous reserves, environmental reserves and quilombos, according to data from the State Secretariat for the Environment (Governo do Amapá 2023), these areas comprehend around 73% of Amapá's territory, contributes to the perception that environmental preservation would be responsible for the local socioeconomic delay. Amapá's political discourse on development is highly guided by the search to attract investments from national and foreign businesspeople, disregarding the importance of incorporating knowledge, traditions, and interests of the local population (Diniz 2016).

One of the activities that allows economic growth combined with environmental preservation and fair distribution of wealth is tourism, as long as it is followed by adequate planning and guided by local characteristics. Despite of Amapá not being a common national touristic destination, the presence of international tourists is a reality in the state, due to its international border.

One other relevant aspect of the Amapá state is the triple border constituted by Brazil, Suriname, and French Guiana (France). While the border with Suriname is formed by dense forest and, therefore not accessible, the border with French Guiana, which constitutes France's longest land border with around 730 km (Silva et al. 2019), is connected by a bridge and by small boats, available at both sides of the Oiapoque river (Map 1).

Passenger boats crossing Oiapoque river (2018).

From the perspective of the traditional populations from Oiapoque (Brazil) and Saint Georges (France), this border is, historically, very fluid—having family and

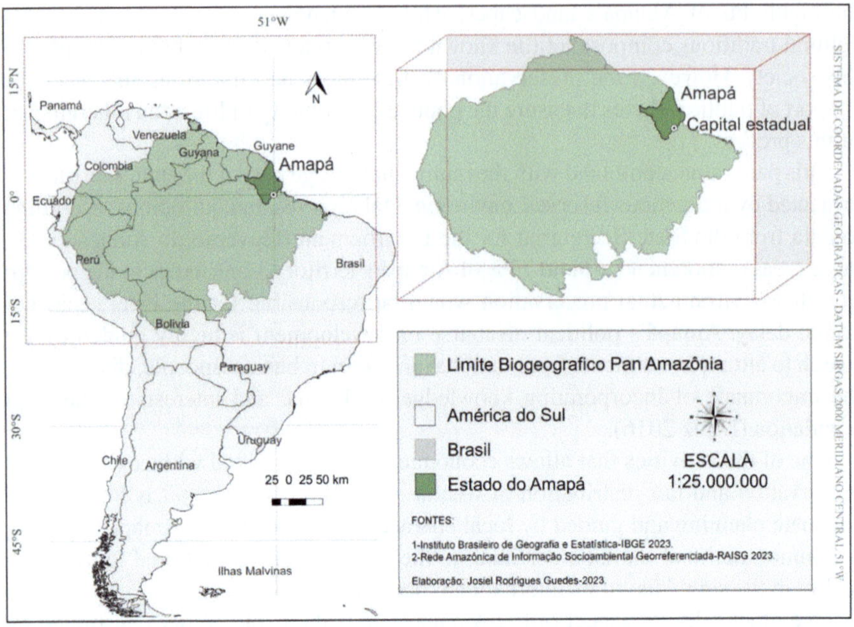

Map 1 Geographic limits of Amapá and Pan-Amazonia. *Source* IBGE databases (2023); RAISG (2023). Credit: Josiel R. Guedes (2023)

business at both sides of the border was and remains very common. From the geopolitical perspective, there was a dispute for this territory between the two countries settled in 1900 and until the 1980's the border cities remained logistically isolated from the regional capitals (Macapá and Cayenne) and the circulation between them was almost forgotten by their central governments (Silva et al. 2019).

Since then, Oiapoque and Saint Georges became more integrated to their political administration centers, and the connection between them more controlled. Cooperation projects also started to be implemented in regional and national levels, culminating with the construction of the binational bridge.

> The land connection brought a greater number of French-Guianese and Europeans to the border, now inclined to discover Brazil from its border with France or even use Macapá as a point of travel to various destinations. Customers also appeared in Oiapoque, now coming more easily from Cayenne and the Guyanese coast, attracted by the cheaper prices for Brazilian goods. Thus, circulation in that geographic space was no longer restricted to indigenous residents and connected the local circulation system to a reticular logic expanded with road connections (Silva et al. 2019, p. 7, author's translation).

Another peculiarity of this border is the fact that it is Brazil's only inverse border in the exchange rate. While in all others the Brazilian real is more valued than the respective foreign currencies, the purchasing power of Guianese, who have the euro as their official currency, is an attraction for tourism in this Brazilian state. Furthermore, another striking aspect of this relationship is the non-application of the principle of diplomatic reciprocity, since Brazilians need a visa to enter French Guiana, while their entrance in Brazil does not have this requirement.

Preliminary research indicates that the flow of international tourism between French Guiana and Amapá is already a relevant element in the state's economy and sociocultural dynamics, specially in the cities closer to the border. In 2023, the Brazilian Federal Police recorded the entry into Brazil, across the border in the municipality of Oiapoque (which has around 27 thousand inhabitants), of 31,747 foreigners, more than 90% of whom are French nationals. However, this research also indicates the fragility in the level of planning and implementation of public policies to coordinate this activity, which compromises control by the government and society over the impacts obtained, favoring the characteristics of predatory tourism.

3 Sustainable Truism and International Collaboration

Nevertheless, the moment of maturation of this research is also marked by political changes in Brazil, both internally and in international relations. After six years of escalation in dysregulations and omissions in the environmental sphere, commitment to the environmental agenda and greater alignment with international organizations favors the intensification of the sustainable tourism agenda in the Amazon, as can be noticed in recent publications and statements of national and international organizations, which align recommendations for member countries to encourage sustainable tourism, aspects also present in many speeches from both federal and Amapá's state government.

In relation to tourism as an activity capable of inducing development, Sachs (2008) emphasizes the difference between the impacts of different types of tourism, also warning against the indiscriminate use of the term ecological tourism when all the aspects of sustainability are not observed. In tourism theories and reports on successful experiences in different regions, the contribution of this activity to the sustainable development of a given location tends to occur only when the population is included in all stages—planning, implementation, and evaluation, favoring that a greater part of the income arising from the activity remains in the community and its circulation favors development without compromising traditions and the local ecosystem. Some of the characteristics of sustainable tourism are:

> [...] understanding of tourism impacts; fair distribution of costs and benefits; generation of direct and indirect local jobs; promotion of profitable businesses; capital injection with consequent diversification of the local economy; interaction with all sectors and segments of society; strategic and logistical development of transport modes; encouraging the productive use of lands considered marginal (tourism in rural areas); subsidies for environmental conservation costs (Beni 2004, p. 14, authors translation).

In a study that analyzed the potential of tourism as a tool for developing the Amazon space and preserving the forest, Ferreira et al. (2021) demonstrated the great latent potential and highlighted several lines of financing available for this purpose. The authors also listed the main obstacles for the activity to be boosted in this region: deficient infrastructure, low qualification of the workforce, low effectiveness of public policies, little or no active participation of the community involved in establishing public policies for the sector, lack of joint actions between States and Countries (partnerships) aimed at tourism development in the Region and low adherence to financing offered to the segment.

Bibliographic research confirms lack of planning and infrastructure as some of the main obstacles for the touristic activity in equipment like the Conservation Units (there are 19 in Amapá, 15 of them allow visiting activities in their regulations). Besides, sensible ecosystems usually require limitation of visitors and complementary activities or funds for the financial viability of conservation (Almeida et al. 2022).

One example of financial support is the Amazon Fund (AF), an initiative created to raise donations for non-reimbursable investments in efforts to prevent, monitor and combat deforestation, as well as to promote the preservation and sustainable use in the Brazilian Amazon. The AF started its operations in 2009, with an innovative arrange that allowed international donations with national administration of the fund, maintaining the sovereignty of the area unchanged. It was discontinued in 2019 due to disagreements of Bolsonaro's government with its regulations and environmental agenda and reactivated on January 1, 2023. The graphic bellow shows the total amount of projects funded by the AF and how Amapá's participation is still low (Brasil 2023).

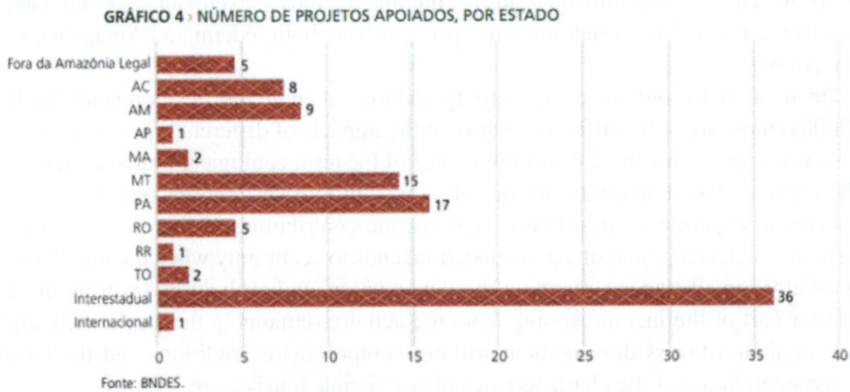

*Source*Amazon Fund Annual Report 2022.

A note published by the Center for Economic Policy and Conjuncture-Cecon (Bastos 2023) argues that one of the obstacles, mainly for small associations of traditional populations of the Amazon, to access such resources are technical, since procedures like elaborating a project and dealing with the banks bureaucracies are unknown territory for them and should be mediated by the state and other institutions, like the universities. This limitation has greater impacts on the axes of support possibilities that directly involve traditional populations, like the projects related to the sustainable use in the Brazilian Amazon, leading to a disparity in the number of initiatives supported between the axes.

Some reports of concluded projects in other Brazilian states, showcase examples of how community-based tourism can be strengthened by fundings like the AF, when supported by local organizations. In the state of Amazonas, two initiatives helped to structure an ecological trail in a Conservation Unit and build a community inn (FAS 2020), increasing the communities 'capacity to complement their income through touristic activities. The responsible presence of tourist in conservation areas can also be a mechanism to help local communities in maintaining the security of their territories and disseminating the relevance of their ways of life in combating climate change.

4 Final Considerations

In recent years, setbacks, and flexibilities in environmental legislation, as well as precarious supervision in this sphere, have promoted the advancement of deforestation and pollution in the Brazilian Amazon. A report released by the Civil Society´s Working Group for the 2030 Agenda (GTSC 2021), points out the serious setbacks in social, environmental, economic, civil and political rights experienced in Brazil in this period. Added to the above, are the efforts that segments of the private sector and of the public authorities in Amapá have been dedicating to activities such as mining, soybean monoculture and wood extraction, with a discourse that conditions the development of the state to the expansion of these activities.

At the same time, increasingly frequent events such as environmental catastrophes, climate change and pandemics highlight the exhaustion caused by the predatory and unequal growth model, which is no longer tolerable. Such evidence, although growing, is not new and has, for several decades, highlighted the need for integrative solutions—rivers, natural phenomena, pollution, and biological agents do not respect borders, therefore, measures restricted to these imaginary limits are not sufficient and generate the need for establishment of international laws and agreements regarding the environment.

But how to conciliate global interest in the conservation of a region with the rights and autonomy of local populations? The perception of the urgency of the environmental crises is not evenly clear to everyone, for some due to ideological factors, and for others due to the daily urgency of survival related matters. Ferdinand (2022) presents a complex and interesting perspective on the conflicts, categorizations and hierarchizations latent in theorizations about today's global challenges—if, on the one hand, climate and environmental calamities prove the urgency of changes at a global level in the environmental sphere, the "Colonial fractures" remain not only present, but aggravating the scenario of which they are the main cause.

According to Ferdinand, environmentalism guided by colonial logic, based on the interests and epistemologies of the Global North, white, Christian, male and heteronormative, is incapable of providing answers to more just and viable ways of inhabiting the world. His work invites the reader to rethink the Eurocentric logic of northern countries as the only providers of knowledge and solutions and revisit traditional knowledge, attacked by coloniality to impose the way and conditions of life that led the planet to its current conditions.

Therefore, it is important to highlight that the current global appeals in defense of the Amazon point to paths very close to those considered primitive and not in line with modernity a few decades ago. The plan for the occupation of the Amazon, more intensively led by Brazilians dictatorial government from the late sixties until the early eighties, assumed the forests as well as its traditional populations as the greatest enemies for the development of the region and the country, and a strong convincing campaign was made to spread this belief, as shown in the documentary "Transamazônica: uma estrada para o passado", directed by Fabiano Maciel and Jorge Bodanzky (2021).

This seems to be a necessary and urgent moment for intensive investment in environmental awareness and sustainable technological solutions to stop predatory growth in Amapá state and prevent the model that devastated so many areas of the Brazilian Amazon from being repeated in the state of Amapá. In our perception, this evolves greater and wider understanding of the paths that brought us here as well as to the traditional knowledge of harmonious coexistence with nature, allied with local and national governance and international cooperation.

References

Almeida de LMLD, da Fontoura AGC, Vasconcelos de IM de, Brito DMC, Hilário RR (2022) Current status, attractions, and obstacles for ecotourism in protected areas of Amapá, Brazil. Ambiente & Sociedade 25:e00061. https://doi.org/10.1590/1809-4422asoc20220006r1OA

Bastos PPZ (2023) Fundo Amazônia e desenvolvimento socioambiental inclusivo: problemas e recomendações int https://www.eco.unicamp.br/images/arquivos/nota-cecon/nota-do-cecon-20.pdf

Beni MC (2004) Globalização do turismo: megatendências do setor e a realidade brasileira. Aleph, São Paulo

BNDES (2023) Fundo amazônia: Relatório de Atividades 2022 Int https://www.fundoamazonia.gov.br/export/sites/default/pt/.galleries/documentos/rafa/RAFA_2022_port.pdf

Brasil (2023) Fundo Amazônia. https://www.fundoamazonia.gov.br/en/home/

Brito DMC, Drumond JAL (2022) Reflexões sobre a gestão ambiental das Unidades de Conservação no estado do Amapá. Confins 55:2022

Diniz RES (2016) Quilombo de São Pedro dos Bois: memória biocultural subvertida nas lógicas de ocupações recentes do Amapá. Tese (Doutorado Desenvolvimento Sustentável do Trópico Úmido)–Programa de Pós-Graduação em Desenvolvimento Sustentável do Trópico Úmido, Núcleo de Altos Estudos Amazônicos, Universidade Federal do Pará, Belém

Ferdinand (2022) Malcom. Uma ecologia decolonial: pensar a partir do mundo caribenho. São Paulo: Ubu Editora.

Ferreira AF, Oliveira L da CO (2021) O desenvolvimento do turismo como uma estratégia de sustentabilidade do espaço amazônico: um instrumento de preservação da floresta amazônica int. https://seer.unirio.br/rdpp/article/view/10457

Fundação Amazonas Sustentável (FAS) (2020) Fomento à bioeconomia e ao turismo de base comunitária na Amazônia: Resultados e lições aprendidas do Edital Floresta em Pé/Fundação

Amazonas Sustentável. Manaus: Fundação Amazonas Sustentável. 102 p. il. Fundo Amazônia (2022) Relatório Annual 2022 Int https://www.fundoamazonia.gov.br/pt/transparencia/relatorios-anuais/

Governo do Amapá (2023). https://www.portal.ap.gov.br/noticia/1807/amapa-abriga-maior-percentual-de-areas-protegidas-do-pais

GTSC (2021) Grupo de trabalho da sociedade civil para a Agenda 2030. https://gtagenda2030.org.br/relatorio-luz/relatorio-luz-2021/

IBGE (2022) Instituto Brasileiro de Geografia e Estatística Int https://cidades.ibge.gov.br/brasil/ap/panorama

Rauber AL, Ferreira ME (2020) A concentração socioeconômica e os ciclos de exploração dos recursos naturais no estado do Amapá. PRACS: Revista Eletrônica de Humanidades do Curso de Ciências Sociais da UNIFAP, 13(4):181–200

Sachs I (2008) Desenvolvimento: includente, sustentável, sustentado. Garamond, Rio de Janeiro

Silva G de V, Granger S, Tourneau F-ML (2019) Desafios à circulação na fronteira entre Brasil e Guiana Francesa (França). Mercator (Fortaleza) Int 18:e18018. https://doi.org/10.4215/rm2019.e18018

Transamazônica: uma estrada para o passado. Direção: Fabiano Maciel e Jorge Bodanzky. HBO, Brasil

Sustainable Recovery in the Amazon: An Analysis of the Main Practices Adopted in Engineering Courses

Diego Marques Cavalcante, Lucas Veiga Ávila, Laís Viera Trevisan, Rosley Anholon, Álvaro Luiz Neuenfeldt Júnior, and Vitor William Batista Martins

Abstract This study aims to identify sustainability practices, Sustainability Development Goals and Innovation (SDGI) in engineering courses in the Legal Amazon, from the perception of specialists. A quantitative and qualitative methodological approach was adopted to conduct this study. Firstly, a research instrument was developed based on a systematic literature review on the topic of engineers, innovation and sustainability, and with the participation of experts, assisted by the Fuzzy Delphi method, the variables were validated. With the participation of 10 experts (professors/researchers) from Higher Education Institutions (HEIs) located in the Brazilian Amazon and with the aid of the Fuzzy Delphi method, both for the validation of the variables in a research instrument developed based on the main evidence from the Systematic Literature Review and the literature, 38 validated variables were

D. M. Cavalcante (✉) · L. V. Ávila · Á. L. N. Júnior
Technology Center, Federal University of Santa Maria (UFSM), Avenida Roraima, Santa Maria, RS 1000, 97105-900, Brazil
e-mail: dieegomc@hotmail.com

L. V. Ávila
e-mail: lucas.avila@ufsm.br

Á. L. N. Júnior
e-mail: alvarojeng@gmail.com

L. Viera Trevisan
School of Administration, Federal University of Rio Grande Do Sul (UFRGS), 855 Washington Luiz St, Porto Alegre, RS 90010460, Brazil
e-mail: lais.trevisan@ufrgs.br

R. Anholon
School of Mechanical Engineering (FEM), State University of Campinas (UNICAMP), Cidade Universitária Zeferino Vaz—Barão Geraldo, Campinas-SP 13083-970, Brazil
e-mail: rosley@unicamp.br

V. W. B. Martins
Department of Production Engineering, State University of Pará (UEPA), R. Do Úna, N° 156-Telégrafo, Belém-PA 66050-540, Brazil
e-mail: vitor.martins@uepa.br

W. Leal Filho et al. (eds.), *Amazon 2030—Sustainability Issues in the World's Largest Rainforest Region*, World Sustainability Series,
https://doi.org/10.1007/978-3-031-81465-5_10

obtained. The findings are relevant for engineering course coordinators, professors, researchers, and academics, as they enable better planning and execution of training to assist engineering courses in the Amazon in the process of integrating innovation and sustainability approaches into their curricula.

Keywords Sustainability · Recovery · Brazilian amazon · Engineering · Education · Innovation

1 Introduction

There are many challenges to be overcome by Higher Education Institutions (HEIs) on their path towards sustainable development (Rampasso 2019). By analysing the demands of society regarding the adoption of innovative approaches to face the current sustainability challenges, there is a considerable need for more studies on the topic (Martins 2019).

In terms of curricula, it is difficult to insert sustainability according to a transdisciplinary view. When taught in specific disciplines, sustainability is usually addressed partially by focusing on environmental aspects. When it comes to engineering courses, teaching about social aspects, particularly, is a major challenge, due to the lack of clarity on this topic (Rampasso 2019). However, sustainability and social well-being are connected concepts since the environment needs to be preserved to provide long-term life to mankind. As a result, sustainable development concepts can be combined with the environment and society in the curriculum (Elsaadany and Helmi 2018). For that, creating forward-looking strategies, connecting key areas, involving stakeholders through processes, and inspiring people are key actions in the process (Leal Filho et al. 2020).

In the development of the Brazilian Amazon Region, the State was the main inducer for the occupation of the region and supported the economic rationality of the industry (Cipolat et al. 2021), leaving aside the issues of socio-environmental sustainability, the interests of the local populations, social relations, and their culture.

The practice of land grabbing is still one of the actions that help to dilapidate the Amazon forest, in addition to plantations on illegal lands, and the extraction of wood that increases the inflammability of the forest, resulting in fires that set in motion a vicious cycle of tree mortality, increased fuel load, re-entry of fire, and finally, destruction of the forest (Vecchia et al. 2021).

Nevertheless, there is still potential for Amazon's wealth and it is necessary to transform it into inputs and products for the industry segments that present a growing demand (Enríquez 2008). In this regard, the discussion emerges about the need to induce economic activities that are compatible with forest preservation, capable of generating work and income for the local population (Gomes et al. 2022) among other elements that are fundamental in a new production model towards regional development (Cipolat et al. 2021).

Therefore, the main objectives of this study are to identify practices of sustainability and innovation in engineering courses and validate the practices and applicability of the Sustainable Development Goals from the perception of experts in engineering courses, through the Fuzzy Delphi method.

These specific results will drive the development of impactful initiatives and resources that directly benefit the engineering course academic community (coordinators, professors, researchers, and academics), enabling better planning and execution of the training of engineers to assist Amazonian engineering courses in the process of integrating innovation and sustainability approaches in their curricula, becoming able to inspire behaviors, knowledge, and technologies that can effectively contribute to the recovery of sustainability and local development.

2 Theoretical Background

Social and environmental degradation has been an issue of growing concern for many stakeholders (Rampasso 2019) and leads to a process that leads to a more rational use of resources, in line with the principles of eco-efficiency, equity and social justice (Antonio et al. 2019).

In this context, Higher Education Institutions (HEIs) are taking on an important role in sustainable development and have been adapting and redirecting their actions and developing their institutional leadership role in promoting sustainability (Leal Filho et al. 2020).

Sustainable development and climate change are inseparable. Sustainable development, which aggregates the economic, ecological, and human spheres, is an essential condition for maintaining the global climate balance (Vianna et al. 2018). Climate change is already happening and is already producing impacts, and the greater the warming, the greater the future impacts and risks that humanity will face, including the possibility of irreversible damage to ecosystems, biodiversity, agricultural production, and the economy and society in general (Baleia and Alana 2018).

With climate change and increased vulnerabilities in ecosystems, climate change currently represents a challenge to humanity and its structures on a global level (Lusz et al. 2021) and the perception of the need for concrete planning and governance actions to face it becomes necessary (Torres et al. 2020).

The Amazon region of Brazil, motivated by the environmental issue, has been the scene of heated debates recently. The polarization produces myths, such as being the "lung of the world," but also brings truths such as the importance of the Amazon Forest to the global climate (Vieira et al. 2019).

Because the region has the greatest biodiversity on the planet, Brazil becomes essential to avoid the catastrophic effects of climate change, and the analysis of the perceptions of engineering students in this region about innovation for sustainability and sustainable recovery is of utmost importance because these are actions and practices that impact different regions of the world.

In general, there is a consensus on valuing knowledge and quality education for all (Almeida 2021). This broadening of the debate on sustainability-oriented education brings contributions to the theoretical field when it increases the continuous exploration of new ideas, methods, products and processes, but now with a focus on solving the social and environmental problems we face, thus combining innovation with sustainability (Fernandes 2022; Fernando and Cotian 2022).

Sustainable innovation is part of a context in which innovating is necessary, but it is equally necessary that we know how to respect our environment (Fernandes 2022). This is reflected in social contributions when engineering courses can give their graduates the ability to assume their role as leaders who are sensitive to sustainability and, at the same time, in tune with society by generating innovative solutions within a new reality (de Andrade 2022).

In theory, innovative practices in the curriculum can develop experiences in engineering education that will give students and future engineers, a learning/experience to undergo cognitive and behavioral changes and be development leaders in their societies (Al-Bahi et al. 2021), as well as contribute from the bottom up to global sustainable development (Stock and Kohl 2018).

There is a need for a proposed approach for engineering courses that contemplates sustainability at the core of the innovations generated and resulting in practical contributions.

Thus, the maturation of the use of knowledge management as a strategy allows for the recognition of the competencies needed to become an innovative thinking and learning organization with the ability to put the principles of sustainability into practice (Martins 2019). It is about caring about your educational system and top management teams, employees, professors and researchers as sustainable leaders and encouraging the development and education of tomorrow's leaders who will assume important positions in companies, NGOs and government (Leal Filho et al. 2020).

Therefore, it is relevant to recognize the competencies and by trying to measure and evaluate more appropriately the innovation systems (training and applicability) responsible for acquiring and transforming knowledge into innovative products and bringing them to society (Llorca-ponce et al. 2021) with a focus on sustainability in the Eastern legal Amazon region.

The movements towards creating a sustainable future require that people fully understand the complexity of the world we live in and appreciate the need for sustainable ways of living (Aginako and Guraya 2021). One way that academia can help in this process is by bringing sustainable solutions to companies that minimize the impact of their activities on the environment, thus becoming more competitive by reducing costs and improving quality (Fernandes 2022).

Starting from the fact that in organizational innovation, change happens in practices, as in the introduction of new education or training programs (Fernando and Cotian 2022), it is relevant to present proposals that help instigate the creation of innovations for sustainability in the training of engineers in the Brazilian Amazon region. By understanding that future engineering professionals need to respond to the challenges faced by organizations (along with their increasing complexity) and, consequently, the paradigm shift in assessing organizational success (considering these challenges) (Ogrean and Herciu 2020). And thus, contribute so that the innovations developed by HEIs contribute to local sustainable development.

3 Method

In the research process, a mixed methods approach was adopted. A qualitative stage, with the development of the Systematic Literature Review (SLR) and theoretical basis, and a quantitative stage, through the application of the Fuzzy Delphi method, which served as a technique to support the decision consensus, observing the practices and stances regarding innovation, sustainability and sustainable recovery raised in the systematic literature review and added to the 17 SDGs.

In the first stage, a SLR was carried out to search for scientific publications that provided a basis for the theme addressed and to survey the practices and attitudes applied to innovation for sustainability in the learning and training of engineers.

The literature review performed is a systematic one. Its methodology involved steps such as defining the research objectives and problem, designing the data collection instruments, and defining approaches for the analysis and interpretation of the results, serving to identify gaps in the literature and thus guiding researchers on the ways forward (Okoli 2019).

For the systematic literature review, two questions were created from the question that motivated the research problem (What/how are/are):

- Integrated innovation and sustainability approaches in engineering education/ curriculum?
- The education and learning of engineers?

Data was retrieved from the database for a ten-year interval, encompassing February 28, 2013, to February 28, 2023, in a ten years line, in all fields using the terms: engineers, innovation, and sustainability.

The terms used in the search of the search strings were defined after preliminary readings and conversations with experts (teachers/researchers). After this initial research, it was possible to better understand the terms that would be most appropriate and most used in the area, thus, the construction of the search chain would become more adherent to the topic of interest.

A search strategy was defined, and inclusion and exclusion criteria were established (open access, English language, document type: article, and research area: engineering, environment, sustainability, and engineering education/teaching) for the articles and, above all, an analysis of the selected literature was performed, always seeking answers to the questions posed previously.

The following shows how the systematic literature review process was structured. Within this process four steps were fundamental to the RSL, these were: (1) formulation of research questions, (2) selection and evaluation of studies, (3) content analysis of the selected articles, and (4) description of the results. The steps followed in the RSL are shown in Fig. 1.

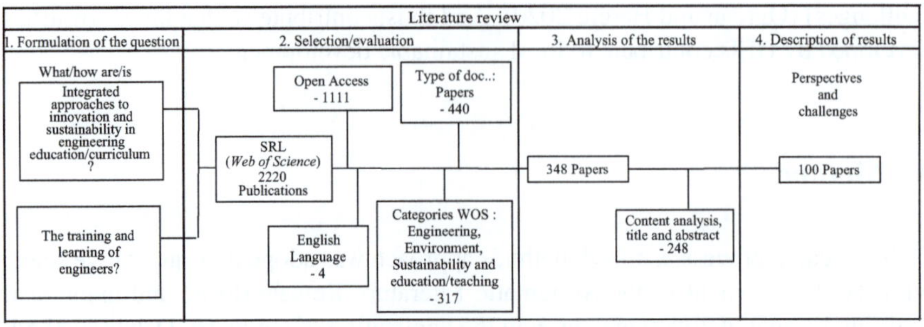

Fig. 1 Fundamental steps of RSL (Adapted from Caiado et al. 2018)

The following table presents the organization of the categories of the statements raised in the RSL, which are outlined in 3 blocks: block 1 General; block University/course and block 3 SDGs, as presented in Table 1.

Based on the literature review, the table presents the main indications of innovation, sustainability, and SDG practices in engineering education, as shown in Table 2.

The fuzzy delphi method is composed of a few methodologies (Ishikawa et al. 1993; Kumar et al. 2019; Kuo and Chen 2008). Aiming to validate the alternatives raised in the qualitative stage (Table 2), the methods of Ishikawa et al. (1993), Kumar et al. (2019) and Kuo and Chen (2008) were used, and the Ishikawa et al. (1993), Kumar et al. (2019) method was defined by technical criteria for this validation, which is presented in the next section.

Table 1 Structure of the data collection instrument of the quantitative stage

Block	Category	Factor
I	General	Innovation
		Sustainability
		Sustainable Recovery
II	University	Innovation
		Sustainability
		Sustainable Recovery
	Course	Sustainability
		Sustainable Recovery
III	General	SDG's

Table 2 Practices and attitudes towards innovation and sustainability

B	C	F	VARIABLE	AUTHOR
I	G	Innov.	R&D practices increase the likelihood that universities will be more innovative	Roman et al. (2018)
			Sustainability and innovation-oriented education promotes economic growth	Li & Liu et al. (2021)
			Partnerships between companies and universities strengthen sustainability and innovation.	Dundas (2011)
			The promotion of conferences, professional and academic events are important for scientific connections for sustainability and innovation	Dyer et al. (2009)
		Sust.	Cooperation between universities strengthens innovation and sustainability in the institution	Vega-Jurado et al. (2017)
			Education for sustainability and innovation is essential for engineering students	Amr Elsaadany e Ahmed Helmi 2022
			The study of sustainability helps to understand the social, economic and environmental issues of organizations and society	Amr Elsaadany e Ahmed Helmi 2022
			Sustainability skills and knowledge are a priority in many companies	Amr Elsaadany e Ahmed Helmi 2022
			The study of sustainability helps change behavior towards the environment	Amr Elsaadany e Ahmed Helmi 2022
			I believe that engineers are a key player in creating sustainable solutions to existing problems	Amr Elsaadany e Ahmed Helmi 2022
		Rec.	The development of extracurricular activities, can be a tool for sustainable recovery	Sady, Żak, Rzepka (2019)
			The reformulation of educational programs can assist in the development of sustainability and sustainable recovery	Sady, Żak, Rzepka (2019)
			Long-term planning can influence the development of new practices aimed at sustainable recovery.	Wiek et al. (2010)
			The lack of legislation forcing organizations to adopt sustainability in their practices influences company performance.	Wiek et al. (2010)
			The creation of awareness campaigns in the form of advertisements, notices, thematic events, etc. can facilitate the change in people's behavior for the recovery of the sustainability.	Allan et al. (2020)
			The emergence of new research and technologies, in areas such as artificial intelligence, can assist in the process of sustainable recovery.	Shittu et al., (2021)

(continued)

Table 2 (continued)

II	U	Innov.	Creativity and Innovation at the university generate continuous improvement in processes	Bessant & Caffyn (2001)
			The university provides technical and scientific support for the creation of new ideas/projects.	Dyer et al. (2009)
			From the experiences at the university, I am interested in developing proposals for innovative and sustainable products and projects	Greco et al. (2016)
		Sust.	The university is agile in solving problems for sustainable development	Wang & Wang (2012)
		Rec.	The interaction between the university and the local community is an important factor for sustainable recovery.	Wiek et al. (2010)
			The participation of students, professors (academic community) in actions in the community, positively influences the sustainable recovery after the pandemic.	Olsen et al., 2005
	C	Sust.	The course curriculum enables the development of education for sustainability and innovation	Duval-Couetil & Dyrenfurth (2012)
			I see innovation and sustainability as an opportunity to increase your course's collaboration with communities, businesses and NGOs	IUSDRP e ESSSR
			The support currently offered by academic staff to learn about sustainability is appropriate.	IUSDRP e ESSSR
			The course provides a systemic (broad) view of society and the goals for a sustainable agenda	Savage, E., Tapics, T., Evarts, J., Wilson, J. and Tirone, S. (2015)
			In the course I am motivated to develop practical tools to advance the sustainability agenda	Savage, E., Tapics, T., Evarts, J., Wilson, J. and Tirone, S. (2015)
			The problems/practices addressed in my course always consider the environmental and economic implications of a decision or process	Savage, E., Tapics, T., Evarts, J., Wilson, J. and Tirone, S. (2015)
			The course promotes and encourages participation in community activities that encourage sustainability (extension)	Rocío Valderrama-Hernández et al 2020
			The course promotes education in values oriented to the formation of responsible, active and democratic citizens.	Rocío Valderrama-Hernández et al 2020
			The course subjects lead to a vision that understands the functioning of natural, social and economic systems, as well as their inter-relationships and problems, both locally and globally.	Rocío Valderrama-Hernández et al 2020

(continued)

Table 2 (continued)

		Rec.	The course is innovative in developing tools that enable practices for the recovery of: bioactives, residues, renewable energies, low carbon practices, development of new materials, better use of natural resources, etc.	Herrero e Ibanez (2018)

Chart 2 with the 32 practices and attitudes indicated was used as the basis for developing the questionnaires applied during the research. There are several techniques used for decision consensus with qualitative variables, but for this research, the fuzzy Delphi method was proposed, since it demands that the interviewees respond to the research instrument only once, so it reduces the number of interviews and the research time and it is estimated that the number of interviewees should vary between five and twenty (Bui et al. 2020).

For the development of the instrument, it was considered the 5-point Likert scale (Likert 1932) focused on measuring the knowledge and experiences of respondents, as to the topic in question, as follows the scale: 5 = strongly agree; 4 = agree; 3 = indifferent; 2 = disagree; 1 = strongly disagree. For the SDG context, a 5-point Likert level scale of importance level was adopted: 5 = very important; 4 = important; 3 = indifferent; 2 = not very important; 1 = unimportant.

For the validation of this step, 15 experts were invited in the period November 2022, which are teachers with doctoral degrees in engineering courses in public HEIs located in the legal Amazon, working for more than 10 years in research.

In return, we obtained the participation of 10 specialists, 7 of whom are professors in production engineering, 2 in environmental engineering, and 1 in agronomy. The modeling is presented below.

To apply the method, following Ishikawa et al. (1993) and Kumar et al. (2019), it is necessary to identify the criteria and the number of experts (decision makers) to determine the importance through a questionnaire using linguistic variables.

Once the linguistic inputs are provided by the experts, the answer for each criterion is converted into fuzzy numbers (Triangular Fuzzy Number—TFN):

Score 1 (0.1;0.1;0.3);

Score 2 (0.1;0.3;0.5);

Score 3 (0.3;0.5;0.7);

Score 4 (0.5;0.7;0.9);

Score 5 (0.7;0.9;0.9).

A fuzzy number corresponding to the j-th criteria provided by the i-th expert is represented in (1) and the fuzzy weights of the criteria are given in (2), (3) and (4).

The fuzzy weights will be converted into crisp values and will rank the elements by real numbers.

$$\tilde{a}ij = (aij,\ bij, cij)\ \text{para}\, i = 1, 2, 3 \ldots\ nj = 1, 2, 3 \ldots m \quad (1)$$

- n = number of experts;
- m = number of criteria.

$$aj = \min\{aij\} \quad (2)$$

$$bj = (\Pi i = 1nbij)1/n \quad (3)$$

$$cj = \max\{cij\} \quad (4)$$

Finally, the fuzzy weight is averaged to calculate the defuzzification (5).

$$aj = \frac{aj + bj + c}{3}, = 1, 2, 3 \ldots m \quad (5)$$

As an example, for variable 1.1 expert 1 assigned a rating of "5", whose TFN is (0.7;0.9;0.9). Similarly, we find the TFN for each rating assigned by all 10 specialists with respect to variable 1.1, as indicated below:

Esp. 1—(0,7;0,9;0,9) [note 5];

Esp. 2—(0,7;0,9;0,9) [note 5];

Esp. 3—(0,5;0,7;0,9) [note 4];

Esp. 4—(0,7;0,9;0,9) [note 5];

Esp. 5—(0,7;0,9;0,9) [note 5];

Esp. 6—(0,7;0,9;0,9) [note 5];

Esp. 7—(0,7;0,9;0,9) [note 5];

Esp. 8—(0,7;0,9;0,9) [note 5];

Esp. 9—(0,7;0,9;0,9) [note 5];

Esp. 10—(0,7;0,9;0,9) [note 5].

Now, the fuzzy weight of variable 1.1 as given in Table 1, is (0.5;0.9;0.9), was obtained as follows:

The first entry of the fuzzy weight, i.e. 0.5, was obtained by taking the minimum value among all the first 10 entries given above, i.e.;

$$\text{Min}\ (0, 7;\ 0, 7;\ 0, 5;\ 0, 7;\ 0, 7;\ 0, 7;\ 0, 7;\ 0, 7;\ 0, 7;\ 0, 7) = 0, 5;$$

The second entry of the fuzzy weight, i.e. 0.9, was obtained by taking the geometric mean of all the geometric means of all 10 entries given above, i.e.;

$$(0,9 \times 0,9 \times 0,7 \times 0,9 \times 0,9 \times 0,9 \times 0,9 \times 0,9 \times 0,9 \times 0,9)^{(1/10)} = 0,9$$

The third entry of the fuzzy weight, i.e. 0.9, was obtained by taking the maximum value among all the 10 third entries given above, i.e.;

$$\text{Max}\ (0,9;\ 0,9;\ 0,9;\ 0,9;\ 0,9;\ 0,9;\ 0,9;\ 0,9;\ 0,9;\ 0,9) = 0,9.$$

Similarly, the diffuse weight was obtained for all 32 variables.

The defuzzification value of each variable was performed by taking the average of the fuzzy weight (Eq. (5)). For example, the defuzzification value variable 1.1 with a fuzzy weight (0.5, 0.9, 0.9) is obtained as (0.5 + 0.9 + 0.9)/3 which is "0.8" as indicated in Table 1. Similarly, the defuzzification value was obtained for all 32 variables.

The threshold ã is defined to select or reject the factors by means of the average of the minimum value of the linguistic variable "important" (0.5) and maximum value of the linguistic variable "indifferent" (0.7), so ã is equal to 0.6 and the principle to select or reject the criterion is: if $\tilde{a}j \geq \tilde{a}$ then the criterion j is selected since if $\tilde{a}j < \tilde{a}$ then the criterion j is rejected (Singh and Sarkar 2020).

The results of the FDM calculation are shown in Table 3.

Of the thirty-two factors raised in the systematic literature review and that composed the first and second block of statements, 21 were selected and in the third block with the SDGs, all were selected. Eleven variables were eliminated because ãj is less than 0.6 (ã).

In the first block (1.8) Sustainability skills and knowledge are a priority in many companies.

In the second block, in the university category, only the variable (2.4) The university is agile in solving problems for sustainable development was declassified.

In the second block, in the course category, the following variables were disqualified (2.7) The course curriculum enables the development of education for sustainability and innovation, (2.9) The support currently offered by academic staff to learn about sustainability is appropriate, (2.10) The course provides a systemic (broad) view of society and the goals for a sustainable agenda, (2. 11) On the course I am motivated to develop practical tools to advance the sustainability agenda, (2.12) The problems/practices addressed in my course always address the environmental and economic implications of a decision or process, (2.13) The course promotes and encourages participation in community activities with a focus on sustainability, (2. 14) The course promotes education in values oriented to the formation of responsible, active and democratic citizens, (2.15) The course subjects lead to a vision that understands the functioning of natural, social and economic systems, as well as their inter-relationships and problems, both locally and globally, and (2.16) The course in which I participate is innovative to develop instruments that enable practices for the recovery of: bioactives, waste, renewable energy, low carbon practices, development of new materials, better use of natural resources, etc. In the third block with the SDGs, none of the statements were rejected.

Table 3 Result of the FDM

Block	Cat.	Factor	Variable	aj	Bj	cj	TFN (corresponding)	Defuzzification value	Decision (selected/rejected)
1	General	Innovation	1.1	0,5	0,9	0,9	(0,5;0,9;0,9)	0,8	Selected
			1.2	0,3	0,8	0,9	(0,3;0,8;0,9)	0,7	Selected
			1.3	0,5	0,9	0,9	(0,5;0,9;0,9)	0,8	Selected
			1.4	0,5	0,9	0,9	(0,5;0,9;0,9)	0,8	Selected
		Sustainability	1.5	0,3	0,8	0,9	(0,3;0,8;0,9)	0,7	Selected
			1.6	0,3	0,8	0,9	(0,3;0,8;0,9)	0,7	Selected
			1.7	0,3	0,8	0,9	(0,3;0,8;0,9)	0,7	Selected
			1.8	0,1	0,4	0,9	(0,1;0,4;0,9)	0,5	Rejected
			1.9	0,5	0,8	0,9	(0,5;0,8;0,9)	0,7	Selected
			1.10	0,1	0,7	0,9	(0,1;0,7;0,9)	0,6	Selected
		Sustainability Recovery	1.11	0,3	0,8	0,9	(0,3;0,8;0,9)	0,7	Selected
			1.12	0,3	0,8	0,9	(0,3;0,8;0,9)	0,7	Selected
			1.13	0,5	0,8	0,9	(0,5;0,8;0,9)	0,7	Selected
			1.14	0,1	0,7	0,9	(0,1;0,7;0,9)	0,6	Selected
			1.15	0,3	0,8	0,9	(0,3;0,8;0,9)	0,7	Selected
			1.16	0,3	0,8	0,9	(0,3;0,8;0,9)	0,7	Selected
2	Univ.	Innovation	2.1	0,5	0,8	0,9	(0,7;0,8;0,9)	0,7	Selected
			2.2	0,1	0,7	0,9	(0,1;0,7;0,9)	0,6	Selected
			2.3	0,1	0,7	0,9	(0,1;0,7;0,9)	0,6	Selected
		Sustainability	2.4	0,1	0,4	0,9	(0,1;0,4;0,9)	0,5	Rejected
		Sustainability Recovery	2.5	0,5	0,8	0,9	(0,5;0,8;0,9)	0,7	Selected
			2.6	0,5	0,8	0,9	(0,5;0,8;0,9)	0,7	Selected
	Course	Sustainability	2.7	0,1	0,6	0,9	(0,1;0,6;0,9)	0,5	Rejected
			2.8	0,5	0,9	0,9	(0,5;0,9;0,9)	0,8	Selected
			2.9	0,1	0,4	0,9	(0,1;0,4;0,9)	0,5	Rejected
			2.10	0,1	0,6	0,9	(0,1;0,6;0,9)	0,5	Rejected
			2.11	0,1	0,5	0,9	(0,1;0,5;0,9)	0,5	Rejected
			2.12	0,1	0,5	0,9	(0,1;0,5;0,9)	0,5	Rejected
			2.13	0,1	0,4	0,9	(0,1;0,4;0,9)	0,5	Rejected
			2.14	0,1	0,6	0,9	(0,1;0,6;0,9)	0,5	Rejected
			2.15	0,1	0,5	0,9	(0,1;0,5;0,9)	0,5	Rejected
		Sustainability Recovery	2.16	0,1	0,6	0,9	(0,1;0,6;0,9)	0,5	Rejected
3	General	Sustainable Development Goals – SDG	3.1	0,5	0,8	0,9	(0,5;0,8;0,9)	0,7	Selected
			3.2	0,3	0,8	0,9	(0,3;0,8;0,9)	0,7	Selected
			3.3	0,3	0,8	0,9	(0,3;0,8;0,9)	0,7	Selected
			3.4	0,3	0,8	0,9	(0,3;0,8;0,9)	0,7	Selected
			3.5	0,3	0,8	0,9	(0,3;0,8;0,9)	0,7	Selected
			3.6	0,3	0,8	0,9	(0,3;0,8;0,9)	0,7	Selected
			3.7	0,5	0,8	0,9	(0,5;0,8;0,9)	0,7	Selected
			3.8	0,5	0,8	0,9	(0,5;0,8;0,9)	0,7	Selected
			3.9	0,5	0,9	0,9	(0,5;0,9;0,9)	0,8	Selected
			3.10	0,3	0,8	0,9	(0,3;0,8;0,9)	0,7	Selected
			3.11	0,1	0,8	0,9	(0,1;0,8;0,9)	0,6	Selected

(continued)

Table 3 (continued)

			3.12	0,5	0,9	0,9	(0,5;0,9;0,9)	0,8	Selected
			3.13	0,5	0,8	0,9	(0,5;0,8;0,9)	0,7	Selected
			3.14	0,3	0,8	0,9	(0,3;0,8;0,9)	0,7	Selected
			3.15	0,3	0,8	0,9	(0,3;0,8;0,9)	0,7	Selected
			3.16	0,1	0,7	0,9	(0,1;0,7;0,9)	0,6	Selected
			3.17	0,5	0,8	0,9	(0,5;0,8;0,9)	0,7	Selected

Now following the method presented in Kuo and Chen (2008) once the linguistic contributions provided by the experts for each of the practices presented in Table 6 according to the following scale: 1 (Very low relevance), 2 (Low relevance), 3 (Medium relevance), 4 (High relevance) and 5 (Very high relevance). Each score is designated by Rij, where the index of practice i is rated by respondent j.

Analyzing the experts' answers if determines the Triangular Fuzzy Numbers (TFN) for each training practice i, composed of three elements (Li; Mi; Ui); where:

Li: indicates the minimum value of the rating of all experts for practice i;

Mi is the geometric mean of the value of the rating of all experts for practice i, determined as follows:

$$(\Pi j = 1k \ Rij)1/k \tag{6}$$

Ui indicates the maximum value of the rating of all experts for practice i;

Thus, the TFNi index (Li; Mi; Ui) is established for each practice i.

Once TFNi is determined, the Center of Area (COA) approach is used to defuzzify the TFNi values. The COAi value, for each practice i, is calculated as follows:

$$\mathrm{COAi} = [(\mathrm{Ui} - \mathrm{Li}) + (\mathrm{Mi} - \mathrm{Li})]/3 + \mathrm{Li} \tag{7}$$

The evaluation criteria are established by setting the threshold value α as follows:

$$\alpha = (\mathrm{U} + \mathrm{L})/2 \tag{8}$$

where:

U indicates the maximum possible rated value;

L indicates the minimum possible rated value.

And the selection principle is defined by:

- if COAi $\geq$ α; then practice is accepted;
- if COAi $\leq$ α; then practice is not accepted.

The results of the FDM calculation are shown in Table 4 as follows:

Of the thirty-two factors raised in the systematic literature review and that composed the first block of statements, 26 were selected and in the second block with the SDGs, all were selected. 6 variables were eliminated because α is less than 3.0.

Table 4 Result of the FDM 2

Block	Cat.	Factor	Variable	aj	bj	cj	TFN (corresponding)	Defuzzification value	Decision (selected/rejected)
1	General	Innovation	1.1	4	4,9	5	(4;4,9;5)	4,6	Selected
			1.2	3	4,5	5	(3;4,5;5)	4,2	Selected
			1.3	4	4,8	5	(4;4,8;5)	4,6	Selected
			1.4	4	4,8	5	(4;4,8;5)	4,6	Selected
		Sustainability	1.5	3	4,3	5	(3;4,3;5)	4,1	Selected
			1.6	3	4,7	5	(3;4,7;5)	4,2	Selected
			1.7	3	4,7	5	(3;4,7;5)	4,2	Selected
			1.8	1	2,5	5	(1;2,5;5)	2,8	Rejected
			1.9	4	4,6	5	(4;4,6;5)	4,5	Selected
			1.10	2	3,9	5	(2;3,9;5)	3,6	Selected
		Sustainability Recovery	1.11	3	4,3	5	(3;4,3;5)	4,1	Selected
			1.12	3	4,4	5	(3;4,4;5)	4,1	Selected
			1.13	4	4,7	5	(4;4,7;5)	4,6	Selected
			1.14	2	4,0	5	(2;4;5)	3,7	Selected
			1.15	4	4,5	5	(4;4,5;5)	4,5	Selected
			1.16	3	4,5	5	(3;4,5;5)	4,2	Selected
2	Univ.	Innovation	2.1	4	4,7	5	(4;4,7;5)	4,6	Selected
			2.2	2	3,8	5	(2;3,8;5)	3,6	Selected
			2.3	2	4,2	5	(2;4,2;5)	3,7	Selected
		Sustainability	2.4	1	2,3	4	(1;2,3;4)	2,4	Rejected
		Sustainability Recovery	2.5	4	4,5	5	(4;4,5;5)	4,5	Selected
			2.6	4	4,7	5	(4;4,7;5)	4,6	Selected
	Course	Sustainability	2.7	1	3,6	5	(1;3,6;5)	3,2	Selected
			2.8	4	4,9	5	(4;4,9;5)	4,6	Selected
			2.9	2	2,7	4	(2;2,7;4)	2,9	Rejected
			2.10	2	3,7	5	(2;3,7;5)	3,6	Selected
			2.11	2	2,8	4	(2;2,8;4)	2,9	Rejected
			2.12	1	2,8	4	(1;2,8;4)	2,6	Rejected
			2.13	1	2,5	4	(1;2,5;4)	2,5	Rejected
			2.14	1	3,5	5	(1;3,5;5)	3,2	Selected
			2.15	2	3,1	4	(2;3,1;4)	3,0	Selected
		Sustainability Recovery	2.16	2	3,4	5	(2;3,4;5)	3,5	Selected
3	General	Sustainable Development Goals	3.1	4	4,7	5	(4;4,7;5)	4,6	Selected
			3.2	3	4,6	5	(3;4,6;5)	4,2	Selected
			3.3	3	4,3	5	(3;4,3;5)	4,1	Selected
			3.4	3	4,6	5	(3;4,6;5)	4,2	Selected
			3.5	3	4,3	5	(3;4,3;5)	4,1	Selected
			3.6	3	4,6	5	(3;4,6;5)	4,2	Selected
			3.7	4	4,7	5	(4;4,7;5)	4,6	Selected
			3.8	4	4,6	5	(4;4,6;5)	4,5	Selected
			3.9	4	4,8	5	(4;4,8;5)	4,6	Selected
			3.10	3	4,5	5	(3;4,5;5)	4,2	Selected
			3.11	2	4,3	5	(2;4,3;5)	3,8	Selected

(continued)

Table 4 (continued)

3	General	Sustainable Development Goals	3.12	4	4,8	5	(4;4,8;5)	4,6	Selected
			3.13	4	4,7	5	(4;4,7;5)	4,6	Selected
			3.14	3	4,6	5	(3;4,6;5)	4,2	Selected
			3.15	3	4,6	5	(3;4,6;5)	4,2	Selected
			3.16	2	4,1	5	(2;4,1;5)	3,7	Selected
			3.17	4	4,7	5	(4;4,7;5)	4,6	Selected

In the first block (1.8) Sustainability skills and knowledge are a priority in many companies.

In the second block, in the university category, only the variable (2.4) 'The university is agile in solving problems for sustainable development' was declassified.

In the second block, in the course category, the variables (2.9) The support currently offered by academic staff to learn about sustainability is appropriate, (2.11) On the course I am motivated to develop practical tools to advance the sustainability agenda, (2.12) The problems/practices addressed in my course always consider the environmental and economic implications of a decision or process, (2.13) The course promotes and encourages participation in community activities focused on sustainability.

In the third block with the SDGs, none of the statements were rejected.

After the validation of the experts, by the two methods, it was decided to adopt the modeling according to Ishikawa et al. (1993); Kumar et al. (2019), resulting in Table 5 that follows.

Main Implications

Some of the practices raised in the RSL of sustainability applicability and innovation in engineering courses, in the current perception of experts (teachers/researchers) working in HEIs in the Brazilian Amazon region, are identified as in disagreement, based on the experts' perception of the presence of such practices in their teaching experiences.

However, the exclusion of the statements does not allow us to say that the experts do not judge the practices of the discarded statements as not important in the training of engineers. It is possible to see that the practices discarded differently in the approach are included in other statements that were selected in the Fuzzy Delphi. Therefore, what can be concluded is that there was a refinement of the practices and attitudes listed in Table 2 submitted to the judgment of the experts.

This current perception of practices in engineering courses needs to be overcome in order to allow, at the first moment, new possibilities for planning and executing the training of engineers and, consequently, to increase the process of integrating innovation and sustainability approaches into the course curricula, allowing their graduates to assume their role as leaders who are sensitive to sustainability and, at the same time, in tune with society by generating innovative solutions within a new reality (de Andrade 2022).

Table 5 Practices and stances of innovation, sustainability, sustainable recovery and SDGs

B.	CAT.	FAT.	VARIABLE	
I	G.	Innovation	1.1	R&D practices increase the likelihood that universities will be more innovative
			1.2	Sustainability and innovation-oriented education promote economic growth
			1.3	Partnerships between companies and universities strengthen sustainability and innovation
			1.4	Promoting conferences, professional and academic events are very important for the scientific and practical connections for sustainability and innovation
		Sustainability	1.5	Cooperation between universities strengthens innovation and sustainability in the institution
			1.6	Training for sustainability and innovation is essential for engineering students
			1.7	The study of sustainability helps to understand the social, economic and environmental issues of organizations and society
			1.8	The study of sustainability helps to change behavior towards the environment
			1.9	I believe that engineers are a major player in creating sustainable solutions to existing problems
		Sustainability Recovery	1.10	The development of extracurricular activities, can be a tool for sustainable recovery in the Amazon
			1.11	The reformulation of educational programs can assist in the development of sustainability and sustainable recovery in the Amazon.
			1.12	Long-term planning can influence the development of new practices aimed at sustainable recovery.
			1.13	The lack of legislation forcing organizations to adopt sustainability in their practices influences company performance.
			1.14	The creation of awareness campaigns in the form of advertisements, notices, thematic events etc. can facilitate the change in people's behavior making them more committed to the sustainability of the Amazon
			1.15	The emergence of new research and technologies, in areas such as artificial intelligence, can assist in the process of sustainable recovery in the Amazon.

(continued)

Table 5 (continued)

II	UNI/ COUR.	Innovation	2.1	Creativity and innovation at the university generate continuous improvement in processes
			2.2	The university provides technical and scientific support for the creation of new ideas/projects
			2.3	Based on my experiences at the university, I am interested in developing proposals for products, innovative and sustainable projects, etc.
		Sustainability	2.4	I see innovation and sustainability as an opportunity to increase course collaboration with communities, businesses, and NGOs
		Sustainability Recovery	2.5	The interaction between the university and the local community is an important factor for the sustainable recovery in the Amazon.
			2.6	The participation of students, professors (academic community) in actions in the Amazon, positively influences the sustainable recovery after the pandemic.
III	G.	Sustainable Development Goals	3.1	1) Poverty eradication
			3.2	2) Zero Hunger and Sustainable Agriculture
			3.3	3) Health and Wellness
			3.4	4) Quality Education
			3.5	5) Gender equality
			3.6	6) Safe Water and Sanitation
			3.7	7) Accessible, Clean Energy
			3.8	8) Decent Work and Economic Growth
			3.9	9) Industry, Innovation and Infrastructure
			3.10	10) Reduction of Inequalities
			3.11	11) Sustainable cities and communities
			3.12	12) Responsible Consumption and Production
			3.13	13) Action against global climate change
			3.14	14) Life in Water
			3.15	15) Life on land
			3.16	16) Peace, justice and effective institutions
			3.17	17) Partnerships and means of implementation

The training of engineers should include practical activities for the application of innovation and sustainability (1.4 The promotion of conferences, professional and academic events are very important for the scientific and practical connections in favor of sustainability and innovation) and activities that help students identify the possibility of the applicability of innovation and sustainability in processes (1.10 The development of extracurricular activities can be a tool for a sustainable recovery in the Amazon). The universities/ engineering courses need to reflect and identify the possible gains with the applicability of the concepts of innovation and sustainability (2.3 From the experiences at the University I am interested in developing proposals for innovative and sustainable products, projects, etc.) and support and promote among the students the interest in implementing and developing the concepts (2.2 The university provides technical and scientific support for the creation of new ideas/ projects).

Subsequently, it is important to break the paradigm that innovation and sustainability practices are two separate things, universities/engineering courses should start working on innovation and sustainability in a related way in curricula (Elsaadany and Helmi 2018) and assume their institutional leadership role in promoting sustainable development (Leal Filho et al. 2020).

This main challenge is related to the best practices to be adopted in the training/ training of engineers, relating the contents covered in the classroom with the demands of society, and bringing innovative approaches in the face of current sustainability challenges (Martins 2019).

Finally, the main challenge related to structuring the theoretical part of the training/ training of engineers to create knowledge and technologies that can effectively contribute to the recovery of sustainability and local development is linked to the approaches integrated into the methodology of active learning (1.3 The partnerships between companies and universities; 1. 10 The development of extracurricular activities, can be a tool for sustainable recovery in the Amazon; 1.14 The creation of awareness campaigns in the form of announcements, notices, thematic events etc.; 2.4 collaboration of the course with communities, companies and NGOs; 2.5 The interaction between the university and the local community; 2.6 The participation of the academic community in actions in the Amazon) that motivate and keep the student in contact with reality.

4 Conclusions

This study aimed to identify the main practices of innovation and sustainability applicability and validate the practices and applicability of the Sustainable Development Goals from the perception of experts in engineering courses, using the Fuzzy Delphi method. Based on the results, it is possible to state that the objectives were met.

The results present thirty-two innovation and sustainability practices identified through the systematic literature review and twenty-one main practices validated by the fuzzy Delphi method, in addition to the seventeen objectives for sustainable development that also had their applicability validated by applying the fuzzy Delphi method. The thirty-eight validated practices were grouped into three blocks: fifteen practices in the first general block; six practices in the second university/course block; and seventeen objectives for sustainable development in the third SDG block.

It should be noted that in the fuzzy Delphi method, as described in the method section, it is estimated that the number of interviewees should vary between five and twenty (Bui et al. 2020). As this is a method that has already been used in previous studies, it is felt that the present study has the necessary foundation so that the academic community of engineering courses (coordinators, teachers, researchers and academics) can consider the results presented to improve their current course programs.

Overall, it is expected that this study will assist in the process of integrating innovation and sustainability and applicability of the SDGs in the curricula of engineering courses in the Brazilian Amazon and, consequently, contribute to inspiring changes in behavior towards the environment and generation of knowledge and technologies that contribute to sustainable recovery and local development in the Brazilian Amazon.

A limitation of this study concerns the opinions of the participants. As a suggestion, it is suggested that the group of participants be broadened, considering that they have different views of the degree course. More specifically, apply it to groups of engineers, civil, agronomy, fisheries, environmental, etc. In other words, with similar basic training, but with experiences and practices that may vary. The aim is to increase the number of pieces of evidence, given that the method used has a validated scientific vision. In addition to the above, it is proposed that future research consider the application of weighting methods (such as Analytical Hierarchy (AHP) to classify the main challenges presented in this research, and it is also suggested that the research be expanded to include students and graduates from all areas of knowledge, to verify how teaching, research and extension are adopting sustainability in HEIs.

References

Aginako Z, Guraya T (2021) Students' perception about sustainability in the engineering school of Bilbao (University of the Basque Country): Insertion Level and Importance. Sustainability, 13:8673. https://doi.org/10.3390/su13158673

Al-Bahi AM, Abd-Elwahed MS, Soliman AY (2021) Implementation of sustainability indicators in engineering education using a combined balanced scorecard and quality function deployment approaches. Sustainability (Switzerland), 13(13). https://doi.org/10.3390/su13137083

Almeida FI (2021) Sustentabilidade: Desafios do Brasil no Século XXI. Política Democrática Revista de Política e Cultura. Ano 21(59)

Baleia R, Alana G (2018) Direitos Humanos Engajamundo. Greenpeace C, Marengo JA, Souza Jr C Mudanças Climáticas: impactos e cenários para a Amazônia

Bui T et al (2020) Identifying sustainable solid waste management barriers in practice using the fuzzy Delphi method. Resour, Conserv Recycl 154:104625

Cipolat C, Vinicius M, Bidarte D (2021) Os polos de Carajás e Juriti na Amazônia oriental: desenvolvimento regional ou industrialização? The Carajás and Juriti poles in the eastern Amazon: regional development or industrialization? http://orcid.org/0000-0002-5545-3969

Elsaadany A, Helmi A (2018) Sustainable development, technological and industrial impacts on engineering education. Interdiscip Descr Complex Syst 16(2):227–237. https://doi.org/10.7906/indecs.16.2.3

Enríquez GEV (2008) Desafios da sustentabilidade da amazônia: Biodiversidade, cadeias produtivas e comunidades extrativistas integradas

Fernandes A (2022) Eco-Inovação: o paradoxo presente na nossa sociedade. https://orcid.org/0000-0002-4172-3833
Alves MFR, Galina SVR, Dobelin S (2018) Literature on organizational innovation: past and future. Innov Manag Rev 15(1):2–19. https://doi.org/10.1108/INMR-01-2018-001
Fernando L, Cotian P (2022) Engenharias. Ciência e Tecnologia 7. https://doi.org/10.22533/at.ed.933193101
Filho WL (2011) About the role of Universities and their contribution to sustainable development. High Educ Pol 24(4):427–438. https://doi.org/10.1057/hep.2011.16
Freitas ARP, Paiva LEB (2019) Revisão da produção científica internacional de brasileiros acerca das mudanças climáticas. Rev Gestão Soc E Ambient, 12(3):95–113. https://doi.org/10.24857/rgsa.v12i3.1615
Gomes AF, Quaresma P, Giraldi P, Santos VF, Porto J (2022) Mestrado em desenvolvimento regional: 15 anos, na busca de sinergias, possibilidades e expectativas de desenvolvimento. In Mestrado em desenvolvimento regional: 15 anos, na busca de sinergias, possibilidades e expectativas de desenvolvimento. Uniedusul Editora. https://doi.org/10.51324/80277988
Ishikawa A, Amagasa M, Shiga T, Tomizawa G, Tatsuta R, Mieno H (1993) The max-min Delphi method and fuzzy Delphi method via fuzzy integration. Fuzzy Sets Syst 55:241–253. https://doi.org/10.1016/0165-0114(93)90251-C
Kumar A, Zavadskas EK, Mangla SK, Agrawal V, Sharma K, Gupta D (2019) When risks need attention: adoption of green supply chain initiatives in the pharmaceutical industry. Int J Prod Res 57: 3554–3576. https://doi.org/10.1080/00207543.2018.1543969
Kuo YF, Chen PC (2008) Constructing performance appraisal indicators for mobility of the service industries using Fuzzy Delphi Method. Expert Syst Appl 35(4):1930–1939
Leal Filho W et al (2020) Sustainability leadership in higher education institutions: An overview of challenges. Sustainability 12:3761
Llorca-ponce A, Rius-sorolla G, Ferreiro-seoane FJ (2021) Is innovation a driver of sustainability? An analysis from a Spanish Region. Sustainability 13:9286
Lusz P, Cristina I, Zaneti BB, Rodrigues Filho S (2021) Situação: O preprint foi submetido para publicação em um periódico Educação ambiental na educação do campo: jovens, pesquisa ação e mudanças climáticas (versão 2) (AAAA-MM-DD). https://doi.org/10.1590/SciELOPreprints.2365
Martins VWB (2019) Knowledge management in the context of sustainability: Literature review and opportunities for future research. J Clean Prod 229:489–500
Ogrean C, Herciu M (2020) Business models addressing sustainability challenges—towards a new research agenda. Sustainability 12:3534
Okoli C (2019) Guia para realizar uma revisão sistemática da literatura. Tradução de David Wesley Amado Duarte; Revisão técnica e introdução de João Mattar. eaD em Foco
Rampasso IS (2019) Analysis of the perception of engineering students regarding sustainability. J Clean Prod, 233:461–467
De Andrade SE (2022) Universidade federal de santa catarina centro tecnológico programa de pós-graduação em engenharia e gestão do conhecimento. Simon Rampasso, i. (n.d.). universidade estadual de campinas Faculdade de Engenharia Mecânica Guidelines for insertion of sustainability in engineering education in Brazil Diretrizes para inserção da sustentabilidade no ensino de engenharia no Brasil. https://orcid.org/0000-0003-1633-6628
Singh PK, Sarkar P (2020) A framework based on fuzzy Delphi and DEMATEL for sustainable product development: A case of Indian automotive industry. J Clean Prod, [s. l.], 246:118991
Stock T, Kohl H (2018) Perspectives for international engineering education: sustainable-oriented and transnational teaching and learning. Procedia Manuf 21:10–17. https://doi.org/10.1016/j.promfg.2018.02.089
Torres PHC, Jacobi PR, Leonel AL (2020) Nem leigos nem peritos: o semeador e as mudanças climáticas no Brasil. Política & Sociedade 19(44):17–38. https://doi.org/10.5007/2175-7984.2020v19n44p17

Vecchia DD, da Costa AM, Conceição FA (2021) Ecocide crimes in the amazon rainforest: the european ecological pact as an alternative against environmental crimes. J Law Sustain Dev, 9(2). https://doi.org/10.37497/REVCAMPOJUR.V9I2.737

Vianna S, Alexandre C, Luis B H, Barioni G, Mozzer G B, Bergier I (2018) Ação contra A mudança global do Clima Contribuições da Embrapa

Vieira PA, Buainain AM, Contini E (2019) Amazônia Um mosaico em construção. Revista Ponto de vista, 28(1344)

Estimation of the Vulnerability of the Coastal Zone of the Amazon to Sea-Level Rise Due to Climate Change

Luan Victor Pereira de Sousa, Fabio Augusto Siqueira dos Santos, Pâmela Beatriz Cantanhede Aguiar, Roseane Costa Diniz, Thomas Vinicius Barros Pacheco, Felipe Martins Sousa, Flavia Rebelo Mochel, Celso Henrique Leite Silva-Junior, Stella Manes da Silva Moreira, and Denilson da Silva Bezerra

Abstract The present research aims to analyse the vulnerability of the coastal zone of the Maranhão Legal Amazon (MLA) to rising sea levels resulting from climate change and the contribution of the local mangrove forest to reducing this risk. The MLA coastline is made up of many ecosystems, such as Ramsar sites and approximately 47% of Brazil's mangrove areas (about 460,745 ha); changes induced by human action can affect essential services provided by these ecosystems in the study

L. V. P. de Sousa (✉) · F. A. S. dos Santos · P. B. C. Aguiar · R. C. Diniz · T. V. B. Pacheco · F. M. Sousa · F. R. Mochel
Department of Oceanography and Limnology, Federal University of Maranhão–UFMA, São Luís, Maranhão, Brasil
e-mail: engamb.luan@gmail.com

F. A. S. dos Santos
e-mail: fabio.siqueira@discente.ufma.br

P. B. C. Aguiar
e-mail: pamela.aguiar@discente.ufma.br

T. V. B. Pacheco
e-mail: thomas.vinicius@discente.ufma.br

F. M. Sousa
e-mail: felipe.martins1@discente.ufma.br

F. R. Mochel
e-mail: flavia.mochel@ufma.br

C. H. L. Silva-Junior
Instituto de Pesquisa Ambiental da Amazônia–IPAM, Belém, Pará, Brazil

S. M. da Silva Moreira
Instituto Internacional Para a Sustentabilidade (IIS), Rio de Janeiro, Brazil

D. da Silva Bezerra
Federal University of Maranhão–UFMA, São Luís, Brazil

W. Leal Filho et al. (eds.), *Amazon 2030—Sustainability Issues in the World's Largest Rainforest Region*, World Sustainability Series,
https://doi.org/10.1007/978-3-031-81465-5_11

area. In this research, two computational models are used to estimate the vulnerability of the Maranhão coastal zone to climate change; the first is the Integrated Valuation of Ecosystem Services and Trade-offs (InVEST) to assess coastal vulnerability, the second is "BR-MANGUE" to simulate the response of the mangrove to rising sea levels. The results highlight the importance of all coastal ecosystems and the possible damage resulting from their losses, especially in the loss of mangrove areas on the MLA coast. The results also indicate that the protection provided by coastal ecosystems is highly relevant to proposing nature-based solutions to combat the impacts of climate change on the Amazon coastal zone.

Keywords Ecosystem services · Nature-based solutions · Mangroves · InVEST model

1 Introduction

The IPCC Sixth Assessment Report (AR6) warns that climate change does not result in a potential future threat but rather a current one (IPCC 2022; Nobrega 2022). The present estimates indicate a much greater climate risk than in previous reports, arising from the considerable increase in atmospheric CO_2 concentration in recent years (Pörtner et al. 2022). However, even with effective climate mitigation, it is no longer possible to completely prevent climate change and avoid its consequences (Nunez et al. 2019), which means that intense actions are needed to reduce the negative impacts of climate adaptation actions.

Actions to de-escalate climate change are thus urgently needed. Climate mitigation, i.e. actions that reduce the concentration of greenhouse gasses in the atmosphere, is imperative to reduce the magnitude of climate change, hoping to keep its impact as low as possible (IPCC 2018). Thus, on top of climate mitigation, the current climate emergency also calls for urgent adaptation actions, i.e. actions capable of reducing, moderating or avoiding potential risks of climate change, increasing natural and human systems' resilience (IPCC 2018).

Sea-level rise (SEL) is one of the consequences of climate change and can pose serious environmental and socioeconomic challenges, especially in coastal environments (Nevermann et al. 2023). The SEL exacerbates the impacts of extreme sea-level events as well as coastal hazards, and has multiple detrimental effects on marine ecosystems and services (Nevermann et al. 2023; Moftakhari et al. 2017; Masson-Delmotte et al. 2021; Martyr-Koller et al. 2021; Hurk et al. 2022). IPCC estimates suggest a continuous and progressive global mean sea level increase for the next few decades. Such a trend can already be observed when, for example, we compare the rate of change of 3.7 [3.2–4.2] mm/year between 2006 and 2018 with the rate of 1.9 [0.8 and 2 0.9] mm/year between 1971 and 2006 (Masso-Delmotte et al. 2021). The global mean sea level will continue to rise throughout the twenty-first century, reaching between 0.28 and 0.55 m under a very low greenhouse gas scenario (SSP1-1.9) up to 0.63–1.01 m at a very high setting (SSP5-8.5), which would transform coasts across the world (IPCC 2021; Manes et al. 2023).

Extreme sea-level events and land subsidence have the potential to significantly affect landscapes, land use, infrastructure, morphology and ecosystem services. Therefore coastal areas are among the most vulnerable regions in the world (Nicholls and Cazenave 2010; Davtalab et al. 2020). The coastline, both in the coastal and marine portions, comprises many ecosystems, including a variety of natural habitats, such as coastal forests, mangroves, wetlands, restingas, fields, dunes, rocky shores, and coral reefs, among others (Mapbiomas 2022; Manes et al. 2023).

The coastal zone of the Legal Brazilian Amazon, to the north, is covered mainly by tropical forests, mangroves and grasslands (Mapbiomas 2022; Manes et al. 2023), in addition to some coral reefs located in deep waters (Moura et al. 2016). It is also important to mention that the mangroves present in the coastal zone of the Legal Brazilian Amazon correspond to approximately 80% of the Brazilian mangroves (Mapbiomas 2022). Sea-level rise can affect essential services performed by estuaries and strongly associated ecosystems, such as mangroves and salt marshes, as habitats for many species, fishing, tourism, navigation, and carbon storage (Donato et al. 2011; Ewel et al. 1998; Lee 2008; Lee et al. 2014). The Amazonian mangroves correspond to a valid natural response to potential sea-level rise events; therefore, their preservation is a valuable adaptation strategy to reduce climate change damage to coastal human populations, infrastructure and economic activities. However, studies on nature's contribution to coastal defence still need to be made available in the national scientific literature (Morris et al. 2019; Manes et al. 2023).

Among the strategies with the most significant potential for climate adaptation are the Nature-based solutions (NbS; Pörtner et al. 2021, Pires et al. 2021, Scarano et al. 2020, Manes et al. 2022), having been applied in several adaptation strategies around the world, favoring the preservation of essential ecosystems (Oliveira 2020), such as mangroves, enabling the maintenance of environmental services and the conservation of biodiversity. NbS comprise actions that protect, recover or sustainably manage natural or modified ecosystems to solve global societal problems, providing benefits for humans and nature alike (Cohen-Shacham et al. 2016; Manes et al. 2022).

As an umbrella concept, NbS can take many forms within different ecosystem-related approaches, including 'Ecosystem Based Adaptation', which refers explicitly to addressing the role of ecosystems to better adapt to climate change, accounting for social, economic and cultural co-benefits for local communities (Scarano 2017); 'Ecosystem-based Disaster Risk Reduction', where nature can reduce the likelihood of natural hazards to become disasters. Nonetheless, all NbS approaches aim to enhance ecosystems' resilience and services to counteract significant societal challenges (global changes) and ensure human well-being (Cohen-Shacham et al. 2016, Manes et al. 2022).

Because of the above, the present study aims to estimate the vulnerability of the coastal zone of the Brazilian legal Amazon, which has the largest continuous area of mangroves on the planet, to potential sea level rise events. In this way, we present an unprecedented experiment to understand the importance of coastal ecosystems for protecting the Brazilian Amazon coastal zone.

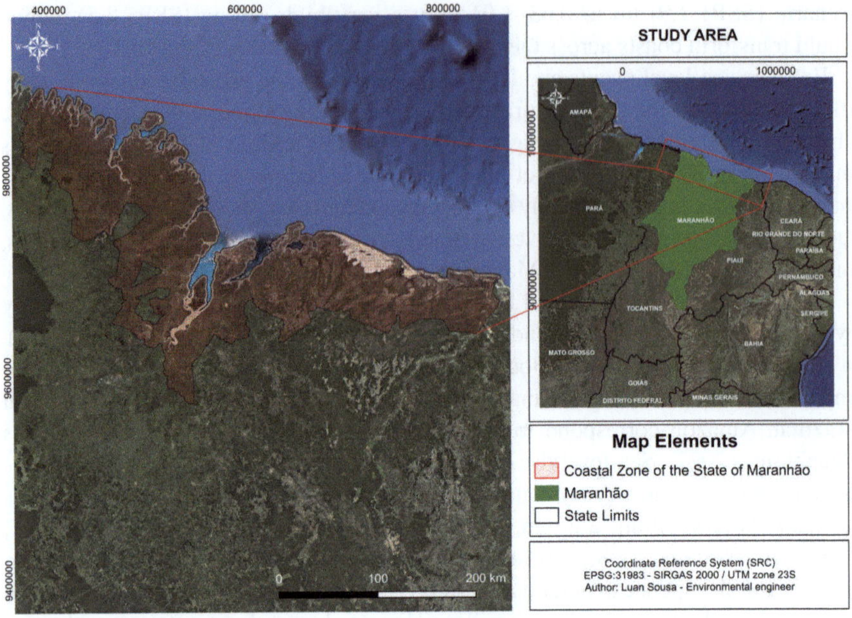

Fig. 1 Location of the study area

2 Methodology

2.1 Study Object Area

The study area for this research corresponds to the coastal zone of the state of Maranhão. This state is part of the coastal zone of the Brazilian legal Amazon. With a length of 640 km, between the mouths of the Gurupi and Parnaíba rivers, the coast of Maranhão is divided into five sectors with very different physiographic characteristics, namely the Golfão Maranhense, the Eastern Coast, the Western Coast, the Baixada Maranhão and the Marine State Park from Parcel Manuel Luís. The state stands out for its significant area of mangroves alone it has approximately 47% of the mangrove areas in the Brazilian coastal zone and 60% of the area where Brazilian salt marshes occur (Fig. 1).[1]

The Coastal Vulnerability model of the Integrated Valuation of Ecosystem Services and Trade-offs (InVEST) software was used to assess the risk for the coastal zone of Maranhão (the area of study for this research). InVEST is a software package widely used in the scientific literature to map and assess ecosystem services (Sharp et al. 2020).

[1] For more information access: https://brasil.mapbiomas.org/2022/11/18/radiografia-da-costa-brasileira-nos-ultimos-37-anos-manguezais-estaveis-praias-e-dunas-em-retracao/.

The coastal vulnerability model generates a risk index along the coast through Eq. 1 based on six biophysical factors: wind speed, wave power, high tide potential, relief, and sea level elevation. Finally, in addition to physical variables, the model incorporates the degree of protection of natural ecosystems to assess the vulnerability of the coastal zone.

:

$$EI = \left(\prod_{i=1}^{n} R_i\right)^{\frac{1}{n}} \tag{1}$$

Thus, to assess the role of nature in coastal protection, we use two scenarios: (i) with current natural ecosystems along the coast of Maranhão and (ii) total absence of all natural ecosystems (i.e., current natural ecosystems no longer provide coastal protection).

The second model used, "BR-MANGUE", was prepared following the theoretical precepts described by Bezerra et al. (2014), who developed a computational model to simulate the response of the mangrove ecosystem to rising mean sea levels.

The model simulates the SEL for the study area in 79 (seventy-nine) elevation steps, from 0.11 m to 1.01 m, according to an arithmetic progression of rate 0.11 m per year, from 2020 to 2100. The highest elevation value corresponds to the most alarming scenario of sea level rise in AR6-IPCC, that is, 1.01 m of global average elevation in 2100.

The implementation is based on the computational model of cellular automata (Wolfram 1983; Bezerra et al. 2014), a logical system that has the concept of the cell as the basic unit: each cell has a neighbourhood of cells and a discrete state that can vary during the simulation, according to its transition rules.

For the model simulation procedure, a geographic database was organised in the TerraView 4.2.0 software, containing land cover data obtained from the most recent Mapbiomas database (https://mapbiomas.org/), with a spatial resolution of 30 m. The database also contains information from a Digital Elevation Model (source: Brazil in Relief—https://www.cnpm.embrapa.br/projetos/relevobr/) and soil classes (source: Ecological-Economic Zoning in Maranhão—http://zee.ma.gov.br/).

3 Analysis of Results and Discussion

The results obtained in this article agree with other global-scale studies (Manes et al. 2023). Our results warn of the severe consequences of the continued loss of coastal ecosystems, which increases the number of coastal areas and populations exposed to risks induced by climate change. According to Manes et al. (2023), all coastal natural habitats are essential, and the damage results from their losses. Therefore, actions aimed at the protection and sustainable use of these habitats should not be reduced and consequently putting people and property on the Brazilian coast at risk. Still, rather than the opposite, protection actions should be increased. Protecting these habitats, considering the relationships between people and nature, is imperative to ensure climate adaptation against coastal hazards and safeguard coastal human populations (Manes et al. 2023).

Prates et al. (2012) state that all Brazilian states have areas above 10% of their mangroves legally protected. However, as APA ensures, a large portion of this protection has low effectiveness. In Maranhão, the percentage of protection reaches 97.2% of the total in the state, mainly due to the portion protected exclusively by APA (85% of the total area under protection in the state).

Notably, Brazil is likely among the countries with the highest concentration of people benefiting from nature-based coastal risk reduction worldwide (Ferrario et al. 2014; Manes et al. 2023). However, although the importance of NbS is being increasingly recognized in Brazil and worldwide, the loss of natural habitats in the last decades indicates that recognition in the paper still has limited practical application. An example was the recent attempt by the Brazilian Ministry of the Environment to reduce the protection status of mangroves (Ottoni et al. 2021; Vale et al. 2021; Bezerra et al. 2022) placing 110,000 km^2 of mangroves at risk (Rosa and Azevedo 2020).

This change would have been disastrous not only for the mangroves themselves but also for the people who depend on their crucial role in the maintenance of fisheries and attenuation of climate-related hazards (e.g. Ferreira and Lacerda 2016; Bernadino et al. 2021; Ottoni et al. 2021). Mangroves also have innate persistence due to their adaptive ecosystem engineering capacity to adjust to changing environmental conditions (Jones et al. 1994, 1997; Gijsman et al. 2021). Their natural regeneration and sediment trapping capacity enables mangroves to (1) stabilise shorelines (Krauss et al. 2014; Woodroffe et al. 2016; Gijsman et al. 2021), (2) recover from impacts of extreme events (Krauss and Osland 2020; Gijsman et al. 2021), and (3) adapt to longer-term changes in environmental conditions (Schuerch et al. 2018; Saintilan et al. 2020; Gijsman et al. 2021). Hence, mangroves have the potential to offer a shoreline-stabilising, impact-resilient and climate-adaptive strategy to reduce coastal flood risk. In doing so, mangroves can provide additional value for society while implementation and maintenance costs could be substantially lower compared to hard-engineering infrastructure (Borsje et al. 2011; Narayan et al. 2016; Gijsman et al. 2021).

To evaluate the services provided by terrestrial and marine ecosystems for coastal protection in the study area, we used three scenarios among the five worked on by the IPCC (SSP scenarios are described in sections TS1.3 and 1.6 and Box 1.4 between chapters of the contribution of Working Group 1.). We divided the resulting risk indices into three categories (organised through inVEST), with areas of lower risk (overall risk indices from 1 to 2.33), intermediate risk (overall risk indices from 2.33 to 3.66), and higher risk (overall risk indices overall risk score from 3.66 to 5).

The partial results obtained during the research show that natural ecosystems provide essential protection against risks induced by the effects of climate change along the coastal zone of the state of Maranhão.

3.1 Coastal Index

Currently considering all ecosystems capable of protecting the Maranhão coast, specifically the mangrove ecosystem, which had the highest degree of protection compared to other coastal and transition ecosystems between terrestrial and marine environments.

Under current conditions, with all remaining natural habitats, it is possible to observe in item A of Fig. 2, that more than half of the coast of Maranhão is at the lowest risk (82.62%, in 3,520 points along the coast) (Fig. 2). The remaining part of the coast is currently divided between intermediate risks (17.37%) and no high-risk points.

In item B of Fig. 2, considering the IPCC SSP2-4.5 scenario and its effect on sea level rise in the calculation, the length of the coast at lower risk would fall to just 49.31%. There is also a significant increase in the length of the coastline under intermediate and high risks, to 50.44% and 0.23%, respectively.

For the SSP5-8.5 scenario and its effect on sea level rise on the calculation, the length of the coast at lower risk would have an even more evident drop to just 34.65%. There is also a significant increase in the length of the coastline under intermediate and high risks, to 63.61 and 1.73%, respectively (Fig. 3).

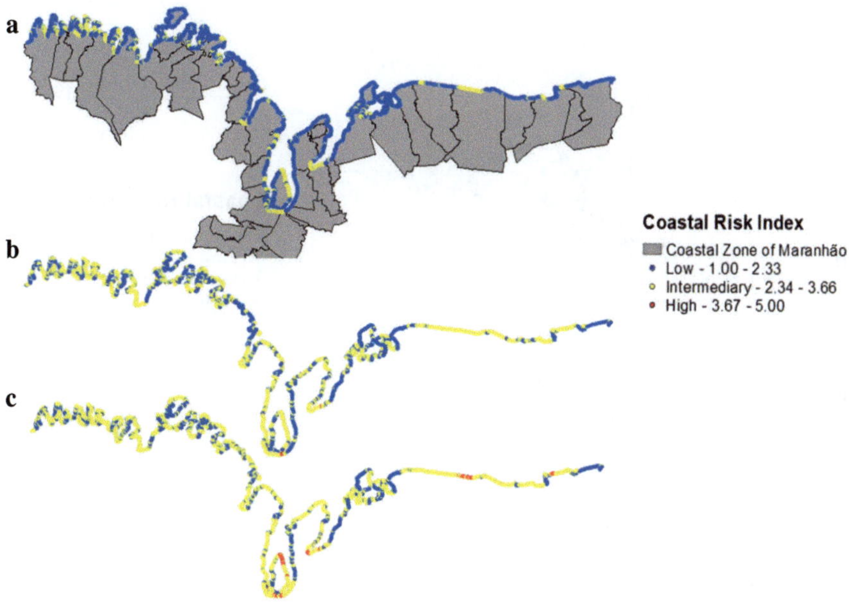

Fig. 2 Coastal risks along the coast of Maranhão under different IPCC scenarios. All habitats are shown for current conditions. **a** Scenario SSP1-1.9 with the uplift rate to 2100 between 0.28 and 0.55 m, **b** Scenario SSP2-4.5 with the uplift rate up to 2100 between 0.43 and 0.76 m, and **c** Scenario SSP5-8.5 with the uplift rate until 2100 between 0.63 and 1.01 m

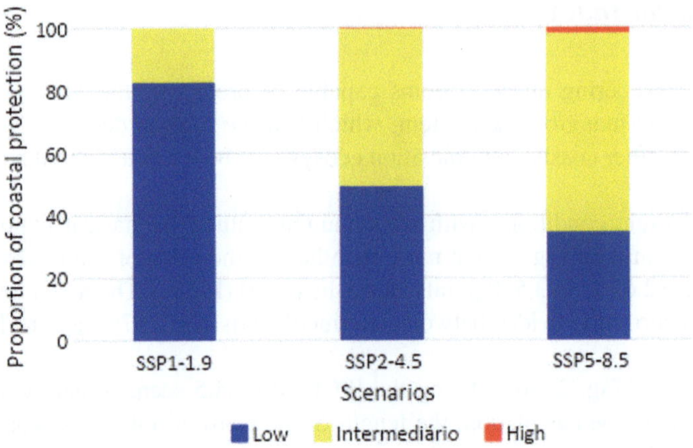

Fig. 3 The proportion of coastal protection associated with different scenarios

Given the current conditions of the habitats, tests were carried out without their presence in the study area to measure the degree of protection against sea-level rise scenarios. In Fig. 4, we divide it again into three scenarios, A, B, and C, and all results with the value of each IPCC scenario are added to the calculation.

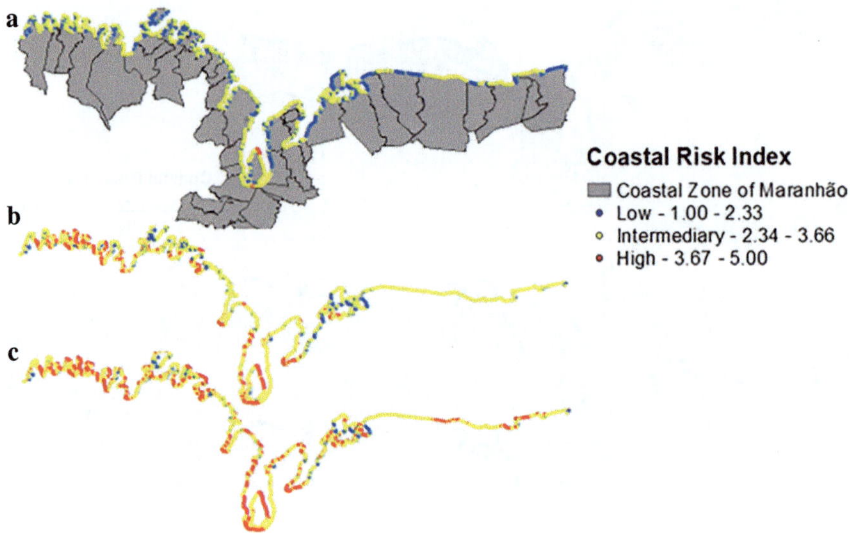

Fig. 4 There are coastal risks along the coast of Maranhão without natural habitats under different IPCC scenarios. **a** Scenario SSP1-1.9 with the uplift rate to 2100 between 0.28 and 0.55 m, **b** Scenario SSP2-4.5 with the uplift rate up to 2100 between 0.43 and 0.76 m, and **c** Scenario SSP5-8.5 with the uplift rate up to 2100 between 0.63 and 1.01 m

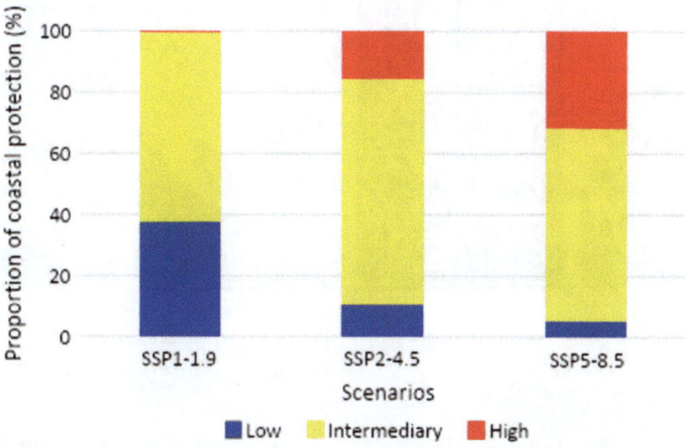

Fig. 5 Proportion of coastal protection associated with different scenarios

It was possible to observe in the simulation of item A in Fig. 4, that more than half of the points are found at intermediate and high levels, and consequently, in items B and C, which correspond to the scenarios we call intermediate and severe, the values of these points increase even further (Fig. 5).

3.2 Coastal Areas with the Most Vulnerability

The region's most at risk from the effects of the SSP5-8.5 scenario are located mainly on the Maranhão Western Coast (item A in Fig. 6), where the "reentrâncias maranhenses[2]" are located, characterized by part of the largest continuous strip of mangroves on the planet (Nascimento et al., 2013). The Maranhão recesses represent the largest continuous mangrove forest in the world, comprising several peninsulas in the various sectors that make it up (Souza-Filho 2005; Spalding et al. 2010; Souza et al. 2018).

Another region of greatest risk identified is the Golfão Maranhense (item B in Fig. 6), which is an important coastal zone with estuaries and inlets, consisting of a complex mangrove environment (Guterres et al. 2022). Rebelo-Mochel (1997) describes this environment highlighting the characteristics of the coastline, high volume of fresh water and tidal ranges. The region is at the center of a golf system, characterized by semidiurnal macrotides, and tidal currents greater than 4 m/s (Guterres et al. 2022).

The Maranhão gulf is characterized by having Maranhão Island (IM) as its center, with a population of over one million inhabitants, the four municipalities (São Luís,

[2] Important Ramsar site located in the coastal zone of the Brazilian legal Amazon, for more information access: https://rsis.ramsar.org/ris/640.

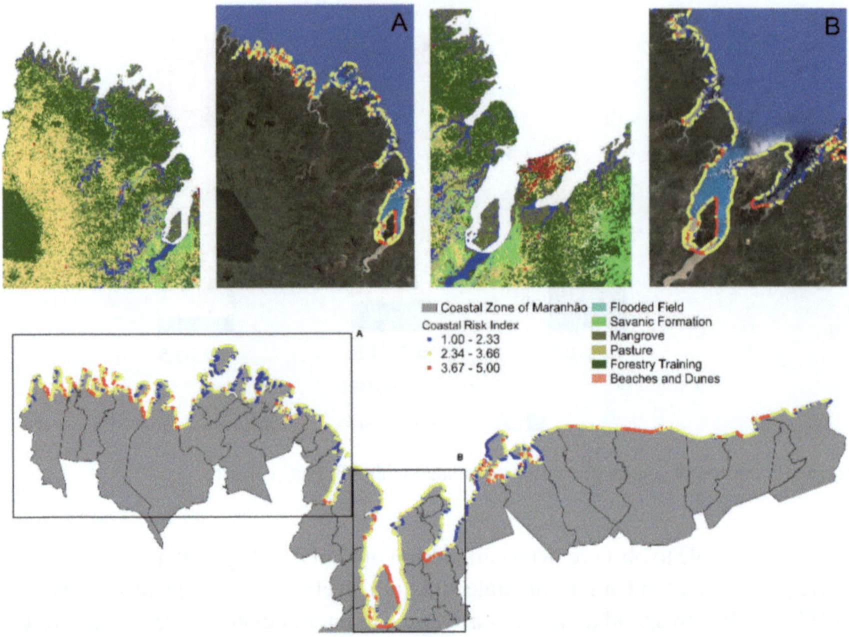

Fig. 6 Comparison between vulnerability indices and natural ecosystems in risk areas along the coast of Maranhão

Ribamar, Raposa and Paço do Lumiar) belonging to IM induce high urban and regional development, consequently the intensity and magnitude of human actions are perceived in the coastal strip of IM and impose great vulnerability to the landscape and aquatic ecosystems.

3.3 Simulation of Mean Sea-Level Rise

The results obtained in the simulation are validated through literature that indicates positive variation in mangrove areas according to sea elevation levels, also portraying potential areas of increase in mangroves as salt water increasingly penetrates rivers and estuaries (Lacerda et al. 2007; Bezerra et al. 2014). In the literature, according to several authors, the investigated impacts of sea level rise and climate change on the Brazilian Amazon coast during the late Holocene confirmed that sea level rise is the most likely factor that can cause the expansion of mangroves (France et al. 2012; Bezerra et al. 2014).

Using spatially explicit modelling can help understand the pattern of mangrove response to climate change and other environmental stressors (Berger et al. 2008; Bezerra et al. 2014). Therefore, our results continue a series of attempts, which will begin with Bezerra et al. (2014), to understand the mangrove patterns of a specific case

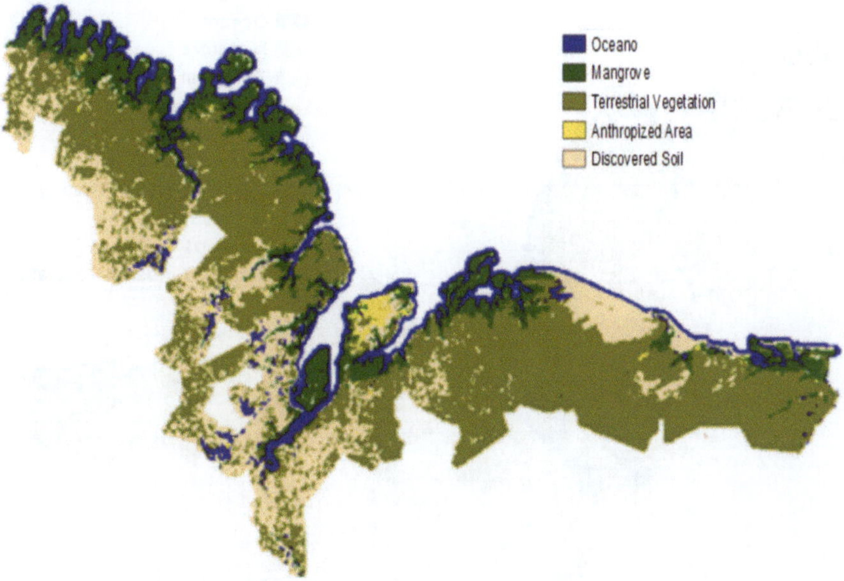

Fig. 7 Land cover classes and spatial distribution of mangrove areas on the Maranhão coast for initial conditions

study associated with simplifying this complex process proposed in our conceptual model.

With the results obtained through the BR-MANGUE computational model, it was possible to observe the behaviour of the mangrove in the face of rising sea levels. In Fig. 7, we have the current condition of the study area, considering it without significant anthropogenic and natural modifications.

In Fig. 8 it is possible to observe the result of the simulation with the current conditions plus the conditions predicted by using the Br-Mangue model. The sea level used in the simulation was 0.013 m/year, a value defined by the IPCC in the SSP5-8.5 scenario; in this scenario, an increase of up to 1.01 m is predicted for the year 2100, this value was divided into 79 years, being the difference between the years 2021 (year of the last report) and the year 2100.

It was identified that the number of mangrove pixels that migrated into the continent were greater than the flooded mangrove pixels. In other words, according to the simulation carried out in the study area, an increase of around 1 m in sea level can induce a mangrove migration process that is more intense than the occurrence of loss of mangrove areas. Such results can highlight the significant potential that Amazonian mangroves have to adapt to climate change; however, human uses adjacent to mangrove areas can act as barriers to migration in eventual sea level rise. This process must be better studied and understood by scientific knowledge.

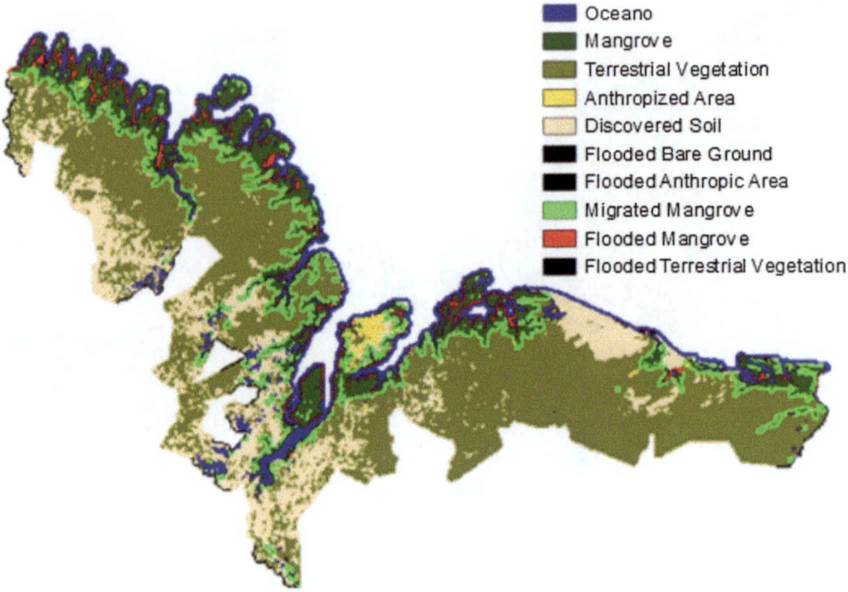

Fig. 8 Land cover classes and spatial distribution of mangrove areas on the coast of Maranhão for the year 2100

4 Conclusion

The coastal zone of the Brazilian legal Amazon corresponds to the largest continuous area of mangroves on the planet. The area under analysis in this research is part of this immense strip of mangroves, comprising approximately 47% of the entire coverage of Brazilian mangrove forests.

The simulation results for the study area indicate that without the presence of mangroves, most of the coastal zone for the study area would have an intermediate risk value (for the most alarming IPCC scenario). This indicates that the mangroves offer a critical environmental service of protecting the coastline against potential sea-level rise events, mainly because simulations carried out without other coastal environments resulted in no evidence of increased vulnerability of the study area to sea rise.

The results presented in this research highlight the importance of all coastal ecosystems and the possible damage resulting from their losses, especially in the loss of mangrove areas on the Maranhão Legal Amazon coast. The results also indicate that the protection provided by coastal ecosystems is highly relevant to proposing nature-based solutions to combat the impacts of climate change on the Amazon coastal zone.

Acknowledgements This research was funded by the Maranhão Research and Scientific and Technological Development Support Foundation—FAPEMA in partnership with the Maranhão Port

Administration Company, through the MASTERS SCHOLARSHIP Notice—FAPEMA RESOLUTION Nº 10, OF 13/06/ 2022—PORTO DO FUTURO PROGRAM—EMAP. C.H.L.S-J. was supported by the National Council for Scientific and Technological Development—CNPq (grant 401741/2023-0).

References

Berger U, Rivera-Monroy VH, Doyle TW, Dahdouh-Guedas F, Duke NC, Fontalvo-Herazo ML, Hildenbrandt H, Koedam N, Mehlig U, Piou C, Twilley RR (2008) Avanços e limitações de modelos baseados em indivíduos para analisar e prever a dinâmica de florestas de mangue, Aquat

Bernadino AF, Nóbrega GN, Ferreira TO (2021) Consequences of terminating mangrove's protection in Brazil. Mar Policy 125:104389

Bezerra DS, Santos AL, Bezerra JS, Amaral S, Kampel M, Anderson LO, Mochel FR, Nunes JLS, Araujo NA, Barreto LN, Pinheiro MSS, Celeri MJ, Silva FB, Viegas AM, Manes S, Rodrigues TCS, Viegas JC, Souza UDV, Santos ALS, Silva-Junior CHL (2022) Brazil's mangroves: Natural carbon storage. Science 375:1239

Bezerra DS, Amaral S, Kampel M, Andrade PR (2014) Simulating sea-level rise impacts on mangrove ecosystem adjacent to anthropic areas: the case of Maranhão Island, Brazilian Northeast. Pan-American Journal of Aquatic Sciences. 9(3):188–198

Borsje BW, van Wesenbeeck BK, Dekker F, Paalvast P, Bouma TJ, van Katwijk MM et al (2011) How ecological engineering can serve in coastal protection. Ecol Eng 37:113–122. https://doi.org/10.1016/j.ecoleng.2010.11.027

Castellanos EJ, Lemos MF, Astigarraga L, Chacón N, Cuvi N, Huggel C, Miranda L, Vale MM, Ometto JP, Peri PL, Postigo JC, Ramajo L, Roco L, Rusticucci M (2022) Capítulo 12: América Central e do Sul. in: Mudanças Climáticas 2022: Impactos, Adaptação e Vulnerabilidade. IPCC, ONU

Celliers L, Rosendo S, Costa MM, Ojwang L, Carmona M, Obura D (2020) Uma abordagem de capital para avaliar a governança costeira local. Ocean Coast Manag 183:104996

Cohen-Shacham E, Walters G, Janzen C, Maginnis S (eds) (2016) Nature-based Solutions to address global societal challenges. Gland, Switzerland: IUCN. xiii + p 97

Cooley S, Schoeman D, Bopp L, Boyd P, Donner S, Ito S, Kiessling W, Martinetto P, Ojea E, Racault M, Rost B, Skern-Mauritzen M, Ghebrehiwet DY (2022) Capítulo 3: Oceanos e ecossistemas costeiros e seus serviços. In: Pörtner HO, Roberts DC, Tignor M, Poloczanska ES, Mintenbeck K, Alegría A, Craig M, Langsdorf S, Löschke S, Möller V, Okem A, Rama B (eds) Climate change 2022: Impacts, adaptation, and vulnerability. Contribuição do Grupo de Trabalho II ao Sexto Relatório de Avaliação do Painel Intergovernamental sobre Mudança do Clima. Cambridge University Press (no prelo)

Davtalab R et al (2020) Sea level rise effect on groundwater rise and stormwater retention pond reliability. Water, [S.L.], 12(4):1129. MDPI AG. https://doi.org/10.3390/w12041129

Donato D, Kauffman JB, Murdiyarso D, Kurnianto S, Stidham M, Kanninen M (2011)

Ewel KC, Twilley RR, Ong JE, Ecology G, Letters B, Mangrove F (1998) Different kinds of mangrove forests provide different goods and services. Glob Ecol Biogeogr Lett 7:83–94

Ferrario F, Beck MW, Storlazzi CD, Micheli F, Shepard CC, Airoldi L (2014) The effectiveness of coral reefs for coastal hazard risk reduction and adaptation. Nat Commun 5:1–9

Ferreira CF, Lacerda LD (2016) Degradation and conservation of Brazilian mangroves, status and perspectives. Ocean Coast Manag 125:38–46

França MC et al (2012) Os últimos manguezais Ilha do Marajó-Amazônia Oriental: Impacto das mudanças climáticas e/ou mudanças relativas do nível do mar. Revisão De Paleobotânica e Palinologia 187:50–65

Gijsman R et al (2021) Nature-based engineering: a review on reducing coastal flood risk with mangroves. Front Mar Sci, 8:702412

Guterres CE et al (2022) Espacialização e análise dos registros de fauna do golfão maranhense nos últimos 300 anos e sua correlação com uso e ocupação do solo. https://doi.org/10.1038/ngeo1123

van den Hurk B et al (2022) Living with sea-level rise in North-West Europe: science-policy challenges across scales. Clim Risk Manag, [S.L.], 35:100403. Elsevier BV. https://doi.org/10.1016/j.crm.2022.100403

IPCC (2018) Summary for policymakers. In: Global warming of 1.5 °C. An IPCC Special Report on the impacts of global warming of 1.5 °C above pre-industrial levels and related global greenhouse gas emission pathways, in the context of strengthening the global response to the threat of climate change, sustainable development, and efforts to eradicate poverty

IPCC (2021, in press) Summary for policymakers. In: Climate change 2021: the physical science basis. In: Delmotte M et al (eds) Contribution of Working Group I to the Sixth Assessment Report of the Intergovernmental Panel on Climate Change. Cambridge University Press.

IPCC (2022) Climate change 2022: Impacts, adaptation, and vulnerability. In: Contribution of working group ii to the sixth assessment report of the intergovernmental panel on climate change [Pörtner H-O, Roberts DC, Tignor M, Poloczanska ES, Mintenbeck K, Alegría A, Craig M, Langsdorf S, Löschke S, Möller V, Okem A, Rama B (eds.)]. Cambridge University Press. Cambridge University Press, Cambridge, UK and New York, NY, USA, p 3056. https://doi.org/10.1017/9781009325844

Jones CG, Lawron JH, Shachak M (1997) Positive and negative effects of organisms as physical ecosystem. Engineers 78:1946–1957. https://doi.org/10.1890/0012-9658(1997)078[1946:PANEOO]2.0.CO;2

Jones CG, Lawton JH, Shachak M, Organisms M (1994) Organisms as ecosystem engineers. Oikos 69:373–386. https://doi.org/10.2307/3545850

Krauss KW, Osland MJ (2020) Tropical cyclones and the organization of mangrove forests: a review. Ann Bot 125:213–234. https://doi.org/10.1093/aob/mcz161

Krauss KW, Mckee KL, Lovelock CE, Cahoon DR, Saintilan N, Reef R et al (2014) How mangrove forests adjust to rising sea level. New Phytol 202:19–34. https://doi.org/10.1111/nph.12605

Lacerda LDD, Menezes MOTD, Molisani MM (2007) Changes in mangrove extension at the Pacoti River estuary, CE, NE Brazil due to regional environmental changes between 1958 and 2004. Biota Neotropica 7:67–72

Lee SY (2008) Mangrove macrobenthos: assemblages, services, and linkages. J Sea Res 59(1–2):16–29

Lee SY et al (2014) Ecological role and services of tropical mangrove ecosystems: a reassessment. Glob Ecol Biogeogr 23(7):726–743

ManesNature-based solutions for climate change adaptation: from global to local scale analysis. S (2022) 133f. Tese (Doutorado em Ecologia) - Universidade Federal do Rio de Janeiro, Rio de Janeiro, p 2022

Manes S et al (2023) Nature as a solution for shoreline protection against coastal risks associated with ongoing sea-level rise. Ocean & Coast Manag, [S.L.], 235:106487. Elsevier BV. https://doi.org/10.1016/j.ocecoaman.2023.106487

Mapbiomas (2022) Coleção 6 de mapas anuais de uso e cobertura da terra no Brasil. Disponível em:. Acessado em: 17 de jun. de 2023

Martyr-Koller R, Thomas A, Schleussner CF, Nauels A., Lissner T (2021) Loss and damage implications of sea-level rise on small island developing states. Curr Opin Environ Sustain 50:245–259

Masson-Delmotte V et al (2021) Climate change 2021: the physical science basis. In: Contribution of working group I to the sixth assessment report of the intergovernmental panel on climate change, vol 2

Moftakhari HR et al (2017) Compounding effects of sea level rise and fluvial flooding. In: Proceedings of the National Academy of Sciences, [S.L.], vol 114, no 37, pp 9785–9790, 28 ago. Proceedings of the National Academy of Sciences. https://doi.org/10.1073/pnas.1620325114

Morris RL, Konlechner TM, Ghisalberti M, Swearer SE (2019) From grey to green: Efficacy of eco-engineering solutions for nature-based coastal defense. Glob Change Biol 24:1827–1842

Moura RL et al (2016) Um extenso sistema de recifes na foz do rio Amazonas. Ciência Adv 2:e1501252. Otonni, FP, Hugnes, RM, Katz, AM, Rangel-Pereira, FS

Narayan S, Beck MW, Reguero BG, Losada IJ, Van Wesenbeeck BK, Pontee N et al (2016) The effectiveness, costs and coastal protection benefits of natural and nature-based defences. PLoS ONE 11:e0154735. https://doi.org/10.1371/journal.pone.0154735

Nascimento WR, Souza-Filho PWM, Proisy C, Lucas RM, Rosenqvist A (2013) Mapping changes in the largest continuous Amazonian mangrove belt using object-based classification of multisensor satellite imagery. Estuar Coast Shelf Sci 117:83–93. https://doi.org/10.1016/j.ecss.2012.10.005

Nevermann H et al (2023) Land loss implications of sea level rise along the coastline of Colombia under different climate change scenarios. Clim Risk Manag, [S.L.], 39:100470. Elsevier BV. https://doi.org/10.1016/j.crm.2022.100470

Nicholls R, Cazenave A (2010) Sea-level rise and its impact on coastal zones. Science 328(5985):1517–1520. https://doi.org/10.1126/science.1185782

Nóbrega MR et al (2022) Impacto das Mudanças Climáticas na Aridez da Bacia Estendida do rio São Francisco, Brasil. Rev Bras Meteorol. https://doi.org/10.1590/0102-7786371003

Nunez S, Arets E, Alkemade R, Verwer C, Leemans R (2019) Assessing the impacts of climate change on biodiversity: is below 2 °C enough? Clim Change 154:351–365. https://doi.org/10.1007/s10584-019-02420-x

Oliveira RA (2020) Contribuições para elaboração de um plano municipal de adaptação dos efeitos das mudanças climáticas. Tese de Doutorado

Otonni FP, Hugnes RM, Katz AM, Rangel-Pereira FS, Bragança PHN, Fernandes R, Palmeira-Nunes ART, Nunes JLS, Santos RR, Piorski NM, Rodrigues-Filho JL (2021) Brazilian mangroves at risk. Biota Neotropica 21(2)

Pires APF, Soto CR, Scarano FR (2021) Strategies to reach global sustainability hould take better account of ecosystem services. Ecosyst Serv 49:101292

Pörtner HO et al (2021) IPBES-IPCC co-sponsored workshop report on biodiversity and climate change. 105281/zenodo4782538

Pörtner HO et al (2022) Climate change 2022: impacts, adaptation and vulnerability

Prates AP, Gonçalves MA, Rosa MR (2012) Panorama da conservação dos ecossistemas costeiros e marinhos no Brasil

Rebelo-Mochel F (1997) Mangroves on São Luís Island, Maranhão, Brazil. In: Kjerfve B, Lacerda LD & Diop EHS (ed.) Mangrove ecosystem studies in Latin America and Africa. UNESCO, Paris, pp 145–154

Rosa MR., Azevedo TR (2020) Nota Técnica sobre o impacto do Despacho n 4.410/2020 do Ministério do Meio Ambiente (MMA). MapBiomas. Available at https://mapbiomas.org/nota-t. Accessed on 20/03/2022

Saintilan N, Khan NS, Ashe E, Kelleway JJ, Rogers K, Woodroffe CD et al (2020) Thresholds of mangrove survival under rapid sea level rise. Science 368:1118–1121. https://doi.org/10.1126/science.aba2656

Scarano FR, Arroyo MTK, Cortés J, Pires APF (2020) Biodiversity. In: Moreno JM, LagunaDefior C, Barros V, Buendía EC, Marengo JA, Oswald U (eds) Adaptation to climate change risks in Ibero-American Countries—RIOCCADAPT Report. Springer, Madrid

Scarano FR (2017) Ecosystem-based adaptation to climate change: concept, scalability and a role for conservation science. Perspect Ecol Conserv 15:65–73

Schuerch M, Spencer T, Temmerman S, Kirwan ML, Wolff C, Lincke D et al (2018) Future response of global coastal wetlands to sea-level rise. Nature 561:231–234. https://doi.org/10.1038/s41586-018-0476-5

Sharp R et al (2020) InVEST 3.10 user's guide. The Natural Capital Project, Stanford University, University of Minnesota, The Nature Conservancy, and World Wildlife Fund

Souza Filho PWM (2005) Costa de manguezais de macromaré da Amazônia: cenários morfológicos, mapeamento e quantificação de áreas usando dados de sensores remotos. Rev Bras Geofísica, 23:427–435

Souza CA et al (2018) Biodiversidade e conservação dos manguezais: importância bioecológica e econômica. Educação Ambiental sobre Manguezais. Unesp, São Vicente, pp 16–56

Spalding M, Kainuma M, Collins L (2010) World atlas of Mangroves. Earthcan, 319

Vale MM, Berenguer E, Menezes MA, Castro EBV, Siqueira LP, Portela RCQ (2021) The COVID-19 pandemic as an opportunity to weaken environmental protection in Brazil. Biol Cons 255:108994

Wolfram S (1983) Statistical mechanics of cellular automata. Rev Mod Phys 55(3):601

Woodroffe CD, Rogers K, McKee KL, Lovelock CE, Mendelssohn IA, Saintilan N (2016) Mangrove sedimentation and response to relative sea-level rise. Ann Rev Marine Sci 8:243–266. https://doi.org/10.1146/annurev-marine-122414-034025

Xavier LY, Guilhon M, Gonçalves LR, Corrêa MR, Turra A (2022) Ondas de mudança: em direção à gestão ecossistêmica para a adaptação às mudanças climáticas. Sustentabilidade 14:1317

Territorial Racism: The Urban Context of the Warao Indigenous People in Transit Through the Brazilian Amazon

Laisse Andressa Nascimento dos Santos and Adnilson de Almeida Silva

Abstract This study sought to discuss the migratory issue and the change in lifestyle of the Warao Indigenous people, from the capital Tucupita, state of Delta Amacuro—Venezuela, and aims to elucidate the social, cultural, linguistic and economic organization within the Brazilian urban context, as well as its acceptances, limitations, difficulties and motivations, given that territorial racism is part of the planning dynamics. The investigation also encompasses actions taken by local governmental and non-governmental organizations, which proposed measures to meet the social, religious, educational, and health demands of the Warao. The field research took place at Cáritas Arquidiocesana of Porto Velho, which serves the aforementioned public through solidarity and emergency actions. The geographic category analysis was based on the concept of territory and its territorial aspects, of territoriality, multiterritoriality and decoloniality; as well as the territorial markers. This is a bibliographical and empirical research, using data from theses, dissertations and articles published in the scientific and academic field, from the application of a semi-structured questionnaire, necessary for the collective construction of the study, with a qualitative nature and phenomenological analysis, in order to punctuate the human events as a result of migratory manifestations.

Keywords Public policy · Environmental insecurity · Growing migration/immigration · Venezuelan indigenous people

Territorial racism: the urban context of the Warao indigenous people in transit through the Brazilian Amazon.

L. A. Nascimento dos Santos (✉) · A. de Almeida Silva
Federal University of Rondônia, UNIR, Porto Velho, RO, Brazil
e-mail: laissecristo@gmail.com

A. de Almeida Silva
e-mail: adnilson@unir.br

W. Leal Filho et al. (eds.), *Amazon 2030—Sustainability Issues in the World's Largest Rainforest Region*, World Sustainability Series,
https://doi.org/10.1007/978-3-031-81465-5_12

1 Introduction

Discussions about the growing immigration within the Brazilian Amazon, driven by several factors, make the characteristics arising from a strong population contingency fit into a context with different social and territorial policies, becoming increasingly necessary. Another essential context discussed in this study is migration within Brazil, where mobility has been on the rise every day, in addition to the public policies intended for the studied population.

In this sense, we propose to address the issue of territorial racism (Haesbaert 2007), in the context of territoriality (Almeida Silva 2010), since immigrants/migrants require, upon arrival in a country, to be welcomed from their most basic needs, such as decent housing, access to health and education, food, water, medicine, clothing, among other necessities. The investigation into the quality of life of these individuals is of interest to us, from a geographic perspective related to the services provided to them, which took place at Cáritas Arquidiocesana of Porto Velho.

Law Number 13.445 of May 24, 2017, grants immigrants rights equivalent to those of native citizens when they establish themselves in the country, either through a refugee request or as migrants.

It deals with the reception of these individuals in an international environment such as Brazil, which, through state and municipal organizations, adopts measures of protection and emergency aid, such as reception projects and financial benefits, or through non-governmental organizations, such as Cáritas Arquidiocesana of Porto Velho, which through the Bem Viver Warao Project, enabled the empirical research of this study.

The Bem Viver Warao Project by Cáritas, in the year 2022, had as its general objective to welcome indigenous Venezuelan immigrants in social vulnerability, has an individualized and safe physical space with food, and promotes self-sustainability for socioeconomic integration. Among the specific objectives aimed at meeting this demand from native peoples, shelters were offered in a care system as a palliative and transitory solution.

Racial traits are used to justify the practice of racism because they are considered "deviating from the norm" and "inferior" from the outset and it is an issue that must be addressed urgently by all social layers. Another significant feature among this population is the practice of their culture in an urban context, where women have organized themselves in front of markets and at traffic signals to collect financial assistance requests, to meet their main needs, and to achieve mobility to other urban centers within the country.

Territorialities are the basis of belonging, a sacred tree considered the source of life, a river that provides food, and the unique vegetation that characterizes the locality in which they are inserted. These elements are found in the region and make the Warao population feel closer to their culture and accumulate experiences from their ancestors. According to Almeida Silva (2010), these characteristics of nature are the very life of this population.

The Warao, with diversity in countless characteristics that make them more susceptible to experiences that are not part of their culture, have lived for many decades far from their territory, driven by the changes that have occurred in their lands, in the national politics of their country (Venezuela) and mainly in the economy, which accelerated the exodus to neighboring Venezuelan cities and later to neighboring countries, with the intention of destination and at times arriving without planning.

2 Cultural and Millennial Path

Figure 1 shows the path taken by indigenous Venezuelans, from Venezuela to the arrival in Brazilian territory. The individuals surveyed in this study reported that they came by bus, via Pacaraima-RR (which also forms the border between the countries), and within the national territory, they traveled through the states of Roraima, Amazonas, until they reached Rondônia and settled in the capital Porto Velho.

The image in the right upper corner shows the Cathedral of the Divina Pastora, also known as Tucupita Cathedral, located in the Delta Amacuro state in the capital Tucupita and in the lower left corner there is an image representing a Warao family served by the Bem Viver Warao Project of Cáritas Arquidiocesana of Porto Velho.

Fig. 1 Mobility Venezuela/Brazil. *Source* SANTOS, Laisse. SANTANA, Ravele. July, 2023

The Warao are the second largest Venezuelan indigenous population, originating from the state of the Orinoco Delta,[1] in Northeast Venezuela. They began to migrate to urban locations within the country from the moment agricultural expansion was installed in their natural territories, as an initiative of the Venezuelan government in the 1950s, which modified the water course of the Mánamo River, considered their main source of raising resources for subsistence through the creation of a dike. This project influenced the adaptations adopted as resistance and new family organizations were formed, based on the needs arising in the urbanized area (Ventura 2020).

Also according to Ventura (2020), all the changes that occurred in the Warao indigenous territory, such as water salinization, acidity and permanent flooding of the soil, the latter of which led to the construction of new houses on stilt, built on piles driven into the marshy soil, generated territorial insecurity and the exodus of the indigenous population to the city. They began to engage in a system of barter through informal work, such as unloading products from trucks or even collecting donations at traffic lights, in front of pharmacies and supermarkets, so that it constituted a new economic activity.

However, what we can see in the Warao's speech, according to Rosa (2021, p. 263) is that these informal jobs generate family insecurity and staying at the signs is an option that can be replaced by crafts, a much more acceptable way, and in the speech of one of the Warao women who lived in Belém—PA, there is dissatisfaction with living of donations, the latter of which can be exchanged for making handicrafts. "Soy una mujer trabajadora, no quiero comida, quiero miçanga [sic] para hacer pulseras. No quiero salir todos los días a pedir dinero, quiero hacer artesanía y sostener a mi familia".[2]

Another economic activity revolves around the buriti fruit, Mauritia flexuosa, sacred to the Warao people, which requires permission from the spirits before it can be taken from nature and utilized, and also various parts of the tree, such as the pulp used for making flour and performing rituals, as well as palm fronds used for crafting, contribute to the collection and income from this work, as a means of raising funds to purchase tickets for future migratory journeys within Brazil (Ventura 2020).

For the ancestral/traditional population, the forest is a place represented by symbolic forms and subjectivity; Every culture is characterized by the representations that govern it, designated by its symbols, history, space in which it is inserted, and its territoriality, to which actions are attributed and which qualifies objects and values, coming from its ancestors through oral tradition and social organizations, in addition to the symbolic function being represented by the object (Almeida Silva 2010).

In accordance with Almeida Silva, the anthropologist Molina (2021) emphasizes that land for indigenous peoples is irreplaceable, as it is as valuable as life itself. It serves as a reference for culture and the community, meaning that it encompasses

[1] From the Warao language: wiri (place) and noco (rowing), or "place where one rows".

[2] This statement by Dolores Warao is found in the thesis Warao mobility in Brazil and the ways of managing a population in transit: reflections from the experiences of Manaus-AM and Belém-PA/ Marlise Rosa. 1st ed. Rio de Janeiro: E-papers, 2021.

their histories and their relationship with nature and addresses the continuity of generations, thereby postulating that there is no possibility of exchanging an indigenous area for any other that does not possess the deep-rooted characteristics integral to the entire construction of values and accomplishments measured by their knowledge and values.

3 National Mobility

Figure 2 depicts the municipality of Porto Velho, the capital of Rondônia, which is located in the Northern Region of the country, and is home to 460,413 inhabitants, according to the 2022 Demographic Census by the Brazilian Institute of Geography and Statistics (IBGE). In this city, indigenous and non-indigenous Venezuelan immigrants arrive daily in search of personal and family improvements. They generally seek assistance from Cáritas Arquidiocesana and receive the initial guidance to organize themselves within the desired expectations. In order to do so, they undergo a short interview about how they are at the moment and how they got to that service.

According to Ventura (2020), the first Warao individuals arrived in Brazil in 2014. Over the years, the number of refugees and migrants has significantly increased, with approximately 6,000 new indigenous people seeking opportunities in neighboring South American countries, of which Brazil is a part.

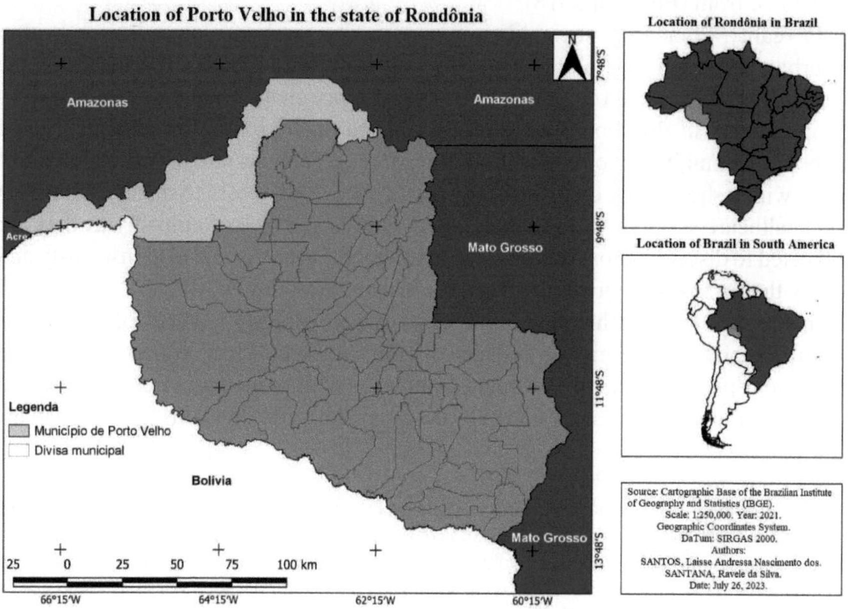

Fig. 2 Municipality of Porto Velho—RO. *Source* SANTOS, Laisse. SANTANA, Ravele. July, 2023

According to Figueira (2018), three factors illustrated the dissatisfaction of Brazilians with the increasing number of Venezuelan immigrants in Pacaraima in 2018. In some cities, violent and intolerant actions were witnessed, in disagreement with their presence in urban centers and were, in some ways, characterized by xenophobia and contempt.

> (1) racism, as a structural component of social relations in Brazil, but also—and especially for the case I analyze here—as a traditionally guiding element of the country's immigration policies, and even as an organizing parameter of a hierarchy of nationalities; (2) the recovery of the idealized image of the nation-state and its association with the idea of property; (3) populism as a discursive trigger that activates the first two factors through the circulation of fears and resentments as mobilizing emotions in the attention and aspirations of the state's population [Our translation].[3]

With the daily increasing influx of Venezuelan immigrants crossing the border and entering Brazil through Pacaraima, some Brazilians have felt threatened in their competition for formal employment and access to public policies (such as education, health, and housing). Racism manifests itself in various contexts and is underpinned by the same approach, whether related to race, gender, privilege, or language. What truly prevails is structural racism, ingrained within many Brazilians as something cultural.

The foreigner, the immigrant, and the refugee would all represent, in this sense, a threat to the territorial domain to which the power of the native community would extend; the relationship between the native citizens and the immigrants, even those who migrate in search of refuge to protect themselves from the dangers of the place they come from (Figueira 2018).

In reality, what may seem like an opportunity for better organization within an urban context that is completely different from the cultural experiences in the Venezuelan Amazon, is a centralization of subsistence characterized by rented houses or apartments in dire conditions, (dis)organized shelters accommodating a large number of immigrants, or even life on the streets. They survive in unstable conditions, with a shortage of clean drinking water, food, and access to quality education and healthcare services. Additionally, they often face a lack of empathy as they are subjected to discriminatory actions and xenophobia, which can lead to insecurity and justify their transitory movement between spaces (Barbosa 2020).

In Porto Velho, the housing configuration also exhibits features of precarious conditions. Despite receiving assistance from the Bem Viver Warao Project, the indigenous people reside in apartment complexes with limited air circulation and no natural light, with an entrance and exit (the door), no room divisions (just a space for

[3] From the original: "o racismo, como um componente estrutural das relações sociais no Brasil, mais também – e sobretudo para o caso que aqui analiso—como um elemento tradicionalmente orientador das políticas de imigração do país, e, inclusive, como parâmetro organizador de uma hierarquia de nacionalidades; (2) recuperação da imagem idealizada do Estado-nação e sua associação com a ideia de propriedade; (3) o populismo como gatilho discursivo que aciona os dois primeiros fatores por meio da circulação de medos e ressentimentos como afetos mobilizadores da atenção e anseios da população do Estado."

sleeping, eating, and studying), lacking privacy, and being subject to vulnerability, that is, they have already settled in substandard housing conditions.

Since their arrival in Brazil, the Warao families have already encountered instability and territorial insecurity, where they often found themselves homeless and had to share the streets with drug users and the homeless Brazilian population, and at times they had their belongings subtracted living in Mercado Ver-o-peso in Belém-PA, several conflicts were the result of this housing interference (Rosa 2021).

4 Results

The research was conducted with five Warao indigenous individuals who responded to a questionnaire regarding their immigration journey leading up to their arrival in Brazil. The guiding questions were presented in both Portuguese and Spanish, they were open-ended questions, and required assistance to be answered. For this investigation, we had three male and two female indigenous participants.

Graph 1 portrays the time each Warao indigenous person has lived in Brazil. Bk has been in the country for 4 years. Among the five employees interviewed, he has been in the Brazilian urban context for the longest time.

In Graph 2, except for R., who was not treated well in Pacaraima, upon her arrival in Brazil, the participants reported being treated well by Brazilians, and one of those surveyed considers them to be good.

Graph 3 illustrates the intention to immigrate and the possibility of Brazil being the intended destination. Bk reported that her family and she were only passing

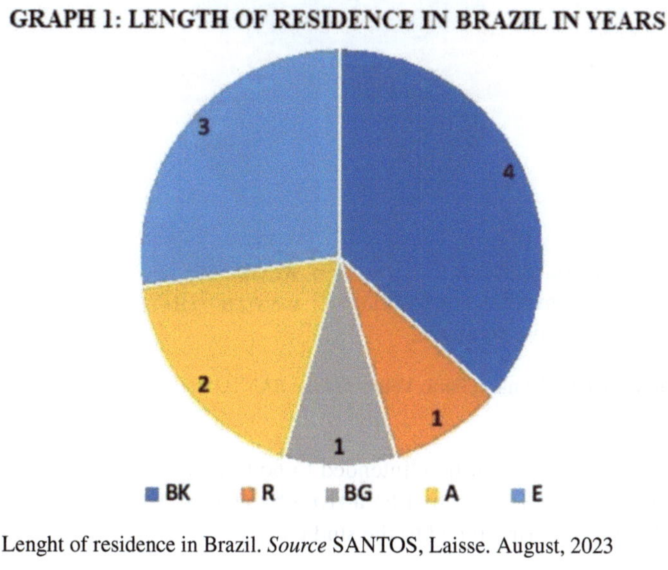

Graph 1 Lenght of residence in Brazil. *Source* SANTOS, Laisse. August, 2023

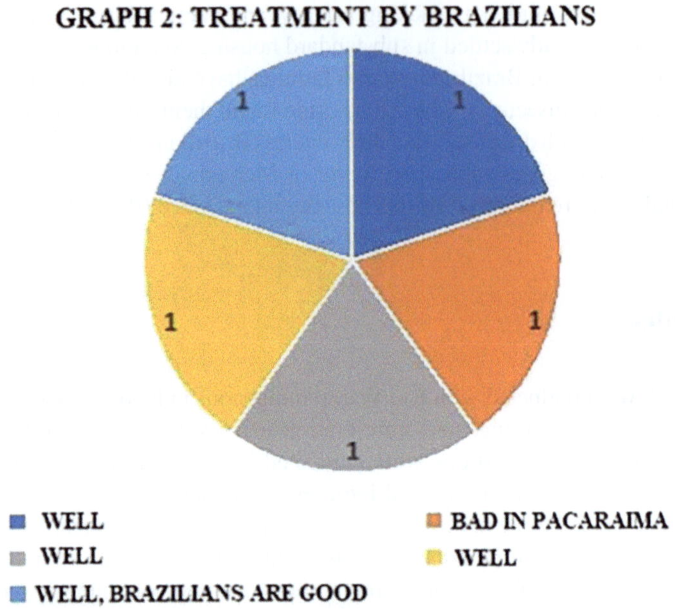

Graph 2 Treatment by Brazilians. *Source* SANTOS, Laisse. August, 2023

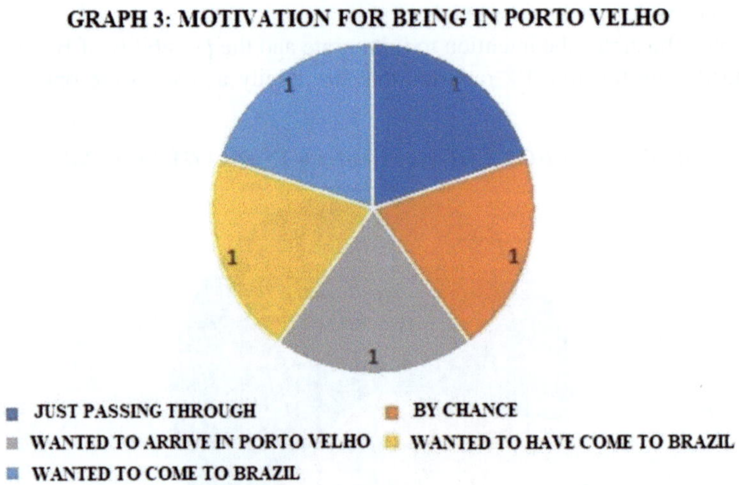

Graph 3 Motivation for being in Porto Velho. *Source* SANTOS, Laisse. August, 2023

through Porto Velho and that they intended to go to another place in Brazil to join other family members. R. said that he arrived by chance, while the others had the intention of coming to Brazil and to the studied city.

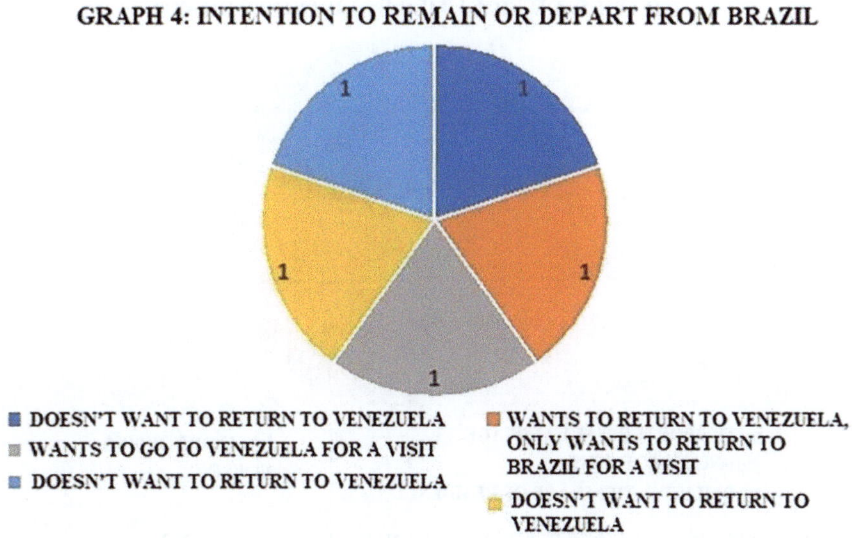

Graph 4 Intention to remain or depart from Brazil. *Source* SANTOS, Laisse. August, 2023

When questioned about whether they would stay in Brazil or return to Venezuela, four participants stated that they had no intention of returning to their home country, while one of them would consider returning for a visit. Returning to Brazil for a visit was the ideal option for the fifth participant in the study, as shown in Graph 4.

Based on the data presented in the study, it is understood that one of the issues that permeates the lives of these people is territory, which has the meaning expressed by Haesbaert (2007):

> (...) every territory is, simultaneously and necessarily, functional and symbolic, as power relations have an inseparable component in space both in the performance of "functions" and in the production of "meanings." The territory is "functional," beginning with its role as a resource, from its relationship with what are known as 'natural resources" [Our translation].[4]

The territorial issue is seen subjectively by the Warao, and family groups, as immigrants, have the objective of settling where they feel best, which justifies the constant national migrations and continuous mobility.

Graph 5 illustrates the difficulties faced by those who participated in the survey. Since communication is fundamental for interaction with others, participant 1 finds the Portuguese language difficult to learn; for participants 2 and 4, lacking a home poses an obstacle to moving forward on other issues, while participant 3 faces the challenge of unemployment, and beyond the complexities of urban life, there is a sentiment of not liking the place of residence.

---

[4] From the original: "(...) todo território é, ao mesmo tempo e obrigatoriamente, funcional e simbólico, pois as relações de poder têm no espaço um componente indissociável tanto na realização de "funções" quanto na produção de "significados". O território é "funcional" a começar pelo papel enquanto recurso, desde sua relação com os chamados "recursos naturais".

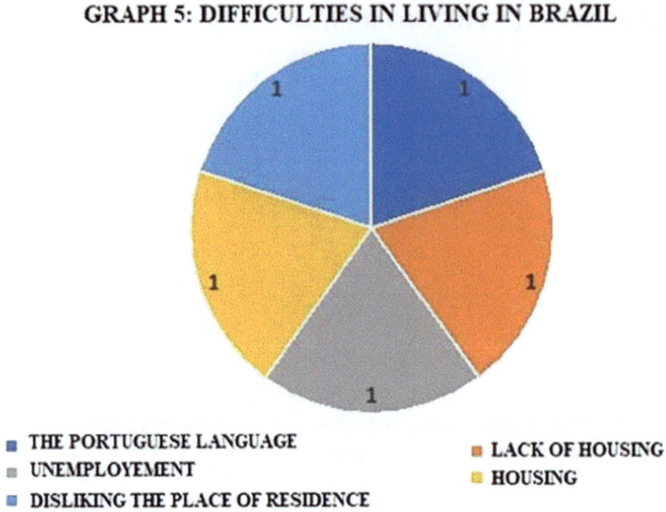

Graph 5 Difficulties in living in Brazil. *Source* SANTOS, Laisse. August, 2023

As the portuguese language is very important for communication between nationals and non- nationals, this was highlighted as a difficulty for one respondent, as it is "difficult to learn" and in Porto Velho, communication in spanish flowed in a way that was understandable by both the parts. In Belém-PA the health proposal was to offer immunization, prevention and guidance; and education was to propose an intercultural and bilingual interaction, where the Warao themselves would be responsible for teaching classes with the aim of maintaining culture, in addition to creating pamphlets guiding health-related issues (Belik 2021).

Graph 6: showed that among the five participants, only respondent 4 felt that he had experienced prejudice and discrimination in Brazil, where at times he believed to have experienced such discomfort. In contrast, respondents 1, 2, 3, and 5 reported that they had never faced any discriminatory situations up to the date of the research.

In Graph 7, respondents were asked if they would recommend Brazil as a place for any relatives, family members, or friends to settle either permanently or temporarily. Only participant BG would not extend an invitation due to the difficulty of interacting with others caused by language barriers when learning the Portuguese language.

As can be observed, the difficulties faced by immigrants, particularly the Warao indigenous people, are immense. Their way of life encompasses collectivity and a culture vastly distinct from that of the Brazilians. They have to overcome numerous obstacles in their daily lives within a country that holds values significantly different from those of their territory of origin. It falls upon Brazilians to comprehend these disparities and foster cultural closeness, particularly considering the social vulnerability experienced by these immigrants.

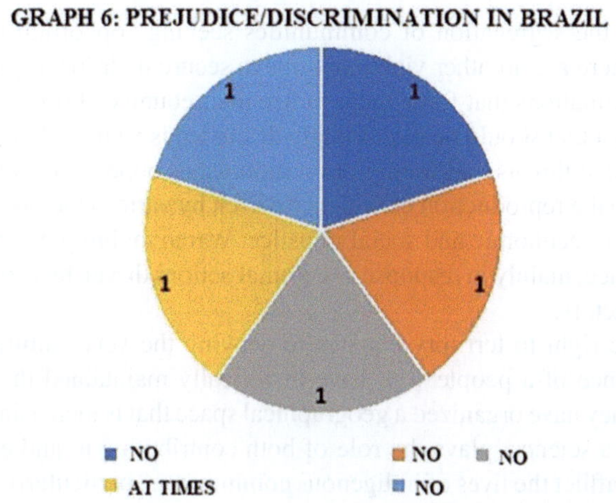

Graph 6 Prejudice/discrimination in Brazil. *Source* SANTOS, Laisse. August, 2023

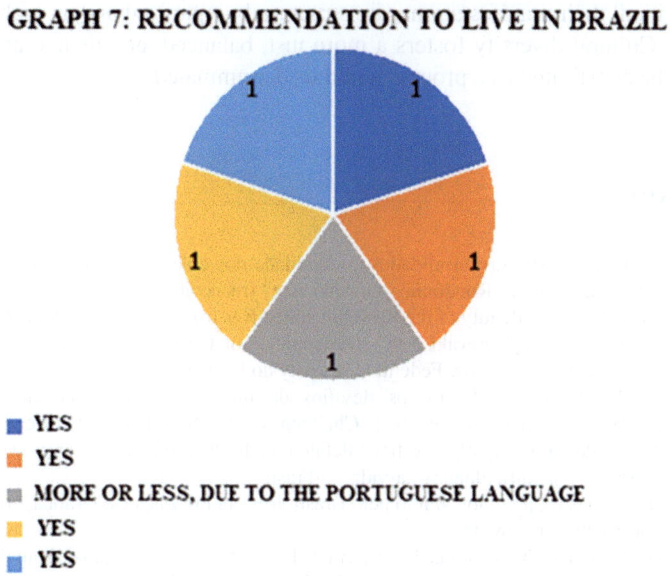

Graph 7 Recommendation to live in Brazil. *Source* SANTOS, Laisse. August, 2023

5 Conclusions

The aim of study on territorial racism, we sought to discuss the migratory practices that have affected indigenous populations in the Amazon, particularly the Warao, whose territory underwent changes due to the construction of a dike, which was compounded by the political and economic crisis that overshadowed their lives. It is evident that the segregation of communities seeking "opportunities" has taken place, and as there are no other viable options to secure their local spaces, they find themselves in situations that force them to flee their country of origin.

Another point that should be increasingly discussed is xenophobia, a practice that comes with veiled threats that immigration diminishes opportunities for native citizens, leading to the reproduction of violence driven by various factors and obscuring the true causes of economic and social injustice. Warao mobility exhibits characteristics of resilience, mainly in response to colonial actions driven by economic, social, and political factors.

Denying the right to territory equates to denying the very cultural, social, and spiritual existence of a people who have historically maintained their lives in the forest, where they have organized a geographical space that is increasingly restricted. Geography, as a science, plays the role of both contributing to and exposing these intrusions that afflict the lives of indigenous communities, particularly those who are forced to leave their territories and relocate to places devoid of any connection to their heritage.

With that said, it serves as a way to emphasize that racism, xenophobia, and prejudice, in all their manifestations, are structurally organized and as such must be countered. Cultural diversity fosters a more just, balanced, and democratic world. This must be constituted as a process and thus disseminated.

References

Almeida Silva A de (2010) Territorialidades e identidade dos coletivos Kawahib da Terra Indígena Uru-Eu-Wau-Wau em Rondônia: "Orevaki Are" (reencontro) dos "marcadores territoriais" [Territorialities and identity of the Kawahib collectives from the Uru-Eu-Wau-Wau Indigenous Land in Rondônia: "Orevaki Are" (reencounter) of the territorial markers]. (Doctoral dissertation, Ph.D. in Geography, Federal University do Paraná)

Barbosa C (2020) Indígenas Warao: os desafios da migração e as dificuldades da vida no Brasil [Warao Indigenous People: Challenges of Migration and Life Difficulties in Brazil]. Brasildefato. https://www.brasildefato.com.br/2020/08/25/indigenas-warao-os-desafios-da-migracao-e-as-dificuldades-da-vida-no-brasil

Belik D (2021) Grupos e Caminhos Warao pelo Brasil. Revista Presença Geográfica, vol 08, núm. 02, Esp.; março-abril; pp 28–46

Brasil (2016) Constituição da República Federativa do Brasil: texto constitucional promulgado em 5 de outubro de 1988, com as alterações determinadas pelas Emendas Constitucionais de Revisão nº 1 a 6/94, pelas Emendas Constitucionais nº 1/92 a 91/2016 e pelo Decreto Legislativo nº 186/2008 [Constitution of the Federative Republic of Brazil: Constitutional text promulgated on October 5, 1988, with amendments determined by Constitutional Amendments No. 1 to 6/94, Constitutional Amendments No. 1/92 to 91/2016, and Legislative Decree No. 186/2008]. Senado Federal, Coordenação de Edições Técnicas, Brasília

Brasildefato. https://www.brasildefato.com.br/2021/08/25/o-que-e-o-marco-temporal-e-como-ele-atinge-os-indigenas-do-brasil

Figueira RR (2018) Razões da xenofobia: ensaio sobre os fatores contribuintes da violência xenófoba contra imigrantes e refugiados venezuelanos em Roraima [Reasons for Xenophobia: An Essay on Contributing Factors to Xenophobic Violence Against Venezuelan Immigrants and Refugees in Roraima]. In: Baeninger R, Jarochinski Silva JC (eds) Migrações venezuelanas , pp 1–400.

Núcleo de Estudos de População "Elza Berquó"–Nepo/Unicamp. https://www.nepo.unicamp.br/publicacoes/livros/mig_venezuelanas/migracoes_venezuelanas.pdf

Haesbaert R (2007) Território e multiterritorialidade: um debate [Territory and multiterritoriality: A debate]. Revista Geographia, 9(17):19–45. https://periodicos.uff.br/geographia/article/view/13531/8731

IBGE (2023) Cidades e Estados [Cities and States]. https://www.ibge.gov.br/cidades-e-estados/ro/porto-velho.html

IHU (2020) Usininos. O êxodo do povo Warao para o Brasil [The Exodus of the Warao People to Brazil]. https://www.ihu.unisinos.br/categorias/598788-o-exodo-do-povo-indigena-venezuelano-warao-para-o-brasil-edicao

Molina L (2021) Para entender o marco temporal [To Understand the Temporal Framework] [Video]. YouTube. https://youtu.be/RamN8J-Yg1s

MPMS (2019) Ministério Público do Mato Grosso do Sul. O povo indígena Warao—um caso de imigração para o Brasil [The Warao Indigenous People—A Case of Immigration to Brazil]. https://www.mpms.mp.br/noticias/2019/01/o-povo-indigena-warao-um-caso-de-imigracao-para-o-brasil

Rosa M (2021) A mobilidade Warao no Brasil e os modos de gestão de uma população em trânsito: reflexões a partir das experiências de Manaus-AM e de Belém-PA/Marlise Rosa. 1 ed. Epapers, Rio de Janeiro

Veiga E (2021) Entenda o marco temporal e saiba como ele atinge os povos indígenas do Brasil [Understanding the Temporal Framework and how it affects indigenous Peoples in Brazil]

Ventura L (2020) Migração e direitos coletivos: a presença dos Warao no Brasil [Migration and Collective Rights: The Presence of the Warao in Brazil]. Relatório: Violência contra os povos indígenas no Brasil—Dados de 2018 [Report: Violence Against Indigenous Peoples in Brazil – 2018 Data]. Conselho Indígena Missionário—CIMI

Natural Arte—Sustainable Use of Wood

Juliana Mendes de Oliveira, Angelo Ricardo Sousa Chaves, Rommel Noce, Aline Ariel Silva Cardoso, and Iris Mayara Leão Mota

Abstract The Amazon Rainforest has been under great pressure from the expansion of Brazilian agricultural frontiers and mining, especially in recent years. However, national and international pressure for forest preservation, climate change, the valorization of traditional and indigenous populations and consumer awareness have been pressuring the market for the sustainable use of natural resources. This work had as its objective the study of the case of Natural Arte Atelier, a company conceived by Environmental Engineer Angelo Chaves, specialized in the production of wooden pieces and furniture from the Amazon Forest.

Keywords Sustainability · Wooden furniture · Social responsibility

1 Introduction

The Brazilian Amazon, in more than 500 years of post-discovery history, has as a result of its endogenous models of economic exploitation a predominant scenario of poverty and inequality. In summary: progress without development (Melo 2015). In

J. M. de Oliveira (✉) · R. Noce · A. A. S. Cardoso · I. M. L. Mota
Biodiversity and Forests Institute, Universidade Federal do Oeste do Pará, Rua Vera Paz, S/N, Santarém, PA, Brazil
e-mail: juliana.oliveira@ufopa.edu.br

R. Noce
e-mail: rommel.noce@ufopa.edu.br

A. A. S. Cardoso
e-mail: aline.cardoso@discente.ufopa.edu.br

I. M. L. Mota
e-mail: iris.mota@discente.ufopa.edu.br

A. R. S. Chaves
Natural Arte, Santarém, PA, Brazil

W. Leal Filho et al. (eds.), *Amazon 2030—Sustainability Issues in the World's Largest Rainforest Region*, World Sustainability Series,
https://doi.org/10.1007/978-3-031-81465-5_13

addition to economic development below expectations within the concept of sustainable development, the Western Region of the State of Pará has to deal with another problem: the advance of soybean monoculture. This monoculture, on the past few years, has been expelling small local farmers from their lands and putting pressure on the indigenous population of the region (Colón 2021).

The agribusiness-monoculture complex, and its relations to the problems and peculiarities of global meat production is factual (Kröger 2022). Weis (2013), argues that we are currently living in a period where the scale and pace of destruction has reached new proportions, as the incessant pressure to compete, grow, and accumulate under capitalism has pushed the expansion and simplification of farming landscapes to radical new extremes.

The monoculture of soybeans, in addition to disrespect premises of sustainable development, especially regarding deforestation to open new arable areas and the use of pesticides, also escapes the premise of endogenous development, because it does not allow local actors to participate in the development process and to enjoy the income generated (Araújo 2014).

The concept of sustainable development emerged during the Brundtland Commission, in the 1980s, where the *Our Common Future Report* was prepared, when the Norwegian Prime Minister, Gro Harlem Brundtland, presented the following definition for the concept: "It is the way in which current generations meet their needs without, however, compromising the ability of future generations to meet their own needs" (Brundtland apud Scharf 2004).

Just a decade after the Commission met, it was seen that addressing environmental issues alone—which were the ones that most afflicted society—would not solve the problems of a sustainable global economy. To achieve sustainability, it would be necessary to achieve other means as well. Sustainability could not be thought only as a matter of pollution control, but rather as an environmental, economic and social issue (Elkington 2001). The goal of sustainable development is the preservation of global wealth, which refers both to financial assets and natural resources, but also to the life quality of the population (Scharf 2004).

A new balance concept arises, the Triple Bottom Line (TBL), which is to simultaneously achieve economic prosperity, with a clean environment and with social responsibility (Othman et al. 2010). The concept works with three pillars: economic, social and environmental.

In the region studied, West of Pará, in the vicinity of the city of Santarém, the low urban infrastructure, precarious electrical network and the lack of industrial activities makes the job opportunities scarce, even more in the context of sustainable development. The lack of opportunities also causes the pressure on traditional communities and small rural producers for the expansion of agricultural monoculture.

2 Purpose of the Paper

The present work aimed to present the case study of Natural Arte Atelier, in order to publicize the experience of a company that seeks to work with the concept of sustainable enterprise in the Amazon Region.

3 Approach

3.1 Place of Study

The study was made on the company Natural Arte Atelier, located in the community of Bom Sossego, municipality of Belterra, in the western region of Pará. The company has as bordering areas the Tapajós National Forest (Flona Tapajós), small traditional communities and large soybean plantations. It is important to note that Flona Tapajós is a Federal Nature Conservation Unit in Brazil. It was created in 1974 and has an area of 527,319 hectares, covering the municipalities of Aveiro, Belterra, Placas and Rurópolis (ICMBio 2023).

3.2 Case Study Analysis

Interviews were conducted with the owner and site visits, in order to subsidize the development of the work. The work made in partnership with the Federal University of Western Pará (UFOPA) also counted with visits of the Forest Engineering Course, during the discipline of Design of Forest Products, to show the reality of the entrepreneur's experience in the Amazon.

4 Results Achieved

Conceived by Environmental Engineer Angelo Ricardo Sousa Chaves, the company, created in 2020, in the midst of the crisis caused by the Covid 19 pandemic, was born with the proposal of working with the pillars of sustainable development. Chaves is native of the community of Surucuá, in the Tapajós-Arapiuns Extractive Reserve, created in 1998, within the Legal Amazon, also in the western region of Pará.

At the age of 13, he was already helping his parents in the extraction of latex in the Extractive Reserve. At 17, he left his community to study in Manaus, capital of Amazonas, where he studied the Technical Course in Agriculture, Management in

Agroforestry and later in Environmental Engineering. He had professional experiences in forest production of seedlings, recovery of degraded areas and community actions with traditional people.

In 2014 Chaves returns to the state of Pará, Santarém, where he begins to work with traditional communities and socio-productive projects using wood. A great connoisseur of local communities and the Amazon Forest, Chaves always saw great potential in the use of forest residues and what the forest could offer as raw material, that would not result on its extinction (like seeds, vine, chestnuts), generating income for the local communities.

In 2020 he decided to create Natural Arte Atelier, a company dedicated to the production of furniture and decoration elements with materials from the forest, exclusively with woods from forest management residue, woods that fell naturally and non-wood forest products.

The *Reduced Impact Forest Management*, implemented in Flona Tapajós and Resex Tapajós-Arapiuns, generates large residues, which are left in the forest for natural decomposition. These residues, mainly of branches and roots don´t compose income for the forest communities. Its use is an income alternative proposed by Natural Arte (Figs. 1 and 2).

It is not uncommon for large trees to fall naturally. Due to the heavy rains, common in the Amazon region, it is constant and part of the natural process of the forest dynamics, the toppling of trees. These trees, due to their high density and wood quality, can take hundreds of years to decompose naturally, hindering the natural regeneration of the area affected by the fall (Fig. 3). Even though it's a natural dynamic, in 1989, Colinvaux (1989) was already concerned about the past and the future of Amazon, pointing that the climatic history of the Amazon rain forest indicates that the ecosystem was well adapted to certain natural disturbances, but it wouldn't have the resilience to tolerate human exploitation. The climate change is something that now is important to concern too.

Fig. 1 Forest residue of *Piquiá* wood (*Caryocar villosum*) being sawed in the Natural Arte Atelier

Fig. 2 Tree branch residue. Often the residues stay on the "*forest floor*" for several years. It was estimate that this residue has remained in the forest for 6 years

Fig. 3 Fallen tree found in the forest. The heavy rains in the Amazon region cause many trees to fall in a natural process of the Amazon forest dynamics

In the social pillar, the company hires and trains people from the local community. Being in a rural area, 40 km away from the city of Santarém, the use of community labor was the initial premise for the creation of the company. With the view that communities can and should be included in production processes, Chaves works with his neighbors who are members of the community itself. Another important factor was the hiring of the same number of women and men. Gender equality is effective both in number of workers, values of wages and functions. All workers in the production sector are trained to perform all functions, from the initial sawing, to finishing.

In 2021, in Brazil, the wage gap between women and men performing the same function increased from 20.7 to 22% (IBGE 2021). Reports from Brazilian and international institutions make the problem very evident: women are paid less than men to perform the same function. When it comes to wage inequality, Brazil is in an uncomfortable position, in one of the last positions in the international ranking of equal pay (Global Gender Gap Report 2020). The Brazilian business community doesn't seem to combat this problem, as it should.

The Atelier was also idealized with the intention of creating an *Amazonian Design*, where the natural elements are protagonists, highlighting the shapes and colors of the natural elements of the region. Instinctively, the company is also working on the concept of ecodesign. Although widely used in academia, it still raises doubts about its meaning; its comprehension generates numerous interpretations. Even though the word is self-explanatory (design = design and eco = ecological criteria), the term synthesizes a large set of design activities that tend to address the themes posed by the environmental issue. The extent of the environmental problem and the transversally of the themes involved in the act of designing, from the choice of raw material, production method, packaging and transportation, makes the term ecodesign to have a very comprehensive use (Manzini and Vezzoli 2016).

Methodologies are fundamental in all design process, tracing guidelines for products development and being characterized by studies of guided principles and procedures. The Designer, over time, guarantee a fundamental part in the process of products creation (Platcheck 2003).

One of the difficulties to work with these concepts is the use of wooden pieces. As a natural material, with the use of several wood species, and without conformity of forms, one of the biggest challenges is to see the final piece. Figures 4, 5 and 6 shows the same piece, from the beginning of its creation to the final product. Visualizing the shape that the piece will have, respecting the natural forms of the tree is also a form of art. Imagining how the piece shown in Fig. 1 will turn into the finished piece in Fig. 6 requires not only talent, but also a lot of training and woodworking skills.

The creations of the Atelier aim to highlight the natural beauty of the Amazonian woods as the chairs shown in Figs. 7 and 8, where the protagonist is the unique figure of the Pequiá wood.

The *piquiá* wood is one of the most used in the Atelier for presenting unique characteristics. The *piquiá*, *Caryocar villosum*, of the Caryocaraceae family, is gaining market share and it is use in the production of furniture. Featuring distinctive design due to its unique features and figure.

The *piquiá* occurs mainly in the Amazon mainland forests, but it is also found in localities of the Northeast region and even in Mato Grosso. It can reach from 20 to 40 m in height and from 0.9 to 1.8 m in diameter. Characterized by yellowish coloration, reverse grain and tangential figure of attenuated fibrous aspect, caused by the prominence of the axial parenchyma, vascular lines and fibers (Brazilian Forest Service 2023).

Fig. 4 Carved piece removed from the residue shown in Fig. 1

Fig. 5 Bowl with final finish

Because it is a large tree and has a variety of figures (wood drawings), the *piquiá* is widely used. Every part of this tree can be used, from the heartwood (the most valued and with the best quality), the sapwood, the branches and even the roots. Figure 9 shows a coffee table made with wood from the roots of the *piquiá*.

The forks of the *piquiá*'s roots reach such large sizes, that during the visit to the Natural Arte Atelier, by the students of the Forest Engineering program of the Federal University of the West of Pará, sixteen people climbed on a root that was being worked, without it even swing the piece (Fig. 10). This root, specifically, was found in a rural area and they had already tried to burn it. It has several marks of carbonization. They just couldn't burn it because of its large dimensions and the humidity of the piece.

It is important to emphasize that the entrepreneur is always available to pass on his knowledge to the new generations and to the traditional communities. Showing cutting techniques, sanding and presenting natural products for finishing (Fig. 11).

Fig. 6 Finished piece in the hands of its creator, Angelo Chaves

Fig. 7 *Piquiá* chair. Wood figure highlighting

Fig. 8 *Piquiá* chair

Fig. 9 Coffee table made with the roots of a *piquiá*

The property where the Natural Arte Atelier was implanted also has a small area of native forest, where several Amazonian forest species can be found. In order to contribute to the preservation of nature, visitors to the Atelier are invited to enter this forest and get to know the Amazon biome up close (Figs. 12 and 13).

Chaves report the pressure that locals have been under in recent years, and his effort to strengthen local communities for the preservation of the forest. A few days

Fig. 10 Students of the forest engineering program of the federal University of Western Pará (UFOPA) visiting the natural arte atelier

Fig. 11 Practical class of the discipline design of forest products. Chaves showed the production and demonstrated how manufacture the pieces

before the visit made by the University, an agricultural tractor had invaded his land, felling a 2 m strip of forest, on the border with the neighbor, a soybean planter. A history that repeats itself in the Amazon region.

It is with great regret that we inform that in November 2023 (after the "Amazon 2030—Sustainability Issues in the World´s Largest Rainforest Region"), the small area of native forest preserved by Natural Arte Atelier caught fire, and was totally destroyed. Due to the great drought that the Amazon is going through this year,

Fig. 12 Visitors are invited to get to know the company's small private forest area

Fig. 13 Chaves showing the existing forest species

enhanced by the El Ninho phenomenon (Lopes 2023), and the fires in the region to "clear" land for agriculture, the traditional communities have been increasingly losing the forest for the large producers of monocultures.

5 Conclusions

Traditional Amazonian communities urgently need viable alternatives to generate income, without the felling of the forest. The dissemination of the "know how" of companies that are working within the precepts of sustainable development is of paramount importance for the communities of the forest. The Amazon Rainforest will only be maintained if the people who live here can make a living from it without having to destroy it or be expelled by big farmers. The strengthening of enterprises that aims the sustainable development of the Amazon is vital for the preservation of the biome.

All images presented belong to the authors.

References

Araújo RC (2014) Análise sobre a monocultura de soja e o desenvolvimento sustentável na Amazônia com base na teoria do desenvovimento endógeno. Revista Economia e Desenvolvimento 26(1):105–118

Colón M (2021) The environmental violence of soy cultivation in the Brazilian Amazon. J Latin Am Geogr 20(3):186–194.https://doi.org/10.1353/lag.2021.0055

Elkington J (2001) Canibais com Garfo e Faca. Makron Books, São Paulo

ICMBio (2023) Instituto Chico Mendes de Conservação da Biodiversidade. Floresta Nacional do Tapajós. https://www.icmbio.gov.br/flonatapajos/

IBGE (2021) Síntese de Indicadores Sociais: Uma Análise das Condições de Vida da População Brasileira. https://biblioteca.ibge.gov.br/visualizacao/livros/liv101892.pdf

Kröger M (2022) Extra activisms, existences, and extinctions monoculture plantations and amazon deforestation. Routledge Taylor & Francis Group, New York

Lopes LT (2023) Mudanças climáticas na atualidade. Universidade Federal do Mato Grosso do Sul. https://repositorio.ufms.br/handle/123456789/7403

Manzini E, Vezzoli C (2016) O desenvolvimento de produtos sustentáveis—Os requisitos ambientais dos produtos industriais. Edusp, São Paulo

Melo AF (2015) Dilemas e desafios do desenvolvimento sustentável da Amazônia: O caso brasileiro. Revista Crítica de Ciências Sociais 107. https://doi.org/10.4000/rccs.6025

Othman MR, Repke J, Güntet W, Huang Y (2010) A modular approach to sustainability assessment and decision support in chemical process design. Ind Eng Chem Res 49:7870–7881

Platcheck ER (2003) Metodologia de ecodesign para o desenvolvimento de produtos sustentáveis. Universidade Federal do Rio Grande do Sul. https://lume.ufrgs.br/bitstream/handle/10183/117875/000396871.pdf?sequence=1&isAllowed=y

Scharf R (2004) Manual de Negócios Sustentáveis. São Paulo, Amigos da Terra

Serviço Florestal Brasileiro—SFB (2023) Laboratório de Produtos Florestais—LPF. Brasília. https://lpf.florestal.gov.br/pt-br/?option=com_madeirasbrasileiras&view=especieestudada&especieestudadaid=55

Weis AJ (2013) The ecological hoofprint: the global burden of industrial livestock. Zed Books, London, UK

World Economic Forum (2020) Global gender gap—report. https://www3.weforum.org/docs/WEF_GGGR_2020.pdf

The Amazon that is News on Regional Television News Broadcasts: An Analysis of the 'Amazônia Viva' Operation Based on the Frame Analysis Theory

Jorgelene dos Santos Oliveira

Abstract Sustainability issues in the largest tropical forest region in the world involve issues related to communication. In this text, we seek to understand the Amazon narrative broadcast on regional television news programs in the State of Pará, broadcasted by affiliates of the main Brazilian TV channel, based on the analysis of coverage of a combat operation against deforestation launched by the Government of the State of Pará in August 2020 in light of the framing theory.

Keywords Television journalism · Amazon · Narrative · Regional

1 Introduction

The general objective of the present investigation is to study the image of the Amazon in contemporary narratives through the analysis of news frames of reports (units of analysis) published in regional television news programs in Brazil. The Jornal Liberal 1st edition and Jornal Tapajós 1st edition are part of this scope, broadcast regionally in the cities of Belém and Santarém, respectively, by the broadcaster with the largest audience in Brazil, Globo Television Network.

The study is focused on television reports broadcasted during the second half of 2020. Conflicts (Smetko and Valkenbug 2000) and environmental consequences are the predominant frames. In these reports was found that the focus is on the event itself and its relationship with ecosystem results whose emphasis lies on what it will bring to groups and/or individuals.

This is an investigation anchored in Framing Theory, based on Pragmatic Narrative Analysis (Mota 2007), with an approach that mixes elements of qualitative (Sousa

J. dos Santos Oliveira (✉)
Fernando Pessoa University, Porto, Portugal
e-mail: jorgelene.oliveira@ufopa.edu.br

Federal University of Rio Grande do Sul, Porto Alegre, Brazil

Federal University of West Pará, Santarem, Para, Brazil

W. Leal Filho et al. (eds.), *Amazon 2030—Sustainability Issues in the World's Largest Rainforest Region*, World Sustainability Series,
https://doi.org/10.1007/978-3-031-81465-5_14

2006) and quantitative (Cervi 2017) data collection. It expands the types of frameworks created by authors Semetko and Valkenburg (2000) and creates a fifth type: the environmental consequences framework.

Operation Amazônia Viva launched by the Government of the State of Pará in August 2020 was warranted in order to combat the deforestation rates in the State, one of the highest in Brazil. This operation integrates the Command and Control pillar of the Plano Estadual Amazônia Agora (PEAA). It is noteworthy that this axis is the only one with a prominent definition on the home page: "The design of the Command and Control Component (C&C) focuses on understanding, based on Pará's own experience over the last 25 years, that loss of forests and biodiversity is a problem with complex variables, which only consistent state action and an efficient rational arrangement with Civil Society and the Business Sector can reverse". To achieve this objective, the Command and Control (C&C) axis has carried out field inspections at least once a month since its creation in August 2020, according to information published on the page at the State Department of Environment and Sustainability (SEMA).

These actions are executed with the support of the Military Police, Army, Environmental Police and government agents from different regions of the state, the actions occurred mainly in the Western region of Pará, where the city of Santarém is located, headquarters of TV Tapajós, from where it is generated the JT1 signal, which reaches 11 cities and covers a region of almost 1.5 million people (Table 1).

The two regional television stations, in which these news programs are broadcasted, belong to the Globo Television Network conglomerate in the state of Pará.

Table 1 States that contributed to deforestation

States	PRODES 2020 (Km^2)	Contribution (%)
Acre	706	6.51
Amazonas	1.512	13.93
Amapá	24	0.22
Maranhão	336	3.10
Mato Grosso	1.779	16.39
Pará	4.899	45.15
Rondônia	1.273	11.73
Roraima	297	2.74
Tocantins	25	0.23
AMZ. Legal	**10.851**	**100.00**

Note The table presents deforestation data in the nine states of ALB in 2020. *Source* INPE (2022)

According to data from the National Institute for Space Research (INPE),[1] the consolidated rate of deforestation due to clear cutting, for the nine states of the Brazilian Legal Amazon (ALB), in 2020, it was 10,851 km.[2]

> This rate is calculated annually based on data generated by the Satellite Monitoring Project for Deforestation in the Legal Amazon (PRODES). The mapping, to record and quantify deforested areas larger than 6.25 hectares, was based on images from the Landsat series of satellites or similar. PRODES defines deforestation as the complete removal of primary forest cover by clear cutting, regardless of the future use of these areas (INPE 2022).

To calculate this rate, 229 images covering the entire ALB were analyzed. In this context, Pará was one of those that contributed most to this deforestation (45.15%), as shown in the table above.

On the PEAA website, the C&C pillar presents results of Operation Amazônia Viva achieved throughout 2020, the time frame of this investigation, compared to the previous year (2019):

> Inspection actions in areas of state interference contributed decisively to a 14% reduction in deforestation in 2020, when compared to the performance in these areas in 2019 (data from Deter/INPE). This data indicates that Operation Amazônia Viva—together with the beginning of land and environmental regularization measures—is already presenting important first results. If we consider the 2019–2020 comparison for the entire state (federal and state areas), the variation points to a reduction of just 1% in 2020, indicating that areas under federal interference have reduced performance and, therefore, increased their share of total deforestation accounted for in Pará territory. (PEAA website, C&C pillar: 2022).

After contextualizing Operation Amazônia Viva and demonstrating its impact on combating environmental crimes in Pará, now it is time to analyze the coverage of the operation on regional news broadcasts in this state, considering one of the categories listed in the codebook: environment.

The topic was analyzed separately, the decision for separate analysis was made with the aim of understanding aspects of regional coverage on said topic, from an internal perspective, that is, from those who live and work in the region. When specifically addressing this topic, elements of coverage of the Legal Amazon, as a whole, were found in reports broadcasted on each of the two regional news programs, as well as on the two national news programs analyzed, as demonstrated in previous topics and in details which will be done in later topics.

The sample of this separate analysis of Operation Amazônia Viva is a sum of seven television reports that correspond to all those that were shown, regarding the aforementioned operation, by the two regional news programs listed, from August to December 2020, the time frame of this investigation. This period of time was chosen for analysis because it coincides with the date of creation of the operation (August 2020), whose main objective is to combat environmental crimes in the state, as previously mentioned.

---

[1] For more information, see: http://www.inpe.br/noticias/noticia.php?Cod_Noticia=5811#:~:text=O%20valor%20consolidado%20da%20%C3%A1rea,os%20nove%20estados%20da%20ALB.

[2] For more information, see: http://www.inpe.br/noticias/noticia.php?Cod_Noticia=5811#:~:text=O%20valor%20consolidado%20da%20%C3%A1rea,os%20nove%20estados%20da%20ALB.

To gather this sample all editions of the news broadcasts shown during six months were observed, comprising: August, September, October, November and December of the year 2020. Considering that the broadcast is carried out from Monday to Saturday, there are 26 editions per month, totaling 130 editions of each of the news programs, in the selected period. In total, between the two news programs, 260 editions were observed, with an average duration of 1 h per edition, that is, 260 h of news programs were observed.

On JL1 (TV Liberal), based in Belém, two reports were shown in the period listed, totaling 2 min and 11 s of total coverage, in five months. The other five reports analyzed were broadcast by JT1 (TV Tapajós), based in Santarém, totaling 32 min and 34 s. The difference in time allocated to cover the topic in one channel and another is already noticeable (Table 2).

When watching each report, images shown (videos and photographs) were selected and classified, as well as the captions included in each of them (a character generator that leaves a summary of the story displayed on the screen throughout the time the report is on air being said), in addition to the text read/spoken by reporters and the images displayed.

It is considered that these two elements, mainly (images displayed and subtitled text), contribute to substantiate what is being broadcasted. Therefore aim to demonstrate the narrative constructed by both television news programs in a clearer tone. However, the way the theme/subject is presented builds the narratives.

Below, it will be presented how each of these broadcasters, based in different regions of the state, covered Operation Amazônia Viva, created by the Government of the State of Pará to combat environmental crimes.

Table 2 Reports about operation Amazônia Viva published in JL1

Date	Unit of analysis	Number	Party tipe	Theme	Framework	Time	Sources
13/10/20	JL1	1	Live link without sound	Operation Amazônia Viva	SEMAS releases the balance of the 4th phase of Operation Amazônia Viva in Pará 2'	2'	No sources in the article
					NOVEMBRO		
28/11/20	JL1	2	Note LOC OFF	Operation Amazônia Viva	Operation Amazônia Viva reached its 6th phase this week 1' No sources in story	1'	No sources in the article

Note Source Own preparation (2021)

Fig. 1 Operation Amazônia Viva. *Note* JL1 Reporter during live participation talking about activities of Operation Amazônia Viva. *Source* TV screenshot by the author (2021)

On TV Liberal, the first record of Operation Amazônia Viva (OAV) is from October 13, 2020, bringing news of the fourth phase of the operation. It is noteworthy that the operation was created in August of that year. In other words, three months after the operation was created, JL1 reported on the topic for the first time, informing viewers of the fourth phase of actions, without previously explaining that there were the first, second and third phases (Fig. 1).

The format used to report this phase was "live entry" with the reporter outside. To illustrate, the image insertion feature was used throughout the speech, with information appearing in the video that the images were provided by the State Secretariat for the Environment and Sustainability (SEMAS). The reporter remained on air for 1min58s, which is considered an adequate time for live appearances, in a news program with an average duration of 1 h, as is the case (Fig. 2).

Despite reporting on the fourth phase of the operation, as already highlighted, there were no reports on the topic in the newscasts previously observed. This leads to a discontinuity of information for viewers, who are left without any knowledge on the previous phases of the operation. Throughout the reporter's participation in the video, the following text appears: "Amazônia Viva: SEMAS releases balance of the 4th phase of the operation".

In JL1, the words operation, Amazon and environmental crimes are used in the captions of the reports and also of the "live" entries as a resource to anchor the "live" participations, as well as the images of Operation Amazônia Viva. The images used were provided by members of the operations. No sources were heard in this newscast (Figs. 3 and 4).

With the caption "Fighting environmental crimes: Operation Amazônia Viva has reached the 6th phase", this was how JL1 presented the second piece of news on the

Fig. 2 4th fase of Operation Amazônia Viva. Legend the picture: "Operation amazon alive". *Note* JL1—Image shown during live participation talking about activities in the fourth phase of Operation Amazônia Viva. *Source* JL1 from TV Liberal.TV screenshot by the author (2021)

Fig. 3 Activities of Operation Amazônia Viva. Legend the picture: "Combating environmental crimes". Note. JL1 –Presenter reads note LOC OFF live entry about operation activities. *Source* JL1 from TV Liberal. TV screenshot by the author (2021)

topic, lasting 53 s. This time, the focus is on combating environmental crimes. The images presented were again provided by a government agency, Agência Pará (State Government Communications Agency), as explained in the article.

Fig. 4 Reading by the LOC/OFF presenter. Legend the picture: "Combating environmental crimes" Note. JL1 Image shown during a reading by the LOC OFF presenter about the operation's activities. *Source* JL1 da TV Liberal. TV screenshot by the author (2021)

The format used was voice-over (LOC OFF), when the presenter reads the note live, from the studio, and continues reading as the speech is then covered by images. The difference between the LOC OFF note and the covered note is that, in the first there is no prior editing of the images, that is, the presenter does not go to the off booth to record the text in advance. As aforementioned, the text is read in the studio. This format is generally used when there is no need for synchronization between image and text, as required in recorded reports, known as VTs. For example, if you are talking about seized vehicles, you cannot show forest on the background. In this case, the images shown are general, about the subject covered. This is a resource that gives the news agility, but it is rarely used in editions. According to the first report presented, no sources were heard in this one either.

In the period from August to December 2020, JT1 (Santarém) aired five reports covering Operation Amazônia Viva, totaling 32 min and 34 s. All images used are the result of records made by members of the field operation. There are no records of images taken by the broadcaster's film crew. Only the sources heard resulted from the work of the news team, whose interviews were conducted after the operations were carried out, at the police station in the city of Santarém (Table 3).

The subtitles used during the exhibition, as well as the images shown during the analyzed reports are repeated. It should be noted that, in all five articles shown, the same sources were heard. There were two reports on phase four of the operation, using the same sources and, consequently, the same information (Fig. 5).

In this report above, although the main focus is the seizure of illegal timber, the fact that workers were found in precarious conditions "similar to slavery" is also highlighted. The delegate highlights the fact in his speech the reporter does the same,

Table 3 Reports about operation Amazônia viva

Date	Unit of analysis	Number	Party tipe	Theme	Framework	Time	Sources (characters)
20/8/20	JT1	1	Live link with sound	inspections	Operation Amazônia Viva: MP catches illegal logging: clandestine sawmill closed	7′36”	Chief delegate, inspection agent, military police lieutenant
24/8/20	JT1	2	Live link with sound	inspections	Operation Amazonia Viva: the balance of efforts to combat illegal deforestation over the weekend	7′05”	Inspection agent
					September		
1°/9/20	JT1	3	Live link with sound	Inspections	Operation Amazônia Viva: 3rd phase of actions to combat illegal logging comes to an end	9′51”	Delegate of the environmental crimes police station
29/9/20	JT1	4	Live link with sound	Inspections	Operation Amazônia Viva targets areas in Mojuí dos Campos in Pará - This is the 4th phase of the operation to combat environmental crimes	7′35”	Sergeant, coordinator of the operation, delegate of the agrarian Conflict police station
					October		

Note Reports about operation Amazônia Viva published in JT1

however, then moves on to another subject. The topic, despite its seriousness, is not discussed again in other reports by the vehicle. Like the other articles shown, this also uses images provided by members of the operation, focusing on the piles of seized wood. The situation and the workers are not shown besides the brief mentions on the speech of the delegate and the reporter herself (Fig. 6).

With the caption “Operation Amazônia Viva: the balance of actions to combat illegal deforestation over the weekend”, JT1 presents another report on the topic

Fig. 5 Activities of Operação Amazonia Viva. Legend the picture: "Operation Amazon alive". *Note* JT1 Reporter during live participation talking about activities of Operation Amazônia Viva. *Source* TV screenshot by the author (2021)

Fig. 6 Activities of Operation Amazona live. Legend the picture: "Operation Amazon alive". *Note* JT1 Image shown during a reading by the about the operation's activities. *Source* JT1 da TV Tapajós, 24 August. TV screenshot by the author (2021)

studied. This time, the focus is inspections' results. The images presented (videos and photos) are the same ones shown in previous reports but the credits have changed. According to what is explained, the images were provided by the Agrarian Conflicts Police Station, as stated in the credits of the article.

The characters (sources) heard previously are repeated in the article. Once again, the same sources are heard: inspection agent Solange Chaves (SEMAS) and the chief delegate of the Environmental Crimes Combat Police Station. The only difference is in the scenario used by the reporter, who, this time around, appears inside a room instead of outside, and the location's identification is not clear to the viewer. The following image shows this reporter's participation. Several times the reporter repeats information about the subject being discussed.

During this participation, the reporter uses the image resource, as shown in Fig. 7. The two news programs have similar characteristics in the sense that they use images provided by the institutions responsible for the operation. Recently, there has been a greater presence of this type of image use, both on regional and national news programs in Brazil. It can be inferred that this attitude is anchored in some factors, including the precariousness and lack of structure in newsrooms. Another factor can be attributed to issues relating to new production routines limited by external factors, for example, the covid-19 pandemic, which hit the world in 2019 and continues in 2021, the year this investigation was concluded. The "live" link format with the support of images and sound is a resource widely used by JT1, during this and other topics (Fig. 8).

The third recording of the operation, aired on September 1st, 2020, lasted 9min51s and used the same format as the previous report: live entry anchored by recorded interviews. The anchor presented, according to the caption displayed during the report, the framework that highlights the closure of the third phase activities of the operation, returning to a previously discussed topic: "Operation Amazônia Viva: ends the 3rd phase of actions to combat the illegal exploitation of wood".

Fig. 7 Operation Amazon alive. Legend the picture: "Operation Amazon alive" *Note* JT1—Live input on operation activities. *Source* JT1 from TV Tapajós, 24 Aug 2021. Screenshot from the TV by the author (2021)

Fig. 8 Operation Amazon alive. Legend the picture: "Operation Amazon alive". *Note*. JT1—Image displayed during live input on operation activities. *Source* JT1 from TV Tapajós, August 24, 2021. Screenshot from the TV by the author (2021)

In this second record, it is worth highlighting two similarities with the reports shown previously. The first is the theme, in this case the third phase of the operation. The second similarity is the source, which had already been heard during the live entry of the first report: delegate Fábio Amaral, the only one to be heard in the report. This time, he was identified as head of the Agrarian Conflicts Police Station (DECA), and no longer as the person responsible for the operation.

Conflicts and environmental consequences are the predominant frameworks. In these reports, it was found that the focus is the event and its relationship with ecosystem results whose emphasis is what it will bring to groups and/or individuals (Figs. 9 and 10).

With a running time of 7min35s, the report broadcast on September 29th follows the same format as the other ones mentioned: live entry, followed by recorded sound. The caption displayed is: "Illegal logging: the balance of the 4th phase of Operation Amazônia Viva in the metropolitan region".

This time, were used as sources the sergeant of the 1st CIPAMB, the coordinator of the Amazônia Viva/Eixo Uruará operation and, once again, the main delegate identified as the person responsible for the "Division of Agrarian Conflicts of the Civil Police"; a table with a summary of activities was another resource used (Fig. 11).

The last recording of the operation on JT1 was made on October 6, 2020. With a time of 2 min 57 s, the report in videotape (VT) format presents the off passage and sound structure, which is quite usual on TV. The caption is "Operation Amazônia Viva: the fourth phase was launched in the municipality of Mojuí dos Campos", summarizing the last phase of the operation.

Fig. 9 Activities of Operation Amazon alive. Legend the picture: "Operation Amazon alive". *Note* Live input on operation activities. Image captured from the TV screen. *Source* JT1 on TV Tapajós, 1 September. 2021. TV screenshot by author (2021)

Fig. 10 Activities of Operation Amazon alive: legend the picture: "illegal loggin". *Note* JT1—Live input on operation activities. *Source* JT1 on TV Tapajós, 29 September. 2021. TV screenshot by author (2021)

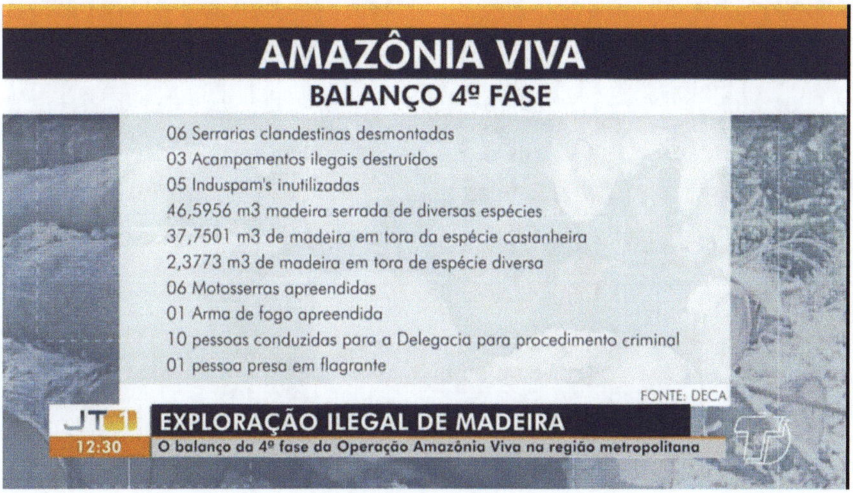

Fig. 11 Operation activities: legend the picture: "Operation Amazon alive" Balance 4a phase. *Note* JT1—Table with a summary of the operation's activities. *Source* JT1 on TV Tapajós, 29 September 2021. TV screenshot by author (2021)

As sources (characters in the news), it presents the SEMAS environmental inspection agent, again the same chief delegate (DECA). The latter is a source present in all reports about the operation shown on JT1. The framing of environmental consequences and their relationship to their ecosystem outcomes continues to be reinforced. The biggest emphasis is on conflict. Next, in the image reproduced from one of the inspections, the close-up of the agents, right at the opening of the VT, reinforces this approach (Fig. 12).

2 Conclusions

When analyzing the coverage of Operation Amazônia Viva on two regional TV news channels, Jornal Liberal 1st edition (Belém) and Jornal Tapajós 1st edition (Santarém), it was found that both channels focused on factual news, i.e. the issue that was happening at the moment, without going into depth. An example is that both stations prioritized the operation itself, staying on the event's surface level. There was no concern in varying the opinions presented. In both vehicles, the images were also repeated in the same proportion, all of which came from the enforcement agents.

There was little space for reflection on the environmental issue and its social, economic and political consequences, etc., in neither TV news programs. Even though we recognize that the short viewing time on TV limits this possibility, it is, however, possible to broaden the debate by diversifying sources and images. At JT1, the time factor cannot be taken into account in this instance. This newscast

Fig. 12 Operation activities Amazon alive. Legend the picture: "Operation Amazon alive". *Note* JT1—image of operation Amazônia Viva shown during the reporter's participation. *Source* JT1 on TV Tapajós, 6 Oct. 2021. TV screenshot by author (2021)

devoted more than 30 min to publicizing the operation, which would have made it possible to broaden the discussion. On JL1, only two minutes were devoted to the operation. In both newscasts, only the images provided by the police officers who were part of the operation were aired, with credits (text) for the government agencies. In some cases, only photos were used, featuring no moving images of what happened. Few of the reports analyzed featured images filmed by the broadcasters' cinematographic reporters.

Despite spending different amounts of time on the subject, the two news programs used the same formats: live coverage using images provided, followed or not by sound bites (recorded interviews). Both newscasts prioritizing terms related to enforcement and, in the focus of the seven subtitles analyzed, repeated terms such as environmental crimes, seizure, arrests, enforcement and balance sheet, among others.

As prior mentioned, the news published about the operation in both vehicles prioritized the factual, considering that the idea is to keep the "Amazon alive", it was expected that in addition to dead forest images—logs, equipment used to commit environmental crimes, information on arrests, figures on the balance of inspections etc.—would be used. The images were expected to be accompanied by such expanded views, through the opinions of specialists, scientists or even residents, family farmers or even rural producers, about the difficulties of keeping the forest standing and even the daily lives of the inhabitants of the largest rainforest on the planet.

Finally, it was expected that the dichotomy of preservation/destruction would not only be highlighted and reinforced by the framing of environmental consequences, always focusing on the event itself, in this case Operation Amazonia Viva, without broadening the focus to its relationship with ecosystem results for groups or individuals.

References

Cervi E (2017) Manual de métodos quantitativos para iniciantes em ciência política. CPOP, UFPR
INPE Instituto Nacional de Pesquisas Espaciais (2022) https://www.gov.br/inpe/pt-br
Mota LG (2007) Análise pragmática da narrativa. In: Lago C, Benetti M (Orgs) Metodologia de Pesquisa em Jornalismo. Vozes
Sousa JP (2006) Elementos de teoria da Comunicação e dos media. Letras Contemporâneas
Semetko HA, Valkenburg PM (2000) Framing European politics: a content analysis of press and television news. J Commun 50(2):93–109

Analysis of News About the Amazon in the Light of the Sustainable Development Goals

Jonathan Oliveira Fernandez, Gabriele de Sousa Araújo, Marcelino Silva da Silva, and Fábio Manoel França Lobato

Abstract The Amazon is the largest tropical forest in the world, with great importance for climate regulation and global biodiversity preservation. Given its relevance and in light of the Sustainable Development Goals (SDGs), it is pertinent to investigate how they are being implemented in the Amazon region. Therefore, this work analyzes Amazon-related news to identify the most discussed topics and their relationships with the SDGs. Text mining and content analysis techniques were used to correlate news with the SDGs. These analyses were based on grounded theory, weak labeling, and topic modeling. The results show that private companies operate in reforestation and environmental protection programs; the government also plays an essential role in combating deforestation and poverty, influencing environmental policies, and strengthening environmental regulation and partnerships with other countries. Such factors are aligned directly with SDGs 13 and 15, which are aimed at forest conservation, natural resources sustainable management, international cooperation, and other strategies to mitigate environmental impacts in the region. We conclude that the SDGs contribute to preserving the Amazon ecosystem, covering not only terrestrial and aquatic life but also the health and well-being of the traditional people who inhabit the region.

Keywords SDG · Amazon rainforest · Text mining · News analysis · Artificial intelligence

---

J. O. Fernandez · M. S. da Silva · F. M. F. Lobato (✉)
Federal University of Western Pará, Instituto de Engenharia e Geociências, Santarém, Pará, Brazil
e-mail: fabio.lobato@ufopa.edu.br

M. S. da Silva
e-mail: marcelino.ss@ufopa.edu.br

G. de Sousa Araújo · F. M. F. Lobato
State University of Maranhão, Engineering and Systems Graduate Program, São Luis, Maranhão, Brazil

W. Leal Filho et al. (eds.), *Amazon 2030—Sustainability Issues in the World's Largest Rainforest Region*, World Sustainability Series,
https://doi.org/10.1007/978-3-031-81465-5_15

1 Introduction

The Amazon is the largest tropical forest in the world, where a rich biodiversity of plants and animals is preserved, offering diverse natural resources and helping combat global warming (Mendoza et al. 2023). According to data from the Instituto Nacional de Pesquisas Espaciais (INPE), this is among the biomes most affected by fires and deforestation, which could lead to a critical point, transforming the forest into a major carbon emitter (Wit 2020; INPE 2023), harming biodiversity and impairing the Sustainable Development Goals (SDGs) achievement.

The United Nations (UN) established the SDGs in 2015 as an extension of the Millennium Development Goals (MDGs), which were in force from 1990 to 2015. In 2015, the UN launched the 2030 Agenda, a global action plan that established 17 SDGs and 169 goals to strengthen the MDGs, including new targets for climate change, innovation, sustainable consumption, and socioeconomic problems mitigation (Villagra and Campos 2019). It is clear that forests, like the Amazon region, play a prominent role in achieving several SDGs. In this context, a non-exhaustive list includes the food resources provision (SDG 2—zero hunger), plants that can be used for medicinal purposes (SDG 3—good health and well-being), the local biodiversity protection, both aquatic and terrestrial life (SDG 14—life below water and SDG 15—life on land) and the promotion of income for families living in the region (SDG 1—no poverty), to mention a few. However, deforestation makes it challenging to carry out these activities (Cardozo et al. 2020).

In particular, the Amazon rainforest is of great importance due to its biodiversity, fertile soils, and among the largest river basins in the world. Consequently, the Brazilian government established priorities such as reducing deforestation and fires, thus decreasing biodiversity loss and the national impact on global warming, among other problems (Cardoso et al. 2021). With technological evolution, social media expansion, and other information platforms, issues involving the SDGs, such as sustainability and social responsibility, are increasingly present in society (Silva Júnior 2023). The same authors state that people are more aware of the products they consume, correlating with aspects related to the SDGs. Thus, information plays an important role in communication and all human activities, including professional and personal life, education, politics, and others (Assunção et al. 2021). Thus, news as an information source is widely used in various areas. From this perspective, there is a need to analyze news regarding how the SDGs contribute to combating the Amazonian environmental problems and promoting sustainable development in the region.

Until now, the works investigating SGDs and news have taken a more holistic approach, as in Fitri et al. (2021), who applied topic modeling to identify relevant SDG news topics in Indonesia. López-Carrión and Martí-Sánchez (2022) conducted a descriptive text analysis of the Spanish digital press about the SDGs and the 2030 Agenda, including semantic analysis and text mining. In this context, no studies were identified that used text mining for analyzing news in light of the SGDs; moreover, there is no work dedicated to correlating SGDs to the Amazon Rainforest based on news analysis. Therefore, this study aims to analyze news related to Amazon to identify the themes disclosed and which SDGs are prevalent, using analyses based on the grounded theory (Baumer et al. 2017) and also through current text mining and news content analysis techniques, mapped by Araújo et al. (2023).

The remainder of this paper is organized as follows. The materials and methods are described in Sect. 2. The results are discussed in Sect. 3. Finally, the final remarks and suggestions for future works are presented in Sect. 4.

2 Materials and Methods

Text mining aims to extract valuable information from a given corpus (text dataset) (Dang and Ahmad 2014). According to Žižka et al. (2019), it can be defined as a knowledge-intensive process where the user interacts with a corpus using analytical tools to discover new and potentially useful patterns. In this sense, it was decided to use a methodological process based on Aleixo et al. (2023) and Rodrigues et al. (2022), which are summarized in Fig. 1.

For the **Data Collection**, "Amazônia" was defined as the keyword to be searched on news portals. Subsequently, a web crawler was developed to collect news data. Regarding the data source, five leading Brazilian news portals were selected considering their relevance: Portal G1, Metrópoles, Folha de São Paulo, Exame, and CNN Brasil. A comparative news metadata analysis is summarized in Table 1.

This data was stored in Comma-Separated Values (CSV) files used in the next step (**Data Preparation**). This step is required to correct inconsistencies in the data,

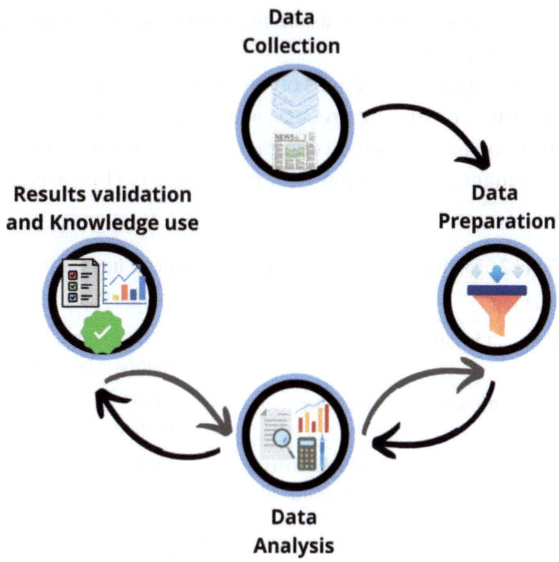

Fig. 1 Methodological process stages summary

Table 1 News meta-data collected

Item	Data type
Headline	Textual
Summary	Textual
Body	Textual
Publication date	Date/time
Region	Textual
Link	Textual
Keyword	Textual
Platform	Textual

such as dispersion, missing values, noise, and various other elements (Mishra et al. 2020). Based on studies by Aleixo et al. (2023) and Cirqueira et al. (2018), the news pre-processing pipeline was built as shown in Table 2.

For the **Data Analysis** stage, filtering was carried out to check which news was within the scope of the study using the grounded theory method for news described by Baumer et al. (2017). This method involved reading the headlines and the article text body, focusing on the relationship between the news and the keyword (search-query). When they did not meet this criterion, the news was discarded. This step was essential to maintaining data congruence and reducing possible bias. For example, on some portals, the recommendation system indicated other news related to the topic, but that did not deal precisely with the topic searched (e.g., Amazon region in our case). In these cases, the news was removed from the data set. The filtered news was then passed to the next stage, where text mining techniques were used to extract information from the data set by automatically labeling the news based on a keywords dictionary associated with the SDGs and identifying the most pertinent topics in the news.

This approach is based on the understanding that manual annotation is a laborious and time-consuming process, which requires a significant investment of time to build an annotated data set. In contrast, weak labeling can be much faster and more efficient than manual annotation (Liu et al. 2022). This strategy allows labeling the dataset

Table 2 Methods applied to pre-processing news headlines and their outputs

Method	Output
Original headline	COP26: ONGs formam aliança para restaurar 4 milhões de hectares no Brasil
Standardization for lowercase	COP26: ONGs formam aliança para restaurar 4 milhões de hectares no brasil
Removal of punctuation, numbering, accentuation, and special characters	COP ONGs formam alianca para restaurar milhoes de hectares no brasil
Stopwords removal	COP ONGs formam alianca restaurar milhoes hectares brasil

automatically, identifying correspondence patterns between a list of terms and the texts needing labeling (Mendes et al. 2021). So, we saw the possibility of adopting news labeling based on a keywords dictionary associated with the SDGs, considering that weak labeling is an alternative to reduce dependence on manual annotation, allowing more data to be easily labeled (Tekumalla and Banda 2021).

To this end, we used a list of terms drawn up from 16 datasets with keywords developed by Elsevier (2023), based on the SDG Research Mapping Initiative, a collaboration between the University of Southern Denmark, Aurora (represented by Vrije Universiteit Amsterdam), the University of Auckland and Elsevier for identifying research that supports the SDGs (Kashnitsky et al. 2022; Wang et al. 2023). The keyword dataset produced by Jayabalasingham et al. (2019) was translated to Portuguese by the Universidade Estadual do Maranhão—UEMA (2021) to be used in academic works, so these terms were incorporated into the present work. As for keywords from other databases presented in the SDG Keywords Dictionary Project developed by the University of Auckland (n.d), they were translated into Portuguese using *Googletrans*, a *Python* programming language library.

Given that these databases consist of queries geared towards academic research, many terms include boolean operators such as AND, OR, or AND NOT. To tackle this problem, some criteria were applied: (i) in cases where there was an AND conjunction between the terms, for example, "(food market AND restriction)", the term with greater emphasis was added to the dictionary, in this case, "food market", and the more generic term was eliminated or commented on in the list. In situations where expressions like "(Physical OR Emotional OR Sexual) AND (Abuse)", the words were rewritten according to cohesion and context, resulting in expressions such as "Physical Abuse", "Emotional Abuse", and "Sexual Abuse"; (ii) when the AND NOT operator was present, the terms were not increased in the list. In this way, these keywords were reproduced in a dictionary in Python. In other words, they were defined as keys (SDG 1–16) and the values (keywords) associated with the SDGs for each key. That said, news was automatically labeled based on the SDGs.

To validate the labeling results, we use these more precise methods since Big Data Analytics has led to the development of advanced machine learning, natural language processing, and topic modeling techniques capable of identifying patterns in the datasets (Egger and Yu 2022); steps were taken to improve interpretability, making it easier for news content experts to annotate topics. Among these methods, BERTopic has been applied in various text-mining works to extract insights from topic modeling and identify relevant topics (Araújo et al. 2023). This technique generates word embeddings from a pre-trained model. Then clusters are identified to generate/aggregate topics based on the words' importance through a variation of Term Frequency-Inverse Document Frequency (TF-IDF) called class-based TF-IDF (c-TF-IDF). This method assesses the term's importance based on its frequency considering the entire corpus (Grootendorst 2022).

Following the model's default configuration, we used the *language = multilingual* parameter, which contains the *paraphrase-multilingual-MiniLM-L12-v2* embedding model, supporting more than 50 languages, including Portuguese. To ensure greater coherence and agreement between topics, we implemented the topic reduction function *reduce_topics*, which performs the recursive merging of topic pairs (Gomes and Attux 2023). In addition to annotating the topics, they were classified and associated with the SDGs by four experts, members of the Rede ODS Brasil (2023).

In the last stage, it is possible to visualize, analyze, and document the knowledge usage, thus disseminating the results obtained during the analyses and presenting both qualitative and quantitative results. For example, in the quantitative aspect, there is the number of collected news articles; in the qualitative aspect, there is the topic modeling, which reveals the main themes found in the news. In the Quanti-Quali dimension, there is the distribution of the SDGs covered by the news and also the temporal distribution. The ground truth technique was adopted, evaluating the results jointly between researchers to validate them and improve the research reliability. For this purpose, the authors read and discussed the results, which involved a detailed analysis of each methodological process step, looking for inconsistencies, incoherence, or incorrect information that could lead to going back to the previous stages. Furthermore, the study was presented at the symposium "Amazonia 2030—Sustainability Issues in the Largest Tropical Forest Region in the World" where comments and questions about the work were essential for validation, ensuring even more reliable results.

3 Results and Discussion

Based on the keyword defined in the data collection stage, 6.607 news items were obtained from the portals Portal G1, Metrópoles, Folha de São Paulo, Exame, and CNN Brasil in the period from 2008 to 2023, resulting in 3.092 articles after filtering. Through exploratory analysis, it was possible to identify which portals had the most significant amount of news collected, as illustrated in Fig. 2.

As observed in Fig. 2, the news portal Exame presented a higher number of collected news articles than the other portals. This may be related to the platform's ability to allow historical data collection, including older information from other sources. Concerning Portal G1, a greater news article volume was anticipated, given its significant relevance as an information source in Brazil. However, many changes in the website's structure over the years may have affected the data collection capability, resulting in a focus on more recent news articles. This fact is evidenced in Fig. 3, which depicts the temporal distribution over the months for each platform. This representation makes it easier to visualize the publications over time.

In Fig. 3, there is an increase in the amount of news on the Exame portal and Portal G1 in July, August, and September. This fact may be related to several factors, for instance, there are news that addressed the political aspects and environmental context, particularly the concern with Amazon's protection. Another prevalent factor is related to news about deforestation and fires occurring in the region. De Faria et al. (2023) highlight that fires occur mainly in August and September, with the latter

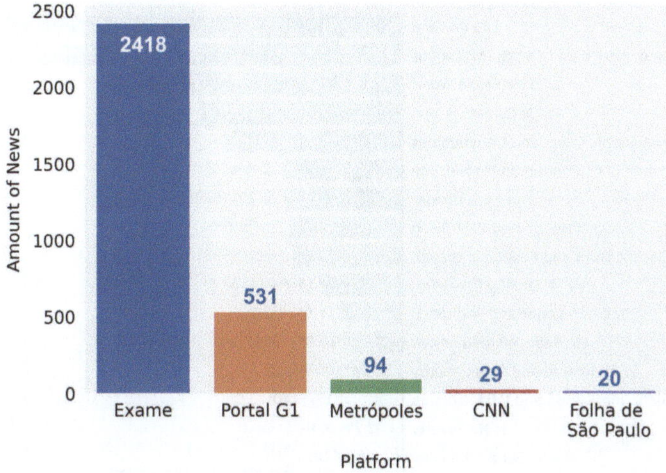

Fig. 2 Number of news published per platform

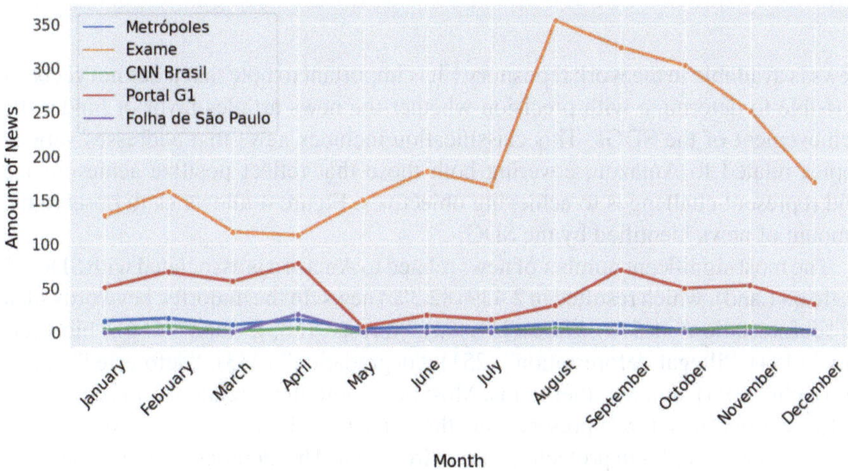

Fig. 3 Monthly news distribution per platform

having a higher incidence precisely due to the dry climate that favors the spread of fire.

3.1 Sustainable Development Goals and News

By applying this method, it was possible to label 2.965 news items using 718 keywords related to the SDGs. The table containing the keywords identified in the

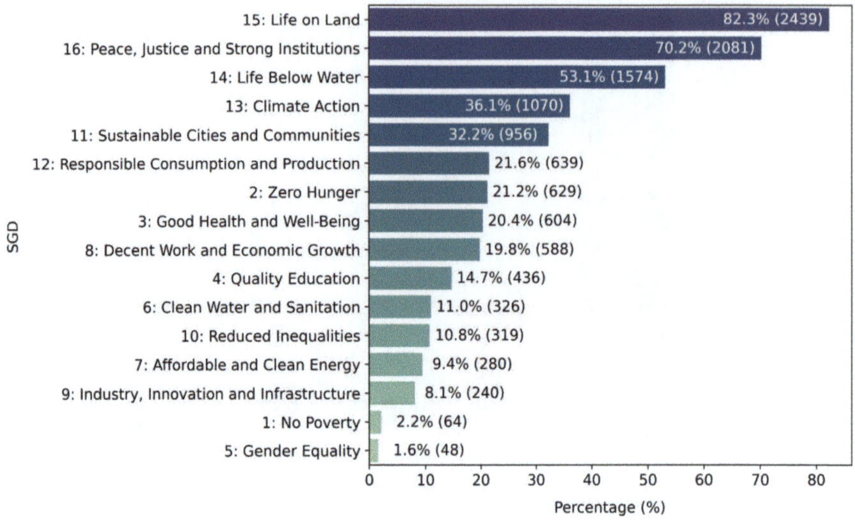

Fig. 4 Distribution of the news related to the SDGs

news is available in the work repository.[1] It is important to note that it has not yet been possible to determine with precision whether the news articles favor or hinder the achievement of the SDGs. This classification includes news that addresses various topics related to Amazon, covering both those that reflect positive achievements and represent challenges to achieving objectives. Figure 4 allows us to observe the amount of news identified by the SDG.

The most significant number of news related to Amazon is associated with SDG 15 (Life on Land), which resulted in 2.439 (82.3%) news. In the majority, keywords such as "deforestation" (1.518), "forest" (1.405), "forests" (683), "land" (616), "biodiversity" (484), "illegal deforestation" (251), "degradation" (233), "deforested" (225), "drought" (194), among other terms. Most news with these related terms deals with deforestation, fires, forest preservation, the impact on biodiversity, and even the role of indigenous peoples in protecting the environment. This demonstrates the relevance of these issues in journalistic coverage and public awareness of the challenges faced in the region, particularly concerning the escalating deforestation and forest degradation that have historically exacerbated people's quality of life while concurrently amplifying poverty, vulnerability, and violence (Santos et al. 2023; Celentano et al. 2022).

In the case of SDG 16 (Peace, Justice, and Strong Institutions), 2.081 (70.2%) news articles were labeled through keywords such as "government" (1.551), "politics" (494), "society" (455), "crimes" (181), "crime" (152), "institution" (141), "human rights" (121), "war" (106), "violence" (105), "institutional" (93). These terms suggest various socioeconomic and political issues affecting the Amazon, such as territorial

[1] https://github.com/fabiolobato/Amazon2030-News_ODS.

conflicts, indigenous rights, corruption related to resource exploitation, and crime and violence impacting indigenous communities and forest preservation. When carefully analyzing the news, these terms are mainly associated with political speeches, especially during the Bolsonaro government at the UN Assembly, and this has influenced environmental and sustainability policies in the Amazon (Peres et al. 2023). Celentano et al. (2022) highlight that better governance depends mainly on reducing illegality and requires coordination between governments, civil society, indigenous peoples, and local communities.

The news articles were also categorized under SDG 14 (Life Below Water), with 1.574 (53.1%) involving keywords such as "forest" (1.405), "coast" (133), "fishing" (68), "sea" (66), "fishermen" (39), "oceans" (37), "seas" (14), "ecotourism" (14), "marine" (12), and "aquaculture" (10). These news articles generally address factors related to forest conservation, fishing practices, preserving marine ecosystems, and developing ecotourism in the Amazon region (Jimenez et al. 2021). Furthermore, the news with the most keywords focuses on international initiatives looking for sustainable solutions. However, the complexity and political impasses are notorious. This global attention underscores the need for international cooperation to preserve the forest, given its significance in addressing climate change and providing natural resources, aligning with SDG 13 (Climate Action), which is also prevalent in the results.

Thus, Osborne et al. (2021) highlight that forming these partnerships allows for sharing the responsibility for restoring ecosystems, considering human well-being, environmental sustainability, and social justice. However, it is imperative that such investments not only focus on biophysical outcomes but also consider essential ecological and social benefits in the Amazon (Celentano et al. 2022).

3.2 Topic Modeling Analysis

Applying BERTopic, described in Sect. 2, identified ten topics with the ten most prevalent terms. Based on these results, the topics were labeled and associated with the SDGs, and this outcome can be viewed in Table 3.

The obtained topics can be seen in Table 3 second column. The first column represents the annotation/labeling by domain experts in news analysis who summarized the topics by reading keywords and the news. The last column denotes the correlation between the news/topics and the SDGs. For example, the topic "Environmental problems and actions in the Amazon" refers to problems that occur in the Amazon that may include issues related to deforestation, biodiversity loss, and climate change, among other challenges that impact the Amazon ecosystem and, by extension, the global environment. The topic also addresses actions and measures to address these problems and protect the ecosystem. These initiatives may include efforts related to forest conservation, reforestation programs, sustainable natural resource management policies, international cooperation, and other strategies that aim to mitigate environmental impacts in the region (De Andrade et al. 2021; Celentano et al. 2022).

Table 3 Topic modeling with BERTopic

Topic label	Terms	SDG
Environmental problems and actions in the Amazon	Amazonia, deforestation, wildfires, fund, brazil, bolsonaro, government, fires, lula, forest	2, 13, 15
Indigenous territories	Indigenous, indians, land, against, leader, yanomami, road, morales, government, belo	1, 2, 10, 16
Combating deforestation	Forest, code, deforestation, environment, forest, government, climate, new, climatic	2, 12, 13, 15
Sustainable development	Sustainability, green, sustainable, company, millions, natura, environmental, food, innovation, environment	8, 9, 12, 13, 15
Environmental and political challenges	Un, bolsonaro, cop, assembly, indigenous chief, speech, raoni, general, emission, goal	10, 13, 16, 17
Carbon emission	Carbon, emission, credit, market, low, gases, greenhouse, effect, company, offset	7, 9, 13
Technology and innovation	Startups, innovation, company, ranking, investment, diversity, technology, sector, google, dilma	4, 8, 9
Investments	Investment, esg, development, financing, management, sustainable, project, bndes	2, 7, 8, 9, 10, 17
COP goals	Brazil, deforestation, Brazilian, cop, carbon, emission, reduction, environment, sustainable, goal	1, 2, 4, 13, 15
Forest fires	Wildfires, pantanal, fires, bolsonaro, ngos, rain, sp, smoke, prohibition, cerrado	11, 13, 15, 17

This is related to SDG 2 (Zero Hunger), SDG 13 (Climate Action), and SDG 15 (Life on land), which focuses on protecting biodiversity and terrestrial ecosystems.

The topic "Indigenous territories" concerns the local people's territory protection, the challenges indigenous communities face regarding infrastructure development, the role of indigenous leaders in defending their rights, and the influence of government policies and public figures (Peres et al. 2023). SDG 1 (No Poverty), SDG 2 (Zero Hunger), SDG 10 (Reduced Inequalities), and SDG 16 (Peace, Justice, and Strong Institutions) are linked to these issues, encompassing justice-related matters, respect for human rights, and community participation in decisions affecting their lives.

The topic "Combating deforestation" addresses issues such as forest management, the impact of deforestation on the environment, climate change, and government and societal efforts to develop codes and policies that promote forest conservation and climate change mitigation. This topic is intrinsically linked to SDG 12 (Responsible Consumption and Production), as highlighted by Peres et al. (2023), pointing out the insignificance of deforestation in small-scale agriculture compared to livestock and soy production. Where these practices, through fires, impact climate change SDG 13 (Climate Action) and drought conditions SDG 15 (Life on Earth). The authors emphasize strategies such as agroforestry systems to mitigate these SDG 2 (Zero Hunger) impacts, mainly through local production and community management, changing communities to change (SDG 15).

Meanwhile, the "Sustainable development" topic, although semantically close to the previous topic, focuses on environmental and economic sustainability, highlighting sustainable practices with Environmental, Social, and Governance (ESG) in companies, innovation in production and consumption, and efforts to protect the environment. It is related to SDG 8 (Decent Work and Economic Growth), SDG 9 (Industry, Innovation, and Infrastructure), SDG 12 (Responsible Consumption and Production), SDG 13 (Climate Action), and SDG 15 (Life on Land).

The topic "Environmental and political challenges" contextualizes issues related to social and political inequalities reduction, as well as the global debate on climate change with the Conference of the Parties (COP), government policies, greenhouse gas emissions targets, and the work of indigenous leaders like Raoni who focuses on defending environmental issues. Therefore, this topic is relevant to SDG 10 (Reduced Inequalities), SDG 13 (Climate Action), and SDG 16 (Peace, Justice, and Strong Institutions).

The "Carbon emission" topic focuses on carbon and greenhouse gas emissions, emphasizing the need to reduce these emissions to combat climate change. For example, companies are looking to reduce their emissions by adopting low-carbon practices and participating in the carbon credit market, where they can offset their emissions by supporting projects that capture or reduce carbon emissions (Celentano et al. 2022). These actions promote a transition to a more sustainable economy, which is aligned with SDG 7 (Affordable and Clean Energy), SDG 9 (Industry, Innovation, and Infrastructure), and SDG 13 (Climate Action).

Among notable topics is "Technology and innovation," emphasizing that the Amazon extends beyond environmental and climate protection. This involves diverse movements in the technology sector, fostering technological innovation, and investing in research and development within the Amazon, including fostering projects related to educational activities, literacy, gender, sustainability, and cultural roots in municipal, quilombola, and indigenous communities (Justino et al. 2023), aligning with SDG 4 (Quality Education). There are also discussions about technology-related public policies, such as those during the Dilma Rousseff government, and their impact on the innovation sector and economic growth. This topic is directly related to SDG 8 (Decent Work and Economic Growth) and SDG 9 (Industry, Innovation, and Infrastructure).

Among the other topics covered, "Investments" is related to SDG 2 (Zero Hunger and Sustainable Agriculture), SDG 7 (Affordable and Clean Energy), SDG 8 (Decent Work and Economic Growth), and SDG 10 (Reduced Inequalities). The topic "COP goals" is linked to SDG 1 (No Poverty), SDG 2 (Zero Hunger and Sustainable Agriculture), SDG 4 (Quality Education), SDG 13 (Climate Action), and SDG 15 (Life on Land). Finally, "Forest fires" are associated with SDG 13 (Climate Action) and SDG 15 (Life on Land).

Each topic addresses specific issues aligned with the SDGs' achievement, and the SDGs' 13 and 15 prevalence in most topics corroborates with the automatic labeling findings. When it comes to the Amazon, the focus is on forest-based ecosystem services and how they contribute to well-being and, consequently, the SDGs' achievement. However, there are still persistent challenges that weaken these objectives. As noted, among the biggest causes is deforestation, which spreads in several areas, such as the reduction of income from forest products (SDG 1), agricultural productivity (SDG 2), pollutants and pathogens that spread, causing diseases (SDG 3 and 6), threats to clean energy (SDG 7), habitat loss (SDG 14), there are several problems that such action can generate (De Andrade et al. 2021).

This research finding corroborates the conclusions presented by De Faria et al. (2023), highlighting that 21% of the total events recorded in the Legal Amazon from April 2020 until 10/2022 are related to deforestation fire events. However, in recent years, the government has demonstrated weakness in planning policies to prevent deforestation (Cardoso et al. 2021). As advocated by Celentano et al. (2022), forest restoration is urgently needed to avoid the critical point of the Amazon's collapse, serving as a vital element of a broader conservation strategy, and this should begin with the proper public policies. Seymour and Busch (2016) describe that tropical forests protect a significant portion of the planet's terrestrial biological diversity (SDG 15) and serve as a secure and natural means of carbon capture and storage. Therefore, they are essential for any strategy aimed at climate stabilization (SDG 13). Reduce deforestation, improve equity and accountability mechanisms, and promote improvements in governance and social justice structures (SDG 16).

4 Conclusion

The work proposes a multifaceted methodological approach to analyze news linked to the Amazon, identifying the main covered topics and relating them to the SDGs. The study was designed considering that news platforms influence sustainability and social responsibility to achieve these SGDs since they help form public opinion. In this sense, the approach combines grounded theory; text mining techniques, such as topic modeling and language models, weak labeling; and the experts' knowledge for annotating topics and validating the research findings.

Classifying news based on the SDG dictionary revealed that the discussed topics are frequently related to SDG 15, reflecting the concern for the Amazon Rainforest's preservation and sustainability and the protection of indigenous human rights. These findings are intrinsically associated with SDG 13, as it presents challenges related to deforestation and wildfires, which weaken the achievement of many other development goals in the region. Hence, the people consuming the news have been aware of the challenges faced in the Amazon. That is confirmed in the topic analysis, as they reaffirm themes like "Environmental problems and actions in the Amazon", "Indigenous territories", and "Combating deforestation" emphasizing SGD 13 and 15 and the related issues. With this study, we hope to shed light on the influence of news in promoting sustainability in Amazon.

We also demonstrate which environmental challenges are highlighted in the news platforms' and how this has affected the region. These insights can catalyze strategic decisions by governments, organizations, and other stakeholders. Moreover, this study initiates an approach based on automatically labeling news aligned with the

SDGs. Although the results are consistent with the topic analysis, we recognize that there are points for improvement. This includes continuously revising the research design, such as needing a larger news dataset for broader and more representative analyses and the SDG dictionary refinement. In future work, we intend to expand our collaboration by including more domain experts in the Amazon region and SGDs, ensuring the research reliability.

Acknowledgements The authors acknowledge the support of the *Conselho Nacional de Desenvolvimento Científico e Tecnológico* (CNPq)—DT-303031/2023-9, PIBIC—132759/2022-5, PIBITI—141598/2022-0; the *Fundação Amazônia de Amparo a Estudos e Pesquisas* (FAPESPA)—PRONEM-FAPESPA/CNPq nº 045/2021.

References

Assunção SS, Ferreira MMC, dos Santos Junior RL, Redigolo FM, Condurú MT (2021) Sistema de informação ambiental como ferramenta para preservação da Amazônia: atuação do Sistema Deter. Informação & Informação 26(3):125–156

Aleixo MJA, de Sousa Araujo G, do Carmo FA, Junior AFLJ, Lobato FMF (2023) Análise Comparativa de Plataformas de Mídias Sociais Turísticas: um Estudo de Caso de Booking e TripAdvisor para o Município de Santarém, Pará. Anais Do Computer on the Beach 14:70–77

Araújo GDS, Leite JBP, da Silva MS, Junior AFJ, Lobato FM (2023) Natural Language Processing and Social Media: a systematic mapping on Brazilian leading events. In Anais do XX Encontro Nacional de Inteligência Artificial e Computacional. SBC, pp 741–755

Baumer EP, Mimno D, Guha S, Quan E, Gay GK (2017) Comparing grounded theory and topic modeling: extreme divergence or unlikely convergence? J Am Soc Inf Sci 68(6):1397–1410

Cardoso TB, dos Santos Chaves D, Sobragi CG (2021) Combate ao desmatamento por queimadas na Amazônia Legal, para o desenvolvimento da ODS no Brasil. Salão Integrado de Ensino, Pesquisa e Extensão da Uergs (SIEPEX) 1(10)

Cardozo LE, Schreiber MDS, Cenci DR (2020) Exploração De Florestas Com Enfoque Na Amazônia E Suas Contribuições Para Os Objetivos Do Desenvolvimento Sustentável. Salão do Conhecimento 6(6)

Celentano D, Moraes M, Ferreira J, Nahur A, Coutinho B, Rousseau GX, Kanashiro M et al (2022) Forest restoration to promote a fair post COVID-19 recovery in the Brazilian Amazon. Land Use Policy 116:106076

Cirqueira D, Pinheiro MF, Jacob A, Lobato F, Santana Á (2018) A literature review in preprocessing for sentiment analysis for Brazilian Portuguese social media. In: IEEE/WIC/ACM international conference on web intelligence (WI). IEEE, pp 746–749

Dang S, Ahmad PH (2014) Text mining: techniques and its application. Int J Eng Technol Innov 1(4):22–25

De Faria DR, Bernini H, de Paula JA, Rivera-Lombardi RJ, Júnior HV (2023) Análise Da Dinâmica Do Intervalo Entre Desmatamento E Fogo: Bases Para Tipificação De Fogo De Desmatamento. Trabalho apresentado em Anais do XX Simpósio Brasileiro de Sensoriamento Remoto

De Andrade LC, Borges-Pedro JP, Gomes MCRL, Tregidgo DJ, do Nascimento ACS, Paim FP, do Amaral JV et al (2021) The sustainable development goals in two sustainable development reserves in central amazon: achievements and challenges. Discov Sustain 2(1):54

Egger R, Yu J (2022) A topic modeling comparison between lda, nmf, top2vec, and bertopic to demystify twitter posts. Front Sociol 7:886498

Elsevier (2023) SDG research mapping initiative. https://www.elsevier.com/about/partnerships/sdg-research-mapping-initiative. Accessed 9 Sept 2023

Fitri H, Widyawan W, Soesanti I (2021) Topic modeling in the news document on sustainable development goals. Int J Inf Technol Electr Eng 5(3):82–89

Gomes GBB, Attux R (2023) Contributions to Social media analysis based on topic modelling. In: Anais do XI symposium on knowledge discovery, mining and learning. SBC, pp 113–120

Grootendorst M (2022) BERTopic: neural topic modeling with a class-based TF-IDF procedure. Preprint at arXiv:2203.05794

Liu M, Luo H, Thong L, Li Y, Zhang C, Song L (2022) SciAnnotate: a tool for integrating weak labeling sources for sequence labeling. Preprint at arXiv:2208.10241

López-Carrión AE, Martí-Sánchez M (2015–2022) Analysis of the coverage and discourse of the sustainable development goals and the 2030 Agenda in the Spanish digital press

INPE, Instituto Nacional de Pesquisas Espaciais (2023) Queimadas: Monitoramento de focos. https://queimadas.dgi.inpe.br/queimadas/portal-static/situacao-atual/. Accessed 29 mai 2023

Jayabalasingham B, Boverhof R, Agnew K, Klein L (2019) Identifying research supporting the United Nations sustainable development goals. Mendeley Data 1(1)

Jimenez ÉA, Gonzalez JG, Amaral MT, Fredou FL (2021) Sustainability indicators for the integrated assessment of coastal small-scale fisheries in the Brazilian Amazon. Ecol Econ 181:106910

Justino A, Nhoatto F, Souza E, Lobato F (2023) Análise orientada a dados como auxílio para tomada de decisão em Gestão de Pesquisa. Revista da CGU 15(27)

Kashnitsky Y, Roberge G, Mu J, Kang K, Wang W, Vanderfeesten M, Labrosse I (2022) Identifying research supporting the United Nations sustainable development goals. Preprint at arXiv:2209.07285

Mendes AR, Passador RV, Caseli HM (2021) Identificando sintomas de depressao em postagens do Twitter em português do Brasil. In Anais do XIII Simpósio Brasileiro de Tecnologia da Informação e da Linguagem Humana. SBC, pp 162–171

Mendoza AYG, Cavalcante FSA, Lima RA, de Lima JPS (2023) Macrofungos Na Amazônia Brasileira E Importância Para O Meio Ambiente: Uma Revisão Integrativa. Revista Valore 8

Mishra P, Biancolillo A, Roger JM, Marini F, Rutledge DN (2020) New data preprocessing trends based on ensemble of multiple preprocessing techniques. TrAC, Trends Anal Chem 132:116045

Osborne T, Brock S, Chazdon R, Chomba S, Garen E, Gutierrez V, Sundberg J et al (2021) The political ecology playbook for ecosystem restoration: principles for effective, equitable, and transformative landscapes. Glob Environ Chang 70:102320

Peres CA, Campos-Silva J, Ritter CD (2023) Environmental policy at a critical junction in the Brazilian Amazon. Trends Ecol Evol

Rede ODS Brasil (2023). https://www.redeodsbrasil.org/. Accessed 9 Sept 2023

Rodrigues L, Prado A, Lobato FMF (2022) Pandemia de covid-19 no brasil: uma análise sobre notícias e comentários de usuários. Culturas Midiáticas 16:26–26

Santos D, Lima M, Wilm M, Seifer P, Veríssimo B (2023) Índice de Progresso Social na Amazônia Brasileira—IPS Amazônia 2023. Belém: Instituto do Homem e Meio Ambiente da Amazônia (Imazon)

Seymour F, Busch J (2016) Why forests? Why now?: The science, economics, and politics of tropical forests and climate change. Brookings Institution Press

Silva Júnior CRGD (2023) Objetivos de desenvolvimento sustentável como estratégia de mercado para organizações: estudo no grupo Neoenergia

Tekumalla R, Banda JM (2021) Using weak supervision to generate training datasets from social media data: a proof of concept to identify drug mentions. Neural Comput Appl:1–9

UEMA, Universidade Estadual do Maranhão (2021) Palavra-chave ODS-SCOPUS. https://ods.uema.br/2021/09/02/palavra-chave/. Accessed 9 Sept 2023

University of Auckland (n.d) SDG Keyword Mapping. https://www.sdgmapping.auckland.ac.nz/. Accessed 9 Sept 2023

Villagra EC, Campos MP (2019) Avances de los objetivos de desarrollo sostenible (ODS) en las universidades: Caso UNAN-Managua. Revista Compromiso Soc 2:25–34

Wang W, Kang W, Mu J (2023) Mapping research to the sustainable development goals: a contextualised approach. Preprint (Version 3) available at Research Square. https://doi.org/10.21203/rs.3.rs-2544385/v3

Wit F (2020) Governança climática e a floresta amazônica: analisando o desenvolvimento sustentável na Amazônia brasileira sob o olhar da geógrafa Bertha Becker. Saberes geográficos e Geografia institucional: Relações luso-brasileiras no século XX

Žižka J, Dařena F, Svoboda A (2019) Text mining with machine learning: principles and techniques. CRC Press

Implementation of University Meliponary

Iris Mayara Leão Mota, Rommel Noce, Juliana Mendes de Oliveira, Graciene Conceição dos Santos, and Aline Ariel Cardoso

Abstract The Meliponiculture, rational breeding of stingless bees, is a very important activity for the preservation of species considering the defense of biodiversity, as well as potential to benefit communities in the composition of income. This study aimed to measure the cost of implanting a meliponary in Santarém, West of Pará, Brazil. The data were based on the structure of the meliponary present at Fatima Mekdece´s Grove of the Universidade Federal do Oeste do Pará (Federal University of Western Pará). It was considered the acquisition of materials, labor for implementation and the period for carrying out the activity.

Keywords Animal production · Meliponiculture · Sustainability · Community economy

1 Introduction

Meliponiculture is the rational breeding of native bees, better known as stingless bees (Gemim et al. 2022) that belong to the tribe Meliponini. It is a common practice for obtaining products that can be used for both gastronomic and medicinal use (Texeira

I. M. L. Mota (✉) · R. Noce · J. M. de Oliveira · G. C. dos Santos · A. A. Cardoso
Biodiversity and Forests Institute, Universidade Federal do Oeste do Pará, Rua Vera Paz, S/N, Santarém, PA 68040-255, Brazil
e-mail: iris.mota@discente.ufopa.edu.br

R. Noce
e-mail: rommel.noce@ufopa.edu.br

J. M. de Oliveira
e-mail: juliana.oliveira@ufopa.edu.br

G. C. dos Santos
e-mail: graciene.santos@ufopa.edu.br

A. A. Cardoso
e-mail: aline.cardoso@discente.ufopa.edu.br

W. Leal Filho et al. (eds.), *Amazon 2030—Sustainability Issues in the World's Largest Rainforest Region*, World Sustainability Series,
https://doi.org/10.1007/978-3-031-81465-5_16

et al. 2022). The activity has been demonstrating importance for human survival and nature preservation. Since bees play a very important role in flowering, performing the pollination of plants that may be being cultivated (Costa et al. 2021). Resulting in the production of seeds of the most varied species that are fundamental for cultivation.

The breeding of stingless bees was not always practiced consciously due to the intense presence of exploitation and inadequate management of hives. This resulted in the loss of dozens of colonies, corroborating the threat of extinction that these species suffer (Rosa et al. 2019). Although the activity of exploration is still very present, the methods and knowledge about the creation have changed with the creation of the Meliponiculture. This allowed an adequate management that facilitates the removal of the products without harming the colonies (Zapechouka and Silva 2021). Such a method allowed the activity to be seen as an educational, recreational, sustainable and economically viable mean. Ensuring improvement in income for communities, benefiting man and nature, with alternatives that minimize negative impacts on the environment (Barbiéri and Francoy 2020).

The activity has been growing due to the awareness of the damage caused to the environment, especially fires and deforestation, which could cause the disappearance of these pollinators and impair the spread of angiosperms in the Amazon (Silva and Paz 2012). In addition, many who perform the activity still do not have knowledge for production and management practices (Ribeiro and Lima 2015). The meliponiculture in the region is mainly traditional, prioritizing family use, without realizing the potential for the composition of family income (Meireles et al. 2018). However, meliponiculture is an economically viable activity, easy to manage, which can be carried out both for the purpose of obtaining products for marketing and for pollination, for producers of agricultural crops (Ribeiro 2023).

2 Purpose of the Paper

This study aimed to measure the cost of implementation of an University meliponary. Given the potential of the activity to contribute to the composition of the income of the communities as an alternative of sustainable production. Favoring the establishment of balance in the interaction between man and nature, so that both can benefit, through the performance in the preservation of the species, guarantee of the pollination of angiosperms coordinated in addition to the financial gain.

Fig. 1 Aerial view of the meliponary of the Universidade Federal do Oeste do Pará

3 Approach

3.1 Place of Study

The work was made at the Fatima Mekdece´s Grove, at the Universidade Federal do Oeste do Pará (UFOPA) which has a meliponiculture sector with an area of 16 × 13 m (Fig. 1). There are two meliponaries structures containing around seven boxes of native bees each and a Hotel for solitary bees (Fig. 2). The place has the presence of angiosperm plants, which favors pollination and facilitates for bees to find their food. In addition, water is available nearby. Through the structures present on site, it was possible to determine the costs for the implementation of a meniliponary in the West Region of Pará, Legal Amazon, Brazil.

4 Results Achieved

4.1 Fence

All prices of the structures were in Real (official Brazilian currency) and converted to Dollar, using the commercial quotation of September 28, 2023 (US$ 1.00 = R$ 5,28).

For the delimitation of the area of the meliponiculture sector that corresponds to 16 × 13 m, the materials for the construction of the fence were identified, as well as prices. The total linear measurement is 58m of fence. The materials and prices are in Table 1.

Fig. 2 Hotel for solitary bees

Table 1 Materials and prices for fence construction, in US$

FENCE				
Materials	Measurement units	Quantity	Unitary price (US$)	Total price (US$)
Galvanized fabric	Linear meters	58	1.57	90.95
Wood	Units	20	3.41	68.18
Galvanized nails	1kg	1	2.84	2.84
Total value				**161.98**

4.2 Collective Shelter

For the construction of the collective shelter it was observed the material that would be used for the implantation and labor for the activity, being necessary to consider the cleaning and management of the place. It is of paramount importance to keep the breeding site clean and proper management to avoid possible accumulations of residues and phorid attacks that may result in the loss of the colony (Fig. 3). The materials and prices to build the meliponaries are in Table 2.

Fig. 3 Meliponaries at clean places

Table 2 Materials and prices for the collective shelter construction, in US$

Collective shelter				
Materials	Measurement units	Quantity	Unitary price (US$)	Total price (US$)
Fiber-cement tiles	2.44 × 0.50 m	4	4.55	18.18
Ringed nails	500 g	1	3.22	3.22
Galvanized nails	1 kg	1	4.92	4.92
Massaranduba's wood	6 m	10	54.92	549.24
Total value				**575.57**

Note The hotel for solitary bees is built with material that was not used in collective shelters such as wood shavings and tiles as well as alternative objects reused like bottle caps

4.3 Labor

For labor, it was considered the cleaning and management of the area, the installation of the fence, the implementation of the collective shelter and construction of the hotel for solitary bees. The established prices were collected from three constructors which can be observed in Table 3.

Table 3 Labor time and costs, in US$

LABOR				
Services	Measurement units	Quantity	Unitary price (US$)	Total price (US$)
Cleaning/management	Day job	1	15.15	15.15
Installation of the fence	Day job	1	22.73	22.73
Implementation of the collective shelter	Day job	4	34.09	136.36
Construction hotel for solitary bees	Day job	2	34.09	68.18
Total value				**242.42**

4.4 Hives

The prices of the hives, or bee box, are established according to the species that will be worked and the size of the desired box. Thus, the prices of four species of native bees most common in the northern region of Brazil and that have a great potential for production and pollination were collected. They are: Melipona Compressipes (Uruçu), Tetragonistica Angustula (Jataí), Tretagona Clavipes (Borá) and Scaptotrigona Depilis (Canudo). The price of the hives already includes the wooden boxes and it was calculated for one collective shelter (Table 4).

Table 4 Materials and prices of the hives, in US$

HIVES				
Species	Measurements in cm	Quantity	Unitary price (US$)	Total price (US$)
Melipona Compressipes (Uruçu)	20 × 20	3	22.73	68.18
Tetragonistica Angustula (Jataí)	19 × 15	3	18.94	56.82
Tretagona Clavipes (Borá)	20 × 20	3	22.73	68.18
Scaptotrigona Depilis (Canudo)	19 × 15	3	18.94	56.82
Total value				**250.00**

Table 5 Result of costs for the implementation of the meliponary, in US$

Costs for the implementation of the meliponary	
Items	Total price (US$)
Fence	161.98
Collective shelters	1,151.14
Hives	500.00
Labor	242.42
Total value	2,055.54

4.5 *Total Costs for Implementation*

The data collected presented the creation of stingless bees an economically viable activity, of easy and practical application, and can be carried out, in a rational way, even without a deep knowledge. The activity can be used for various purposes, whether for the economy, education, sustainability and recreation. Meliponiculture has the potential to generate income for several communities, either for the sale of products derived from creation, or for the use of bees as pollinators in angiosperm creations. In addition, the activity has a relatively low investment for its implementation, as it was measured. Table 5 shows total value for one meliponary with two collective shelters and one hotel of solitary bees.

The opportunity cost is not contemplated in the work. We assume in a brief digression that the UFOPA's meliponary, installed within the forest area, exceeds in benefit the opportunity cost of the 208 m^2 of the delimited area. The pollination capacity of bees contributes significantly to biodiversity. Still from the institutional perspective, today the meliponary receives the visit of schools and other actors of civil society, contributing to the transversal effort to bring the community into the university. In addition of hosting practical classes, research and extension projects. Considering the context of family farming, 208 m^2 is equivalent to 0.0208 hectares, certainly the gains in improving the environment and the complement of income would exceed the opportunity costs of the immobilized area for the implementation of the meliponary.

5 Conclusions

It was concluded that the practice of meliponiculture is an activity that has been establishing space of consciousness in accordance with the sustainable economy. Encouraging preservation practices and improving the financial situation of several families, with a relatively low investment. In addition of being the object of analysis for education, research for traditional socio-environmental activities.

Acknowledgements All images and tables presented belong to the authors.

References

Barbiéri C, Francoy TM (2020) Modelo teórico para análise interdisciplinar de atividades humanas: A meliponicultura como atividade promotora da sustentabilidade. Ambiente Sociedade 23

Costa GB, Ferreira MAC, Silva L, Meneghetti GA (2021) Meliponicultura no Amazonas: desafios para a manutenção dos serviços ambientais, sustentabilidade e bem-estar das comunidades rurais. In: 59º Congresso da Sociedade Brasileira de Economia, Administração e Sociologia Rural—SOBER e 6º Encontro Brasileiro de Pesquisadores em Cooperativismo—EBPC, 02–06 agosto

Gemim BS, Silva FAM, Schaffrath VR (2022) Aspectos Socioambientais da Meliponicultura na Região do Vale do Ribeira, São Paulo, Brasil. Guajú, Matinhos 8:1–25

Meireles AM, Tavares FB, Lopes ELN, Cordeiro YEM (2018) Aspectos econômicos, sociais e ambientais inerentes a atividade de meliponicultura no baixo Tocantins, estado do Pará. Revista Observatorio de la Economía Latinoamericana

Ribeiro GC (2023) Desafios da cadeia produtiva da meliponicultura no interior da Amazônia: produção de mel na Comunidade Palmari, Atalaia do Norte, Amazonas, Brasil. 2023. 43 f. Trabalho de Conclusão de curso (Bacharelado em Administração)—Universidade Federal do Amazonas, Instituto de Natureza e Cultura, Benjamin Constant (AM)

Ribeiro MF, Lima CBS (2015) Avaliação da criação de abelhas-sem-ferrão em Fernando de Noronha após 30 anos de sua introdução. Magistra, Cruz das Almas 27(3/4):474–481

Rosa JM et al (2019) Desaparecimento de abelhas polinizadoras nos sistemas naturais e agrícolas: existe uma explicação? Revista De Ciências Agroveterinárias, Centro De Ciências Agroveterinárias—UDESC Lages 18(1):154–162

Silva WP, Paz JRL (2012) Abelhas sem ferrão: muito mais do que uma importância econômica. Natureza Online 10(3):146–152

Texeira IRV, Silva GV, Leite ICS, Oliveira L, Marques LC (2022) Mulheres, Abelhas e Sustentabilidade: O Caso de um Curso de Meliponicultura. Res Soc Dev 11(9):e55511931807, 1–11

Zapechouka AJ, Silva FF (2021) Uma análise da teoria sobre a ação humana e suas consequências para as abelhas nativas sociais. Meio Ambiente (Brasil) 3(5):81–93

Access and the Quality of Drinking Water in Quilombola Communities in the Amazon, the World's Largest Tropical Rainforest

Francisca Oliveira de Jesus, Valéria Bentes, Josiane dos Santos Carmo, Veridiana Barreto Nascimento, Alinne Mara Pinheiro Mota, Beatriz Smidt Celere, Susana Inés Segura-Muñoz, Luciana Gonçalves de Carvalho, and Marina Smidt Celere Meschede

Abstract This study sought to understand how access to and the quality of water used for human consumption occurs in quilombola communities located in the Amazon region. Four quilombos in Santarém (PA), Amazon region, were selected and a questionnaire was applied to collect data about access and, subsequently, samples ($n = 31$) of water were collected for microbiological and physical–chemical analysis. The results obtained regarding the potability of the water were compared with the recommendations of Resolution 888/2021 of the Ministry of Health and the analysis of significance was carried out using the t-test ($p < 0.005$). Access to water for human consumption occurred in different ways through surface collection, artesian wells and springs in the quilombos investigated. The microbiological analyses showed contamination in 90.2% and 61.3% of the samples for total coliforms and Escherichia Coli, respectively. The physico-chemical assessment showed acidic water in 58% of the samples. The parameters of turbidity and apparent color were inconsistent with the recommendations only in the quilombola community that uses surface water for human consumption. Nitrate concentrations in 100% of the samples were within acceptable limits, but the data shows a warning for the concentrations obtained in two of the quilombos investigated, indicating sanitary deficiencies. The water in the quilombos investigated is of compromised quality, especially in terms of microbiological contamination, and could therefore pose a risk to human health, requiring continuous monitoring.

F. O. de Jesus · V. Bentes · J. dos Santos Carmo · A. M. P. Mota · L. G. de Carvalho · M. S. C. Meschede (✉)
Federal University of Western Pará (UFOPA), Santarém, Pará, Brazil
e-mail: marina.meschede@ufopa.edu.br

V. B. Nascimento
Federal University of Amapá (UNIFAP), Oiapoque, Amapá, Brazil

B. S. Celere · S. I. Segura-Muñoz
Universityof São Paulo (USP), Ribeirão Preto, São Paulo, Brazil

W. Leal Filho et al. (eds.), *Amazon 2030—Sustainability Issues in the World's Largest Rainforest Region*, World Sustainability Series,
https://doi.org/10.1007/978-3-031-81465-5_17

Keywords Amazon ecosystem · Quilombola communities · Water quality · Water uses

1 Introduction

There are around 6 thousand quilombola communities in Brazil, but only 3 thousand are officially recognized, according to the Palmares Cultural Foundation (FCP 2017). The region with the largest number of quilombos is the Northeast of Brazil (3,171 communities), followed by the Southeast (1,300 communities), North (873 communities), South (319 communities) and Midwest (250 communities). In the Amazon region, in the state of Pará, according to the Coordination of Associations of Remaining Quilombo Communities of Pará (Malungu), there are around 527 remaining quilombo communities in the territory, only a third of which have obtained land titles, and just over half of which have received the certification issued by the Palmares Cultural Foundation, which is an initial stage in the titling process (Malungu 2021). The region of Santarém, located in the west of the state of Pará, has a Federation of Quilombola Organizations of Santa-rém (FOQs) which includes 12 communities distributed in different spaces: urban, floodplain and dry land regions.

Quilombos are recognized as ethnic-racial groups with their own well-defined historical trajectory and specific territorial relations linked to black ancestry and the oppression they suffered (Brazil 2003). In Brazil, the "Quilombola Brazil Program" developed by the Ministry of Health (MS) has been working since 2007, through Decree No. 6.261/2007, with a set of actions aimed at improving living conditions and expanding access to public goods and services for people living in quilombola communities, among them those focused on Axis 2, which concerns the promotion, quality and access to drinking water (Brazil 2007).

In the literature, studies on access and quality of water for human consumption in quilombola communities located in the Amazon region, particularly those belonging to the city of Santarém, are non-existent. However, studies available in quilombos in other Brazilian regions, such as the northeast and center-west, point to major problems regarding access and potability of water and that actions aimed at Axis 2 (mentioned above) are compromised (Amorin et al. 2013; Queiroz and Oliveira 2018).

Adequate water supply, in terms of quantity, access and quality, is essential for local socio-economic development, with direct repercussions on the health and well-being of the population (Razzoli and Günther 2008), however, according to Ferreira et al. (2017) the water supply has been uneven between urban and rural areas (where most quilombola communities in Brazil are located). The main sources of water supply in rural areas are shallow wells, springs, streams and rivers, which are often susceptible to contamination by sanitary waste (Ferreira et al. 2017). The precariousness of actions aimed at water security may compromise the quality of water intended for human consumption, favoring the appearance of waterborne diseases, such as acute diarrheal diseases, outbreaks of Hepatitis A, intestinal parasites, among others (Meschede et al. 2018).

The Santarém region has significant weaknesses in terms of access to systems considered satisfactory for basic sanitation, including water distribution (Trata Brazil 2022). Geographical barriers in the Amazon rainforest often make it impossible to install treatment plants and periodic maintenance of supply systems in rural locations, and the presence of microbiological contaminants in water samples from communities located on the banks of the Tapajós River has already been reported in the literature (Batalha et al. 2014). In these regions remote from the basic services, the lack of access to safe sources of water can lead to the consumption of water of dubious sanitary quality and in insufficient and irregular volume to meet basic daily needs (Razzolini and Günther 2008). Therefore, regular access to safe drinking water, although a basic human right, has not been extended to the entire population, especially those found in peri-urban areas forgotten by public sanitation and health policies (Razzolini and Günther 2008).

Based on the context and recognizing that inadequate water supply services are related to an increase in water-borne diseases, this study aimed to assess the accessibility and potability of water in quilombola communities located in the rural and urban regions of Santarém, in the Brazilian Amazon, state of Pará.

2 Methodological Procedures

2.1 Study Site

For this study, four quilombola communities were selected from Santarém, west of Pará, from different geographical locations in the Amazon region (Fig. 1). The first quilombo community selected was Saracura, which is located in a floodplain region on the left bank of the Amazon River and has around 125 resident families. Access to the Saracura quilombo is by riverboat and the community's main economic activity is agriculture and the production of manioc by-products and fishing. The second quilombo chosen for this study was Pérola do Maicá because it is the only urban quilombo in Santarém. It has around 30 families and was formed by the displacement of quilombola families who left the countryside and settled in the city at the end of the 1980s due to the phenomenon known as "TerrasCaídas" (accentuated erosion that damages riverside and floodplain areas). The third quilombo was Murumuru, which has around 40 resident families and is located around 55 km from Santarém and its main economic activity is the production of Açaí. Lastly, we selected the Tiningúquilombo, which has around 40 families and is located around 70 km from Santarém. Tiningú is the only quilombo community in Santarém with a Basic Health Unit (BHU) in its territory and its economic activities are based on the artisanal production of manioc flour.

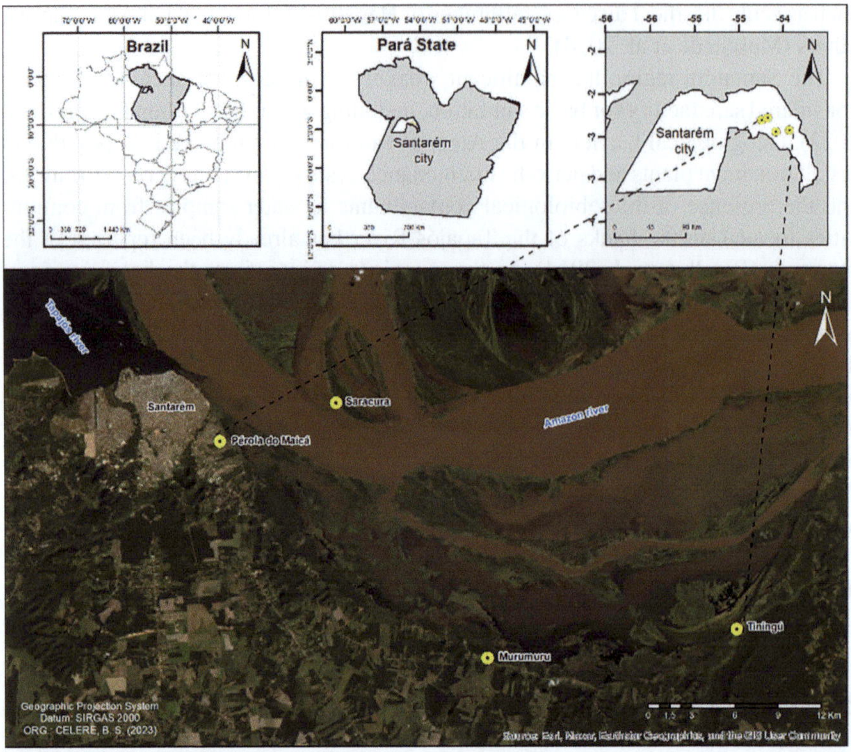

Fig. 1 Map showing the geographical location of the quilombos that took part in the research, Santarém, Amazon biome

2.2 *Data Collection and Analysis*

At random, in order to cover a heterogeneous distribution in the quilombos, 30 households were selected and a questionnaire with closed questions was applied to collect data on access to water. Subsequently, in the homes of the research participants, water samples were collected from the kitchen tap most commonly used for human consumption during the first half of 2023. The samples collected (n = 30; 13% of the total population of the quilombos investigated) were analyzed for microbiological and physical–chemical parameters.

The methods used in the laboratory evaluation, quality control and equipment calibration procedures followed the protocols recommended by the Standard Methods for the Examination of Water and Wastewater (APHA 2023). The micro-biological analyses were carried out using the Colilert® chromogen substrate to determine the presence or absence of total coliforms and *Escherichia coli* (*E.coli*). For the physical–chemical analyses, hydrogen potential (pH) and conductivity were determined by potentiometry and evaluated over a range of 0–14 and 1–1500 μS cm, respectively. Turbidity was determined using the nephelometric method with a scale of

0.20–750 Turbidity Units (UT). Apparent color was determined using the platinum cobalt technique and the Aquacolor® equipment calibrated within the reading range of 2.5–500 Color Units (uH). Nitrate analysis was carried out using photometry, with a reading range of 1.0–100 mg L^{-1}.

2.3 *Ethical Issues*

This study was approved by the Research Ethics Committee (CEP) of the Federal University of the West of Pará (UFOPA) by report number 5.826.150 and obtained authorization from the FOQS of Santarém for data collection.

2.4 *Analysis of Results*

The results were tabulated in an Excel spreadsheet (order number: 91d8c934-41e1-4753-8d54-c216491f124e), after which the parameters obtained were compared with the Potability Standards of Resolution No. 888/2021 of the Brazilian Ministry of Health, which establishes some maximum tolerated limits for water for human consumption (Brazil 2021). In addition, the variations obtained in the physical–chemical parameters in each quilombo were compared using statistical analysis. The Shapiro-Wilks test ($p < 0.05$) was applied to verify normality. The t-test was used for normal distribution and a probability of 0.05 or less was considered significant ($p < 0.05$). Statistical analyses were carried out using Prism—GraphPad (version 5.1, San Diego, CA, USA).

3 Results

The access to water is understood as obtaining it from a source, defined in a spatial and temporal context (Razzolini and Günther 2008). For Howard and Bartram, there is access to waterwhen the source of supply is located up to 1 km away and the time taken to reach it is a maximum ofmaximum of 30 min. According to the authors, the recommended minimumper capita supply is 20 L/per day.

In the communities investigated, the source for obtaining water for human activities (drinking, personal hygiene, use in domestic practices, etc.) differed (Table 1).

In the Saracura, water comes from a surface source, from the Amazon river, which is collected by the residents themselves using hoses, stored in a water tank and then used for human consumption. Residents used to collect water by hand, but this required physical effort and long walks to the source, especially during the region's dry season. Nowadays, residents use a "motor pump" that uses fuel

Table 1 Summary of the characteristics of access to water in the quilombos evaluated, Brazilian Amazon

Quilombo evaluated	Access of water and distribution	Water treatment
Saracura	Surface from the Amazon river by the population through "motor pump". Distribution is done by hoses	It is not dealt with by the public authorities. The population uses individual measures using sodium hypochlorite and household filters
Pérola do Maicá	Private shallow groundwater wells. Distribution is done by Pvc pipes	It is not dealt with by the public authorities. The population uses individual measures using sodium hypochlorite
Murumuru	Shallow groundwater wells. Distribution is done by Pvc pipes	It is not dealt with by the public authorities. The population uses individual measures using sodium hypochlorite
Tiningú	Forest water springs. Distribution is done by Pvc pipes	It is not dealt with by the public authorities. The population uses individual measures using sodium hypochlorite

and a motor to generate energy to collect water directly from the river. Hoses then distribute the water to a household water tank for use by the residents. It is worth noting that the Saracura community is still waiting for the Brazilian public authorities to provide them with continued access to electricity for water collection and other activities that are fundamental to the population. In the Pérola do Maicá and Murumuru communities, access to water is via underground wells with an average depth of 20 to 32 m, which intercept the voluminous Alter do Chão Aquifer. Once the water has been collected, it is used for consumption through hydraulic pipes. In the Tiningú quilombo, water is collected from springs coming directly from within the Tapajós National Forest (Fig. 2a), which is first sent to a reception box and then through hydraulic pipes to the population (Fig. 2b).

As for the presence of microbiological contaminants, the samples collected in the Saracura quilombo showed that 100% (n = 13) grew total coliform bacteria and of these 92.3% (n = 12) grew *E.coli*. Pérola do Maicá was the quilombo with the lowest rate of bacterial contamination, with 30% (n = 07) and 20% (n = 02) of the samples showing growth for total coliforms and *E.coli*, respectively. In the Murumuru quilombo, 100% (n = 03) of the samples were positive for total coliforms and none were found to contain *E.coli*. In Tiningú, 100% (n = 05) of the samples were contaminated by coliforms and *E.coli* (Fig. 3). The results showed inconsistencies with Brazilian legislation, which recommends the absence of growth of these bacteria in water samples for human consumption (Brazil 2021).

The physico-chemical results showed that in the Pérola do Maicá, Murumuru and Tiningú quilombos the water is acidic (pH between 3.9 and 5.6) and outside the range (6.0 to 9.0) recommended by MS ordinance no. 888/2021 (Brazil 2021). On the other

Fig. 2 **a** Representative model of the water flow (1) from springs in the Tapajós National Forest and **b** presentation of the receiving box (2) and pipes for supplying the population (3) in the Tiningú quilombo, Santarém, Pará (*Source* authors of the research)

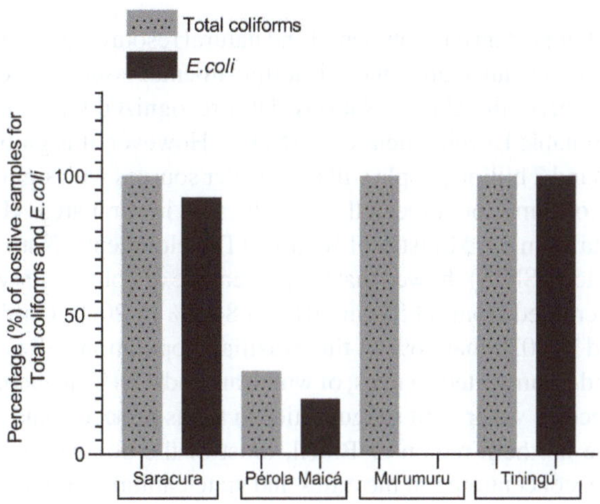

Fig. 3 Percentage (%) of water samples contaminated for total coliforms and *E.coli* in Santarém's ki-lombos (n = 30)

hand, in the Saracura quilombo, the surface water collected from the Amazon river and used for human consumption, the pH values were range adequate (6.9 and 7.7), with values statistically different from the other quilombos evaluated ($p < 0.005$).

The turbidity in the samples from the Saracura quilombo was significantly higher (average value of 14 UT) ($p < 0.005$) when compared to the other quilombos evaluated, and above the maximum recommended value (up to 5 UT) according to Brazilian legislation (Brazil 2021). In the other quilombos evaluated, turbidity remained adequate (values between 0.1 and 1.7 UT). Similarly, the apparent color parameter was only altered in the Saracura quilombo (values between 9.2 and 219 uH), with a recommended maximum value of up to 15 uH for water samples for human consumption. In the other quilombos evaluated, apparent color remained within the recommended range (values between 0.2 and 3.2).

Conductivity expresses the amount of dissolved ions in the water, and when >100 μS/cm indicates an impacted environment (Meschede et al. 2018). The Brazilian resolution does not recommend maximum values for this parameter and the samples evaluated in this study showed low conductivity values ranging from: 22.7 to 90.7; 29.2 to 93.7; 18.8 to 21.3 and from 15.8 to 18.1 μS/cm for the samples from the Saracura, Pérola do Maicá, Murumuru and Tiningú quilombos, respectively. Figure 4 shows the results for the physical and chemical parameters assessed in the water samples from the quilombos investigated.

4 Discussion

Water is the most important environmental and natural resource available to humanity and its access with adequate guarantees is a fundamental issue for the promotion of public health. In 2010, the United Nations (UN) recognized that water is a human right and a Sustainable Development Goal (SDG). However, it is globally estimated that in 2020, around 2 billion people still used water sources with significant sanitary restrictions for consumption, especially those located in rural areas (UN 2022).

In Brazil, data from the Ministry of Regional Development's National Sanitation Information System (SNIS) showed that the percentage of households with a drinking water supply increased from 83.5% in 2017 to 84.0% in 2020. On the other hand, it was calculated in 2020 that 16% of the Brazilian population still consumes water from dubious and/or untreated sources, of which around 41.1% are people who live in the North and receive water with irregularities in terms of access and, consequently, potability. In the northern region of Brazil, most of the population with access to drinking water is concentrated in the cities and state capitals (Gomes et al. 2022).

Access to groundwater in the Pérola do Maicá and Murumuru quilombos was similar to that in the urban context of Santarém and congruent with other studies carried out in Brazilian quilombos (Amorin et al. 2013; Ferreira et al. 2014, 2017; Amorim et al. 2017; Magalhães-Filho et al. 2017), showing that drilling artesian wells could be an option for water supply in rural Amazonian communities due to the presence of the voluminous Alter do Chão aquifer that favors the entire region. In addition, the supply of water through wells is an economically viable alternative, which provides water of good natural quality (requiring little sanitary treatment) and at a low cost (Meschede et al. 2018). However, the WHO has warned that

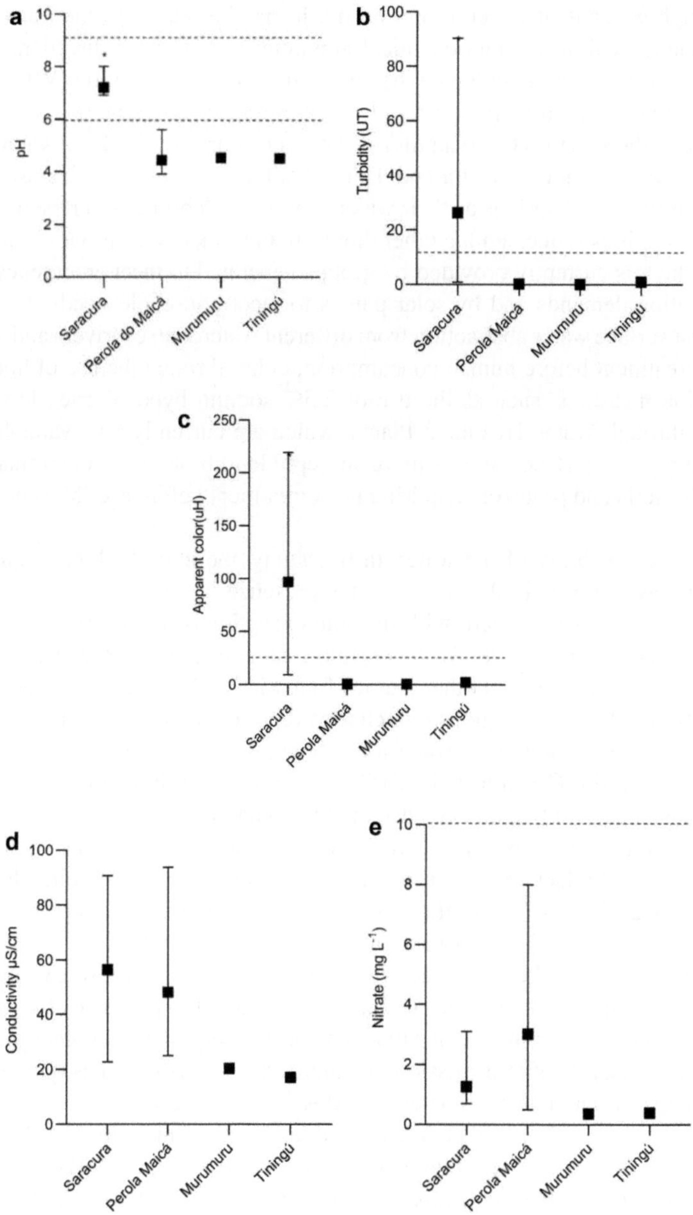

Fig. 4 Physico-chemical parameters in water samples according to the quilombo evaluated in Santarém (n = 30): **a** pH; **b** Turbidity in UT; **c** Apparent color in (uH); **d** Conductivity in µS/cm and **e** Nitrate in mg/L. *t-test: $p < 0.05$, indicates a significant difference between the samples evaluated. Red line indicates reference values according to Ministry of Health Ordinance No. 888/2021 (Brazil 2021)

even though water is of underground origin, it may be related to the transmission of microbiological agents and chemical compounds that cause disorders and are determinants in the epidemiological health profiles of the population (WHO 2017), requiring continuous surveillance.

Although the Alter do Chão aquifer is the main source of drinking water supply in the state of Pará, the Saracura and Tiningúquilombola communities use surface water from rivers and springs as their water supply. The absence of artesian wells in these communities is due, among other things, to the lack of a continuous electricity supply, which is currently provided by groups generated to meet emergency health and education demands and by solar panels to meet household needs. It is worth noting that surface water abstraction from different watercourses (rivers and springs) requires treatment before human consumption, either through the use of household disinfection measures, such as the use of 2.5% sodium hypochlorite (Jesus et al. 2023) or through Water Treatment Plants, which are currently not available in the Santarém region. Surface water is more susceptible to biological contamination by viruses, bacteria and protozoa, requiring prior treatment before use (Meschede et al. 2018).

As for the potability of the water, in this study, the microbiological quality of the water was compromised in terms of the presence of total coliforms and *E.coli* in all the quilombos evaluated, with the sole exception of the Murumuru community, which showed no growth for *E.coli* in the samples investigated. The results of this research are in line with other studies found in the literature that were carried out in different Brazilian regions, which also reported alarming levels of coliform and *E.coli* contamination in the water used for consumption in Brazilian quilombos (Ferreira et al. 2017; Oliveira et al. 2015). The presence of these microorganisms in the samples in this study could be attributed to contamination of the water resource by sanitary waste, by impurities acquired on the way the water reaches the point of consumption, by lack of proper maintenance of household reservoirs, due to the shallow depth of wells (where they exist), by the proximity of supply sources to septic tanks and lack of water treatment. Total coliforms and *E.coli*, when present in drinking water, should be viewed with caution, as they can cause gastroenteritis and other gastrointestinal disorders (Bentes et al. 2020). The presence of E. coli is an indicator of fecal contamination and may indicate the coexistence of other pathogens in the fecal oral cycle of transmission of waterborne diseases such as typhoid fever, cholera, hepatitis and intestinal parasites (Meschede et al. 2018).

As for the physicochemical parameters, some inconsistencies in the pH, turbidity and apparent color values were found in the quilombos investigated. The acidic water obtained in the quilombos, with the exception of Saracura, is in line with other studies in the Santarém region which have also indicated low pH values for water for human consumption (Bentes et al. 2020; Mendes et al. 2017; Meschede et al. 2018). One possible explanation is the presence of siliciclastic materials in the rocks of the Alter do Chão formation, which contribute to the determination of acidity in the aquifer's water naturally (Mendes et al. 2017). The direct health effects of consuming acidic water are still uncertain, but it has been reported in the literature that drinking acidic beverages in general increases the likelihood of changes and decreases in duodenal pH and a significant reduction in periods of duodenal alkalization. In addition, an indirect effect on human health may be the greater solubilization of metals (e.g., aluminum and cadmium) in acidic water (Bentes et al. 2020).

Turbidity was higher in Quilombo Saracura when compared to the other communities investigated. This was to be expected, since surface waters tend to have higher levels of turbidity due to the presence of a greater quantity of suspended solid materials (e.g., organic matter). In addition, in particular, the Amazon River (Fig. 5) and its tributaries, which originate in the Andean and pre-Andean regions, carry large quantities of sediment which are deposited on the banks of the floodplain or form islands within the rivers themselves, contributing to the natural increase in turbidity (Filizola and Guyot 2011). Similarly, the apparent color exceeded the recommended values only in the Saracura quilombo, also due to the characteristic of the Amazon River in carrying large quantities of clay material during its transport (Filizola and Guyot 2011).

The other parameters (conductivity and nitrate) showed values within the normal range according to Brazilian legislation (Brazil 2021). Although within expectations, it should be noted in this study that nitrate concentrations were higher in the Pérola do Maicá and Saracura quilombo, showing significant differences when compared to the other samples ($p < 0.05$). The presence of nitrate in water indicates unsatisfactory sanitary conditions, since it is a by-product of the decomposition of organic matter (Meschede et al. 2018), and may be related to the shallow depth of the wells and population density in the Pérola do Maicá urban quilombo and the presence of sanitary sewage in the Saracura quilombo. It is important to note that consuming too much nitrate can lead to undesirable effects on human health, due to the formation

Fig. 5 Amazon River used as a means of accessing water for human consumption in the Saracura community, Santarém, Pará (*Source* research authors)

of nitrosamines and nitrosamides in the human body after excess intake, which are potentially carcinogenic substances (Alaburda and Nishihara 1998).

5 Conclusion

Taken together, the results showed that the different forms of access to water in the kilometers evaluated (artesian wells, surface collection from the Amazon River and from springs) are subject to the incorporation of pollutants such as coliform bacteria and nitrates. In addition, physical–chemical parameters such as turbidity and apparent color were altered in the water collected from the Amazon River and the pH was acidic in the samples from wells and springs that are used for human consumption.

The consumption of water with total and fecal coliforms, favors gastric-intestinal illnesses. Ingestion of nitrate through water can be associated with harmful effects on humans at excessive levels, such as carcinogenic effects. The parameters turbidity and apparent color do not represent direct health effects, but they become complementary measures of water potability because they indicate the presence of solid materials in suspension in drinking water. It should be borne in mind that other studies in the Amazon have pointed to the consumption of acidic water, which does not seem to be an exclusive characteristic of the Santarém region. The health effects of prolonged consumption of acidic water are still uncertain, but attention should be paid to the fact that the low pH of water increases the solubilization of certain minerals that are harmful to human health.

The health risks associated with the consumption of water with compromised quality can be minimized through periodic monitoring in quilombola communities and through the adoption of satisfactory sanitation and water supply practices by the Brazilian state. To alleviate this situation, household and palliative measures such as chlorinating water with 2.5% sodium hypochlorite and boiling, building piping systems to homes with adequate sealing, periodic maintenance of water reservoirs and health education actions for the population on water safety could be an alternative.

References

Alaburda J, Nishihara L (1998) Presença de compostos de nitrogênio em águas de poços. Rev Saude Publica 32(2):160–165. https://doi.org/10.1590/S0034-89101998000200009

Amorim PM, Silva RA, Sato MT (2017) Latas d'água nas cabeças: Percepções sobre a água na comunidade quilombola de Mata Cavalo. REMEA—Revista Eletrônica Do Mestrado Em Educação Ambiental 34(3):130–146. https://doi.org/10.14295/remea.v34i3.7204

Amorin MM, Tomazi L, Silva RAA, Gestinari RS, Figueiredo TB (2013) Evaluation of housing and health conditions of Boqueirão afro descendant community, state of Bahia, Brazil. Biosci J, Uberlândia 29(4):1049–1057. https://docs.bvsalud.org/biblioref/2018/10/946462/avaliacao-das-condicoes-habitacionais-e-de-saude-da-comunidade-_t7lSayi.pdf. Accessed 08 Sept 2023

APHA—American Public Health Association (2023) Standard methods for the examination of water and wastewater, 24nd edn. Washington DC

Batalha SS, Mortorano LG, Biase AG, Morales GP, Pontes NA, Santos L (2014) Condições físico-químicas e biológicas em águas superficiais do Rio Tapajós e a conservação de Floresta Nacional na Amazônia, Brasil. Revista Ambiente & Água 9(4):647–663. https://doi.org/10.4136/ambi-agua.1304

Bentes VS, Almeida GW, Meschede MSC (2020) Qualidade da água utilizada para consumo humano proveniente do aquífero álter do chão em Santarém (Oeste do Pará) e sua relação com a saúde pública. Geo Brasiliensis 34 (1):101–109. https://doi.org/10.21715/GB2358-2812.2020341101

Brazil (2003) Decreto n. 4.887 de 20 de novembro de 2003. Regulamenta o procedimento para identificação, reconhecimento, delimitação, demarcação e titulação das terras ocupadas por remanescentes das comunidades dos quilombos de que trata o Art. 68 do Ato das Disposições Constitucionais Transitórias. Presidência da República, Brasília. www.planalto.gov.br/ccivil_03/decreto/2003/D4887.htm. Accessed 8 Sept 2023

Brazil (2007) Lei n.º 6.040, de 07 de fevereiro de 2007. Institui a Política Nacional de Desenvolvimento Sustentável dos Povos e Comunidades Tradicionais. DOU, Brasília, DF. https://www.planalto.gov.br/ccivil_03/_ato2007-2010/2007/Decreto/D6040.htm. Accessed 8 Sept 2023

Brazil (2021) Portaria GM/MS nº 888 de 04 de maio de 2021. Dispõe sobre os procedimentos de controle e de vigilância da qualidade da água para consumo humano e seu padrão de potabilidade. https://bvsms.saude.gov.br/bvs/saudelegis/gm/2021/prt0888_24_05_2021_rep.html. Accessed 8 Sept 2023

FCP—Fundação Cultural Palmares (2017) Informações quilombolas. https://www.gov.br/palmares/pt-br/departamentos/protecao-preservacao-e-articulacao/informacoes-quilombolas/. Accessed 8 Sept 2023

Ferreira FS, Queiroz TM, Silva TV, Andrade ACO (2017) À margem do rio e da sociedade: a qualidade da água em uma comunidade quilombola no estado de Mato Grosso. Saúde E Sociedade 26(3):822–828. https://doi.org/10.1590/S0104-12902017166542

Ferreira EP, Ferreira JP, Pantaleão FS, Ferreira YP, Albuquerque KN, Ferreira TC (2014) Abastecimento de água para consumo humano em comunidades quilombolas no município de Santana do Mundaú—AL. Revista Brasileira de Geografia Física 7(06):1119–1125

Filizola N, Guyot JL (2011) Fluxo de sedimentos em suspensão nos rios da Amazônia. Braz J Geol 41(4):566–576. https://doi.org/10.25249/0375-7536.2011414566576

Gomes MCRL, Andrade LC, Nascimento ACS, Pedro JPB, Mota CR (2022) Conditions of use and levels of household access to water in rural communities in the Amazon. Ambiente & Sociedade 25:e01782. https://doi.org/10.1590/18094422asoc20210178r12vu2022L4OA

Jesus FO, Bentes V, Segura-Muñoz SI, Meschede MSC (2023) Eficácia das medidas domiciliares de desinfecção da água para consumo humano: enfoque para o contexto de Santarém, Pará, Brasil. Cad Saúde Pública 39(2):e00205322. https://doi.org/10.1590/0102-311XPT205322

Magalhães FJC, Paulo PL (2017) Abastecimento de água, esgotamento doméstico e aspectos de saúde em comunidades Quilombolas no Estado de Mato Grosso do Sul. Interações (campo Grande) 18(2):103–116. https://doi.org/10.20435/inter.v18i2.1435

Malungu. Comunidades quilombolas do Pará. https://malungu.org/comunidades-1/. Accessed 8 Sept 2023

Mendes A, Galvão P, Sousa J, Silva I, Carneiro RN (2017) Relations of the groundwater quality and disorderly occupation in an Amazon low-income neighbourhood developed over a former dump area, Santarém-PA, Brazil. Environ Dev Sustain 21:353–368. https://doi.org/10.1007/s10668-017-0040-8

Meschede MSC, Figueiredo BR, Alves RI, Segura-Muñoz SI (2018) Drinking water quality in schools of the Santarém region, Amazon, Brazil, and health implications for school children. Revista Ambiente & Água 13(6):e2218. https://doi.org/10.4136/ambi-agua.2218

Oliveira R, Oliveira FGS, Santos MAB, Almeida S, Moura GJB (2015) Avaliação da Potabilidade da Água Consumida por Quilombolas em Juazeiro, Ba, Brasil. Opará: Etnicidades, Movimentos Sociais e Educação 3(4):45–58. https://revistas.uneb.br/index.php/opara/article/view/OPR3.4.4

Queiroz TM, Oliveira LCP (2018) Qualidade da água em comunidades quilombolas do Vão Grande, município de Barra do Bugres (MT). EngSanit Ambient 23(1):173–180. https://doi.org/10.1590/S1413-41522018166375

Razzolini MTP, GüntherWMR (2008) Impactos na saúde das deficiências de acesso a água. Saúde e Sociedade 17(1):21–32. https://doi.org/10.1590/S0104-12902008000100003

Trata Brazil (2022) Organização da Sociedade Civil de Interesse Público. Ranking do Saneamento. https://tratabrasil.org.br/ranking-do-saneamento-2022/. Accessed 8 Sept 2023

UN—United Nations (2022) Global issues water. https://www.un.org/en/global-issues/water. Accessed 8 Sept 2023

WHO—World Health Organization (2017) Guidelines for drinking water quality: fourth edition incorporating the first addendum. Library Cataloguing in Publication Data, Geneva. https://www.who.int/publications/i/item/9789240045064. Accessed 8 Sept 2023

REDD+ – Where is the Money? Funding Protection of Tropical Forests

Fiona Ryan

Abstract Tropical deforestation is a major threat to global biodiversity and produces 11% of global greenhouse gas emissions. Tree cover in Brazil alone fell 13% from 2001 to 2022. The global public good of preserving forests has spawned a number of global initiatives including; Reducing Emissions from Deforestation and forest Degradation (REDD+), commodity certification, proposed unilateral restrictions on deforestation supply chains, and the Leaders Declaration on Forests. Nevertheless, the conversion of forests for agriculture provides a pathway for development and poverty reduction for many developing countries. Funding for REDD+ results based payments (RPB) is designed to provide an alternative development pathway to forest conversion. This study collects and analyses data on the range of sources of funding for REDD+ and other forest conservation activities, with the aim of identifying their potential to meet the development aspirations of people and countries while avoiding the further conversion of forests. These include; compliance emissions trading, official development assistance (ODA) for climate and biodiversity, voluntary Carbon markets and philanthropy. Currently, conservation activities including jurisdictional REDD+ activities, financed from public sources, are the largest finance flows despite only partially covering the net opportunity cost. We conclude emissions trading, such as would be possible under Article 6 of the Paris Agreement, is the only funding category likely to provide sufficient funding for REDD+ to reduce forest emissions.

Keywords Tropical deforestation · Sustainable development · Emissions trading · Climate finance · REDD+

F. Ryan (✉)
Forests Development and Climate, 14/58 Woodward St, Edge Hill, Cairns, QLD 4870, Australia
e-mail: fiona.ryan@forestclimate.net
URL: https://www.forestclimate.net/

W. Leal Filho et al. (eds.), *Amazon 2030—Sustainability Issues in the World's Largest Rainforest Region*, World Sustainability Series,
https://doi.org/10.1007/978-3-031-81465-5_18

1 Introduction

Tropical deforestation is a major threat to global biodiversity and produces 11% of global greenhouse gas emissions (World Bank 2023) The global public good of preserving forests has spawned a number of global initiatives, including, Reducing Emissions from Deforestation and forest Degradation (REDD+), commodity certification, proposed unilateral restrictions on deforestation supply chains such as the recently passed European Commission (2023) Deforestation Free Products regulations, and the Leaders Declaration on Forests. Land use change such as deforestation in tropical countries has been identified as the greatest threat to biodiversity by the Intergovernmental Science/Policy Platform on Biodiversity and Ecosystem Services (IPBES n.d.). Reducing deforestation in tropical countries has also been identified by Nabuurs et al. (2022) as the land based emissions mitigation option that has the largest economic mitigation potential. With a potential to reduce emissions from deforestation and degradation by 3.4 $GtCO_2$/yr by 2050, this is much greater than the next ranked land use option, afforestation, reforestation and restoration at 1.6 $GtCO^2$/yr by 2050. Protecting forests has the greatest potential for reducing CO_2 emissions. Tropical deforestation is concentrated in developing countries where the conversion of forests for agriculture provides a pathway for development and poverty reduction. In the United Nations Framework Convention on Climate Change (UNFCCC) (1992) equity issues between developing and developed countries resulted in the principle of common but differentiated responsibilities and respective capabilities (CBDR-RC) between countries. In addition, developed countries should meet the agreed full incremental cost of mitigation by developing countries. To reduce emissions from deforestation based on these principles, developed countries must provide adequate financial support to developing countries to fulfill the equity provisions contained within the UNFCCC. Ryan (2022) found that if insufficient funds were delivered then the forest protection activity was likely to lose country ownership. That is, the activity would lose political and social support within the country where the activity was conducted and thus, would likely fail.

Few studies have compared funding the potential from different sources of north/south flows such as philanthropy, environmental non-government organisations and international emissions trading. Deutz et al. (2020) have done a comprehensive study of global financial flows for nature but did not discriminate between global and north/south flows. Many of the estimates for this paper are in the form of a price of Carbon emissions. They provide indications of relative cost effectiveness rather than implying the use of a Carbon trading or Carbon pricing mechanism. However, the study does differentiate between price and non-price based mechanisms. Various financing sources such as publicly funded climate finance or international emissions trading will have different drivers and constraints which can influence the ultimate volumes of north/south climate finance these mechanisms deliver, in addition to their relative cost effectiveness.

The aim of this study is to identify the most likely sources of finance that can be expanded to provide sufficient funding to reduce deforestation and provide alternative development pathways to people in forested developing countries. Effective international and national governance frameworks are needed to foster these finance flows to transform the current financial and economic systems that result in large scale deforestation.

2 Scope and Key Definitions

Much of the data found in the literature includes related or overlapping terms and definitions such as nature-based-solutions, forest restoration, ecosystems based adaptation, natural climate solutions and many others. This study focuses on funding sources that provide nature-based solutions to climate change which was defined by the International Union for the Conservation of Nature (IUCN) in 2017 as:

> Actions to protect, sustainably use, manage and restore natural or modified ecosystems, which address societal challenges, effectively and adaptively, providing human well-being and biodiversity benefits (IUCN 2020):

This is a very broad definition which allows for the capture of data from narrower categories. Where the data is available, this study focuses on protection which is most closely represented by terms such as avoided deforestation and forest protection which includes funding for forest protection that may or may not include a results based payment component, which characterise REDD+ projects. There is limited data available on the more restricted definitions. The most used term related to protection or avoided deforestation is REDD+. However, even the term REDD+ includes reduction in forest degradation as well as sustainable management of forests and enhancement of forest carbon stocks. Ecosystem or forest protection also comes under the definition of protect. Many other terms could include forest protection such as nature based solutions (NBS) or ecosystems based solutions (EBS) and ecosystem based approaches (EBA).

This study focuses on north/south flows of finance. Where no separate data of flows from developed to developing countries are available, figures were included that include both north/north and north/south flows. This study does not evaluate non-financial policy options such as deforestation free supply chains or how they interact with climate finance. It does not examine concessionality or grant equivalence of the finance disbursed. It also does not compare the various potential sources of finance with the emissions gap that needs to be overcome to reach the Paris Agreement 1.5 degree goal or other goals such as reaching the Forests Leaders Declaration goal of ending deforestation by 2030.

The focus on forest protection is in alignment with the mitigation hierarchy for biodiversity and ecosystems services and sustainable management of living natural resource proposed by Arlidge et al. (2018) which prioritises avoidance of environmental damage over minimisation which itself is prioritised over restoration with offsetting being the lowest priority. As it is an ecosystem service this hierarchy has been applied to climate mitigation. A list of terms used and their ranking in the mitigation hierarchy is found in Table 1.

This study examines distributional issues between developed and developing countries. The sharing of proceeds between various entities such as levels of government, corporations, indigenous peoples and local communities within any particular country is beyond the scope of this study. Nabuurs et al. (2022) found that protection of forests was much more cost effective than restoration. Financing avoidance of deforestation with offsets against mitigation targets for other mitigation sectors such as fossil fuel is actually shifting effort from one avoidance category—reducing fossil fuel use to another avoidance category—reducing deforestation emissions. In such cases the focus is still avoidance. In contrast, offsetting fossil fuel mitigation with restoration of forests shifts the mitigation effort down the mitigation hierarchy from mitigation or avoided emissions to a lower step in the hierarchy—remediation. The study focuses on the period up to and including 2030 which is policy relevant to the Glasgow Leader's Declaration on Forests and Land Use and most of the first Nationally Determined Contributions (NDC) in the Paris Agreement.

This study focuses on definitions of sources such as climate finance (mostly from ODA) or philanthropy. We do not examine finance totals from individual institutions such as the World Bank as a source of finance. Its finance would normally be counted as climate finance (ODA). Likewise finance from WWF would be included in totals for environmental non-government organisations (ENGO).

Table 1 Relevant categories of nature based climate solutions used in the literature

Code	Activity category	Mitigation hierarchy steps ✱
AD	Avoided deforestation	1
P	Protect forests—avoided deforestation and forest degradation	1
REDD+	Reducing Emissions from Deforestation and Forest Degradation, enhancement of forest carbon stocks and sustainable management of forests: note: most projects under this category have been under mitigation hierarchy step one	1, 2, 3, 4#
GFW	Grasslands, forest and wetlands	1, 2.3.4
A	AFOLU—agriculture, forestry and land-use	1. 2. 3. 4
AFF	Agriculture, fisheries and forestry	1. 2. 3. 4
G	Global natural climate solutions—domestic and north south combined	1, 2, 3, 4

✱ The steps based on Arlidge et al. (2018): 1. avoid, 2. minimise, 3. remediate, 4. offset
Nabuurs et al. (2022) and Bloomberg NEF (2023) indicate that the majority of REDD+ activities implemented are for step 1

3 Methodology

This study begins with a search of the literature to identify the current and potential sources of funds for the various definitions of nature based climate finance. Some of the sources include voluntary carbon markets, public official development assistance (ODA) climate finance and compliance carbon markets. The literature includes peer reviewed journals, reports from non-government organisations, United Nations or other multilateral bodies and market research firms. A comparison is made between the level of funds currently being supplied by various funding sources which is in accordance with the first aim of this study. The carbon prices of current and potential sources are examined and compared with estimates of the net cost of action. Using this method, we identify those activities that both deliver sufficient funds to reduce deforestation and also provide a net benefit to the host country. Similarly, the prices per tonne are also compared to the cost to the global community such as estimates of the global social cost of Carbon to determine if the funding has increased the net benefit to the globe (Stern and Stiglitz 2017). The carbon prices identified do not necessarily represent an actual exchange of value between two parties to a transaction.

We also review the literature to identify the determinants of prices and volumes that may apply to each funding source. And make recommendations on what policy tools can be used to radically reduce deforestation with maximum benefit to developing countries and lead to a more equitable world. Where possible the classes of activities funded were ranked according to Arlidge et al. (2018) nature conservation mitigation hierarchy, see Table 1.

4 Assessment of the Literature

4.1 Theoretical Issues Identified in the Literature

A list of the categories of climate finance identified in the literature is contained in Table 2. Nabuurs et al. (2022) defines economic mitigation potential as mitigation that can be achieved at USD100/tCO_2-eq per year. The EU, UK and USA have stated in their NDCs that they will meet their Paris Agreement mitigation targets domestically (UNFCCC 2023) which strongly indicates they will not use Article 6 emissions trading to meet their Paris targets. If this stance is maintained, this leaves funding to come from climate finance (mostly ODA and private finance mobilised by public funding), philanthropy, environmental NGOs or voluntary carbon markets and other sources. Data on the magnitude of such sources is found in Tables 2 and 4.

The concepts of financial additionality and concessionality were not discussed to any extent in the literature. Additionality refers to the extent to which funding is increased rather than existing funding that has been reallocated with no net increase. Concessionality is the extent to which finance is provided on better terms than could be found in the commercial finance market. Current and recent carbon prices are

Table 2 Identified funding categories and recent financial flows

	Definition	Period	Annualised amount $ billions	Billion tC the O_2e	Source
(AFOLU) voluntary FTEM	Forestry and landuse	2021	1.33	0.227	Forest Trends' Ecosystem Marketplace (2022)
(AFOLU) voluntary FTEM	REDD+	2021	0.86		Forest Trends' Ecosystem Marketplace (2022)
Voluntary Bloomberg NEF	Mostly forest based and REDD+	2022		2	Bloomberg NEF (2023)
REDD+ voluntary market		2022	1.70	0.285	UNEP (2023)
IPCC voluntary AFOLU credits		2010 to 2018	0.47	0.026	Nabuurs et al. (2022)
Private mobilsed through MDBs. Includes all such climate investment	Private finance through MDBs and bilateral aid for biodiversity, as private finance mobilised		0.54		EBRD (2021)
Philanthropy	NBS		0.30		Casado-Asensio et al. (2022)
Philanthropy OECD		2020	0.60		Deutz et al. (2020)
Conservation NGOs	NBS		3.00		EBRD (2021)
MDB forestry ag/For/LU/Fsh	Low & MICs, no separate category for REDD or forestry		2.40		EBRD (2021)
MDB private finance mobilised	Private finance through MDBs for biodiversity, as private finance mobilised		0.72		OECD (2023)
AFW-private finance mobilised Ag fisheries and forestry	OECD DAC		2.30		OECD (2023)
(P) Private finance mobilised environmental rotection	OECD DAC		0.40		OECD (2023)

(continued)

Table 2 (continued)

	Definition	Period	Annualised amount $ billions	Billion tC the O_2e	Source
Mobilised private finance	OECD DAC	2016–2020	2.23		https://www.oecd.org/dac/environment-development/rioconventions.htm
Biodiversity DAC rules-significant	OECD DAC significant	2016–2020	4.90		https://www.oecd.org/dac/environment-development/rioconventions.htm
Total DAC biodiversity-principle	OECD DAC principle	2020	2.35		https://www.oecd.org/development/financing-sustainable-development/development-finance-topics/climate-change.htm
33% of climate finance	33% of climate finance	2000–2025	18.67		https://www.oecd.org/development/financing-sustainable-development/development-finance-topics/climate-change.htm
Cumulative REDD	Olesen et al. (2018) not specifically labelled REDD+	2008–2016	2.70		Olesen et al. (2018)
Cumulative REDD+';	Norman and Nakooda (2014) specifically labelled REDD+	2006–2014	1.10		Norman and Nakooda (2014)
Deutz ODA biodiversity high	This looks at ODA only and includes climate and biodiversity put together	2008–2015	9.70		Deutz et al. (2020)
Deutz ODA biodiversity low			4.00		Deutz et al. (2020)
ODA climate biodiversity UNEP	Includes climate and biodiversity		2.40		UNEP (2023)
Private finance with ODA UNEP	Includes climate and biodiversity		0.54		UNEP (2023)
CBD 2025 Kunming target	Convention or Biological Diversity		20.00		CBD (2022)

(continued)

Table 2 (continued)

	Definition	Period	Annualised amount $ billions	Billion tC the O_2e	Source
Illustrative avoided nature loss and sequestration	TSVCM does not distinguish between developing and developed countries in this source	Natural climate solutions account	0.30		TSVCM (2021)
CORSIA annualised cost high	CORSIA—Carbon Offset Reduction Scheme for International Aviation	2021–2026	0.56		ICAO (2022)
CORSIA annualised cost high	CORSIA—Carbon Offset Reduction Scheme for International Aviation	2021–2026	0.02	1.92	ICAO (2022)

Table 3 Current and recent carbon prices

	Definition	Year	USD/ tCO_2-e
Social cost of carbon low 2020 HLC (Stern and Stiglitz 2017)		2020	40.0
Social cost of carbon high 2020 HLC (Stern and Stiglitz 2017)		2020	80.0
EU ETS		2022–2023	85.0
REDD + net cost - high (UNEP 2023)	REDD+ Net cost	2023	50.0
REDD + net cost - high (UNEP 2023)		2023	30.0
Forest Trends (2022) voluntary carbon market price forest and land use	Forest and land use	2021	5.8
Forest Trends (2022) voluntary carbon market price forest and land use		2020	2.5
Voluntary carbon market (UNEP 2023)		2021	4.7
Jurisdictional REDD+ (Ryan 2022)		2019	10.0

Table 4 Funding projections for 2030

	2030 amount $ billions
(G) Global NBS carbon markets low (Deutz et al. 2020)	24.90
(G) Global NBS carbon markets high (Deutz et al. 2020)	39.90
Voluntary Bloomberg NEF (2023)	15.00
33% climate finance allocated to NBS (method in OECD (2021))	22.80
ODA biodiversity high (Deutz et al. 2020)	19.40
ODA biodiversity low (Deutz et al. 2020)	8.00

found in Table 3 and projected 2030 funding volumes found in the literature are listed in Table 4.

4.1.1 Compliance Carbon Markets

Article 6 under the Paris Agreement makes provision for the trading of Internationally Traded Mitigation Options (ITMO). Implementation of trading under Article 6, cooperative approaches would allow for Carbon offsets under the Paris Agreement. The use of Article 6 has been limited and delayed for a number of years by the slow resolution of agreed rules. (World Bank 2023), while Switzerland has already reported trades to the UNFCCC under article 6.2. Nevertheless, Switzerland has said it will not use forest based credits under article 6.2. Japan and South Korea are also exploring emissions trading contracts with developing countries (Hynes et al. 2023). Despite the lack of international emissions trading, domestic Carbon prices are covering an ever greater proportion of the world's emissions. Over 20% of global emissions are covered by a Carbon price, either a Carbon tax or an emissions trading scheme (World Bank 2023). The Clean Development Mechanism (CDM) was an international project level emissions trading mechanism for the Kyoto Protocol from 2008 to 2020. The inclusion of CDM projects in the Paris Agreement NDCs is currently being negotiated in the UNFCCC (World Bank 2023). Afforestation and reforestation can receive credits under CDM but protection of forests is not eligible.

A crucial element in the operation of carbon markets whether compliance or voluntary is the price of Carbon. The price of Carbon is established by a market price at which the supply of emissions abatement credits for sale is equal to the amount that buyers would like to purchase at a particular price. Other factors such as biodiversity and social costs and benefits of the emissions reduction credits will affect individual credits but the focus here is on the price of carbon. While there are few transactions in the international compliance carbon market, optimal pricing can be estimated by equalizing the marginal cost of abatement with the global social cost (Stern and Stiglitz 2017). One predictor of a future Carbon price is a calculation of the global social cost of Carbon which many researchers have estimated. The Stern and Stiglitz (2017) estimate is used in this paper, in addition we note Nabuurs et al. (2022) "economic potential" of mitigation which limits costs to USD 100/tCO2. There is much political opposition within many countries to the introduction of Carbon taxes and emissions trading schemes. Another compliance mechanism, the Carbon Offsetting and Reduction Scheme for International Aviation (CORSIA) credits prices have been very low due to the disruptions associated with COVID (ICAO 2022; World Bank 2023).

4.1.2 Voluntary Carbon Market

Bloomberg NEF (2023) asserts the market failed to grow in 2022 due to adverse media attention on offsets. Bloomberg NEF (2023) considered REDD+ credits to be characteristic of a low quality market. Emissions reductions from voluntary carbon markets are considered by many studies to be complementary or transitional to compliance Carbon markets. It is expected that as compliance markets expand voluntary markets will decline (TSVCM 2021: UNEP 2023). Public good theory suggests (Samuelson 1954) that voluntary carbon markets will under supply net emissions reductions. Jacobs (2014) found that pledges by companies had a more positive effect on the stock market than announcements of emissions reductions achieved. Corporations respond to both the norms of social behaviour and material incentives. Chrun et al. (2016) found that the literature on corporate environmentalism was plagued by causality issues. As the price of Carbon goes up, the material incentive to reduce emissions decline. Fisher-Vanden and Thorburn (2011) found that firms that join voluntary environmental programs suffer the disincentive of a decline in shareholder wealth.

4.1.3 Climate Bonds/Forest Bonds/Green Bonds

World Bank (2023) indicate that only 1% of green bonds were targeted at forestry or agriculture. None of the literature examined attempted to separately identify the quantity of north/south loan instruments funding to protect forests and ecosystems or nature based climate solutions. In this study it is assumed bond financing is counted within private finance directly mobilised or philanthropy.

4.1.4 Non-Wood Forest Products

FAO (2014), estimates that the total value of non-wood forest products as USD 17 billion. FAO (2020) reported USD 7.1 billion which was judged as an underestimate because many countries did not report the economic value of their non-wood forest products. Including measures of non-wood forest products would reduce the estimated net benefits of forest conversion.

4.1.5 Private Investment

Climate and biodiversity have aspects of global public goods which means much of the benefits of climate action accrue to the entire world (Samuelson 1954 and Kaul et al. 2015). Private investment has mostly gone to renewable energy and transport because electricity and transport services can be sold to generate revenue, rather than forest protection.

4.1.6 Public Finance

Current levels of public finance are driven by collective promises undertaken by developed countries in the UNFCCC negotiations to provide USD100 billion per year to developing countries for climate mitigation and adaptation for the period 2020 to 2025. Later COP decisions have specified that 50% should be targeted to adaptation. Developed countries have not been able to meet the collective commitment for the USD 100 billion goal although they have stated they will meet it by 2023 (Climate Home News 2023). A post 2025 goal is currently under negotiation. In the absence of good data on forest protection from north/south funding, this study has used the illustrative example of a 33% of mitigation climate finance recommended by Griscom et al. (2017) to be allocated to natural climate solutions. We projected that some climate finance for adaptation will be used for NBS activities, bringing the total to 37% of climate finance available nature based activities as shown in Fig. 1. OECD (2021) states that predicting future levels of public climate finance is inherently uncertain due to a host of factors such as macro-economic conditions, the short term nature of budgets and electoral cycles and MDBs are reliant on their ability to raise funds on the capital market which is dependent of their capital base and

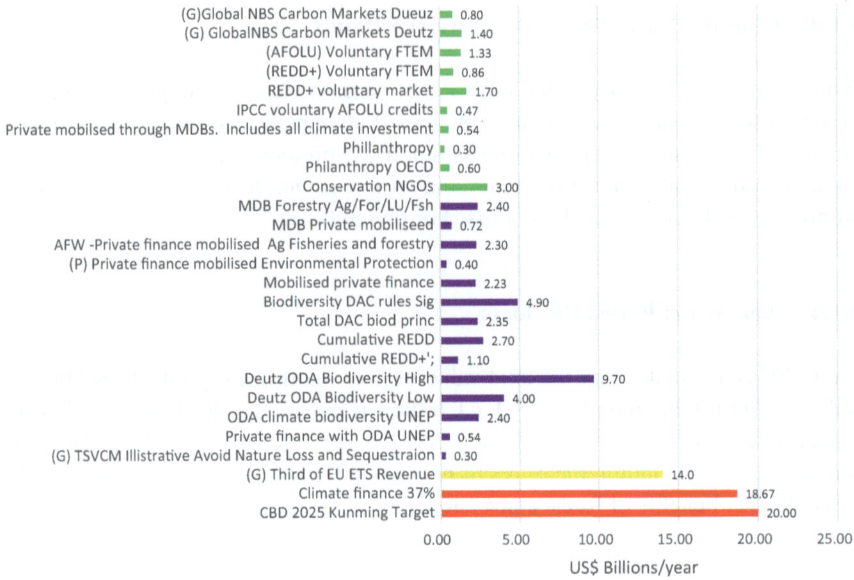

Fig. 1 Recent forest and ecosystem climate funding and comparison with some illustrative potentials

global financial conditions. The Kunming-Montreal Global Biodiversity Framework contained a promise of USD 30 billion per year by 2030 in north/south biodiversity funding. This has a list of possible sources of funding sources including ODA (Convention on Biodiversity 2022). OECD (n.d.), recommends not to "add up" the figures from the biodiversity and climate change markers as this could lead to double counting.

The architecture for REDD+ Transactions (ART) is a mechanism to combine public and private funding to finance jurisdictional REDD+. As such, it can channel private offsets into jurisdictional REDD+ to reduce problems with additionality, leakage and permanence than project based REDD+ (Seymour 2020). Much jurisdictional REDD+ funding buys emission reductions at a fixed price of USD 5/tCO_2or USD 10/tCO_2 (Ryan 2022).

5 Results

The results are contained in Figs. 1, 2 and 3, which detail the volume of funds and Carbon prices from a range of estimates of both public and private funding sources found in the literature. Background data for these Figures is found in Tables 2, 3 and 4.

Green; represents estimates of private funding, Mauve; is disbursed public funding

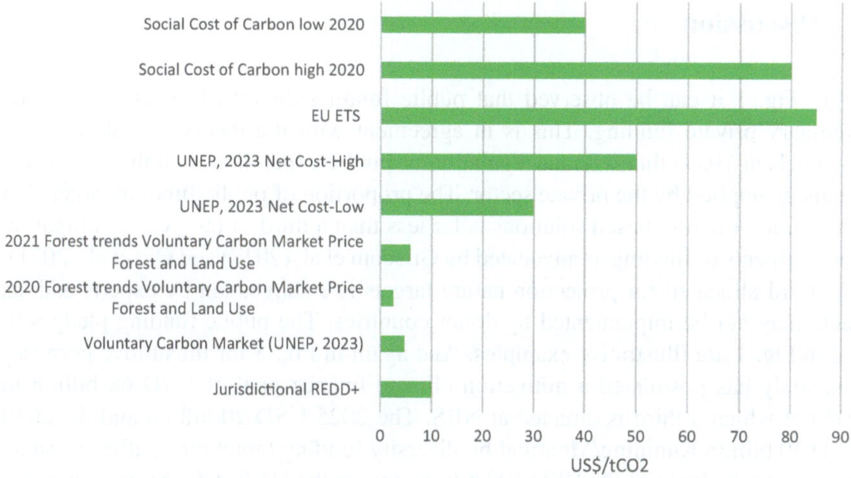

Fig. 2 Prices and costs of carbon for NBS

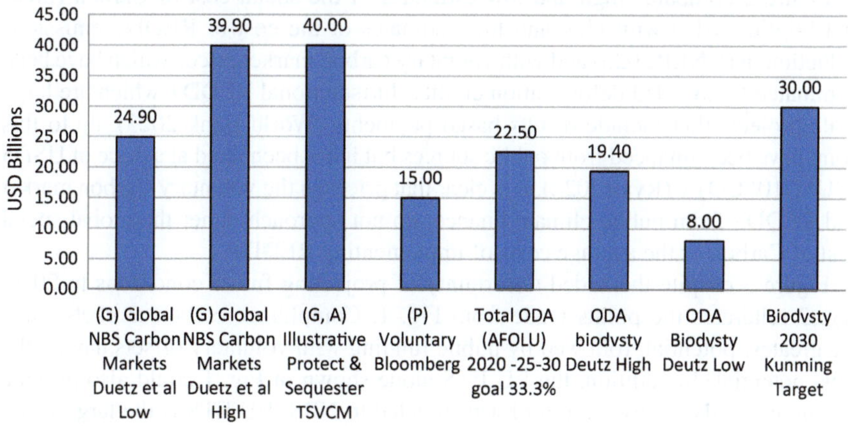

Fig. 3 NBS funding projections for 2030

Red are current targets for all public climate finance north/south targets, adjusted to a third targeted to mitigation for nature based projects (Griscom et al. 2017).

Yellow; represents a third of the public revenue from the EU Emissions Trading Scheme which could also be disbursed to protection or other nature based mitigation if allowed.

(Table 2 lists the data and references from which Fig. 1 is sourced).

6 Discussion

From Fig. 1 it can be observed that public funding disbursed generally exceeds voluntary private funding. This is in agreement with the theory of public goods (Samuelson 1954) that without regulation or public expenditure, public goods will be undersupplied by the private sector. The proportion of public funding targeted at protection or nature based solutions is far less than a third of the overall mitigation commitments of funding as advocated by Griscom et al. (2017) and Qin et al. (2021). The third allocated for protection nature targets is a suggestion by experts and, as such, may not be implemented by donor countries. The public funding pledges in red in Fig. 1 are illustrative examples. And again in Fig. 3 for illustrative purposes this study has postulated a mitigation climate finance goal of USD 68 billion in 2030 of which a third is directed at NBS. The 2025 USD 20 billion and the 2030 USD 30 billion Kunming/Montreal biodiversity funding target may suffer the same statistical rebadging as the USD 100 billion goal in the UNFCCC (Michaelowa and Michaelowa 2011). It is also likely that the same funds will be accounted for against both targets (OECD n.d).

Figure 2 compares high and low estimates of the social cost of Carbon (Stern and Stiglitz 2017) with high and low estimates of the cost of REDD+ emissions reduction in (UNEP 2023) and with voluntary carbon market prices which have been dominated by avoided deforestation credits. Jurisdictional REDD+ which are large scale projects that include results based payments (World Bank 2023), up to this point have been financed from public sources but it has been fixed at a price of USD5 or USD10/tCO_2-e (Ryan 2022). It is clear that prices in the voluntary Carbon market and REDD+ from public climate finance, do not approach either the global social cost of Carbon or the average costs of implementing REDD+.

Figure 3 despite the added uncertainty of projecting future conditions to 2030, Fig. 3 reiterates the points made from Fig. 1. Compliance carbon markets show the greatest potential, followed by public funding with voluntary offsets having the least potential. In addition, the EU ETS alone shown in Fig. 1 could also provide enormous funds to NBS if a third was directed to NBS. EU ETS is the largest of a number of compliance Carbon pricing systems that do align with the global social cost of Carbon (World Bank 2023), even with the current absence of international compliance markets

7 Conclusion

At this point in time, publicly funded climate finance is the major component of nature based funding for climate change to developing countries, exceeding private finance such as voluntary carbon markets and philanthropy. This is in line with traditional economic theory that the private sector will under-supply public goods such as climate change mitigation.

The greatest uncertainty regarding future sources of funds is the present lack of an international compliance carbon market. This is compounded by uncertainty surrounding CORSIA which creates demand for mitigation credits. Conversely, the CDMs inclusion in Article 6 could lead to a greater supply of emissions reduction credits to compete with forest protection. In addition, major developed economies have signaled in their NDCS that they will not use international emissions trading. Of course, these NDCs could be amended at some time before 2030. Many of the projections that voluntary Carbon markets assume that private entities will pay Carbon prices many times current prices and volumes of trading. The approximately \$5/tCO_2 levels in the latest estimates may be the result of the current supply and demand for credits. However, they do not reflect the levels of finance needed to significantly reduce deforestation by 2030. Protection of ecosystems has a lower Carbon cost of mitigation than many other AFOLU activities such as restoration. Therefore, a market mechanism for AFOLU activities favors protection over other activities that are lower on the mitigation hierarchy.

In contrast to this untested assumption for the expansion of voluntary markets, is the presence of domestic compliance Carbon markets, of which the EU ETS is the largest, which although not currently linked to north/south emissions trading has prices above USD 80/tCO2 and raises revenue of around USD 27 billion. The outstanding conclusion of this study is that international compliance emissions trading, the financial mechanism that is most likely to provide funding needed for REDD+ and forest protection in particular, is largely absent from the current financial architecture. If developed countries actively participated in REDD+ NBS emissions trading, deforestation and emissions could be radically reduced and developing countries could find an alternative development pathway that avoids deforestation.

References

Arlidge WNS, Bull JW, Addison PFE, Burgass MJ, Gianuca D, Gorham TM, Jaco C, Shumway N, Sinclair SP, Watson JEM, Wilcox C, Milner-Gulland EJ (2018) A global mitigation hierarchy for nature conservation. Bioscience 68(5):336–347. https://doi.org/10.1093/biosci/biy029

Bloomberg NEF (2023) Carbon offset market could reach \$1 trillion with right rules. https://about.bnef.com/blog/carbon-offset-market-could-reach-1-trillion-with-right-rules/. Retrieved 28 July 2023

Casado-Asensio J, Blaquier D, Sedemund J (2022) "Biodiversity and development finance: Main trends, 2011–20". OECD Development Co-operation Working Papers, No. 110. OECD Publishing, Paris. https://www.oecd-ilibrary.org/development/biodiversity-and-development-finance_b04b14b7-en

Chrun E, Dolsak N, Prakash A (2016) Corporate environmentalism: Motivation and mechanisms. Annu Rev Environ Resour 41:341–362

Climate Home News (2023) Rich countries 'confident' \$100bn climate finance delivered in 2023. https://climatechangenews.com/2023/09/19/rich-countries-confident-100bn-climate-finance-delivered-in-2023

Convention on Biodiversity (2022) Kunming-montreal global biodiversity framework. CBD/COP/DEC/15/4. https://www.cbd.int/doc/decisions/cop-15/cop-15-dec-04-en.pdf. Retrieved September 22, 2023

Deutz A, Heal GM, Niu R, Swanson E, Townshend T, Zhu L, Delmar A, Meghji A, Seth, Tobin- de la Puente J (2020) Financing nature: Closing the global biodiversity financing gap. The Paulson Institute. The Nature Conservancy, and the Cornell Atkinson Center for Sustainability

EBRD (2021) Joint report on multilateral development banks' climate finance. https://www.ebrd.com/documents/comms-and-bis/mdb-climate-finance-report-2021.pdf

European Commission (2023) Green deal: New law to fight global deforestation and forest degradation driven by EU production and consumption enters into force: Media Release, Directorate General for the Environment. https://environment.ec.europa.eu/news/green-deal-new-law-fight-global-deforestation-and-forest-degradation-driven-eu-production-and-2023-06-29_en#:~:text=Paralaxis%2C%20Getty%20Images. Retrieved July 30, 2023

FAO (2014) State of world's forest genetic resources, Commission on Genetic Resources for Food and Agriculture—Food and Agriculture Organizition of the United Nations, Rome. Retrieved 22 July 2023 from https://catalogue.unccd.int/368_a-i3825e.pdf

Fisher-Vanden K, Thorburn KS (2011) Voluntary corporate environmental initiatives and shareholder wealth. J Environ Econ Manag, 62 (3):430–445

Forest Declaration Assessment Partners. (2022) Finance for forests: Theme 3 assessment. Forest Declaration Assessment: Are we on track for 2030? Climate Focus (coordinator and editor). Retrieved July 18, 2023 from https://forestdeclaration.org/wp-content/uploads/www.forestdeclaration.org

Forest Trends' Ecosystem Marketplace (FTEM) (2022) The art of integrity: state of the voluntary carbon markets 2022 Q3, insights briefing. Forest Trends Association, Washington DC

Fuss S, Golub A, Lubowski R (2021) The economic value of tropical forests in meeting global climate stabilization goals. Global Sustain 4:E1. https://doi.org/10.1017/sus.2020.34

Global Forest Watch (2023). https://www.globalforestwatch.org/

Griscom BW, Adams J, Ellis PW, Houghton RA, Lomax G, Miteva DA, Schlesinger WH, Shoch D, Siikamäki JV, Smith P, Woodbury P, Zganjar C, Blackman A, Campari J, Conant RT, Delgado C, Elias P, Gopalakrishna T, Hamsik MR, Fargione J (2017) Natural climate solutions. Proc Natl Acad Sci USA 114(44):11645–11650

Griscom BW, Busch J, Cook-Patton SC, Ellis PW, Funk J, Leavitt SM, Lomax G, Turner WR, Chapman M, Engelmann J, Gurwick N P, Landis E, Lawrence D, Malhi Y, Schindler Murray L, Navarrete D, Roe S, Scull S, Smith P, Streck C, Walker WS, Worthington T (2020) National mitigation potential from natural climate solutions in the tropics. Philos Trans R Soc London Ser B, Biol Sci, 375 (1794):20190126

Hynes D, Hall M, Machnik D (2023) Implementing Article 6—An overview of preparation in selected countries, Gold Standard, Adelphi and BMWK. https://www.goldstandard.org/sites/default/files/implementing_article_6-an_overview_of_preparations_in_selected_countries.pdf

ICAO (2022) Analysis in support of the 2022 CORSIA periodic review. https://www.icao.int/environmental-protection/CORSIA/Documents/2_CAEP_CORSIA%20Periodic%20Review%20(C225)_Focus%20on%20Costs.pdf

IIntergovernmental Science/Policy Platform on Biodiversity and Ecosystem Services (IPBES) Theme 1 the main threats facing biodiversity globally, (retrieved September 28, 2023). https://www.ipbes.net/node/38152

IUCN (2020) Global standard for nature-based solutions. A user-friendly framework for the verification, design and scaling up of NbS, 1st edn. Gland, Switzerland. https://portals.iucn.org/library/sites/library/files/documents/2020-020-En.pdf

Kaul I, Davies R, Glasser R, Gerber M, Etter L (2015) Policy debate|financing the SDGs: Global vs local public goods. Int Dev Policy|Rev Int Polit Développement, (6.2)

Michaelowa A, Michaelowa K (2011) Coding error or statistical embellishment? The political economy of reporting climate aid. World Dev 39(11):2010–2020

Nabuurs G-J, Mrabet R, Abu Hatab A, Bustamant M, Clark H, Havlík P, House J, Mbow C, Ninan KN, Popp A, Roe, S, Sohngen B, Towprayoon S (2022) Agriculture, Forestry and Other Land Uses (AFOLU). In: IPCC, 2022: Climate Change 2022: Mitigation of climate change. Contribution of working group III to the sixth assessment report of the intergovernmental panel on climate change [P.R. Shukla, J. Skea, R. Slade, A. Al Khourdajie, R. van Diemen, D. McCollum, M. Pathak, S. Some, P. Vyas, R. Fradera, M. Belkacemi, A. Hasija, G. Lisboa, S. Luz, J. Malley, (eds.)]. Cambridge University Press, Cambridge, UK and New York, NY, USA. https://doi.org/10.1017/9781009157926.009

Norman M, Nakhooda S (2014) The state of REDD+ finance. Center for Global Development, Washington DC

OECD (n.d.) OECD DAC rio markers for climate: Handbook. Retrieved July 17 2023. https://www.oecd.org/dac/environment-development/Revised%20climate%20marker%20handbook_FINAL.pdf

OECD (2021) Climate finance and the USD 100 billion goal forward-looking scenarios of climate finance provided and mobilised by developed countries in 2021–2025 technical note: Technical note. OECD Publishing. https://doi.org/10.1787/a53aac3b-en

OECD (2023) Private finance mobilised by official development finance interventions. OECD Publishing, Paris, Development Co-operation Directorate

Olesen A, Böttcher H, Siemons A, Herrmann L, Martius C, Roman-Cuesta RM, Atmadja S, Hansen DS, Andersen SP, Georgiev I, Bager SL, Schwöppe C, Wunder S (2018) Study on EU financing REDD+ related activities, and results-based payments pre and post 2020: Sources, cost-effectiveness and fair allocation of incentives. European Commission, DG Environment/Climate Action, Luxembourg

Qin Z, Griscom B, Huang Y, Yuan W, Chen X, Dong W, Li T, Sanderman J, Smith P, Wang F, Yang S (2021) Delayed impact of natural climate solutions. Glob Change Biol 27(2):215–217

Rennert K, Errickson F, Prest BC, Rennels NRG, Pizer W, Kingdon C, Wingenroth J, Cook R, Partham B, Smith D, Cromar K, Diaz D, Moore FC, Muller UK, Plevin RJ, Raftery AE, Sevcikova H, Sheets H, Stock JH, Tan T, Watson M, Wong TE, Anthoff D (2022) Comprehensive evidence implies a higher social cost of CO_2. Nature 610(7933):687–692

Roe S, Streck C, Beach R, Busch J, Chapman M, Daioglou V, Deppermann A, Doelman J, Emmet-Booth J, Engelmann J, Fricko O, Frischamann C, Funk J, Grassi G, Griscom B, Havlik P, Hanssen S, Humpenöder F, Landholm D, Lomax G, Lehmann J, Mesnildry L, Nabuur G, Popp A, Rivard C, Sanderman J, Sohngen B, Smith P, Stehfest E, Woolf D, Lawrence D (2021) Land-based measures to mitigate climate change: Potential and feasibility by country. Glob Change Biol 27(23):6025–6058

Ryan F (2022) Is a low and fixed price for mitigation credits effective in reducing deforestation emissions? In: Chaiechi T, Wood J (eds) Community empowerment, sustainable cities, and transformative economies. Springer, Singapore. https://doi.org/10.1007/978-981-16-5260-8_12

Samuelson P (1954) The pure theory of public expenditure. Rev Econ Stat 36(4):387–389. https://doi.org/10.2307/1925895

Seymour F (2020) INSIDER: 4 Reasons why a jurisdictional approach for REDD+ Crediting is superior to a project-based approach. May 5. https://www.wri.org/insights/insider-4-reasons-why-jurisdictional-approach-redd-crediting-superior-project-based. Retrieved July 30 2023

Standing Committee on Finance (SCF) (2022) First report on the determination of the needs of developing country Parties related to implementing the Convention and the Paris Agreement, UNFCCC

Stern N, Stiglitz JE (2017) Report of the high-level commission on carbon prices. World Bank, Washington, DC

Taskforce on Scaling Voluntary Carbon Markets (TSVCM) (2021) Taskforce on scaling voluntary carbon markets, January 25th, 2021: Summary pack, https://www.iif.com/Portals/1/Files/TSVCM_Summary.pdf

UNFCCC (2023) NDC Registry. https://unfccc.int/NDCREG. Retrieved September 24, 2023

United Nations Environment Programme (UNEP) (2023) Pricing forest carbon. Nairobi

World Bank (2023) State and trends of carbon pricing 2023. World Bank, Washington, DC. https://doi.org/10.1596/978-1-4648-2006-9

World Bank (2017) Shadow price of carbon in economic analysis: Guidance note. https://pubdocs.worldbank.org/en/911381516303509498/2017-Shadow-Price-of-Carbon-Guidance-Note-FINAL-CLEARED.pdf

Carbon, Communities and Conservation Projects: A Case Study in Rondônia State, Brazil

Fábio Olmos, Adriana Martins, and Miguel Milano

Abstract The Rio Cautário State Resex Forest Conservation Project covers 146,400 ha in Rondonia state. It aims to protect native ecosystems and safeguard traditional communities' rights to an area under constant threat from illegal activities that have reduced or degazetted many protected areas in Rondonia over the last decades. In 2020, Permian Brasil, a nature conservation private company, in association with 92 local families, started a 30-year forest carbon project. Since then, the company has directed upfront investments of R$ 5.5 million/year to the reserve, benefiting families through Payment for Environmental Services (PES) and community development programs. As a result, and compared to the situation prior to the project, the project has created job opportunities for local people, empowered women and the youth, increased local income and improved family life quality with infrastructure such as internet access, solar energy, wells, etc. The project employs local staff dedicated to protection, community engagement, biodiversity monitoring and restoration, investing in building local capacity. The project at Rio Cautário, although in an initial stage, shows carbon finance can be a viable model of sustainable development for the region, benefiting communities and protecting the Amazon rainforest and its biodiversity. It is expected to generate carbon credits in a large scale from 2024 on, and its model can be replicated and customized for other protected areas under pressure in the Amazon to use carbon markets to improve local livelihoods and help local communities to conserve native ecosystems with the needed resources, means and safety.

Keywords Amazon · Carbon projects · Extractive reserves · Communal lands · Forest conservation

---

F. Olmos · A. Martins · M. Milano
Permian Brasil, Av. Paulista 1765/121, São Paulo, SP 01311-930, Brazil
e-mail: adriana.martins@permianglobal.com

F. Olmos (✉)
Perfin Climate, Av. Brig. Faria Lima - 2277/301, São Paulo, SP 01452-000, Brazil
e-mail: folmos@perfinclimate.com.br

W. Leal Filho et al. (eds.), *Amazon 2030—Sustainability Issues in the World's Largest Rainforest Region*, World Sustainability Series,
https://doi.org/10.1007/978-3-031-81465-5_19

1 Introduction

Long-term funding for conservation projects has proved to be elusive and a major bottleneck for the protection of Amazonian forests (Borner et al. 2020; Silva et al. 2022; Soares-Filho et al. 2023). Finance from carbon markets trading carbon credits generated by projects of reducing emissions from deforestation and degradation (REDD+) have been touted, where core carbon integrity principles are verified, as an additional and effective way to channel private capital into conservation (for example, Paoli et al. 2010; Lima et al. 2013).

The state of Rondônia has experienced the highest increase in deforestation rates in the Brazilian Amazon; from 1985 to 2021 its forest area went from 20.2 to 13.2 million ha, a loss of 35% (data from Map Biomas[1]). Forest loss increased with the opening of the Cuiabá-Porto Velho Road (BR 364) in 1964, exploding from 1984 on with the World Bank's POLONOROESTE project, which funded the paving of the road and resettlement programs implemented by the Federal Government through INCRA (see Rosendo and Brown 1998; Pedlowski et al. 1997).

Public outcry over the resulting deforestation and violations of the rights of traditional communities (including recently contacted indigenous groups) led the World Bank to foster the PLANAFLORO[2] project, under which the newly created state (formerly a federal territory) was to implement a system of protected areas based on an ecological-economic zoning (ZEE) as a condition for receiving resources for further infrastructure and agricultural development (Keck 1998; Millikan 2001; Browder 2002).

Rondônia has been a textbook example of protected area downgrading, downsizing and degazettement (PADDD) supported by local politicians who see public lands as open for grabbing (see Pedlowski et al 2005; Araújo and Barreto 2010; Bernard et al. 2014; Pack et al. 2016; Tesfaw et al. 2018; Keles et al. 2020). As a result, state (as opposed to federal) protected areas in Rondônia have low impact in curbing deforestation/degradation (Herrera et al 2019).

Organised social movements have been active actors in the creation of protected areas across the Brazilian Amazon, namely extractive reserves (see Brown and Rosendo 2000 for an overview). Nevertheless, the lack of long-term funding sources has hindered the success of several extractive reserves, resulting both in an increase in cattle-ranching and associated deforestation (Pantoja et al. 2010) and vulnerability to invasions and land-grabbing. This led to situations like Jaci-Paraná State Extractive Reserve, which has lost 147,365 ha between 2000 and 2022[3] (Scherer 2021; Garzon et al. 2022) and Rio Preto-Jacundá State Extractive Reserve, with a loss of 1,901 ha just in 2022 (MapBiomas 2023).

Permian Brasil[4] is a conservation-oriented company developing forest carbon projects to provide finance for avoided deforestation and restoration initiatives.

[1] https://mapbiomas.org/estatisticas.

[2] See also https://acervo.socioambiental.org/acervo/documentos/plano-agropecuario-e-florestal-de-rondonia-planafloro.

[3] https://uc.socioambiental.org/pt-br/arp/1319.

[4] See www.permianbrasil.com.

Stating in 2020, Permian has carried such a project at Rio Cautário State Extractive Reserve that provides a case study on how carbon finance can improve both local livelihoods and the conservation outcomes of an Amazonian protected area located in a very problematic setting. Here we describe key aspects of this project.

2 The Rio Cautario Forest Carbon Project

Details on the project's methodologies and outcomes are presented in the Project Document (PD) and Monitoring Report (MR) lodged in VERRA's project registry[5] and available for public consultation. Below is a summary.

Rio Cautário State Extractive Reserve (RCSER) covers 146,400 ha in the southwestern part of the state of Rondônia, Brazil (Fig. 1). The reserve is managed by the local communities, currently with 92 families (315 people) organised in 7 villages, under the auspices of SEDAM, the state environment agency. RCSER was created in 1995 based on Rondonia's ZEE and to comply with the requests from the organised rubber-tapper (*seringueiros*) movement, then fighting land-grabs from cattle ranchers who would eventually occupy and destroy the adjoining São Domingos State Forest (GTA 2008). For a history of the area see its management plan (SEDAM 2016).

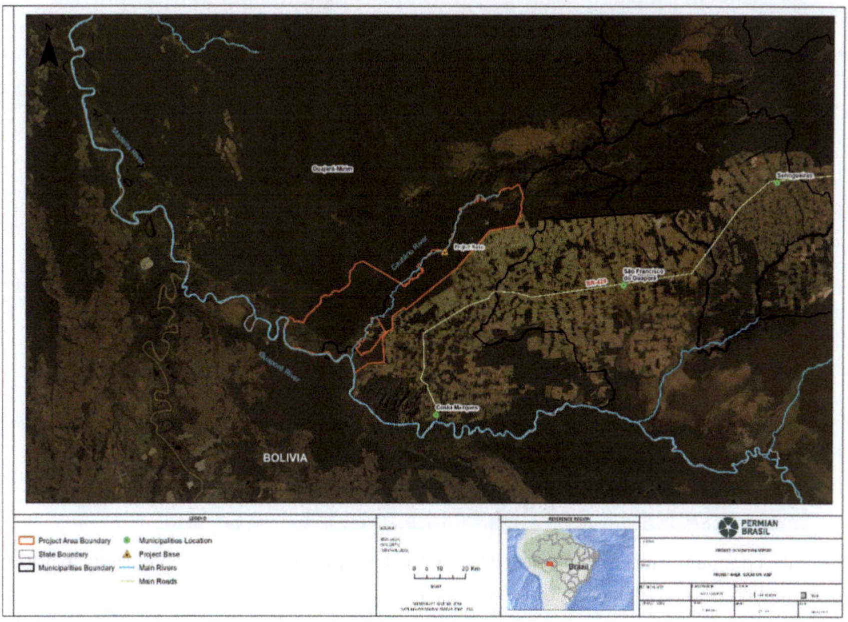

Fig. 1 Rio Cautário state extrative resex in Southwest Rondônia, Brazil. The reserve is a barrier against the deforestation front stemming from the BR 421 road and protects the largest remaining forest block in this part of the Amazon

[5] https://registry.verra.org/app/search/VCS/All%20Projects.

The Cautário cutting through the forests of Rio Cautário State Extractive Reserve

One of the community meetings promoted by the Rio Cautário project for the community to decide where funding is to be applied for common benefit

The project's fire brigade in action

One of the tree nurseries sponsored by the project. These are attended by local families and support restoration initiatives

The local community descends from rubber-tappers who were already established in the area in the 1930s and more recent migrants. The main sources of income are gathering Brazil nuts and natural rubber and small-scale agriculture, manioc flour being the main product sold outside the reserve. Monthly income per family varies considerably but, before the project (2020), 33% of families earned less than a minimum wage (c. US$ 250), 33% up to a minimum wage and 34% up to two minimum wages.

Prior to the project there was no material gain for the community to comply with the management plan and logging was promoted as an alternative to generate more income, what failed to materialise and resulted in forest degradation and debts incurred by the local association (Dias 2014; SEDAM 2016).

The project began as a consultation by SEDAM to the community, following former previous informed consent procedures, leading to a public call for proposals for developers interested in carrying carbon projects in the reserve. The proposals were screened by SEDAM and, in March 2020, presented to the community for their choosing. Permian Brasil's proposal was selected, with contracts signed by Permian with each family and SEDAM in June 2020 and the project effectively starting the following September.

The primary objectives are to reduce carbon emissions through forest protection and restoration in the reserve, stop illegal deforestation and encroachment, control fire hazards, restore areas currently covered by pasture and create economic alternatives for the local community through a Payment for Ecosystem Services (PES) program, which also seeks to improve the sustainability of agricultural and extractive practices promoted in the State Extractive Reserve Management Plan (SERMP). The project also aims to improve the conservation outlook of globally threatened species and local ecosystems, and promote environmental education among residents, particularly the youth. The project has potential to generate over 20 million verified carbon units (VCUs) over its lifetime.

3 Benefits for the Communities

The project is based on the compliance of the reserve's management plan by the beneficiaries, as well as on their agreement to cease any commercial logging operations in the reserve, as defined in a contract that also specifies the mechanism for benefit sharing after credits are sold. In return, the project has implemented the following activities:

PES: every family enrolled in the project receives monthly payments of R$ 1,375. The value is readjusted yearly following an index adopted by Rondônia state. This has doubled the average family income and provided financial security, meaning that the local producers can sell their products when prices are higher. It has also allowed improvements in the houses and access to credit, among other benefits linked to guaranteed basic income programmes.

Communal Development: R$ 250,000, adjusted yearly according to the same PES index, are available every year for investments in goods and services for the common benefit, the communities deciding what they wish through a process of direct democracy. So far, the project has provided internet access to all the communities, paid for mechanical preparation of agricultural plots, wells and reservoirs for water supply, solar panels, vehicles, garages and social gathering furniture, etc.

Agricultural and Rural Extension Free Assistance: the Project has a team dedicated to assist the Community to improve agricultural techniques, increasing productivity while replacing the use of fire and the need of further forest clearing for more productive and sustainable methods. Special focus has been given to empower women by training and assistance that engage then with vegetable gardens to improve food security and variety, health and additional income.

Direct jobs and training: the team at the project site currently includes 29 community members, plus four technical staff. Local people are able to combine their duties in protection, monitoring or rural extension with traditional activities like Brazil nut harvesting. Staff has been trained in using tools from drones to GPS, firefighting, biodiversity monitoring and other skills necessary for their jobs.

Protection: Since its start, the project has curbed two organised invasions by groups claiming property rights in the area, the most recent case in July 2023 involving a group moving into the reserve, by bus, after an advance party had entered the area to mark plots with the intention to claim them. The local project staff organised a response with law enforcement officers, successfully avoiding the invasion, with the ringleaders now facing legal action.

RCSER has had almost zero deforestation since the project start, with just 4.6 ha in 2020, zero ha in 2021 and 2.21 ha in 2022, all secondary forest where clearing was approved by SEDAN under the reserve´s management plan.

Biodiversity monitoring: the Project has installed a network of 92 permanent forest plots for biomass mapping and long-term monitoring. So far, the area has 312 tree species, with six globally Endangered (EN) and eight Vulnerable (VU). Monitoring using standard transects and camera-traps—to be carried on an annual basis—has found 44 species of mammals (two EN, seven VU, five Near-threatened), while the bird list stands at 485 species, with nine VU. One reptile is VU.

Restoration: the project has started the restoration of 235 ha of abandoned pastures left in the reserve. This has included training locals to harvest seeds of native trees and creating a production chain for seedlings, with 16 families now enrolled, up from eight in 2022. Both assisted natural restoration techniques and agroforestry systems are being used, with an initial 22 ha of the latter implemented in 2022.

Support to local institutions: the project also sponsors the meetings of the reserve's deliberative council, the main body governing it, and has donated R$ 3.6 million to the Rondônia Climate Fund so SEDAM, if willing, has funding available for the protection of other areas. Prior to the project there was no tagged funds for council meetings and those lacked regularity.

Permian Brasil has a budget of R$ 27.2 million over the first five years of the project as upfront investment prior to any sales of carbon credits, these being scheduled for later in 2025. In its first 2.5 years, the project has already directed R$ 3.5 million for PES, R$ 825,000 to community development and R$ 2 million to forest restoration.

This budget for a single area compared to the 2023 budget of R$ 11.9 million SEDAM had for all the state's 40 protected areas, most of it dedicated to daily stipends by personnel and law enforcement, shows the potential of such kind of project to implement protected areas and really avoid deforestation at same time improves local population´s quality of life through private funds.

4 Perspectives for Amazonian Communal Areas

Carbon markets are one, among several tools, for decarbonising economies and companies (Bayer and Aklin 2020; Sylvera 2023; Ecosystem Marketplace 2023a). Further, when incorporating carbon credits from avoided deforestation/restoration projects, carbon markets do provide finance to prevent forest loss and degradation (Busch and Ferretti-Gallon 2023; Ecosystem Marketplace 2023b).

REDD+ projects selling credits to voluntary carbon markets have proved to be effective in reducing deforestation (Guizar-Coutiño et al. 2022; Demarchi et al. 2023; Mitchard et al. 20,323) while premium projects, like Cautário, see trees beyond mere carbon storage (see Burivalová et al. 2023) and do include biodiversity conservation and community benefits as intrinsic to their business model, validated by certifications such as CCB (Climate, Community and Biodiversity[6]) that add to market value.

Despite that, forest carbon/REDD+ projects have been criticised by some researchers as failing to achieve credible results, but these studies have also been criticised for cherry-picking and choosing poorly performing controls against real-world deforestation rates and pressures (see Pauly 2023 for a summary and Mitchard et al. 2023 for details). Other critics come from NGOs against any private sector/market-based solutions on conservation or climate-change and are beyond a rational dialogue.

The authors' own experience shows that local people, who actually live in areas at risk of being lost to deforestation and face their hardships, are keen to embrace market-based economic alternatives providing well-being and forest/biodiversity conservation, as long as their rights are respected, and projects do deliver. This point has been made by others (see Pairojmahakji and Ganz 2023) but missed by many campaigners.

As it is too common in the conservation arena, local realities often clash with ideological views and political interests. The widely recognised shortfall of governments and philanthropy to provide the necessary finance for forest conservation, including formally protected areas, and improve the livelihoods of local populations demands a pragmatic approach.

For example, extractive reserves were once portrayed as the best way to safeguard the Amazon, but the reality of economics, demography and government support has resulted in significant deforestation not only by invaders (see Introduction) but also by the very people supposed to be protecting the forest (as in the case of Chico Mendes Extractive Reserve, see Mascarenhas et al. 2018; Fittipaldy and Castello 2021).

[6] https://verra.org/programs/ccbs/.

5 Conclusions

Carbon markets may not be the silver bullet some portrait, but it is a tool not to be discarded in the fight to curb disastrous climate change while recognising the role played by natural ecosystems and traditional communities. Rather, improvements both in the science and the policies grounding forest carbon projects are already being carried in a moving landscape project developers must cope with.

This is shown by the late 2023—early 2025 changes in methodologies adopted by certification bodies like VERRA[7], the lively debates on legislation creating a domestic carbon market in Brazil (ennacted in late 2024) and agreements allowing carbon trading between countries.

The Rio Cautário project shows carbon finance can deliver measurable benefits to local communities and to the local biodiversity while making a contribution to climate resilience even before carbon credits are actually sold. This model can be adapted and improved to allow private capital to flow into the diversity of protected areas in the Amazon, from formally peopleless biological reserves to indigenous lands, contributing to their economic viability and future resilience.

Governments have a role not only in creating and regulating carbon markets but, specifically in Brazil's case, also in creating the opportunities for public–private partnerships aimed at the conservation of public forests and in mediating the relationship between project developers and communities owning forest areas.

As we have seen at Cautário, the path has many bumps, but the rewards make them worth.

References

Araújo E, Barreto P (2010) Formal threats to protected areas in the Amazon. The State of the Amazon, 16. https://imazon.org.br/en/publicacoes/formal-threats-to-protected-areas-in-the-amazon/

Bayer P, Aklin M (2020) The European Union emissions trading system reduced CO_2 emissions despite low prices. PNAS 117(16):8804–8812. https://doi.org/10.1073/pnas.1918128117

Bernard E, Penna LA, Araújo E (2014) Downgrading, downsizing, degazettement, and reclassification of protected areas in Brazil. Conserv Biol 28(4):939–950. https://doi.org/10.1111/cobi.12298

Börner J, Schulz D, Wunder S, Pfaff A (2020) The effectiveness of forest conservation policies and programs. Annu Rev Resour Econ 12:45–64. https://doi.org/10.1146/annurev-resource-110119-025703

Browder JO (2002) Conservation and development projects in the Brazilian Amazon: Lessons from the community initiative program in Rondônia. Environ Manage 29:750–762. https://doi.org/10.1007/s00267-001-2613-3

Brown K, Rosendo SR (2000) Environmentalists, rubber tappers and empowerment: the politics and economics of extractive reserves. Dev Chang 31(1):201–227. https://doi.org/10.1111/1467-7660.00152

[7] https://verra.org/verra-launches-new-era-of-forest-protection-with-transformative-redd-methodology/.

Buřivalová Z, Yoh N, Butler RA, Sathya Chandra Sagar HS, Game ET (2023) Broadening the focus of forest conservation beyond carbon. Curr Biol 33(11):R621–R635. https://doi.org/10.1016/j.cub.2023.04.019

Busch J, Ferretti-Gallon K (2023) What drives and stops deforestation, reforestation, and forest degradation? An updated meta-analysis. Rev Environ Econ Policy, 17(2):217–250. https://www.journals.uchicago.edu/doi/abs/10.1086/725051

Demarchi D, Subervie J, Catry T, Tritsch I (2023) Using publicly available remote sensing products to evaluate REDD projects in Brazil. Glob Environ Chang. 80, (102653). https://doi.org/10.1016/j.gloenvcha.2023.102653

Dias RS (2014) Reserva Extrativista do Rio Cautário: o manejo florestal como uso sustentável no território. MSc. Dissertation, Universidade Federal de Rondônia. https://www.ri.unir.br/jspui/handle/123456789/2022

Ecosystem Marketplace (2023a) All in on climate: the role of carbon credits in corporate climate strategies. https://www.ecosystemmarketplace.com/publications/2023-em-all-in-on-climate-report/

Ecosystem Marketplace (2023b) 2023 state of the voluntary carbon markets report: Paying for quality. https://www.ecosystemmarketplace.com/publications/state-of-the-voluntary-carbon-market-report-2023/

Fittipaldy MCPM, Castelo CEF (2021) Há boi pastando, há desmatamento e outras coisas mais: A retrato da RESEX Chico Mendes. Conexões, 9(2). https://periodicos.ufpa.br/index.php/Conexoes/article/viewFile/13749/9577

Garzon LFN, da Silva DS, Ribeiro MS (2022) Desmanche de territórios indígenas e de unidades de conservação no novo arco de desmatamento em Rondônia/Brasil e Beni/Bolívia (2018–2020). Desenvolv E Meio Ambient, 60. https://doi.org/10.5380/dma.v60i0.80233

GTA/Grupo de Trabalho Amazônico (2008) O fim da floresta. A devastação das Unidades de Conservação e Terras Indígenas no Estado de Rondônia. Regional Rondônia, 46. https://fase.org.br/wp-content/uploads/2010/01/O_fim_da_floresta.pdf

Guizar-Coutiño A, Jones JPG, Balmford A, Carmenta R, Coomes DA (2022) A global evaluation of the effectiveness of voluntary REDD+ projects at reducing deforestation and degradation in the moist tropics. Conserv Biol 36(6):e13970. https://doi.org/10.1111/cobi.13970

Herrera D, Pfaff A, Robalino J (2019) Impacts of protected areas vary with the level of government: Comparing avoided deforestation across agencies in the Brazilian Amazon. Proc Natl Acad Sci 116(30):14916–14925. https://doi.org/10.1073/pnas.1802877116

Keck ME (1998) Planafloro in Rondônia: the limits of leverage, pp 181–218. In: Fox JA, Brown LD (eds) The struggle for accountability: the World Bank, NGOs, and grassroots movements. MIT Press, Cambridge, Mass

Keles D, Delacote P, Pfaff A, Qin S, Mascia MB (2020) What drives the erasure of protected areas? Evidence from across the Brazilian Amazon. Ecol Econ 176:106733. https://doi.org/10.1016/j.ecolecon.2020.106733

Lima RF, Olmos F, Dallimer M, Atkinson PW, Barlow J (2013) Can REDD+ help the conservation of restricted-range island species? Insights from the endemism hotspot of São Tomé. PLoS ONE 8(9). https://doi.org/10.1371/annotation/f3b7b8dc-a2d6-43e8-945a-566a7082cb12

MapBiomas (2023) Relatório Anual de Desmatamento 2022. São Paulo, Brasil, p 125. http://alerta.mapbiomas.org

Mascarenhas FS, Brown IF, da Silva SS (2018) Desmatamento e incêndios florestais transformando a realidade da reserva extrativista chico mendes. Desenvolv E Meio Ambient 48. https://doi.org/10.5380/dma.v48i0.58826

Millikan B (2001) O Painel de Inspeção do Banco Mundial e o pedido de investigação sobre o Planafloro. In: Vianna Jr, A (ed) Banco mundial, participação, transparência e responsabilização: A experiência brasileira com o painel de inspeção. Rede Brasil, Brasília

Mitchard E, Carstairs H, Cosenza R, Saatchi SS, Funk J, Nieto Quintano P, Brade T, McNicol I, Meir P, Collins M, Nowak E (2023) Serious errors impair an assessment of forest carbon projects: A rebuttal of West et al. (2023). https://papers.ssrn.com/sol3/papers.cfm?abstract_id=4661873

Pack SM, Ferreira MN, Krithivasan R, Murrow J, Bernard E, Mascia MB (2016) Protected area downgrading, downsizing, and degazettement (PADDD) in the Amazon. Biol Cons 197:32–39. https://doi.org/10.1016/j.biocon.2016.02.004

Pairojmahakji R, Ganz D (2023) Carbon markets: Time to listen to indigenous peoples and local communities. https://www.ecosystemmarketplace.com/articles/carbon-markets-time-to-listen-to-indigenous-peoples-and-local-communities/

Pantoja MC, Costa EL, Postigo A (2010) A presença do gado em reservas extrativistas: Algumas reflexões. Rev Pós Ciências Sociais, 6(12). https://periodicoseletronicos.ufma.br/index.php/rpcsoc/article/view/60

Paoli GD, Wells PL, Meijaard E, Struebig, Marshall MJ, Obidzinski K, Tan A, Rafiastanto A, Yaap B, Slik JWF, Morel A, Perumal B, Wielaard N, Husson S, D'Arcy L (2010) Biodiversity conservation in the REDD. Carbon Balanc Manag 5:7. https://doi.org/10.1186/1750-0680-5-7

Pauly M (2023) Recent challenges in the retrospective evaluation of forest carbon credit baselines. Integr Environ Assess Manag 19(5):1394–1397. https://doi.org/10.1002/ieam.48

Pedlowski MA, Dale VH, Matricardi EA, da Silva Filho EP (1997) Patterns and impacts of deforestation in Rondônia, Brazil. Landsc Urban Plan 38(3–4):149–157. https://doi.org/10.1016/S0169-2046(97)00030-3

Pedlowski MA, Matricardi EA, Skole D, Cameron SR, Chomentowski W, Fernandes C, Lisboa A (2005) Conservation units: A new deforestation frontier in the Amazonian state of Rondônia, Brazil. Environ Conserv 32(2):149–155. https://doi.org/10.1017/S0376892905002134

Rosendo S, Brown KM (1998) Strategic alliances, partnerships, and collective action: rubber tappers and extractive reserves in Rondonia, Brazil. In: Crossing boundaries, The seventh annual conference of the international association for the study of common property, Vancouver, BC, Canada. https://dlc.dlib.indiana.edu/dlc/handle/10535/1501

Scherer I (2021) O dano ambiental difuso ou coletivo derivado da grilagem de terra e outras vicissitudes sofridas pela reserva extrativista Jaci-Paraná. Revista da Emeron (29)

SEDAM (2016) Plano de manejo da reserva extrativista estadual do rio cautário. Secretaria de Estado de Desenvolvimento Ambiental. https://cuc.sedam.ro.gov.br/reserva-extrativista-do-rio-cautario/

Silva JM, Barbosa LC, Topf J, Vieira IGG, Scarano F (2022) Minimum costs to conserve 80% of the Brazilian Amazon. Perspect Ecol Conserv. https://doi.org/10.1016/j.pecon.2022.03.007

Soares-Filho BS, Oliveira U, Ferreira MN, Marques FFC, Oliveira AR, Silva FR, Börner J (2023) Contribution of the Amazon protected areas program to forest conservation. Biol Cons 279:109928. https://doi.org/10.1016/j.biocon.2023.109928

Sylvera (2023) The state of carbon credits 2023: how the market can move forward. https://www.sylvera.com/resources/the-state-of-carbon-credits-report

Tesfaw AT, Pfaff A, Golden Kroner RE, Qin S, Medeiros R, Mascia MB (2018) Land-use and land-cover change shape the sustainability and impacts of protected areas. Proc Natl Acad Sci 115(9):2084–2089. https://doi.org/10.1073/pnas.1716462115

PNAE and Integrative Policies in Boa Vista, Roraima: Levels of Learning, Development, and Sustainability

Eliane Alves da Silva, Eugenio Avila Pedrozo, and Tania Nunes da Silva

Abstract Amazon Rainforest became urbanized forest, because the population living today in the Legal Amazon is approximately 29,396,954 inhabitants. And though the abundance of natural resources is high, 48.3% of its population suffers severe food and water insecurity. The National School Feeding Program (PNAE), mainly after the adoption of Law nº 11,947, is a public policy that could help solve these problems, because it fosters food and nutritional sovereignty and safety, and makes feasible fair trade and rural development. During the program execution, learning can occur involving natural and human domains to ensure a sustainable future. By taking into account this context, the Chapter aims to analyze the levels of learning about sustainability and the sustainable development fostered by PNAE in Boa Vista, Roraima. In order to investigate it, a qualitative-descriptive research and abductive logic were adopted, using as strategy a single case study. It was observed that there are other integrative policies in addition to the interconnection of municipal departments that catalyze PNAE results; however, Roraima state's capital is in "transition" rather than "transformation" process of learning.

Keywords Brazilian Amazon · Roraima · Rural development · Food safety · Sustainability

---

Eugenio Avila Pedrozo: In Memoriam.

E. A. da Silva (✉)
Universidade do Estado de Mato Grosso (UNEMAT), Cáceres, Brazil
e-mail: eliane.alves.silva@unemat.br

E. A. Pedrozo · T. N. da Silva
Universidade Federal do Rio Grande do Sul (UFRGS), Porto Alegre, Brazil
e-mail: tnsilva@ea.ufrgs.br

W. Leal Filho et al. (eds.), *Amazon 2030—Sustainability Issues in the World's Largest Rainforest Region*, World Sustainability Series,
https://doi.org/10.1007/978-3-031-81465-5_20

1 Introduction

The Amazon Rainforest is the largest tropical forest in the world, according to INPA (National Institute of Amazonian Research). It covers around 7.5 million square kilometers, of which, 5.5 million constitute the tropical forest. It is worth mentioning that 68% of the entire Amazon territory is in Brazil (Leal 2019). However, part of the Amazonian Region became urbanized forest, and the population living today in the Legal Amazon is approximately 29,396,954 inhabitants (IBGE 2021a, 2021b). Therefore, the Amazonian region houses around 13.50% of the Brazilian population.

Though there is high abundance of natural resources, it is in the North Region, where most of the Amazonian Region is located, that 48.3% of the population suffers severe food and water insecurity (Rede PENSSAN 2022). In this context, the State should formulate and strengthen public policies to ensure changes in the dietary pattern and rearrangement of the food system, in addition to fostering agricultural production, thus ensuring access to healthy food, and making feasible fair trade and sustainable rural development (Marques and Triches 2022).

A public policy guided by food and nutritional sovereignty and security (SSAN) that can solve the problems of the region, is the National School Feeding Program (PNAE), whose main objective is to partially supply the nutritional needs of the beneficiary students by offering them at least one daily meal, in order to meet nutritional requirements during the period when they are at school. So, school feeding must cover at least 15% of the students' daily needs (FNDE 2019b).

PNAE originated in the Social Security Feeding Service (SAPS), founded in August 1940 (Peixinho, 2013), and underwent several transformations, until 1976, when the integration of the National School Feeding Campaign (CNAE) to the II National Food and Nutrition Program (II PRONAN) occurred. In 1979 it became the National School Feeding Program (PNAE) (Peixinho 2013; Silva 2020). However, it was in June 16, 2009, with approval of Law nº 11.947, that one of the most significant innovations occurred in PNAE. This law determines that at least 30% of food purchases for school feeding must come from family farmers.

For contemplating local products, paths are open to advance in the sustainable development. So, the actions performed by social actors of the community due to the program, can generate higher level of learning, changing the individual/community reference framework. By learning collectively, individuals attempt to predict and control how their actions affect the natural and human domains to ensure a sustainable future (Lee 1993; Röling and Wagemakers 1998; Maarleveld and Dangbegnonpág 1999; Lankester 2013).

When the issue of individual behavior change is considered, it is concentrated in the role that individuals, as social actors, can play by contributing to reduce their impact on the environment, without necessarily taking into account economic penalties or incentives. These environmental concerns can be classified in a sustainability scale that comprises individuals that value natural resources, and believe that they can't be replaced, and those that argue that these resourced can be replaced. Based on the position about the type of capital to be prioritized, the perception of sustainability can be classified as Very Weak (Cornucopia), Weak (Complacent), Strong (Communalism), and Very Strong (Deep Ecology).

Starting from these discussions, the following research question emerges: What is the relation between levels of learning about sustainability and sustainable development, based on the adoption of Law nº 11.947, PNAE, in Boa Vista/RR? In order to characterize the largest number of learnings, and having more space to discuss sustainable development and sustainability in the Amazonian Region, Boa Vista, Roraima state capital, was chosen. So, this Chapter's objective is to analyze the relation of levels of learning about sustainability and the sustainable development fostered by PNAE, in Boa Vista, RR.

This study is justified for showing the importance of PNAE in the North Region, through the promotion of family farming. The study of this integration of food feeding and family farming enables other ways of increasing the insertion of small farmers as suppliers of healthier food. And that will provide incentives for products from local economies, directed to sustainable development, in addition to the supply of non-industrialized foods to students. For documenting it, the present study provides contributions to municipalities in the Amazonian Region, because the data collected contribute to decision making, and enable learning for the formulation of new strategies, to comply with Law nº 11.947.

To achieve the proposed objective, qualitative-descriptive research (Sampieri et al. 2013; Kumar et al. 2018) and abductive logic (Mezirow 1991; Charreire and Durieux 2003; Cruz 2007) were adopted, using as strategy a single case study (Yin 2018). Interviews were semi-structured (Bishop 2005) and observation was non-participant (Stake 2011; Kumar et al. 2018), documented through field notes (Lofland 1974; Lofland and Lofland 1995; Weber 2009). The analysis of findings was made based on Bardin (2011).

The Chapter is broken down into 5 sections; the first is introduction; the second describes PNAE history and operationalization; the third addresses the literature review used; the fourth describes the methodological procedures used; and the fifth presents results and discussion; it also presents final remarks and references that grounded the study.

1.1 PNAE History and Operationalization

According to Peixinho (2013), the National School Feeding Program (PNAE) originated in the Social Security Feeding Service (SAPS), founded in August 1940. In 1945, the National Feeding Commission (CNA) was implemented. In January 1946 the National Nutrition Institute (INN) was founded, which was divided in four sections: biological research, social research, food education, and clinical pathology (Silva 1995; Silva et al. 2022, 2023a, b).

In 1952, the Food Situation and Nutrition Problems in Brazil plan was prepared, which contemplated nutritional matters, expansion of school meals, food assistance to adolescents, regional programs, basic food enrichment, and support to the food industry. The School Meal Campaign (CME) resulted from this project, and was instituted through Decree n° 37.106, of March 31, 1955, by Getúlio Vargas president (FNDE/PNAE 2019a; Silva 2020).

In 1976 the integration of the National School Feeding Campaign (CNAE) to the II National Food and Nutrition Program (II PRONAN) occurred. In 1979 it became the National School Feeding Program (PNAE). From 1976 to 1984, PNAE was constituted as one of II PRONAN guidelines, coordinated by the National Institute for Food and Nutrition (INAN), autonomous entity linked to the Ministry of Health (Peixinho 2013; Silva 2020).

Law nº 11.947, of June 16, 2009 was a result of an intersectorial process in the Federal Government and the participation of the civil society by means of the National Food and Nutrition Safety Council (CONSEA) (Peixinho, 2013), in addition to union mobilizations, like Grito da Terra Brasil. The union movement, occurred in May 2009 (13–27) with 52 hearings, involving over 30 government bodies, 14 ministers and several executive secretaries, led to the approval, in the Federal Senate, of Provisional Measure nº 455/09 (MP 455/09). This MP determined the purchase of food for school meals should be at least 30% from farmers and family farmers (Broch 2009; Silva et al. 2022, 2023a, b).

Intended to develop interministerial actions, Article 14 of Law nº 11.947, of 2009, instituted the Management Committee, composed of representatives from the Federal Government through Ordinance nº 450, of 2010; and the Advisory Group, with participation of civil society representatives, in addition to representatives from the National Council of State Secretaries of Education (CONSED) and the National Union of Municipal Directors of Education (UNDIME), to advise the Management Committee. This law established a new composition for the School Food Councils (CAE), contemplating more representatives of teachers, students or workers in the field of education, appointed by respective class associations, expanding the participation of organized civil entities representation, and excluding the participation of Legislative representatives (Peixinho 2013; Silva et al. 2022, 2023a).

FNDE Resolution nº 26, of June 17, 2013, corroborated Law nº 11.947, solving doubts about who could meet the school food demand. This resolution established that, in addition to the Declaration of Aptitude to PRONAF (DAP), to be considered family farmer, the individual should practice activities in the rural area, have area of up to four tax modules, labor from their own family, family income linked to their own establishment and management of the establishment or enterprise by the family. It also considers as family farmers: forestry workers, fish farmers, extractivists, fishermen, indigenous, quilombolas and settlers of the Agrarian Reform (FNDE 2019b; Silva et al. 2022, 2023a).

On May 8, 2020, FNDE Resolution nº 6 was approved, bringing novelties and standardizations of measures that have already been adopted by some implementing units. A change that Resolution nº 6 brought involved the criterion of selection of formal groups of family farming, through which cooperatives having in their composition higher percentage of participation of members with individual DAP will have priority. In this case, what influences for tiebreaker is not the number of members, but the percentage of members with DAP against the total number of members (Brasil 2020b; Silva 2020).

Due to the coronavirus pandemic, before Resolution nº 6, on April 7, Law 13.987 was approved that changed Law n° 11.947, to authorize, on exceptional basis, during the period when classes were suspended, the distribution of groceries acquired with PNAE resources to guardians of students from basic education public schools. The FNDE instituted, on April 9, 2020, Resolution nº 2 that regulated the distribution of groceries at the discretion of the local public authorities (Silva 2020).

The year 2020 was challenging, because schools in the whole country remained closed or under hybrid regime, where there was rotation of students per days of week, avoiding agglomerations in schools.

During 2021, on the other hand, some significant changes occurred in PNAE management. One important change was FNDE Resolution nº 21, of November 16, 2021, which changed Resolution nº 06, because it increased the individual limit of sale of family farmers and rural family entrepreneur within PNAE. From then on, the transaction must respect the maximum value of R$ 40,000, observing the following rules: individual suppliers and informal groups, the contracts entered into must respect the maximum value of R$ 40 thousand, per Family DAP, and for trading with formal groups, the maximum amount to be hired must be the result of the number of family farmers with Family DAP enrolled in the Legal DAP multiplied by the individual limit of commercialization.

After knowing PNAE history, it is important to understand how it works, and what institutions contribute to its management. Students from the entire basic education (early childhood education, elementary school, high school, and education of youth and adults) enrolled in public schools, philanthropic schools and community entities (under agreement with public authorities) are served by the program. The federal government passes on to states and municipalities, and federal schools, 10 monthly installments to cover 200 school days, according to the number of students enrolled (FNDE/PNAE 2019a; Silva 2020; Silva et al. 2022, 2023a, b). The values are standardized and set according to the levels and modalities of teaching, and indigenous and quilombolas are also contemplated. The daily values corresponding to each teaching modality follow FNDE Resolution nº 6, of 2020.

From the operational point of view, the Federal Government, through FNDE, is responsible for defining the program rules. The implementing entities are the departments of education of states, Federal District, and municipalities, and include federal schools. The implementing units are civil societies with legal private person status linked to the school, non-profit, that can be instituted by initiative of the school, the community or both. They can be called "school fund", "parent-teacher association", "parent-teacher circle" (FNDE 2019b; Silva et al. 2022, 2023b).

The School Food Council (CAE) is responsible for monitoring the acquisition of products, the quality of the food offered to students, the hygienic-sanitary conditions under which the food is stored, prepared, and served, the distribution and consumption, the financial execution and the assessment of accountability of the implementing entities, and issue of conclusive opinion (FNDE 2019b; Silva et al. 2022, 2023b).

The Federal Court of Accounts, and the Ministry or Transparency and Inspection, and Comptroller General of the Union are bodies of the federal government that also inspect the program. The Federal Prosecution Service, along with FNDE, receives and investigates complaints of mismanagement of the program. Health and Agriculture Departments of the states, Federal Districts, and municipalities can collaborate with PNAE through sanitary inspections, by attesting the quality of products used in the food offered and articulating the family farming production. The CAE is responsible for inspections in schools, and education departments (FNDE 2019b; Silva et al. 2022).

2 Theoretical Reference

2.1 Transformative Learning for Sustainability

For Habermas (1984, 1987), learning is an inherently motivational, cognitive, affective-emotional, and social process. Critical theorization, in this regard, is also explicitly evaluative as its purpose is the emancipation of individuals from domination and exploration in a world increasingly individualized, commodified, and reified (O'Donnell 1999; Silva et al. 2023b).

Habermas (1984, 1987) considers the first human interests as representing distinct learning domains; the instrumental and communicative learning domain, respectively; while the emancipation interest involves a learning dimension of critical reflection with implications for the other two. Each of the two interests also requires a different mode of personal learning, with different learning needs and different implications to facilitators of adult learning. For the author, there is a fundamental distinction between the dynamics of learning to control and handle the environment (instrumental learning) and the dynamics of learning and understanding others (communicative learning) (Mezirow 1991; O'Donnell 1999; Lankester 2013; Boucher 2021; Silva et al. 2023b).

The instrumental action always involves predictions of observable events, physical or social, that may prove to be correct or incorrect. Such action is based on empirical knowledge and is governed by technical rules. The choices in the instrumental action process involve strategies based on this knowledge and deducted from rules of a system of values or investigation rules. Such strategies can be correctly or incorrectly deducted. The choice of the best strategy depends on the correct assessment of alternatives. Criteria for effective control of the reality determine which choice is appropriate or inappropriate. The instrumental learning domain involves centrally the determination of relations of cause and effect and learning through task-oriented problem solving (Mezirow 1991; O'Donnell 1999; Lankester 2013; Boucher 2021; Silva et al. 2023b).

The second form, according to Habermas (1984, 1987), occurs by means of socially coordinated activities of their members. For the author, this coordination must be established through communication in certain central spheres of life, satisfying the conditions of rationality inherent in the communicative action. And that can help reintroduce the dialectic and self-reflexive component of the learning process and the language as activity, since the Theory of Communicative Action establishes internal relation between practice and rationality (Habermas 1984, 1987; O'Donnell 1999; Silva et al. 2023b).

In addition to the technical and practical, Habermas identifies the emancipatory as the third general area of human interest that generates knowledge. The emancipatory interest drives individuals, through reflection, to identify and challenge distorted perspectives of meaning. It is the interest in knowledge resulting from self-reflection, including interest in how history and biography expressed the way individuals see themselves, their suppositions about learning, the nature and use of knowledge, but also their roles, social expectations, and repressed feelings that influence them. The emancipatory knowledge is the knowledge acquired through critical self-reflection, different from the knowledge obtained from the "technical" interest in the objective world, or "practical" interest in social relations (Mezirow 1991; O'Donnell 1999; Lankester 2013; Silva et al. 2023b).

So, the interests constitutive of knowledge are linked to the functioning of an ego that adapts to external conditions by means of learning processes. The individual is inserted in the communication system of the social life through self-formative processes and builds an identity in the conflict between instinctive objectives and social restrictions. The implication of this intention, in light of the discussion of developmental psychology, brings the idea of the three "human interests constitutive of knowledge" of Jurgen Habermas (Boucher 2021).

The social learning provides an alternative to the traditional resource management perspective, the collective and collaborative learning that connects the biophysical to the social, cultural, and political spheres (Finger and Verlaan, 1995). As such, a perspective of social learning can be a guiding structure to continuously adapt to systems of resources. The modified understanding of the conditions in evolution, and, therefore, adaptations in systems of managed resources, basically depends on four questions: "Who learns?"; What is learnt?"; "How is it learnt?"; and "Why is it learnt?" (Maarleveld and Dangbegnonpág 1999; Silva et al. 2023b).

A theory of learning focused on meaning could provide a solid base for a philosophy of adult education from which appropriate practices of establishment of goals, assessment of needs, development of programs, instruction, and research could emerge (Mezirow 1991). Through individuals' capacity to become critically reflexive, reducing the "pre-judgment power" of the world of life on the communicative practice in daily life. Once their experiences are translated into speech acts, they are transformed into judgments and are linked to a claim of validity. People become empowered to "possess" their mutual understanding because they come to them by means of their own interpretations and adult positions about objectionable claims of validity. If the meaning is understood when it is known under which conditions an expressed idea is true, we should pay attention to the process of validity test to learn (Habermas 1984, 1987).

According to psychologist Oskamp (2000), Social Sciences provide evidences that a purely economic approach does no receive the same level of emphasis when the issue of change in individual positioning is considered. The author focused on the role that individuals can play to contribute to reduce their impact on the environment without necessarily considering economic penalties or incentives. His study suggests that the critical problem is no related to convincing individuals of the need to change their positions, but a series of complex structures and psychological changes necessary to enable individuals to perceive these environmental concerns.

In a transformative learning, social "transformations" originate in more plural, emerging, and undisciplined political realignments, involving social and technological innovations driven by diversely immeasurable knowledge, challenging incumbent structures and pursuing conflicting purposes. In fact, they may be due more to the critical practice of other values than to the utilitarian search of purposes (Hulme 2014; Grant 2011; Slote 2007; Wong 2006; Stirling 2014).

Some examples may include countless changes in the energy distributed, practices involving service innovations, radical ecoefficiencies, change in culture, and the massively more extensive good use of renewable resources (Jacobson and Delucchi 2009, 2011). Likewise, there are multiple innovations in the culture and practice of food production and use, involving agriculture, local supply chains, collective property, and more integration with other activities (Pretty 2005; Feenstra 1997; Altieri 2012; White and Stirling, 2013; Stirling 2014).

2.2 *Sustainable Development and Sustainability*

One starting point to discuss sustainable development and sustainability is the growing body of science that has warned that the world is facing environmental threats. And that has led to serious implications to many communities, populations, livelihood, and types of societies (Stirling 2014). Behind that, the roots of environmental challenges are increasingly located in the "behavior" of common people, rather than in powerful structures and interests that so actively restrict the condition associated to the growing individualism, consumerism, and materialism (Ransom and Baird, 2009). The diagnosis increasingly distances from the explicitly political struggle, towards less visible psychological and communicative techniques to ensure the "management of behavior" (Stirling 2014).

For such, researchers sought alternative and more sophisticated forms for the range of premises included in sustainability policies. That led to the development of what is widely known as "weak" and "strong" sustainability approaches. These terms were originally developed by environmental economists like Pearce (1993) and Turner (1993), though they were adopted and adapted by other social scientists (Baker 2006).

The positioning about the capital is at the center of this approach towards sustainable development. A strong approach to sustainable development is based on the wish to keep levels of natural capital, defined by Pearce (1993), as functional or valued.

A strong approach to sustainable development argues that any type of natural capital can be negotiated and replaced by human capital, as long as the total capital passed on to the next generation is constant or growing.

Though Turner (1993) provides an account based on economy, Pearce (1993) developed a continuum where he positions weak and strong forms of sustainable development in terms of environment, political values, and ethics. An adapted version of that is provided in Table 1, with additions of Baker (2006) and Barr (2008).

To understand the sustainability spectrum, Dunlap and Van Liere (1978) seminal study on environmental values and behavior, conducted in the United States, is necessary. The authors argue that "biocentric" and "anthropocentric" perspectives of the environment were significant to guide the environmental action of individuals. These values must be seen in a continuum, with the extreme biocentric position representing values that position humans as equal to nature, in addition to not making distinctions based on the hierarchy between human and not human. An extreme anthropocentric

Table 1 Spectrum of sustainability

	Technocentric		Ecocentric	
	Cornucopia	Complacent	Communalist	Deep ecology
Green indicators	Resources explored	Resources conserved, in management attitude	Resources under preservationist positioning	Extreme preservationist positioning
Type of economy	Anti-green economy, free market, without restrictions	Green economy and market guided by incentives	Deep green economy, 'no' to economy and population growth	Deep green economy, strongly regulated to minimize withdrawal of resources
Management strategies	Primary economy policy, turned to economic growth (Gross Domestic Product—GDP). Considered as undisputed, without restrictions. Markets in conjunction with technical progress that will ensure infinite replacement. Possibilities capable of mitigating all limits of scarcity. Restrictions to environment sources that may disappear	Adjusted economic growth (Uses green accounting to measure Gross Domestic Product – GDP) Important dissociation, because infinite substitution is rejected. Sustainability rules follow the constant capital rule	Dissociation, but no increase in scale systems Perspective of health to the whole ecosystem is important. Gaia hypotheses and implications	Imperative reduction of scale from the extreme to a literal interpretation of Gaia, personalized as agent with moral, obligations, and duties

(continued)

Table 1 (continued)

	Technocentric		Ecocentric	
	Cornucopia	Complacent	Communalist	Deep ecology
Ethics	Support to traditional ethics, discourse on contemporary individual human rights and interests, instrumental value of nature	Extension of the ethical discourse to "take care of others", intra and inter-managerial equity, instrumental value of nature	More extension of the ethical discourse; collective interests outweigh individual interests	Acceptance of Bioethics. Intrinsic value of nature (valuable in its own right)
Sustainability indicators	5 Very weak: Change in existing structures; superficial appearance, and minimum changes	Weak: Change in some processes; negotiates with less tangible problems	Strong: Change in the full system, examining the "system" as an element	Very strong: "Cultural" change; both internal and external elements of the system are changed

Source Based on Barr (2008, p. 46) and Davies (2013, p. 16)

position sustains that humans are distinctly differentiated, and, in hierarchical terms, of nature. In fact, the use of such spectra inevitably means that each individual can be place in a different point along the spectrum. For example, many individuals can keep a general anthropocentric position, believing the humans are distinct from the nature, but would not necessarily consider non humans as "inferior" (Barr 2008, p. 46; Davies 2013; Neumayer 2013).

As argued Pepper (2003), there are countless "positionalities" related to the environment. However, it's important to notice the importance of appreciating the role that underlying values play while framing visions regarding the environment. In short, it is about how the points of view about the use of the natural environment for human development are related to the environment.

3 Methodological Procedures

The present study is qualitative-descriptive (Sampieri et al. 2013; Kumar et al. 2018) with abductive logic (Mezirow 1991; Charreire and Durieux 2003; Cruz 2007) and uses as strategy a single case study (Yin 2018). It is worth reminding that interviews were semi-structured (Bishop 2005) and observation was non-participant (Stake 2011; Kumar et al. 2018), documented with field notes (Lofland 1974; Lofland and Lofland 1995; Weber 2009).

The interviews occurred in PNAE municipal management. In total 16 people were interviewed in Boa Vista/RR. First, the nutritionist, technical responsible for the program, was contacted. She was the first to be interviewed, and indicated the key actors that participated in PNAE management in her municipality. Data collection in Boa Vista occurred in December 2021. Table 2 presents the characterization of interviewed in Boa Vista.

After transcriptions of interviews, content analysis was conducted turned to thematic analysis (Bardin 2011). At a second moment, formulation of propositions and objectives was made to propose a theoretical and/or pragmatic framework where the results obtained were used (Bardin 2011). The codification process occurred during the material exploration according to previously formulated rules, based on

Table 2 Characterization of interviewed in Boa Vista-RR

Institution	Interviewed code	Interviewed characteristics
SMEC—Municipal department of education and culture Nutrition sector	ESME01RR	1. Nutritionist responsible for PNAE
CAE—School food council	ECAE01RR	2. President
	ECAE02RR	3. Secretary
COOPHORTA—Boa vista horticultural cooperative	ECOOHO01RR	4. Administrative manager
COOPANA—Nova amazônia agricultural cooperative	ECOOPA01RR	5. Production manager
	ECOOPA02RR	6. Cooperative member
	ECOOPA03RR	7. Treasurer
	ECOOPA04RR	8. Cooperative member
	ECOOPA05RR	9. Administrative manager
COOPERCINCO—Cinco polos agricultural cooperative	ECOOPCI01RR	10. President
	ECOOPCI02RR	11. Financial director
SEAPA—State department of agriculture, livestock and supply	ESEA01RR	12. Assistant agriculture secretary
Municipal elementary school in rural zone 483 students	EMFBW01RR	13. Pincipal
Municipal early childhood education school (Núcleo Casa Mãe) 50 students	ENCMVC01RR	14. Principal
AAGRIPRORR—Association of Family Farmers and Rural Producer of Roraima State	EAGRIP01RR	15. Administrative and financial director
SMAAI—Municipal department of agriculture and indigenous affairs	ESMA01RR	16. Assistant agriculture secretary

Source Research data

the literature review. At the end, during the treatment and interpretation of results, significant results were selected to propose inferences and anticipate interpretations connected to the objective proposed.

4 Results and Discussion

Boa Vista municipality belonged to Amazonas state until 1944, and became capital of the recently created Federal Territory of Rio Branco. In 1962, this territory was renamed Federal Territory of Roraima, and Boa Vista became the state capital. Its population, estimated in 2021 by the Brazilian Institute of Geography and Statistics (IBGE) was 436,591 inhabitants. With that, the municipality is considered the most populated in the state and the sixty-fourth most populated in Brazil, with population density of 70.2 inh/km^2. It is worth mentioning that Boa Vista has 5,687.037 km^2 and concentrates 63.11% of Roraima population (IBGE 2021b).

Boa Vista presents some characteristics that contribute to PNAE management after the Programa Família Que Acolhe (PFQA). This program promotes follow-up of children from the mother pregnancy until six years old, ensuring integrated access to healthcare, education, and social development (Boa Vista 2018). It was observed that this program is an instrumental action, because it was based on empirical knowledge and is governed by rules and techniques that direct the employee to work complying with established targets. That became even clearer when PNAE technical responsible nutritionist was interviewed.

According to interviewee ESME01RR, the team has 4 nutritionists and 1 administrative assistant, and they all took office in February 2020, by public tender. These public servants, by means of PNAE centralized management, serve around 44 thousand students, distributed across 126 schools, 17 of them indigenous/rural schools.

The current group of nutritionists was greeted by 2 nutritionists working in the city hall through work contract (they were not public servants), one nutritionist worked there for 7 years, and the other for 3 years. It was not a simple process, because, after being instructed by the previous nutritionists, at the moment of going to field, the coronavirus pandemic started.

But the experience with ESME01RR interviewee contributed a lot, as can be observed in the following fragment:

> I'm from Pará. I studied in Manaus and came. Now that the first group of nutritionists of this state left, in our staff, everybody is from outside. In fact, everybody has a degree. But two of them are from Roraima. [...] I came from indigenous healthcare. [...] Mainly from indigenous and rural menu. It is as we tried our best there, closer to that population. But we know that we are in the North region, and here we have items that we seek to meet, according to the purchases, we know the students' profile. So, we have the use of tapioca starch, which would fully meet, right? Not only the rural indigenous population has this habit, but the whole population in the North has this habit (ESME01RR interviewee).

The coronavirus pandemic, in 2020, was a challenging period, the recently-formed team of the Nutrition Sector had to meet with the Planning Sector of the Municipal Department of Education and Culture (SMEC), to which it is subordinated, to prepare reference terms for an emergency public call for acquisition of items to distribute food kits to students from the municipal education network. At that time, nutritionists were informed that SMEC was oriented to spend FNDE resources in family farming, and the other items of school meals should be acquired with the municipality's own funds.

During this period, instrumental learning of nutritionists evolved to communicative learning, practical, with allocation of funds for school meals during the pandemic. The next fragment shows how was the interaction of the Nutrition Sector with school managers and farmers for the delivery of food kits.

> We assembled the kits, but deliveries started in 2020, deliveries with family farming. Some kits contained eight items from family farming, and the kit with protein. In the case of protein, a slightly smaller, but helped a lot. And then, in the last of this delivery, of 2021, we made two deliveries of family farming, the protein kit, and one of cold cuts. We know the importance and needs of students. And then, we have the support of the law resolution to provide food for these students. Through kits. So we sought to do it as best as we could. [...] It was during the whole year of 2020, practically. Thus, we used to make a round of delivery, and this involved great logistics, and we count on managers a lot. The support of managers is indispensable, right? And so, there are many schools across the city and there is the supplier to deliver to these several schools, a demand very different from what we had before, a new logistics is necessary (ESME01RR interviewee).

With the pandemic, the Nutrition Sector's team was taken by surprise, because there wasn't, at first, a normative to follow. The strategy adopted was deducted based on the communication between school managers and farmers, so that they could, in fact, deliver food to the students.

The team and SMEC efforts resulted in monthly delivery of approximately 2,700 baskets destined to children enrolled in nurseries and preschools, and other 1,900 were destined to indigenous and rural schools. All families under social projects, like Programa Família Que Acolhe (PFQA), Visitação Domiciliar, or CadÚnico, which had children enrolled in nurseries or preschools, received baskets only through SMEC. It is important to emphasize that SMEC reused baskets with non-perishable products acquired by the Municipal Social Management Department (SEMGES) and bought meats and dairy compounds to deliver protein kits and inserted products from family farming to deliver other separate kits.

It is estimated that 16 thousand kits were monthly distributed with family farming products and proteins. During 2020 and 2021, it is estimated that SMEC spent around R$15,388,090.25 in the purchase of these kits (CAE Boa Vista 2021). The non-perishable products basket's unit value was R$138.25, and SEMGES spent approximately R$19,522,000.00 in the acquisition of 200 thousand baskets from 2020 to 2021 (CAE, Boa Vista, 2021).

As previously mentioned, these non-perishable food baskets were not delivered only to SMEC, they were also destined to the program "Cesta do Bem", where the city hall delivered food complementation baskets until December 2021. Families in social vulnerability situation enrolled in the Federal Government's CadÚnico received the baskets.

In this interconnection between SMEC and SEMGES, it was observed that secretaries, along with the city hall opted for extracting common characteristics to formulate a response to the problem. There was even sharing of commitment, because non-perishable items' baskets were acquired with SEMGES funds. Here the experience of one the departments was prioritized, in a moment of crisis, to catalyze innovations in the service rendered. It was also observed the importance of implementing the process sustainability model involving social values, ethical capital, and implementation of participation mechanisms across departments.

However, despite the great effort made by the departments, not all students were contemplated, because only families registered in CadÚnico were served, and many others that were not registered, as in cases of some immigrants, were not contemplated. The president of CAE (School Food Council), interviewee ECAE01RR draws attention to this in the following fragment.

> Some people who really needed could not be contemplated. We are in a state that today 40%, 30% of the population is immigrant. There is also this case of Venezuelan immigrants. So, municipal and state schools have a group of twenty students with only immigrants. Therefore, until this situation is adapted ... and so many immigrants are in real social vulnerability situation. Because there is the Venezuelan shelter. Those in the shelter receive all meals, but what about those Venezuelans not in shelter? (Interviewee ECAE01RR).

In addition to the coronavirus pandemic, Boa Vista faces intense migration flow of Venezuelan immigrants since 2014. To shelter part of this population, 11 shelters were created, administered by the Army and UN High Commissioner for Refugees (UNHCR). It is estimated that almost 32 thousand Venezuelans live in the capital (UNICEF 2019).

This immigration issue requires higher integration between municipal, state, and federal governments, but there is no governance structure to coordinate these power spheres. It was verified that actions are being carried out, but no integration of them. Learning occurred amid more diverse and comprehensive political realignments. The growth of Amazonian population is a real agenda, and discussions on sustainable development and sustainability should consider it.

With the increase in population density, PNAE management must also evolve. The principal of the rural school visited presented a report on these changes.

> I'm the school manager [...] a school located in the rural area. I work in this school management since 2013, when I was invited to be the school manager. At first I didn't want o accept, but I was convinced to accept and I'm here until today. Almost 10 years. We have already faced many difficult moments. There was a time when we had to club together (with the mayor) to provide meals to the children. When I arrived here in the school [...] we had 183 students, that was in 2013. Today we have 483 students. It means that the number of students rose. Today we have quality meals, because the mayors entered into agreement with family farming. (Interviewee EMFBW01RR).

This "agreement with family farming" mentioned by interviewee EMFBW01RR is in fact part of other project that promotes the municipality farmers, the Municipal Agribusiness Development Plan (PMDA). PMDA was created in 2018 and is managed by the Municipal Department of Agriculture and Indigenous Affairs (SMAAI). This program provides inputs like limestone, phosphorus, NPK, urea, chloride and seeds for planting of winter culture, grains, and summer and irrigated cultures. The interested producer must associate to a cooperative, individually, or be a member of association linked to a cooperative (Boa Vista 2022).

According to interviewee ESMA01RR, SMAAI assistant secretary, to join the program, the farmer must present active DAP, RG and CPF, property ownership document, or lease agreement. The DAP in the municipality is issued by INCRA, SEAPA (State Department of Agriculture, Livestock and Supply), or Union of Rural Workers.

Each farmer can hire R$20,000 in inputs in the first half of the year and equal amount in the second half, according to availability of products in SMAAI. For associations contracts with the cooperative, the amount will be up to R$60,000. Upon receiving the documentation, delivered by the cooperative in SMAAI, the department sends a technician to inspect the planting area. If the technician approves, the farmer will be qualified to receive inputs. To receive the inputs, the farmer must sign the contract at SMAAI and withdraw the inputs, but the transport is his responsibility. The cooperative which the farmer joined must have machines and implements under caution by the municipal government, because it will be responsible for ensuring preparation and planting in the area hired by the producer.

With regard to costs, the cooperative will charge the farmer a determined amount for hour worked of the tractor, referring to diesel consumed, the value paid to the machine operator and machine maintenance fee (change of oil, filters, parts, and machine repairs). At the time of signature of contract at SMAAI, a collection document in the value of the inputs provided to the farmer will be issued. This document, known as "costing", includes the following due dates: contracts signed in the first half are due on 12/15; contracts signed in the second half are due on 04/30. The producer that pays his debt on the due date will be entitled to 30% discount on the document value. After this date he will lose the discount and will pay fines and corrections according to municipal legislation.

PMDA, somehow, influences PNAE in the municipality, as can be observed in Table 3, from 2016 to 2019, increase in participation of family farming with FNDE funds was really expressive.

Another fact that links PMDA to PNAE is its origin. Interviewee ESEA01RR, former agriculture secretary of the municipality, and in 2021 he was assistant agriculture secretary of Roraima state, and told how the program was created, institutionally. See the fragment below:

> We built the PMDA along with other two or three friends, within a legal structure, legally supported. It is worth saying that in the first round we were 100% wrong. We got there... and in my mind, I came with a definition "look, I will take this producer, and I will dress him, in my understanding", but it didn't work, there was no result, nothing. That's when I was sure; I could understand that each farmer has their own particularity. The producer was here, I talked to him and brought him to a condition with economic result. All activities had to have an equilibrium point in terms of results. It had to make profit. Before that there should be good a good crop, with high yield. So, I never let it go. And then we could appear. Then, in our second action, we came to this condition here. It must be around 150 producers, all

Table 3 FNDE transfer and amount invested in family farming from 2011 to 2021 in Reais for school food

Year	Federal fund	Family farming	% Family farming (%)
2021	*3,184,928.00	**2,033,119.86	63
2020	*2,631,549.20	**1,792,753.19	68
2019	*3,598,008.40	**4,569,497.14	127
2018	*2,043,228.00	**2,616,298.67	128
2017	* 932,413.20	2,182,887.94	234.11
2016	2,392,953.75	2,392,953.75	100.00
2015	3,397,778.00	2,672,577.32	79
2014	2,129,872.00	1,685,715.56	79
2013	3,451,668.00	590,037.49	17
2012	3,105,012.00	592,503.00	19.08
2011	2,459,340.00	–	0.00

Source FNDE family farming participation spreadsheet 2011–2017, * SigPC, 2017–2021 and ** Boa Vista City Hall Transparency Portal 2018–2021 (http://131.255.233.222:8079/Transparencia/)

> of them could reach very high level of production. [...] Then things started to work. Today farmers are very well, planting. I'm talking about planting watermelon and maxixe, and other things. Yes. Now came melon. They are making too much money with melon, watermelon, corn. So, we created a concept that nothing is for free, nothing, and this is a natural condition (Interviewee ESEA01RR).

It is worth mentioning that interviewee ESEA01RR was vice-president in a large agricultural company in Mato Grosso state. Therefore, he participated in the development of new technologies, new equipment, in the experimental farm where we worked. His development perspective was completely different from that of the family farmer in Nova Amazônia Settlement Project. That, initially, generated conflicts, but led to the construction of a public policy. By reading the statement of interviewee ECOOPA01RR, Nova Amazonia Agricultural Cooperative (COOPANA) production manager, one can identify the sequence of this construction.

> And in 2016 I had the pleasure, honor, God's bless, to meet the secretary here [...], he came, looked at our beans, looked at the whole plantation, along with two other guys whose name I don't remember. He looked at the whole plantation and said: "president, pass the machine here, the plow, and transform it into fertilizer". And I said to him: "I will tell you something, here we have fifteen farmers that put their money here, and you look at that, and for you it is worthless. Man, for us, for them, I tell you... I'm sure that they prefer to see that in their house. For you it is nothing, but for us, having that at home, these beans, man, you will see with tears, let me tell you, because he (the farmer) could harvest something!", And he said: "I'm going to help you", and he brought the harvester, and Mr. [...] financed the oil, the diesel oil for the harvester, and harvested. Man, that was so much joy, because until then I had never seen a large harvester doing this job for us from the family farming. [...] So it was encouraging! Something that he knows, and gave us a huge starting point, I know that what I told him touched him, and so he came and did that, and harvested. He said look, I will come back here and we will produce here a beautiful plantation. I will do that with you. Some years later the mayor [...] won and called him to be secretary and then he came back

> here and said, "do you remember me? Today I am secretary and I'm here to work with you. Do you have any project?" at the moment there was nothing. "Then I will bring a team to talk and you will write, do, and we will make it work". I said "really?" Since it was politics, I didn't give much importance. And he came back. When he came back he said "is the project ready?, where can we start talking?", and I said "here", because it was in […] house that he showed, and we started to talk there. I said "the project is very easy, simple, and is good for us all; it will be a wonder for us farmers, for you as secretary and the mayor, as manager". I said to him "here we don't want alms, nothing for free. We only want dignity here, the project is simple. We already have plans, but I'm going to tell you. We will create a name; let's say that this is part of you. We want inputs, seeds, we want everything here. But we don't want it for free, the department must find a way to finance, and we want to pay with our production, with what we will harvest, and everything documented". Then he said "man, it's so simple that no one made it here, right?", then he left. Another year, a month and he came back, and came to our office, [...] Then they started to write everything, and everything will work all right, and so on. [...] They spent a good time doing and redoing and doing until it worked. They bought limestone, bought inputs, sent cautions, machines for us, tractors, harrows, everything they had, and so it started. What you see here today, all that, is this PMDA, it was the beginning (interviewee ECOOPA01RR).

PMDA creation is emancipatory learning, because it made individuals from different realities, through reflection, face their distorted perspectives to solve a real problem. Interviewee ESEA01RR made a self-reflection, taking into account his history and that of those settled in Nova Amazônia PA, in addition to considering his role as agriculture secretary, and the social expectations of a community with regard to his position. In this case, the critical knowledge overcame technical and practical learning. It is worth reminding that PMDA structuring process was a "bottom up" strategy where farmers suggested a proposal according to their reality and that resulted in change in the social and economic systems of the municipality.

To understand the real impact of public policies like PMDA and PNAE, it's necessary to contextualize them in Nova Amazônia Settlement Project (PA), which houses farmers from the main cooperatives of the municipality. Nova Amazônia PA originated in a farm from Bamerindus Group that was retaken by the Union due to debts. In April 2003, families camped at BR 174's left margin, aiming at acquiring lands in the area. In November the distribution of lots started, in 5 Poles (Carvalho 2018).

Initially, families camped in Poles II and III were settled. As of January 2004, the settlement of the rest of the families camped, workers from the former Bamerindus Farm that still lived there, horticultural farmers from Operário district, in Boa Vista, and rural workers from the surroundings of Bamerindus Farm and Monte Cristo was made, totaling 262 families. They received lots in areas defined as Poles I, II, III, and IV. In 2005 Pole V was installed, with families from Rio Grande do Sul that were pioneers in cultivation of grains (soybean and rice) and fruit growing. In 2015, farmers removed from Raposa Serra do Sol Indigenous Land, were sent to the location (Carvalho 2018).

The first cooperative to emerge in Nova Amazônia PA was the Five Pole Agriculture Cooperative (COOPERCINCO). COOPANA arose from Pole I Family Farmers Association, from Elifas Levi Veloso Filho Nova Amazônia settlement project, along with a project by the Federal University of Roraima, in 2015. Details were presented in the following excerpt of interview:

> I'm here in Nova Amazônia PA since 2003, when the project started and lots were divided. And so we are here, in this struggle. At the beginning we faced rough times, under those cashew trees, canvas, and a bloody mess. But that's it, every beginning, if things were easy, it's no good, it's not funny. And so I have this lot here, I'm from the very beginning, I have this lot for a long time, working to build it, and in 2005 this association appeared. Nova Amazônia Settlement Project's Family Farmers Association. Elífas Levi Veloso Filho. This Guy fought a lot for this settlement here. He fought a lot, he was pioneer here. So, we, in 2005, decided to name the association after him, because he was a fighter. He died in one of these travels to here [...] Here, and we already had, in 2008, or 2010, we started to build here. [...] we made Semente Crioulo Project, through the Federal Government, and then each farmer that participated in this project got forty kilos of bean to save money to start building the association, which is that first small building there. And then, from then on, I started to like to work here together. And that's when they found out PNAE, started to work with PAA, CONAB and formation of stock of Semente Crioulo Project (Interviewee ECOOPA01RR).

COOPERCIONCO, on the other hand, was created in 2006 by a group of farmers from the South region of the county, that settled in Nova Amazônia PA. At first, there were 21 farmers, and today it counts on 650 members distributed across Roraima state. Commercialization through PNAE and PAA is its main focus. Interviewee ECOOPCI02RR reports how the start of the cooperative was.

> I'm settled there in Nova Amazônia PA, but we were camped, participated in that whole process of camping, outside the farm, in small canvas shacks, waiting for the opportunity o enter. Then the opportunity came, we could enter the settlement process, in 2003. Three years later a group of 'gaúchos' (from Rio Grande do Sul) arrived. We had never left Roraima, and never heard of cooperative. And they arrived with this incentive from Rio Grande do Sul, with their expertise in working in cooperative, understanding how it worked, so they brought this baggage. When they arrived, there was divergence between them, and the group of 'gaúchos' split. And so, from this split, they founded COOPERCINCO. At the time it wasn't COOPERCINCO, it had another name, and it became COOPERCINCO because of the five Poles. Then they started to participate in social programs like PAA. And so, in this PAA formation they needed to obtain higher resources. They started to open the cooperative. Then the name was changed to COOPERCINCO because the land in Nova Amazônia PA, Murupu region, has 5 Poles. So, they chose COOPERCINCO to contemplate the five Poles. They opened it for everybody. Then, in 2012, I joined the cooperative just as partner and we kept on working (interviewee ECOOPCI02RR).

COOPERCINCO creation was a form of collective learning in which farmers, both from Roraima and Rio Grande do Sul, seek meanings for their reality, in this case survival and construction of a more stable life in Nova Amazônia PA. The opening of communication between those settled was fundamental to establish a dynamics of learning to understand others. In this case, the social learning enabled collective and collaborative learning to create the cooperative.

During the visit to COOPERCINCO, interviewee ECOOPCI02RR told the story that preceded FNDE Resolution nº21, of November 16, 2021. According to this interviewee, cooperatives in Boa Vista delivered a letter to President Jair M. Bolsonaro, on October 26, 2021, asking for increase in the transaction value of purchase of family farming products, and that catalyzed FNDE Resolution nº 21. This resolution changed FNDE Resolution nº 06/20, because it increased the individual limit of sales of the family farmer and the rural family entrepreneur inside PNAE. From then on, the transaction must respect the maximum value of R$ 40,000.00, observing the following rules: for individual suppliers and informal groups, individual contracts must respect the maximum value of R$ 40 thousand, per Family DAP, and for trading with formal groups, the maximum amount to be hired must be the result of the number of family farmers with Family DAP enrolled in the Legal DAP, multiplied by the individual commercialization limit. With that, once again the "bottom up" strategy is observed in the municipality. However, FNDE Resolution nº 21 affected the whole country. Farmers in the country already wanted to increase this value, which was made possible by the cooperatives initiative, in Boa Vista/RR.

Boa Vista Horticultural Farmers Cooperative (COOPHORTA) presents a different story. It was founded in 2003, with assistance from Estufas Project, by the city of Boa Vista. The project was created by the mayor at the time with horticultural producers in the region of the green belt, Operário, and Boa Vista Industrial District. They were producers of greenery to send to Manaus. Thus, COOPHORTA building was loaned out by the city. In 2011, through Law nº 11.947, they started to participate in PNAE. At the time they had 40 active partners. And to start to supply they were contemplated by projects from Banco do Brasil Foundation, Amazon Fund, and could buy trucks, furniture, and vehicles. Today they have 700 members. Interviewee ECOOH01RR talks about PNAE importance to the cooperative, and devastating was the coronavirus pandemic period.

> And today we have over 700 members. They could leverage; 99% until 2015-2016 was PNAE resources that moved the cooperative. Some separate small sales of PAA project that was too little at the time. But 99% of the cooperative invoicing was from PNAE. Today it still represents a lot; today we have around 75 to 80% still from PNAE. But we already have a destination to cooperate through the Army, PAA, state PAA and retail sales. [...] Nobody expected the pandemic. We were in our best year. We had important contracts with the state, the city. It was the year that we expected to earn more, 2020. Then the pandemic came to show us that we were on the path, walking the right path in our vision. However, it was wrong, because we depend a lot on the government, and the government today, it this PNAE ends, for the cooperative today it is 90 or almost 80% of our billing (interviewee ECOOH01RR).

Here it is worth retaking Table 3 analysis, because in 2011 there was no participation of family farming in PNAE, it was through measures of incentive to farmers that this percent started to gradually increase. Everything was developed from the municipal government concern in absorbing the city agricultural production. That led to social, organizational innovations, political and institutional displacements that worked progressively, challenging the concentrations of privilege and power.

Another interesting fact observed in visits to cooperatives is their proximity, both in their origin and with regard to their work developed in the municipality. For example, Roraima State Family Farmers and Rural Producers Association (AAGRIPRORR), headquartered in Rorainópolis, 290 km from Boa Vista, which also serves PNAE in the city, was founded by a former employee of COOPHORTA. Moreover, there is a partnership involving the 3 main cooperatives visited (COOPHORTA, COOPERCINCO and COOPANA), they enter into agreements in the division of institutional sales values, and many times, depending on the region served, one of them brings the products of the other along with theirs, thus optimizing logistic costs.

So we can see how cooperatives, even as competitors, develop strategies for integration of members in their value chain. This is the importance of the municipal government and its programs auxiliary to PNAE that work on the needs of communities to develop a more inclusive economy.

With regard to sustainability, technocentrism prevails. However, in a more complacent view (weak sustainability). There is the recognition of the environment as valuable for human development, promoting economic growth, in addition to the use of technology to improve agricultural efficiency, acknowledging a specific level of natural capital stock. Boa Vista municipality is in "transition" rather than "transformation" process.

It was observed very strong presence of technological innovation in the agricultural area, because SMAAI was a department created in 2017, intended to contain rural exodus. In 2018, PMDA was created to develop Nova Amazônia PA production, and in some indigenous areas, in this case approximately 52 hectares (Boa Vista 2022). It is worth mentioning that SMAAI counts on 38 servants, of which almost half work in the Rural Extension, providing technical services to farmers.

SEMEC organized a new team in the Nutrition Department to continue this intersectorial integration across departments (SMEC/SEMGES/SMAAI), and keep a structured control over municipal public policies. There is a multidisciplinary process guided by management processes that characterizes instrumental learning. It is worth noting that there are tacit and incorporated processes involving emerging political realignments.

About green indicators, there was the concern of creating a settlement in Lavrado area to reduce the existing pressure in the state forest areas (Carvalho 2018). Directing agriculture to Cerrado area occurred, where there was already the structure of three expropriated farms (Cauamé, Murupú and Truarú) from Bamerindus Group (Carvalho 2018).

The type of economy is incentive-driven market economy. Interviewee ESEA01RR report addresses the municipality positioning: "It had to make profit. Before that it had to be a good crop, with high yield". As management strategies, Boa Vista seeks a rule of constant capital in the city, either through incentive to agriculture production or through the social programs implanted. The ethical discourse focus on care of population, particularly early childhood, school meals, and farmers' quality of life. There are changes in some processes and the city government that start o address some less tangible aspects, mainly at social level.

Weak sustainability is based on the idea that human innovation will solve natural and social capital problems in a place and not in a broader global context. Considering that the municipality is in the Amazonian Region, thinking at this level is an advance, however, it is not ideal. Valuing nature depends on the individual and collective view of the world. And it is exactly these views of the world that influence interactions of humanity and nature.

Final Remarks

The Amazon became urbanized forest, and with that it started to incorporate problems like rural exodus, lack of basic sanitation, violence, irregular income distribution, lack of structure in healthcare and child malnutrition. In such a complex biome, in order to preserve the forest, the idea of thriving in equilibrium becomes agile tool to start changes. Focusing on growing nutritious food, and sufficient to all while preserving soils and water, valuing life, and taking care of this population whose identity is under construction, constituted of indigenous people, riparians, immigrants from several regions of Brazil and other nations, is a start.

Uniting this environmental complexity to PNAE execution depends, initially, on individual learning that, through observation or interaction with the environment expands to social learning. And this social learning works as guiding principle for deliberation on problems and solutions to be explored. Correlating a developmental economic subsystem to a sustainability vision originated in learning was the theoretical contribution of this Chapter.

As practical contribution, this Chapter shows the importance of public policies and Law n°11.947 guidelines. In the productive aspect, they foster local economies, generate more jobs, and promote cooperativism. Many cooperatives and agro-industries arouse to meet PNAE in the city. This law is important for farmers, because through it they could remain in their properties, avoiding rural exodus. It was through its insertion in PNAE that these farmers sought to diversify their production to serve schools. And that contributed to foster innovations in the field, like agro-industries, but also led to the domain of new technologies. This effort is converted in investments in their property, equipment, vehicles, among other improvements, that contribute to health, time management, self-esteem, and their quality of life. While complying with Lei n° 11.947, instrumental learning is outstanding, because servants follow technical rules, and the strategies adopted depended on the correct assessment of alternatives.

However, in Boa Vista/RR, PNAE was favored by other public policies, as the Municipal Plan for Agribusiness Development (PMDA). This program is managed by the Municipal Department of Agriculture and Indigenous Affairs (SMAAI), and facilitates the acquisition of inputs like limestone, phosphorus, NPK, urea, and seeds for winter and summer culture planting, grains, and irrigated cultures for farmers. The PMDA creation was considered emancipatory learning, because it resulted in a critical reflection process involving farmers and the municipality managers. All those that participated in the construction of this public policy were influenced by their knowledge, their roles, social expectations, and feelings.

Another relevant aspect was the interconnection between the Municipal Department of Education and Culture (SMEC) and the Municipal Department of Social Management (SEMGES) for acquisition of food parcels to be distributed during the coronavirus pandemic to students from the municipal education network, in vulnerability situation. This SEMGES assistance to SMEC enabled the complementary monthly distribution of 16 thousand kits with family farming products and proteins to these families. This relationship between departments characterizes communicative learning, which occurred through socially coordinated activities of their members.

Despite the efforts, Boa Vista is in "transition" rather than "transformation" process. The municipality is characterized by weak sustainability, because it is based on the idea that human innovation will be able to solve natural and social capital problems. Considering that the municipality is in the Amazonian Region, thinking at this level is an advance, however, it is not ideal.

Since data collection occurred in December, last school month of the year, relevant information was obtained. However, the period was a limiting factor, because it was not possible to visit more schools or even visit an indigenous school to see the innovations occurred in these communities. For the future, we suggest studies on the existing relations between levels of learning and sustainable development, based on the adoption of PNAE Law nº 11.947 in indigenous communities.

Acknowledgements We would like to thank Eugenio Avila Pedrozo, coordinator of the research project Learning for sustainable development: Cases from the Brazilian PNAE (National School Feeding Program). He was responsible for theory conception and supervision of the project that involved five regions of Brazil. It was a pleasure work by his side.

References

Altieri MA (2012) Convergence or divide in the movement for sustainable and just agriculture. In: Lichtfouse E (ed.) Organic fertilisation, soil quality and human health. Sustainable Agriculture Reviews. Springer Netherlands, Dordrecht, pp 1–9. https://doi.org/10.1007/978-94-007-4113-3_1

Bardin L (2011) Análise de conteúdo. Edições 70, São Paulo, Brazil

Barr S (2008) Environment and society: sustainability, policy and the citizen. Ashgate, Aldershot

Baker S (2006) Sustainable development. Routledge, Abingdon

Bishop R (2005) Freeing ourselves from neocolonial domination in research a Kaupapa Maori approach to creating knowledge. In: Denzin NK, Lincoln YS The Sage handbook of qualitative research, 3rd ed. Sage Publications, London. https://doi.org/10.1007/978-94-6091-415-7_1

Boa Vista (2013) Lei Municipal nº 1.545 de 20 de dezembro de 2013. Diário Oficial [do] Município de Boa Vista, Boa Vista, RR. https://sapl.boavista.rr.leg.br/norma/68

Boa Vista (2018) Proposta curricular municipal para a educação infantil. Prefeitura de Boa Vista, Boa Vista

Boa Vista (2022) Observatório de Boa Vista: conheça o agronegócio de Boa Vista. https://observatorio.prefeitura.boavista.br/paginas/agronegocio-3c70ae60-cb3d-4a6b-a7f0-551b888039f3

Boucher G (2021) Habermas and literature: the public sphere and the social imaginary. Bloomsbury Academic, New York

Brasil (2009) Lei Federal nº 11,947 de 16 de junho de 2009. Diário Oficial [da] República Federativa do Brasil, Brasília, DF. https://www.planalto.gov.br/ccivil_03/_ato2007-2010/2009/lei/l11947.htm

Brasil (2020) Lei Federal nº 13.987 de 7 de abril de 2020. Diário Oficial [da] República Federativa do Brasil, Brasília, DF. https://legislacao.presidencia.gov.br/atos/?tipo=LEI&numero=13987&ano=2020&ato=0b6ETQ61EMZpWT52d

Brasil (2013) Resolução nº 26, de 17 de junho de 2013. Diário Oficial [da] República Federativa do Brasil, Brasília, DF. https://www.gov.br/fnde/pt-br/acesso-a-informacao/legislacao/resolucoes/2013/resolucao-cd-fnde-no-26-de-17-de-junho-de-2013

Brasil (2020a) Resolução nº 2, de 9 de abril de 2020. Diário Ofiial [da] República Federativa do Brasil, Brasília, DF. https://www.gov.br/fnde/pt-br/acesso-a-informacao/legislacao/resolucoes/2020/resolucao-ndeg-2-de-09-de-abril-de-2020/view#:~:text=Disp%C3%B5e%20sobre%20a%20execu%C3%A7%C3%A3o%20do,novo%20coronav%C3%ADrus%20%E2%80%93%20Covid%2D19

Brasil (2020b) Resolução nº 6, de 8 de maio de 2020. Diário oficial [da] República Federativa do Brasil, Brasília, DF. https://www.gov.br/fnde/pt-br/acesso-a-informacao/legislacao/resolucoes/2020/resolucao-no-6-de-08-de-maio-de-2020/view

Broch AE (2009) Congresso Nacional aprova projetos importantes para o MSTTR. J Contag Brasília 6(3). http://www.contag.org.br/imagens/f1620contagmaiojunho.pdf

Carvalho LJM (2018) Histórico do assentamento PA Nova Amazônia: a conquista da terra em Boa Vista—RR. Dissertação, Universidade Federal de Roraima. Boa Vista, RR: Programa de Pós-graduação em Sociedade e Fronteira, 149. http://repositorio.ufrr.br:8080/jspui/bitstream/prefix/244/1/Hist%C3%B3rico%20do%20assentamento%20PA%20Nova%20Amaz%C3%B4nia%20A%20conquista%20da%20terra%20em%20Boa%20Vista-RR.pdf

Charreire S, Durieux F (2003) Explorer et tester: Deux voies pour la recherche. In Thietart RA. et al (Org.) Méthodes de recherche en management. Dunod, Paris

Conselho de Alimentação Escolar do Município de Boa Vista [CAE Boa Vista] (2021) Relatório de entrega de gêneros alimentícios da agricultura familiar do ano de 2021. CAE, Boa Vista

Cruz LB (2007) O processo de formação de estratégias de desenvolvimento sustentável de grupos multinacionais. Thesis, administration, Federal University of Rio Grande do Sul. School of Business Administration, Porto Alegre, RS, p 444. https://lume.ufrgs.br/handle/10183/12416

Davies GR (2013) Appraising weak and strong sustainability: searching for middle ground. Cons: J Sustain Dev, 10(1):111–124. https://doi.org/10.7916/consilience.v0i10.4635

Dunlap RE, Van Liere KD (1978) The new environmental paradigm. J Environ Educ 9:10–19. https://doi.org/10.1080/00958964.1978.10801875

Feenstra GW (1997) Local food systems and sustainable communities. Am J Altern Agric 12(1):28–36. https://doi.org/10.1017/S0889189300007165

Finger M, Verlaan P (1995) Learning our way out: a conceptual framework for social-environmental learning. World Dev 23(3):503–513. https://doi.org/10.1016/0305-750X(94)00128-L

Fundo Nacional de Desenvolvimento da Educação (FNDE). Programa Nacional de Alimentação Escolar (PNAE) (2019a) Dados da Agricultura Familiar: Aquisições da Agricultura Familiar no período de 2011 a 2016. Brasília/DF. https://www.gov.br/fnde/pt-br/acesso-a-informacao/acoes-e-programas/programas/pnae/consultas/pnae-dados-da-agricultura-familiar

Fundo Nacional de Desenvolvimento da Educação (FNDE). Programa Nacional de Alimentação Escolar (PNAE) (2019b) Sobre o PNAE. Brasília/DF, 2019. https://www.gov.br/fnde/pt-br/acesso-a-informacao/acoes-e-programas/programas/pnae/historico

Grant RW (ed) (2011) In search of goodness. University Chicago Press, Chicago

Habermas J (1984, 1987) The theory of communicative action. Vol. 1: Reason and the rationalization of society. Vol. 2: Lifeworld and system: a critique of functionalist reason. (Trans. Thomas McCarthy.) Beacon Press, Boston

Hulme M (2014) Climate change and virtue: An apologetic. Open Access Human 3:1–13. https://doi.org/10.3390/h3030299

Instituto Brasileiro de Geografia e Estatística (IBGE) (2021a) Lista de municípios da Amazônia legal. https://www.ibge.gov.br/geociencias/cartas-e-mapas/mapas-regionais/15819-amazonia-legal.html?=&t=downloads

Instituto Brasileiro de Geografia e Estatística (IBGE) (2021b) Estimativas de população enviadas ao TCU. https://www.ibge.gov.br/estatisticas/sociais/populacao/9103-estimativas-de-populacao.html?=&t=resultados

Jacobson MZ, Delucchi MA (2009) A plan to power 100 percent of the planet with renewables. Scientific American, November:1–5. https://doi.org/10.1038/scientificamerican1109-58

Jacobson MZ, Delucchi M (2011) Providing all global energy with wind, water, and solar power, Part I: Technologies, energy resources, quantities and areas of infrastructure, and materials. Energy Policy 39(3):1154–1169. https://doi.org/10.1016/j.enpol.2010.11.040

Kumar V, Leone RP, Aaker DA, Day GS (2018) Marketing research, 3rd edn. John Wiley & Sons Inc., Danvers

Lankester AJ (2013) Conceptual and operational understanding of learning for sustainability: A case study of the beef industry in north-eastern Australia. J Environ Manage 119:182–193. https://doi.org/10.1016/j.jenvman.2013.02.002

Leal JLS (2019) A Amazônia brasileira e o seu caráter transnacional: o aproveitamento do seu patrimônio estratégico para garantia do desenvolvimento. Thesis, legal science, Universidade do Vale do Itajaí—UNIVALI. Itajaí, SC: Doctorate in Legal Science, p 377. https://www.univali.br/Lists/TrabalhosDoutorado/Attachments/259/TESE%20-%20JORGE%20LUIZ%20DOS%20SANTOS%20LEAL.pdf

Lee K (1993) Compass and gyroscope: integrating science and politics for the environment. Island Press, Washington, DC

Lofland J (1974) Styles of reporting qualitative field research. Am Sociol:101–111. http://www.jstor.org/stable/27702128

Lofland J, Lofland LH (1995) Analyzing social settings: A guide to qualitative observation and analysis, 3rd edn. Wadsworth, Belmont, Calif

Maarleveld M, Dangbegnon C (1999) Managing natural resources: A social learning perspective. Agric Hum Values 16:267–280. https://doi.org/10.1023/A:1007559903438

Marques ABGM, Triches RM (2022) Aquisição de alimentos orgânicos pelo. Programa:502–520. https://doi.org/10.5380/dma.v60i0.79120

Mezirow J (1991) Transformative dimensions of adult learning. Jossey-Bass Inc., Publishers, San Francisco, California

Neumayer E (2013) Weak versus strong sustainability: Exploring the limits of two opposing paradigms, 4th edn. Edward Elgar, Cheltenham

O'Donnell D (1999) Habermas, critical theory and selves-directed learning. J Eur Ind Train, 23 (4/5): 251–261. O'Donnell D (1999) Habermas, Critical theory and selves-directed learning. J Eur Ind Train 23(4–5):251–261. Available at SSRN: https://ssrn.com/abstract=818364

Oskamp SA (2000) Sustainable future for humanity? How can psychology help? Am Psychol 55(5):496–508. https://doi.org/10.1037/0003-066X.55.5.496

Pearce D (1993) Blueprint 3: measuring sustainable development. Earthscan, London

Peixinho AML (2013) A trajetória do Programa Nacional de Alimentação Escolar no período de 2003–2010: relato do gestor nacional. Ciência Saúde Coletiva 18:909–916. https://doi.org/10.1590/S1413-81232013000400002

Pepper D (2003) Modern environmentalism. Routledge, London

Pretty J (ed) (2005) The earthscan reader in sustainable agriculture. Earthscan, London

Ransom D, Baird V (eds) (2009) People first economics. Oxfam, Oxford

Rede Brasileira de Pesquisa em Soberania e Segurança Alimentar (Rede PENSSAN) (2022) II Inquérito nacional sobre insegurança alimentar no contexto da pandemia da COVID-19 no Brasil. Fundação Friedrich Ebert, São Paulo, SP

Röling NG, Wagemakers MAE (1998) Facilitating sustainable agriculture. Cambridge University Press, Cambridge, United Kingdom

Sampieri RH, Callado CF, Lucio MPB. (2013) Metodologia de pesquisa, 5ª edn. Penso, Porto Alegre

Silva AC (1995) De Vargas a Itamar: Políticas e programas de alimentação e nutrição. Estud. Avançados 9(23):87–107. https://doi.org/10.1590/S0103-40141995000100007

Silva EA (2020) Teorias de aprendizagens e a sustentabilidade no programa Nacional de Alimentação Escolar (PNAE): caso do município de Porto Velho/RO. Dissertação em Administração, Universidade Federal de Rondônia. Porto Velho, RO: Programa de Pós-Graduação em Administração, 305. https://www.ri.unir.br/jspui/handle/123456789/3384

Silva EA, Pedrozo EA, Silva TN (2022) PNAE (National School Feeding Program) and its events of expansive learnings at municipal level. World 3(1):86–111. https://doi.org/10.3390/world3010005

Silva EA, Pedrozo EA, Silva TN (2023a) The PNAE (National School Feeding Program) activity system and its mediations. Front Environ Sci 10:1–17. https://doi.org/10.3389/fenvs.2022.981932

Silva EA, Pedrozo EA, Silva TN (2023b) The PNAE (National School Feeding Program) activity system and its mediations National School Feeding Program (PNAE): A public policy that promotes a learning framework and a more sustainable food system in Rio Grande do Sul, Brazil. Foods 12:1–24. https://doi.org/10.3390/foods12193622

Slote M (2007) The ethics of care and empathy. Routledge, Oxford

Stake RE (2011) Pesquisa qualitativa: estudando como as coisas funcionam. Penso, Porto Alegre

Stirling A (2014) Emancipating transformations: from controlling "the transition" to culturing plural radical progress. STEPS Working Paper, 64. STEPS Centre, Brighton. https://steps-centre.org/publication/emancipating-transformations-controlling-transition-culturing-plural-radical-progress/

Turner RK (1993) Sustainable environmental economics and management: principles and practice. Belhaven, London

United Nations Children's Fund [UNICEF] (2019) Crise migratória venezuelana no Brasil:o trabalho do UNICEF para garantir os direitos das crianças venezuelanas. https://www.unicef.org/brazil/crise-migratoria-venezuelana-no-brasil#:~:text=Com%20o%20apoio%20de%20parceiros,formal%2C%20incluindo%20a%20educa%C3%A7%C3%A3o%20infantil

Weber F (2009) A entrevista, a pesquisa e o íntimo, ou por que censurar seu diário de campo? Horizontes Antropológicos 15(32):157–170. https://doi.org/10.1590/S0104-71832009000200007

White R, Stirling A (2013) Sustaining trajectories towards sustainability: dynamics and diversity in UK communal growing activities. Glob Environ Chang 23(5):838–846. https://doi.org/10.1016/j.gloenvcha.2013.06.004

Wong DB (2006) Natural moralities: A defense of pluralistic relativism. Oxford University Press, Oxford

Yin RK (2018) Case study research and applications: Design methods, 6th edn. Cosmos Corporation—SAGE, London

Integrated Diagnosis of Biodiversity, Water Quality, and Aquatic Resources as an Insight for Conservation Strategies and Sustainability of Artisanal Fishing on the Tapajós River

Edenise Garcia, André L. C. Canto, Charles H. Faria Junior, Dávia M. Talgatti, Diego M. Zacardi, Frank R. V. Ribeiro, Heloísa N. M. Meneses, Iracenir A. Santos, Priscila S. Miorando, and Tony M. P. Braga

Abstract The Tapajós is a main tributary to the Amazon River, harboring a large biological and cultural diversity. However, despite its socioecological importance, the Tapajós experiences multiple anthropogenic pressures arising from deforestation and forest fires, illegal artisanal gold mining, and agribusiness and infrastructure development. As part of the *Águas do Tapajós* (Water of Tapajós) project, an integrated diagnosis of water quality and biodiversity, as well as of artisanal fishing and subsistence agriculture practiced in eight communities located along the Lower and Middle Tapajós was conducted. Results indicated a decrease in water transparency from 54 to 73% of the levels observed in a study from 1982, probably associated with the input of sediments from deforestation and mining. These activities were also potentially related to the high levels of mercury observed in the riverine population sampled, in average 4.6 times higher than the limit recommended by the World Health Organization (WHO). Furthermore, the presence of microcystin, a toxin that can cause liver failure and death, was detected in the vicinity of seven communities. In one of the sampled points, the level measured was higher than WHO's recommended limit for drinking water. Microplastic waste was recorded at all collection stations and sampling periods. High densities of fish eggs and larvae deriving provided evidence that the Tapajós River is an important reproduction site for ichthyofauna. This can

---

The original version of the chapter has been revised. A correction to this chapter can be found at https://doi.org/10.1007/978-3-031-81465-5_35

E. Garcia (✉)
The Nature Conservancy (TNC) Brasil, São Paulo, Brazil
e-mail: edeniserutzatzgarcia@gmail.com

A. L. C. Canto · C. H. Faria Junior · D. M. Talgatti · D. M. Zacardi · F. R. V. Ribeiro · H. N. M. Meneses · I. A. Santos · P. S. Miorando · T. M. P. Braga
Universidade Federal Do Oeste Do Pará (UFOPA), Santarém, Brazil

W. Leal Filho et al. (eds.), *Amazon 2030—Sustainability Issues in the World's Largest Rainforest Region*, World Sustainability Series,
https://doi.org/10.1007/978-3-031-81465-5_21

be reinforced by the large number (326) of species of fish reported, 190 of which with recognized potential for use in commercial and subsistence fishing. Capture of chelonians, although illegal, occurred in all the communities studied, either for consumption or selling, and a decline in the abondance of turtles was reported. As expected, fish was declared the main source of protein for the families interviewed, 86% of which, in addition to fishing, practice subsistence agriculture. However, there was a general perception of decline in the abondance of 31 species of fish. Overall, the study indicated a tendency towards deterioration of water quality conditions and aquatic biodiversity in the Tapajós River, particularly fishing resources. The worsening conditions can affect the health and water and food security of riverine communities. This scenario reinforces the importance of continued monitoring, as well as effective actions to contain the root causes of the observed degradation.

Keywords Tapajós river · Fishing · Aquatic biodiversity · Water quality

1 Introduction

The Tapajós River is a main tributary to the Amazon River, contributing with 6.5% of the Amazon water. The Tapajós Basin occupies an area of almost 500 thousand square kilometers (8% of the Amazon Basin) and is a confluence zone of infrastructure development, agribusiness expansion and artisanal mining, but still surrounded by large tracts of dense rainforest in indigenous lands and protected areas.

The Lower and Middle Tapajós region, the focus of this study, extends from the municipality of Itaituba, upstream, to the municipality of Santarém, where the crystal-clear waters of the Tapajós meet the muddy, sediment-laden waters of the Amazon. This features a diversity of ecosystems, such as alluvial forests and perennial and/or seasonal streams, or *igarapés*, sandy beaches, rocky river sections, rapids etc., which contributes to the existence of great biodiversity, with significant rates of endemism and highly related to the Tapajós hydrological cycles.

The Tapajós is also home to a large cultural diversity. Indigenous and other traditional and local communities live along the Tapajós and its tributaries and depend on the aquatic resources for their subsistence.

Despite its important sociobiodiversity, there is a lack of more advanced studies with an integrate perspective of the Tapajós complex socio-ecological systems. To fill part of this gap, the *Águas do Tapajós* (Water of Tapajós) project promoted an integrated diagnosis of river water quality; ichthyoplankton (eggs and larvae of fish), ichthyofauna (the fish—juvenile and adult—of the study region), chelonians (turtles) and microplastic occurrence and distribution during the four hydrological cycles in the Lower and Middle Tapajós. The study also surveyed fisheries activities and communities' perceptions about fish stocks trends, subsistence agriculture in riverine communities, and people's exposure to mercury and to microcystins, a toxin produced by blue-green algae. Here it is presented an overview of the project and a summary of the main results found, as well as their implications not only for the conservation of aquatic habitats and biodiversity, but also for maintenance of fish stocks and the health and livelihood of the riverine communities that intensively use the waters of the Tapajós, rely on fish as a main source of protein, and who are highly dependent on the environmental services the river provides.

2 Study Area

The study was conducted in the lower and middle reaches of the Tapajós River, state of Pará, Brazil (Fig. 1), along approximately 300 km, across sampling sites located nearby eight riverine communities: Suruacá, Parama, Cauaçu-Epá, Cametá, Apacê, Nazaré, Barreiras e São Luís do Tapajós. Samples were collected during the four regional hydrological cycles: flooding (June 2021), receding water (September 2021), drought (November 2021) and rising water (February 2022).

To collect fish at different stages (eggs, larvae, juveniles, and adults), microplastic waste and water, a river spatial gradient was considered, with the stations being divided into three sectors (Oliveira et al. 2023). Sector 1, the most upstream, had an average distance between its banks of 2.4 km, with turbulent waters due to the presence of rapids and under the influence of the discharge of muddy waters from tributaries that drain artisanal mining areas. Sector 2, surrounding the urban area of the city of Itaituba, had an average distance between its banks of 3.4 km, and was marked by the presence of sandbanks and alluvial islands along the main river channel. Sector 3 was the widest portion of the Tapajós River, with an average width of 14 km, without sandbanks or alluvial islands, and had a slower flow, being surrounded by environmental conservation areas on both banks, with numerous marginal lagoons connected to the main river and influenced by flood pulses.

3 Methodology

Only a brief description of the adopted methodology for the various components of the project is presented here.

Samples of surface and subsurface waters were collected for analysis of the following parameters: chlorophyll *a*, ammonia, nitrate, and total phosphorus. Temperature, pH, conductivity, and water transparency were measured in situ. A Secchi disk was used to measure water transparency. This is an 8-inch (20 cm) disc with alternating black and white quadrants. It is lowered into the water until it can no longer be seen by the observer. This vanishing depth indicates the transparency of the water (Preisendorfer 1986).

Water samples were also collected for qualitative and quantitative analyses of the phytoplankton community, and particularly to detect the presence of cyanobacteria, a blue-green algae that produces toxins called cyanotoxins, among which microcystins are the most common and highly toxic to the liver.

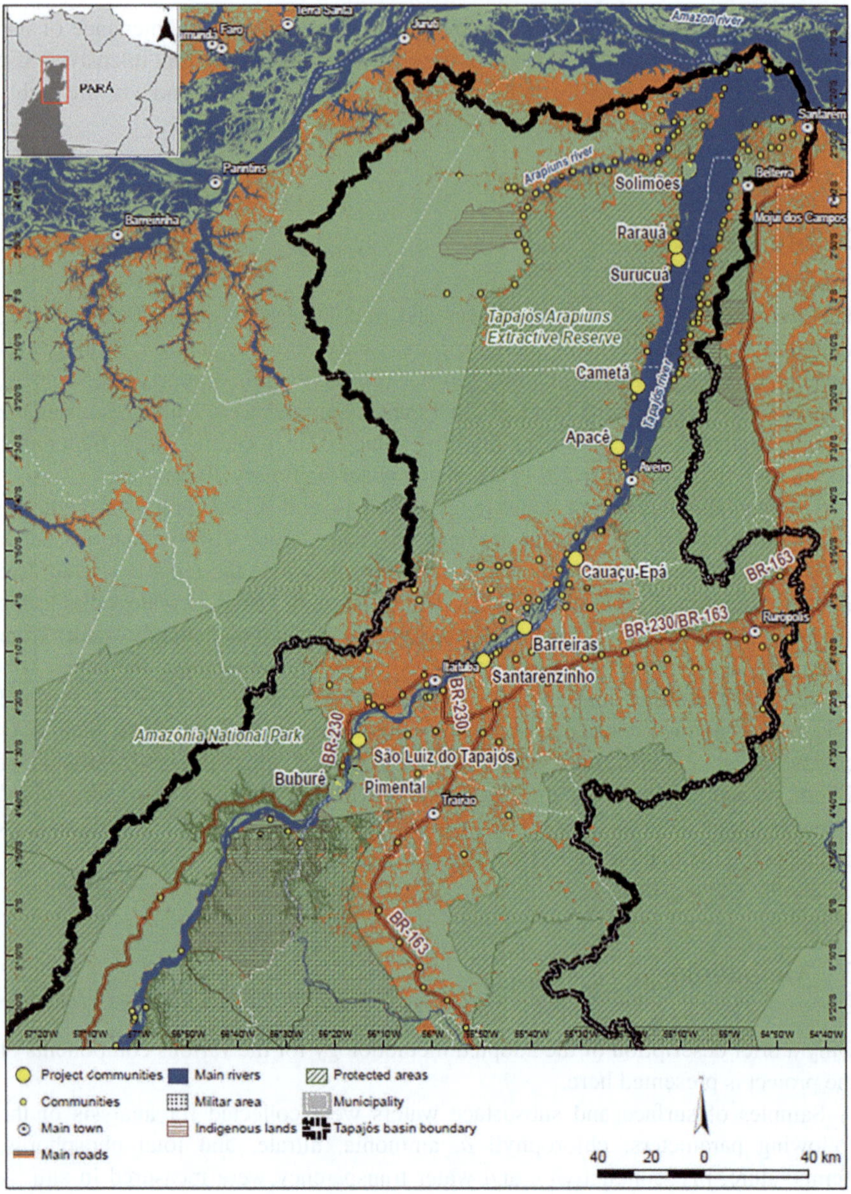

Fig. 1 Map of the study area in the Lower and Middle Tapajós River, showing the surveyed communities. Elaborated by the authors

Ichthyoplankton and microplastic waste were collected through horizontal trawls on the subsurface of the water column using a 300-μm plankton net. Details of ichthyoplankton collection, sample processing and data analysis are presented in (Oliveira et al. 2023).

Fish sampling was carried out using gill nets and human-powered trawl nets, with standardized effort for each type of device, to allow quantitative comparisons between sampling locations. Random collections were also carried out úsing human-powered trawl nets.

Questionnaires composed of objective and subjective questions about i. occurrence, population status and use of chelonian species living in the Tapajós River; ii. fishing activities; and iii. the structure of agricultural production, were applied to the eight riverine communities listed above, in Sect. 2.

To evaluate exposure to mercury in riverine communities, a cross-sectional, descriptive, and quantitative observational study was carried out, including collection of hair samples from 205 adult participants to measure mercury and application of a questionnaire covering age, sex, fish consumption, level of education and involvement in mining activities, in addition to socioeconomic issues and health conditions. More detailed information of the mercury component of the project are presented in a companion chapter (Rego et al. this publication).

4 Results

4.1 Water Quality

The results obtained demonstrate that the waters of the Tapajós River have less transparency, are slightly more basic and have higher electrical conductivity values compared to 1982 data collected in the same period of the year (Schmidt 1982). One of the most remarkable results is related to the transparency of the water, with the maximum transparency recorded being 172 cm, in the month of September, much lower than the that observed in the same month in 1982 (390 cm). Furthermore, in the present study, the average transparency observed between June and December was only 81 cm; for the same period the average in 1982 was 296 cm. The causes for the decrease in water transparency of the Tapajós River are related to land use on the banks of the river and its tributaries, especially deforestation and illegal artisanal gold mining (Lobo et al. 2016; Arias et al. 2018).

The decline in water transparency has important implications for the diversity and productivity of aquatic systems. As light availability decreases, primary production falls, affecting a large part of the aquatic food chain. Consequently, there may be changes in benthic and fish communities, which, in turn, in addition to affecting biodiversity itself, may have impacts on fishing activity and on communities that depend on fish for their subsistence. The increase in suspended sediment can also directly affect water use by the riverine populations (Kjelland et al. 2015; Mol and Ouboter 2004).

Microcystins LR were detected in 12 of the 38 surface water samples analyzed. In general, levels did not exceed the World Health Organization's provisional value for drinking water of 1.0 μg/L (WHO 2020), as observed in previous studies in the Tapajós River (Silva et al. 2019). However, in a study site close to the community of Santarenzinho the observed level (1,353 μg/mL) was higher than WHO's recommended limit. This is particularly concerning given that microcystins are hepatotoxic toxins that can cause liver failure and death (Kamogae and Hirooka 2008). Furthermore, contamination can occur not only by cyanotoxins released into the water, but also by toxins that are still present intracellularly in cyanobacteria (Tsuji et al. 2001), which were not accounted for in the present study, as measurements were carried out in filtered water. Therefore, exposure may be even greater, given that riverine dwellers in the Amazon region, including Tapajós, frequently ingest raw water, unlike residents of large cities, where treatment plants filter the water and retain cyanobacteria biomass in the filters. The high consumption of fish, also exposed to microcystins in the water, represents another source of contamination of riverine communities. Finally, the number of samples in this study containing cyanotoxins raises concern, as they were collected from close to seven of the communities studied.

These findings highlight the need for year-round monitoring and for greater investment in riverside communities' access to drinking water and sanitation.

4.2 *Ichthyoplankton*

Egg and larvae drift is a key mechanism for early fish stages to colonize nursery and growth areas and provides insights into ichthyofauna spawning times and spawning locations, a crucial information for understanding fish biology. This was a pioneer study in the Tapajós River that investigated the drift of ichthyoplankton along the hydrological phases as well as the occurrence and dispersion of microplastic waste associated with plankton, and is presented in detail in Oliveira et al. (2023).

Fish larvae belonging to 10 taxonomic orders and with an average size of 4.98 mm ± 3.14 mm were captured. Clupeiformes larvae occurred in all hydrological phases and reproductive peaks for other orders were recorded. Microplastic waste was classified into 20 types according to color and shape, showing average size of 1.55 mm ± 1.27 mm. Green fragments and blue filaments were the most abundant. It was found the highest density of fish eggs and larvae drifting during the dry season in stretches in Sector 2, whereas microplastics were more abundant during the receding water phase in more turbulent stretches that have impacts from illegal artisanal gold mining (Sector 1). No correlation was recorded between the density of ichthyoplankton with microplastics, however, drifting of these two was observed in all hydrological phases and sampled sectors. This is worrying because, although we did not analyze the ingestion of microplastics by fish larvae, microplastic waste can enter the nursery areas of the Tapajós River and reduce the ecological function of these areas for fish.

The high densities of fish eggs and larvae deriving provides evidence that the Tapajós River is an important reproduction site for ichthyofauna, and that the presence of different landscape configurations is a key factor in the distribution, retention, and development of ichthyoplankton, and microplastic waste dispersal (Oliveira et al. 2023). From the results obtained and considering the river flow, it was possible to identify sections of the river where the largest spawning of the species occurred. This highlights the importance of ichthyoplankton in supporting the mapping of critical areas for the reproduction of sexually mature fish and, therefore, for actions aimed at maintaining the integrity of the river, the ichthyofauna and artisanal fishing.

4.3 Ichthyofauna

The collected samples were used to characterize the fish assemblage of the Tapajós River in terms of composition, richness, diversity, and abundance; as well as for the identification of fish species of greatest economic interest, endemic, threatened with extinction, indicators of environmental quality, as well as those that are scientifically unknown.

The results indicated that the Tapajós River is home to a significant part of the richness and abundance of the Amazonian fish fauna in its middle and lower reaches. It was recorded 326 species of fish inhabiting the main bed alone. Many of these recorded species are important sources of income and animal protein for local populations.

Around 190 species captured had recognized potential for use in commercial and subsistence fishing, of which we highlight those from the families Serrasalmidae (represented by pacus and piranhas), Pimelodidae (catfish), Cichlidae (tucunaré, carás, acaratinga and chaperema), Curimatidae (whitefish), Anostomidae (aracus), Osteoglossidae (aruanã) and Sciaenidae (whitefish). Analyzes of the ornamental fish fauna, even at a preliminary stage, allowed us to recognize the significant contribution of this fauna to the fish assemblage of the middle and lower Tapajós River. Species of the genera *Ancistrus*, *Baryancistrus* and *Peckoltia*, belonging to the Loricariidae family, contributed significantly to the qualitative and quantitative composition of this fauna and their managed use can enhance the local economy.

The similarity analysis demonstrated a significant dissimilarity in the qualitative composition of fish species between the lower and middle reaches of the Tapajós River ($p = 0.0001$; $r = 0.28$) and between the studied periods ($p = 0.0001$; $r = 0.37$). The comparison between the rainy and dry periods allowed us to recognize that the richness of fish species in the middle and lower Tapajós River has been greater during the low water period, while the greatest abundance was recorded during the high water period. This dynamic reinforced the current consensus in river ecology that emphasizes that the hydrological regime is the main driver of ecological dynamics in river-floodplain systems, affecting the taxonomic and functional structure of fish assemblages (Humphries et al. 2014). In the Amazon basin, seasonal movements of fish fauna are directly associated with the annual flood pulse, resulting in significant changes in the density and composition of fish assemblages throughout the year.

The distribution of species in the longitudinal gradient was ordered based on the distance of the collection points in relation to the mouth of the Tapajós River. Through this ordering it was possible to observe the existence of longitudinal gradients in the composition of fish species. The ordering did not reveal a pattern of fauna distribution, but it allowed recognizing homogeneous distributions and fragmentation in the distribution of populations of certain species. Some species seemed to be well adapted to environmental variations in the longitudinal gradient and were widely distributed along the stretches, such as: *Moenkhausia celibela, Moenkhausia gracilima, Triportheus albus, Geophagus altifrons, Brycon pesu* and *Hemiodus unimaculatus.*

Part of the recorded fish fauna required high taxonomic efforts, with recognition of potential new species from different groups of fish to be confirmed. Some recorded species are receiving provisional identification, but still need more refined analyses. Recent taxonomic studies of the Tapajós ichytyofauna, especially small species, have allowed the recognition of new species of fish in the Tapajós (Ribeiro 2020; Birindelli al. 2008; Canto et al. 2022; Ribeiro et al. 2022; Silva-Oliveira et al. 2023). These species belong to the orders Characiformes and Siluriformes, which are the most biodiverse orders in the Amazon basin.

4.4 Chelonians

Amazonian chelonians are icons of Amazonian aquatic biodiversity, not only for the ecological functions they perform, but also for their cultural and social relevance. Deforestation, navigation of large vessels, mining, hydroelectric construction and hunting today threaten several parts of the Amazon basin (Fagundes et al. 2021), including the Tapajós River and its biodiversity (Alarcon et al. 2016).

In this study, the ethnoknowledge approach was used, through 216 interviews, as a more accessible way of obtaining diverse information about the functioning of ecosystems and the ecology of the chelonian species of interest, in addition to being a means that records and values the knowledge acquired by local populations, and which is passed down through generations (Huntington 2000). Priority was given to information on capture, consumption, trade, and population trends of the main species of chelonians used by riverside dwellers—those of the genus *Podocnemis* (*P. expansa*, *P. unifilis* and *P. sextuberculata*), in addition to other species that were cited as important for these themes.

The results of this research indicated that the capture of chelonians, although illegal, occurred in all the communities studied, being an important activity in the relationship between riverine dwellers and the aquatic fauna of the Tapajós River. The study showed that chelonians are part of riverine productive activities: in addition to consumption, the captures were also intended for trade, showing that there is an economy, albeit local, linked to the exploitation of turtles. The highest capture yield was observed in the community of Barreiras and is very possibly related to its proximity (about 6 km) to Tabuleiro de Monte Cristo, a spawning area protected by IBAMA—Brazilian Institute of the Environment and Renewable Natural Resources, where around 10 thousand females spawn annually (Fonseca 2022). The market of turtles reported followed a pattern already known for the Amazon, focusing on larger animals, especially females with eggs, which have higher commercial values. Thus, for each species, females are worth more than males; while among species the Amazon turtle, *Podocnemis expansa*, reaches higher values due to its larger size.

Perceptions of population trends indicated that despite the greater capture of *P. expansa*, this species is considered to have an increasing population; while the situation of greatest concern in terms of viability was indicated for *P. sextuberculata*, considered to be decreasing or disappearing in all communities, except Surucuá. Similarly, *P. unifilis* was also considered to be decreasing in all except Surucuá. The Surucuá community is located within a Sustainable Use Conservation Unit, the Tapajós-Arapiuns Extractive Reserve, that was reported as having a conservation initiative by a large number of interviewees. Finally, it is worth highlighting that the capture of chelonians is a subject considered taboo, as it is summarily prohibited by law, and therefore the numbers gathered in this study can be considered underestimated.

4.5 Fishing Activities, Importance and Perception of Fish Stocks Trends

Fishing resources are used in the Tapajós region for both sale and subsistence not only by fishermen, but by other residents who have other occupations as well. The estimated average daily fish consumption in the 8 riverine communities interviewed was 343 g.person^{-1}.day^{-1} (SD $\pm$ 180). This value is close to that found in other regions of the Amazon (Cerdeira et al. 1997; Batista et al. 1998; Braga et al. 2016; Coelho et al. 2017), and is around seven times greater than the global average consumption, reflecting the strong relationship between residents of the banks of rivers in the region with this resource (Braga et al. 2016; Corrêa et al. 2019). The results demonstrate the importance of fish as one of the main proteins of animal origin consumed by fishermen who complement their meals mainly with rice and cassava flour (37.8%), only with flour (28.9%) or with rice, flour, and beans (20.7%).

Currently, fishing in the surveyed communities is carried out mainly by adult men in partnership with a relative, where the son stands out, besides brother, great father, and brother-in-law. Although the wife also appears as a partner in the fishing activity, it is evident that the male presence predominates in this partnership. The prominence of the male presence as a partner in fishing, also described in other works (Batista et al. 1998; Alcântara et al. 2015; Silva and Braga 2016), indicates that the social actors responsible for teaching fishing activities are mainly parents or a male relative, allowing fishermen to accumulate and pass on detailed traditional knowledge about the environment and exploited resources over generations (Braga and Rebêlo 2014; Silva et al. 2020).

They travel to the fishing spot in wooden canoes, with stern-drive motor propulsion, and carry out fishing using various equipment, especially mesh nets. During fisheries they capture in average 47 ethnospecies of fish, which represent 35 species or groups of fish species described as being the most caught at different times of the year, in addition to alligators and chelonians. Considering that some of these ethnospecies represent more than one species, such as aracus, yams and branquinhas, the number of scientific species is much greater.

Regarding the abundance of resources, 36 types of animals were mentioned that would have decreased in quantity in recent years. Of these, only five (5) were not fish (manatee, chelonians in general, alligator, capybara and deer). Among fish, those that were most cited as decreasing in quantity in fishing environments were maparás (17.6%), jaraquis (15.4%), tambaqui (10.3%) and pirarucu (7.4%). Regarding the reasons that led to this decrease in abundance, different opinions were provided. However, four pieces of information deserve to be highlighted because they were the most declared and because they were related to each other: presence of large boats with glaciers and trawlers (27.7%), large number of people fishing (19.1%), excessive fishing (10, 6%) and the action of certain types of gear used (6.4%) that catch large quantities of fish.

Twenty-five animals were also mentioned that would have increased in abundance in the Tapajós region. Of these, five were not fish (dolphins, capybara, agouti, chickens, and chelonians). Among the most remembered and mentioned animals, three were fish, namely the caratinga (12.9%), tucunaré (11.9%) and the whitefish (8.9%); and one was a mammal, that is, porpoises (10.9%). Regarding the reasons that could be influencing this increase in abundance, the greatest congruences were for the dolphins, which were increasing due to "the large supply of fish/food for them in the region" (32.3%). For the caratinga and tucunaré, the reasons listed would be mainly due to their type of reproduction "with a lot of offspring and quick reproduction" (19.4%) and the whitefish due to being "little captured and consumed" (6.5%).

Finally, the results showed the importance of fishing in the study region, in addition to agriculture, both in the economy and for the subsistence of a large portion of the population, given it is carried out by several residents and not just those who declared themselves fishermen. This information demonstrates the need to monitor fish landings, both in communities and at the nearby urban areas, to complement the declared data and generate information that can be used in fishing planning in the Tapajós.

4.6 Subsistence Agriculture

In addition to fishing, subsistence agriculture is an extremely important activity for the riverine communities of the Lower and Middle Tapajós and is responsible for ensuring food security for these people. It is based on agriculture that families can obtain quality, safe and diverse food to make up their nutrient-rich diets. In addition to making up the food base for families, agriculture provides part of the income and work for family members. Besides food, income and work, family agriculture generates soil coverage with a diversity of plant species, avoiding erosion of riparian areas and becoming an element of forest conservation and reduction of impacts on biodiversity of the Tapajós region.

Approximately 85% of families in the studied communities carry out agricultural activities. The production system most used in family farming are gardens with a variety of cultivated species. In general, beans, corn, potatoes, cassava, and other species are commonly produced. Fruit species as banana, orange trees, lemon trees and *cupuaçu* (*Theobroma grandiflorum*) are also frequent. Large trees are highly valued by the communities, as several people stated that the objective of planting large trees would be "reforestation", "beautifying the land" and "having shade". However, they also plant trees for commercial purposes, such as andiroba (*Carapa guianensis*) and copaiba (*Copaifera langsdorffii*), as the oils extracted from these two species are widely used by locals and by people from nearby cities.

Overall, there is a need for technical assistance to the agriculture practiced by riverine communities in the Lower and Middle Tapajós so that the communities can maintain their diversity of production and improve the quality of cultivated products through the adoption of practices that facilitate their work while producing with low environmental impact.

4.7 Mercury Exposure

Mercury (Hg) is considered one of the most dangerous toxic metals for human and other animals. High levels of Hg can induce different types of toxic effects on human health and the main target is the central nervous system. The metal is released into the environment through natural sources and/or anthropogenic actions, undergoes complex transformations and circulates between the atmosphere, waters, and land (Wright and Welbourn 2002).

In the Tapajós, mercury is widely used in artisanal and illegal mining activities to amalgamate gold. In addition to mining, activities related to land use change, such as deforestation and fires, and the construction of hydroelectric dams contribute to the contamination of aquatic ecosystems in the region (Fearnside 2003), as mercury present in the soils is mobilized and enters surrounding waters. Mercury can be transformed into methylmercury (MeHg), an organic form that has the ability to bioaccumulate in organisms and to biomagnify throughout the trophic chain, thus contaminating aquatic biota and humans.

Exposure to Hg in the Tapajós region is mainly due to the ingestion of fish contaminated with methylmercury and affects indigenous and non-indigenous communities and residents from local urban areas (Costa Júnior et al. 2015; Achatz et al. 2021; Meneses et al. 2022).

As part of this study, mercurial exposure was assessed in 205 adult participants from eight riverine communities in the Lower and Middle Tapajós. Total Hg levels were measured in a hair sample and individuals with levels above 2 μg/g of total Hg were considered exposed, as indicated by the World Health Organization. A questionnaire was administered to describe the health status of research participants, fish consumption, level of education and involvement in mining activities, besides general demographic characterization (age and sex).

Of the participants with total Hg results, 43.6% were above the WHO recommended limit, with an average level of 9.20 μg/g mercury in hair. Among the exposed people, 64% were male; just 10% had worked in gold mining; 62% used to eat fish daily, 30% frequently and 2% not frequently. The highest level of mercury was observed in a male individual, between 21- and 40-year-old, who used to eat fish frequently, but not daily, and who had not worked with gold mining.

Among symptoms of mercurial exposure, "muscle and joint pain" was reported by participants from all communities and "decreased motor coordination", by participants from half of the communities visited. Although these are not exclusive symptoms of mercurial exposure, and can be influenced by other factors, the occurrence of these symptoms draws attention, as they are very characteristic of exposure and reinforces that to understand the real effects of exposure, healthy professionals working with riverine communities need to have a deeper understanding of potential problems provoked by mercury.

5 Conclusion

Although the Amazon basin is recognized as having the largest and most diverse freshwater ichthyofauna in the world, knowledge of its biodiversity has slowly advanced in many of its main tributaries. Due to the basin dimension, studies in many rivers have been neglected. For example, the Tapajós River is one of the largest clear water tributaries on the right bank of the Amazon River and has significant biodiversity and regional socioeconomic importance, but it is still one of the least studied drainages in the basin.

Meanwhile, the Tapajós River has suffered several impacts along its entire length, such as illegal mining, large scale deforestation and removal of riparian vegetation, and risks from hydroelectric construction. As presented in this study, changes in water transparency, presence of cyanotoxins and high exposure of communities to mercury are already threatening the diversity of biological communities and the livelihood of hundreds of riverine communities.

Additionally, no study investigating the spatio-temporal dynamics of ichthyoplankton and microplastic waste was carried out in the Tapajós River basin before the present study. Ichthyoplankton and microplastic waste distribution processes presented here indicated that the Tapajós River is an important breeding ground for fish species, with multiple spawning sites and nursery and growth areas. However, it is becoming a plastic dump, as evidenced by the presence of microplastic waste recorded at all sampling stations and periods. What is more worrisome is that during the receding water phase there is a greater drift of microplastic waste than fish eggs and larvae. This suggests that microplastic waste can reach nursery and ichthyofauna growth areas and be incidentally ingested by fish larvae. This can generate negative impacts on the ichthyoplankton community, and can compromise the recruitment of fish populations.

On a positive note, the richness of species and the new fish species identified recently in this study and others (Ribeiro et al. 2022; Ribeiro 2020; Calegari et al. 2014; Canto et al. 2022) highlight the lower and middle stretch of the Tapajós River as areas of high endemism and reinforce the perception that the Tapajós ichthyofauna is far from being completely known, which increases the need for more scientific investigations in the basin. On the other hand, monitoring the ichthyofauna of the Tapajós River is extremely necessary to evaluate the continuous decline in the fish population perceived by riverine communities, fluctuations in the species' occurrence, distribution areas, habitat quality and loss and presence of subpopulations. Once a monitoring approach is put in place, it will be possible to determine the conservation status of fish fauna and establish effective approaches to protect them and their habitat.

Likewise, despite its socioeconomic importance for the region, fishing is an activity that is still poorly documented, with fragmented information and without a long series of data collection, which leads to imprecise estimates of the total catch and gaps in the sustainability of the sector. Understanding the dynamics of fishing in a region like Tapajós represents an enormous challenge. The great diversity of fish, combined with social and cultural fishers' practices diversity, in a scenario of conflicts between artisanal and commercial fishers, make fishing monitoring a very complex activity. For this reason, monitoring need to go beyond the academic sector and directly involve fishing communities, bringing together scientific and traditional knowledge. This is particularly important to evaluate the effectivity of new Fishing Agreements that are being implemented in the region. These are formal shared management mechanisms agreed by communities in a given region, aiming to establish fishing rules that ensure the sustainability of fishing stocks. The Fishing Agreements can become valuable instruments for co-protection and conservation of aquatic biodiversity, including chelonians and other non-fish species, and essential habitats.

In this context, integrated studies like the ones presented here can fill knowledge gaps and provide subsidies for decision makers and for the definition of public policies. However, special attention is needed so that these results also reach local populations and are complemented by their traditional knowledge, generating effective actions for the conservation of the health of the aquatic biodiversity and the well-being of communities that depend upon the Tapajos River.

References

Achatz RW, de Vasconcellos ACS, Pereira L (2021) Viana PVdS, Basta PC (2021) Impacts of the goldmining and chronic methylmercury exposure on the good-living and mental health of munduruku native communities in the Amazon basin. Int J Environ Res Public Health 18(17):8994

Alarcon DT, Millikan B, Torres M (org) (2016) Ocekadi: hidrelétricas, conflitos socioambientais e resistência na Bacia do Tapajós. International Rivers Brasil, Brasília, DF. Programa de Antropologia e Arqueologia da Universidade Federal do Oeste do Pará, Santarém, PA

Alcantara NC, Gonçalves GS, Braga TMP, Santos SM, Arau Jo RL, Lima JP, Aride PHR, Oliveira AT (2015) Avaliação do desembarque pesqueiro (2009–2010) no município de Juruá, Amazonas, Brasil. Biota Amazônia 5(1):37–42

Arias ME, Lee E, Farinosi F, Pereira FF, Moorcroft PR (2018) Decoupling the effects of deforestation and climate variability in the Tapajos river basin in the Brazilian Amazonia. Hydrol Process 32:11648–11663

Batista VS, Inhamuns AJ, Freitas CEC, Freire-Brasil D (1998) Characterization of the fishery in river communities in the low-Solimões/high-Amazon region. Fish Manage Ecol 5:419–435

Birindelli JL, Sousa LM, Pérez MHS (2008) New species of thorny catfish, genus *Leptodoras Boulenger* (Siluriformes: Doradidae), from Tapajós and Xingu basins, Brazil. Neotrop Ichthyol 6:465–480

Braga TMP, Rebêlo GH (2014) Conhecimento tradicional dos pescadores do baixo rio Juruá: aspectos relacionados aos hábitos alimentares dos peixes da região. Interciência 39(9):1–7

Braga TMP, Silva AA, Rebêlo GH (2016) Preferências e tabus alimentares no consumo de pescado em Santarém, Brazil. Novos Cadernos NAEA 19(3):189–204

Calegari BB, Reis RE, Vari RP (2014) Miniature catfishes of the genus *Gelanoglanis* (Siluriformes: Auchenipteridae): monophyly and the description of a new species from the upper rio Tapajós basin, Brazil. Neotrop Ichthyol 12(4):699–706. https://doi.org/10.1590/1982-0224-20130233

Canto ALC, Hercos AP, Ribeiro FRV (2022) Description of a new species of miniature catfish of the genus Ammoglanis (Siluriformes: Trichomycteridae) from rio Tapajós basin, Brazil. Neotrop Ichthyol, 20(2). https://doi.org/10.1590/1982-0224-2021-0122

Cerdeira RGP, Ruffino ML, Isaac VJ (1997) Consumo de pescado e outros alimentos pela população ribeirinha do Lago Grande de Monte alegre, PA—Brasil. Acta Amazonica 27 (3):213–228. Manaus-AM

Coelho ACS, Faria-Junior CH, Sousa KNS (2017) Fatores que influenciam na compra de peixes por classe social no município de Santarém-PA. Agroecossistemas 9(1):62–83.

Corrêa JCSL, Braga TMP, Laurido SF (2019) Usos de recursos faunísticos pelos moradores da comunidade Boca do Arapiri, Assentamento Agroextrativista Atumã, em Alenquer, Pará, Brasil. Amazônica—Revista Antropol 11(2):741–769

Costa Junior JMF, Lima AAS, Rodrigues Junior D, Silva MCF, Souza GS, Silveira LCL, Pinheiro MCN (2015) Evaluation of mercury exposure levels in riverside communities in the Tapajós region from 2012 to 2014. Rev Pesq em Saúde 14:29–34

Fagundes CK, Fath F, Côtes LG, Uhlig V, Andrade PCM, Vogt RC, Pezzuti JCB, Marcos Júnior PA (2021) Large scale analysis of threats to the nesting sites of *Podocnemis* Species and the effectiveness of the coverage of these areas by the Brazilian action plan for amazon turtle conservation. J Nat Conserv 61:125997. https://doi.org/10.1016/j.jnc.2021.125997

Fearnside PM (2003) A floresta amazônica nas mudanças globais. INPA, p 134

Fonseca RA (2022) Ecologia e Conservação de Podocnemis expansa (Testudines, Podocnemididae) no médio Tapajós, Brasil. Ph.D. Thesis, Universidade Federal do Oeste do Pará, Santarém, Pará

Humphries P, Keckeis H, Finlayson B (2014) The river wave concept: Integrating river ecosystem models. Bioscience 64(10):870–882. https://doi.org/10.1093/biosci/biu130

Huntington HP (2000) Using traditional ecological knowledge in science: methods and applications. Ecol Appl 10:1270–1274

Kamogae M, Hirooka EY (2008) Microcystins: contamination risks in eutrophic waters. Acta Sci Technol 22:1189–1200. https://doi.org/10.4025/actascitechnol.v22i0.3010

Kjelland ME, Woodley CM, Swannack TM, Smith DL (2015) A review of the potential effects of suspended sediment on fishes: potential dredging-related physiological, behavioral, and transgenerational implications. Environ Syst Decis 35:334–350

Rego et al. This book—TBD

Lobo F, Costa M, Novo E, Telmer K (2016) Distribution of artisanal and small-scale gold mining in the Tapajós River Basin (Brazilian Amazon) over the past 40 years and relationship with water siltation. Remote Sens 8(7):22. https://doi.org/10.3390/rs8070579

Meneses HNM, Costa MO, Basta PC, Morais CG, Pereira RJB, Souza SMS, Hacon SS (2022) Mercury contamination: A growing threat to riverine and urban communities in the Brazilian Amazon. Int J Environ Res Public Health 19:2816

Mol JH, Ouboter PE (2004) Downstream effects of erosion from small-scale gold mining on the instream habitat and fish community of a small neotropical rainforest stream. Conserv Biol 18:201–214. ISSN: 0888–8892

Oliveira LS, Oliveira-Junior J, Cajado RA, Silva FK, Zacardi DM (2023) Ichthyoplankton and plastic waste drift in a river in the Amazon Basin, Brazil. Front Environ Sci 11:1068550. https://doi.org/10.3389/fenvs.2023.1068550

Preisendorfer RW (1986) Secchi disk science: Visual optics of natural waters1. Limnol Oceanogr 31:909–926. https://doi.org/10.4319/lo.1986.31.5.0909

Ribeiro RV (2020) A new species of Knodus (Characiformes: Characidae) from the rio Cupari drainage, lower rio Tapajós basin, Brazil. Zootaxa 4747(3):575–584

Ribeiro FRV, Silva-Oliveira SACF, Canto ALC (2022) New species of driftwood catfish of Tatia (Siluriformes: Auchenipteridae) from rio Tapajós, Brazil. Neotropical Ichthyology 20:e210164. https://doi.org/10.1590/1982-0224-2021-0164

Schmidt GW (1982) Primary production of phytoplankton in the three types of amazonian water. Amazoniana 7(3):335–348

Silva JT, Braga TMP (2016) Caracterização da pesca na comunidade de Surucuá (Resex Tapajós Arapiuns). Biota Amazônia 6(3):55–62

Silva IF, Braga TMP, Sousa KNS (2020) Caracterização da atividade pesqueira das comunidades do entorno do Parque Estadual Monte Alegre, Pará, Brasil. Biota Amazônia Macapá 10(2):50–59

Silva SCF, Peleja JRP, Melo S (2019) Flutuação temporal de cianotoxinas (Microcistina) no rio Tapajós (Santarém, Amazônia-Brasil). Scientia Plena 15. https://doi.org/10.14808/sci.plena.2019.082402

Silva-Oliveira C, Canto ALC, Ribeiro FRV (2015) *Bryconops mundur*uku (Characiformes: Characidae), a new species of fish from the lower Tapajós River basin, Brazil. Zootaxa 3994(1):133–141

Tsuji K, Masui H, Uemura H, Mori Y, Harada KI (2001) Analysis of microcystins in sediments using MMPB method. Toxicon 39:687–692. https://doi.org/10.1016/S0041-0101(00)00196-3

World Health Organization—WHO (2020) Cyanobacterial toxins: microcystins. Background document for development of WHO Guidelines for drinking-water quality and Guidelines for safe recreational water environments. Geneva. (WHO/HEP/ECH/WSH/2020.6)

Wright DA, Welbourn P (2002) Environmental toxicology. Cambridge University Press, p 630

Analysis of Fire Hotspot Numbers, Fire Radiative Power and Air Pollutant Concentrations in the Legal Amazonia Between 2016 and 2021

Débora Souza Alvim, Bin Chen, Dirceu Luis Herdies, Gabriela de Alencar Arraes, Luana Santamaria Basso, Bushra Khalid, Gabriel. O. Oyerinde, Simone Marilene Sievert da Costa, Gabriel Brito Costa, Hannah A. Affum, Sergio Machado Corrêa, and Silvio Nilo Figueroa

Abstract Over the past few years, the Amazonia region has experienced a notable increase in fire occurrences, primarily attributed to agricultural management and territorial possession. This increase has, in turn, led to as rise in atmospheric concentrations of pollutants. This chapter presents an observational analysis of the number of

---

D. S. Alvim (✉) · G. de Alencar Arraes
Lorena School of Engineering (EEL), University of Sao Paulo (USP), Lorena, SP 05508-050, Brazil
e-mail: debora.alvim@eel.usp.br

D. S. Alvim · D. L. Herdies · S. M. S. da Costa · S. N. Figueroa
Center for Weather Forecasting and Climate Studies (CPTEC), National Institute for Space Research (INPE), Cachoeira Paulista, SP 12630-000, Brazil
e-mail: dirceu.herdies@inpe.br

S. M. S. da Costa
e-mail: simone.sievert@inpe.br

S. N. Figueroa
e-mail: nilo.figueroa@inpe.br

B. Chen
Institute of Atmospheric Physics, Chinese Academy of Sciences, Beijing 100029, China

Collaborative Innovation Center on Forecast and Evaluation of Meteorological Disasters, Nanjing University of Information Science and Technology, Nanjing 210044, China

B. Chen
e-mail: chen_bin@mail.iap.ac.cn

L. S. Basso
Max Planck Institute for Biogeochemistry, Jena, Germany

B. Khalid
Institute of Geographic Sciences and Natural Resources Research, Chinese Academy of Sciences, 11A Datun Road, Chaoyang District, Beijing 100101, China

W. Leal Filho et al. (eds.), *Amazon 2030—Sustainability Issues in the World's Largest Rainforest Region*, World Sustainability Series,
https://doi.org/10.1007/978-3-031-81465-5_22

fire hotspots and fire radiative power (FRP) across the Legal Amazonia region, alongside the concentration of pollutants in the atmosphere and near the surface, including nitrogen dioxide (NO_2) from OMI sensor on board the AURA satellite; and carbon monoxide (CO), black carbon (BC), organic carbon (OC), and sulfur dioxide (SO_2) from MERRA-2 reanalysis. The primary objective is to assess the changes in spatial and temporal patterns of these air pollutants from 2016 to 2021. Throughout the study period in the Legal Amazonian region, Para State recorded the highest number of fires (193,218), followed by Mato Grosso (177,641), Maranhao (84,760), Amazonas (78,557), Tocantins (73,925), Rondonia (65,447), Acre (45,428), Roraima (15,521), and Amapa (8,450) States. The highest numbers of fire hotspots, FRP and concentrations of NO_2, SO_2, OC, BC, and CO are observed from July through October, aligning with the months of lowest precipitation and relative humidity. The results also reveal that burning biomass raises gas concentration across the region, particularly in the southern part of Mato Grosso near the city of Cuiabá. Gas concentrations increase threefold throughout the entire studied region from March, April, and May to August, September, and October for NO_2 (2.5 to 7.5 10–14 mol cm^{-2}), BC (0.13 to 0.45 micro g m^{-3}), and OC (3 to 9 micro g m^{-3}), indicating the deterioration of the air quality during the fire season.

Keywords Atmospheric pollution · Biomass burning · Legal Amazonia

1 Introduction

The legal Amazonia has been the scene of an intense process of human occupation, resulting in the intensification of economic activities and, consequently, an increase in the exploration of natural resources in the region (Brebbia and Martin-Duque 2002).

Department of Environmental Sciences, International Islamic University, Islamabad, Pakistan

The Abdus Salam International Centre for Theoretical Physics, Trieste, Italy

Gabriel. O. Oyerinde
Physics Advanced Research Centre, Sheda Science and Technology Complex, Federal Capital Territory, Nigeria
e-mail: go.oyerinde@shestco.gov.ng

G. B. Costa
Biodiversity and Forests Institute, Federal University of Western Para, Santarém, PA, Brazil
e-mail: gabriel.costa@ufopa.edu.br

H. A. Affum
National Nuclear Research Institute, Ghana Atomic Energy Commission, Accra, Ghana

Department of Nuclear Science and Applications, University of Ghana, Accra, Ghana

S. M. Corrêa
Rio de Janeiro State University, Resende, RJ, Brazil
e-mail: sergiomc@uerj.br

The rate at which these processes unfold has implications for both human health and the environment. According to Brebbia and Martin-Duque (2002), the primary environmental impacts include the loss of biodiversity, deforestation, fires, erosion, among others. showed that atmospheric pollution negatively affects the quality of life, leading to an increase in hospital admissions and mortality rates. Ignotti et al. (2007), based on epidemiological studies, reported a notable rise in respiratory and cardiovascular diseases and mortality due to atmospheric pollution.

The process of burning vegetation is the most cost-effective and rapid method of soil clearing. This slash-and-burn practice is mainly used for land clearance after deforestation and for agricultural management, serving as a primary driver of fires in the Legal Amazonia. Despite a dry season between July and October in the Amazonia biome, creating conditions conducive to fires with abundant dry vegetation as fuel, the region's humidity prevents combustion from occurring without a human agent to initiate it. This contrasts with the Amazon biome Cerrado, also part of the Legal Amazonia, where fires can occur without necessarily being caused by humans. Therefore, there is a direct relationship between the increase in deforestation, facilitated by the vast expansion of territorial use, and the increase in accidental or intentional anthropogenic fires (IPAM 2020).

According to reports released by the Amazonia Environmental Research Institute (IPAM), the year 2019 had a dry period within the average recorded in recent years. However, 81% more fires were recorded than the average in 2018, explaining the increase in deforestation in the region. According to data from the National Institute for Space Research (INPE), 2019 had 89,000 fire outbreaks and in 2020 there was a significant increase, with 103,000 outbreaks, the highest number since 2017 as highlighted in the Fire Estimation System report. According to Greenhouse Gas Emissions and Removals Estimation System (SEEG), in 2019 the amount of greenhouse gases released into the atmosphere increased by 23%, totaling 968 million tons of carbon dioxide (CO_2) emitted, mostly from deforestation, which is responsible for 44% of emissions in Brazilian territory (IPAM 2020).

The Legal Amazonia lost 10,476 km^2 of forest between August 2020 and July 2021, during which the deforestation season is measured. This rate is 57% higher than the previous season, marking it as the worst in the last ten years, as reported by the Amazonia Institute of People and the Environment (IMAZON). According to IMAZON monitoring, the Amazonia has witnessed its most devastating deforestation rate in the past decade. Notably, within the year 2019–2020, the forest lost an area equivalent to nine times the size of the city of Rio de Janeiro, covering 1200 km^2. This accumulated value is 57% higher than the deforestation recorded in the previous calendar year, from August 2019 to July 2020, which amounted to 6,688 km^2. Para State faced the highest deforestation state in the last 12 months, with an area totaling 4,147 km^2, a 43% increase compared to the previous calendar. Additionally, in July 2021, the registered deforestation was 80% higher than observed in July 2020.

In addition to AMAZON's monitoring system i.e, the Deforestation Alert System (SAD), which uses satellite and radar images, the biome is also monitored by INPE, by the Brazilian federal government. INPE employs two monitoring systems: Deter, which provides real-time monitoring of deforestation and degradation in the

Amazonia for issuing daily alerts; and PRODES, which conducts an annual inventory of forest loss.

With the significant increase in fires, the population is consequently exposed to the polluting gases, mainly through inhalation and ingestion, leading the body to absorb toxic substances (Kampa and Castanas 2008).

Environmental effects are observed soon after, as the chemical composition of the atmosphere is modified by polluting gases. During the burning period, air quality reaches concentrations of up to 500 $\mu g\ m^{-3}$ of particulate matter, approximately three times the concentrations observed in the city of Sao Paulo. In the Amazonia atmosphere, a high concentration of heavy metal particles such as CO and ozone (O_3) can be found, all of which are toxic to human health (Oliveira et al. 2015). The effects of these emissions cause impacts on local, regional and global scale, as they alter the chemical composition and physical properties of the atmosphere, generating a radiative imbalance, absorbing terrestrial radiation and warming the atmosphere (Freitas et al. 2005).

The main goal of this study is to analyze the fire outbreaks, radiative power (FRP) from the MODIS sensor, and the concentration of atmospheric pollutants, including Nitrogen Dioxide (NO_2) from the Ozone Monitoring sensor Instrument (OMI) on board the AURA satellite, Carbon Monoxide (CO), Sulfur Dioxide (SO_2), Black Carbon (BC) and Organic Carbon (OC) from MERRA-2 reanalysis data, near the surface. These pollutants are emitted from the biomass burning in the Brazilian Legal Amazonia in the years 2016 to 2021.

2 Methodology

2.1 Data

A survey of fire radiative power (FRP) data and an analysis of the pollutant gases concentrations were carried out for the legal Amazonia region from 2016 to 2021. The FRP data were obtained through the MODIS sensor on the AQUA satellite, available on the Data Bank platform, and fire data from the INPE website. For NO_2, data were obtained from the OMI sensor on the AURA satellite, while CO, SO_2, BC and OC data were obtained from MERRA-2 reanalysis and downloaded from NASA's Giovanni website. Atmospheric reanalysis produces records of high global spatial and temporal resolution of meteorological fields and the chemical composition of the Earth's atmosphere, which is essential for the analysis in this study (Jesus et al. 2021).

Maps of monthly mean concentrations of NO_2, CO, SO_2, BC and OC in the atmosphere near to the surface of the Legal Amazonia region were generated using NCAR's NCL software. Fire spots and FRP, downloaded from the NCAR website, were plotted using INPE Excel and Power BI software.

2.2 Study Area

The study area is the Legal Amazonia, which represents around 59% of the Brazilian territory and includes the states of Acre, Amapa, Amazonas, Mato Grosso, Para, Rondonia, Roraima, Tocantins and the western part of the 44 ºW meridian of Maranhao (Oliveira et al. 2020), as represented in Fig. 1.

Table 1 shows the land extent and forest coverage areas for all nine states in the Legal Amazonia.

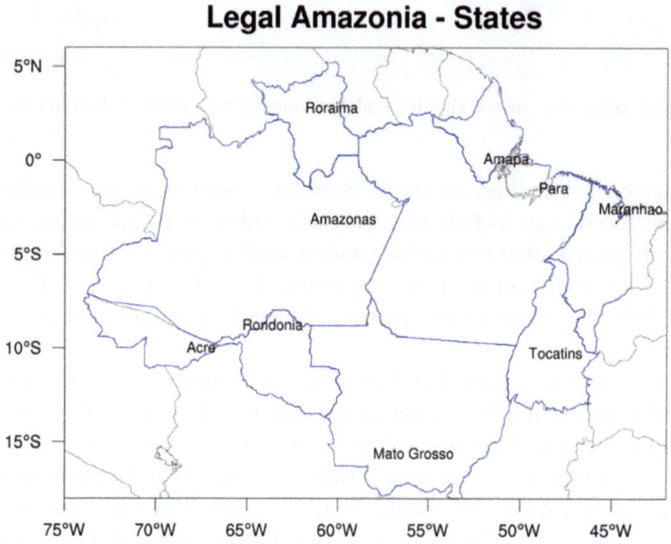

Fig. 1 Division of Legal Amazonia states

Table 1 Land coverage by state of the Legal Amazonia (km^2)—2014

Land coverage by state (km^2)—2014			
State	Total area (km^2)	Forest area (km^2)	Forest coverage (%)
Acre	164,101.9	143,003.5	87.14
Amazonas	1,558,967.4	1,423,798.6	91.33
Amapa	142,765.5	110,686.5	77.53
Maranhao	262,201.9	33,945.7	12.95
Mato Grosso	903,246.7	311,030.2	34.43
Para	1,247,638.7	871,653.6	69.86
Rondonia	237,498.1	136,855.0	57.62
Roraima	224,214.5	151,689.7	67.65
Tocantins	271,773.4	9,726.9	3.58

Source INPE (2018)

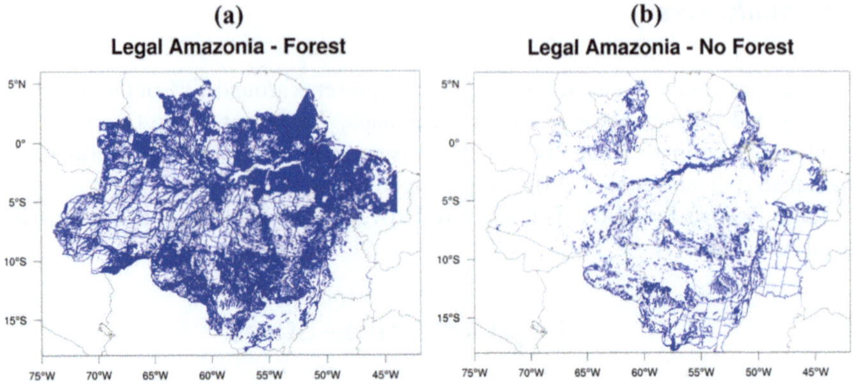

Fig. 2 **a** Legal Amazonia—forest area **b** Legal Amazonia region—area without forest

As presented below, Fig. 2a and b show how these areas are distributed. The area without forests can include urban areas, flooded areas, secondary vegetation, deforested areas, and other types of vegetation, such as pastures, shrubs, grasses and savannas—the so-called non-forest ecosystems. According to IMAZON data, until the year of 2021 the total deforested area in the legal Amazonia corresponds to 21% of this region.

The State with the largest forest coverage is Amazonas, closely followed by Acre, both located in the Amazonia biome, known for their predominantly dense rainforests. Conversely, Tocantins and Maranhao have the least forest coverage, situated in the geographic transition zone between the Amazonia Forest biomes and the Cerrado. In Tocantins, only 9% of the State is occupied by the Amazonia Biome, while in Maranhao 35% of the area is within the Amazonia Biome (Araújo et al. 2017).

With the advancement of agricultural frontiers and other human activities, the establishment of Protected Areas, including Conservation Units (UCs) and Indigenous Lands (TIs) or Quilombo Remaining Territories, has become necessary. Figure 3 shows how UCs are distributed in the Legal Amazonia territory in 2022. Table 2 presents the number of UCs per state in the Legal Amazonia.

The other group of Protected Areas are Indigenous Lands or Quilombo Remaining Lands (TI), territory occupied by remnants of quilombo communities and those used to guarantee their physical, social, economic and cultural reproduction, which until December 2010 represented 21.7% of protected areas in the Legal Amazonia, totalizing 414 TIs. Among these, approximately 15% are in the process of recognition, 74% are already approved and there are still several lands in which indigenous people are seeking official recognition from the State. From article 20 of the 1988 Brazilian Federal Constitution, indigenous lands are recognized and the indigenous right to possession and use is guaranteed (Veríssimo et al. 2011).

According to data from Crisostomo et al. (2015), Brazil currently has 424 TIs in the Legal Amazonia, which means a total area of 115,344,445 ha, representing 23%

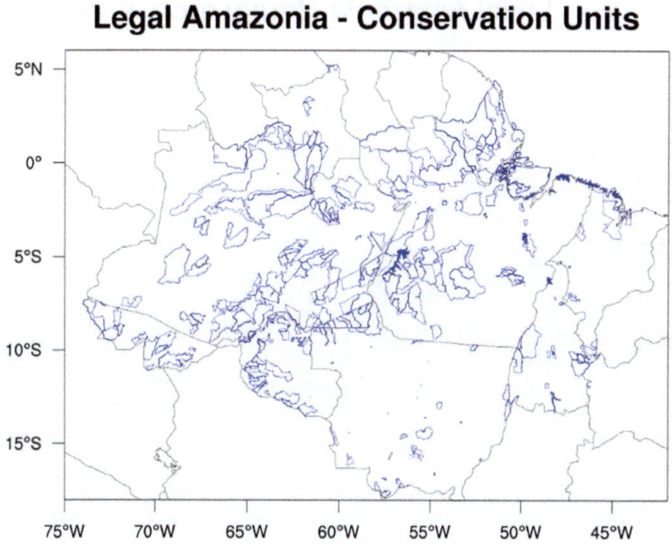

Fig. 3 Distribution of conservation units within the Legal Amazonia region

Table 2 Number of UCs per state in the Legal Amazonia

State	Number of UCs
Acre	20
Amazonas	83
Amapa	12
Maranhao	26
Mato Grosso	46
Para	77
Rondonia	59
Roraima	10
Tocantins	19

Source Adapted from the Socio-Environmental Institute—ISA, 2022

of the Amazonia territory and 98.2% of all TIs on Brazilian territory. According to the 2010 IBGE census, these lands have a population of 433,363 indigenous people. Figure 4 shows how ITs are currently distributed in the Legal Amazonia territories.

Indigenous Lands and Quilombola Remaining Lands are extremely important for the maintenance and conservation of the environment, but even though they act as barriers to inhibit the advancement of deforestation, it is not enough (Crisostomo et al. 2015).

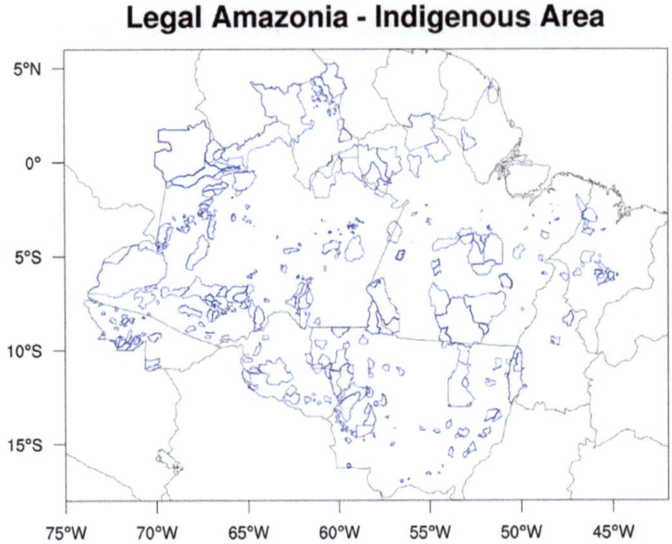

Fig. 4 Current distribution of Indigenous Lands or Quilombo Remaining Lands (Tis) in the Legal Amazonia

The Amazonia is the largest biome in Brazilian territory, according to data from the Brazilian Agricultural Research Corporation (EMBRAPA) and the Brazilian Institute of Geography and Statistics (IBGE). This biome occupies an area of approximately 5 M km^2 in Brazil, which represents around 49% of the territory. It is located in the states of Acre, Amazonas, Amapa, Para, Rondonia, Roraima, Mato Grosso, Maranhao, Tocantins, and also extends in smaller proportions to other countries in South America, such as Bolivia, Colombia, Ecuador, Guyana, French Guiana, Peru, Suriname and Venezuela (Murer and Futada 2022). The Amazonia biome houses the largest tropical forest and biodiversity in the world, harboring an extraordinary richness of fauna and flora that has not been completely documented, with animals and plants that are still unknown.

Deforestation is distributed by the so-called "arc of deforestation", which begins in the southeast of Maranhao, north of Tocantins, south of Para, north of Mato Grosso, south of Amazonas, southeast of Acre and throughout the state of Rondonia (Ferreira et al. 2005).

In general, Amazonian vegetation can be classified into dryland forests and flooded land forests (GEWANDSZNAJDER 2013). Eighty percent of the Amazonia is dryland forest, located in higher regions that are not flooded by rivers. High Forest, Cipó Forest, Campina Forest, forests in areas of transition to the Cerrado are dryland forests, characterized by large trees such as Brazil nut, rubber and palm trees. Flooded Land forests, also called Igapó forests, are located in lower regions and, depending on the region, can remain constantly or seasonally flooded (Murer and Futada 2022). In this type of vegetation, it is possible to find species of water lilies, vines, mosses, and shrubs (Fernando Gewandsznajder 2021).

Considering the tropical geographic location of the Amazonia, the main contributing factor to the Amazonia climate is the solar incidence on the surface. According to studies carried out in the Central Amazonia, the highest levels of solar incidence occur in the period from September to October and the lowest are in the period from December to February. The temperature in the region varies between 24 and 28 °C, characterizing the humid equatorial climate in which this biome is located.

Precipitation is responsible for influencing temperature and relative humidity. The average volume of precipitation in this region is around 2,300 mm $year^{-1}$, increasing slightly near the Andes Mountains. In general, the highest rainfall in the Amazonia is during the period from November to March and the lowest rainfall is between May to September, but this time could vary in the different Amazonia regions. The Amazonia Forest must be understood as controlling the climate regime of the most densely populated regions of Brazil. Without the forest, practically the entire country would have a semi-arid climate, with scarce rainfall, rarely more intense, but associated with storms. Such a scenario would certainly be harmful to a large population and to the environment, with serious socio-environmental and economic problems.

The Amazonia region is peculiar regarding the reception of air masses from the Atlantic Ocean. Once over the forest, these masses condense and form the torrential rains, typical of the region. With the intense evapotranspiration of the forest, increased by the high temperature, humid masses are formed in large quantities that move in the north-south orientation of the Andes Mountains, which act as a buffer, until reaching the States in the central-southern region. Part of these masses is also exported to the Caribbean and the Pacific Ocean, which places the Amazonia Forest in a condition of global importance in terms of its influence on the rainfall regime over a large territorial extension of Latin America.

3 Discussions of Results

The number of fire outbreaks is presented in this study from 2016 to 2021. Fire radiative power (FRP) starts in 2018, since there was no data available on the INPE Fire Bank between the years 2016 and 2017.

Figure 5 shows the total number of fires per month in the Legal Amazonia region from 2016 to 2021. For all years of this study, the highest number of fire events occurred during the months of July to October.

The same seasonal pattern is observed in Fig. 6 regarding the average FRP in the Legal Amazonia region from 2018 to 2021, and an increase in FRP in the months from July to October, corresponding to the dry season in the region, and consequently the largest number of fires recorded.

Figure 7 shows the highest annual number of fires in 2020, 2017 and 2019, respectively, with an increase in fires largely due to the intensification of the region's deforestation rate. Deforestation in this region has increased significantly in the last three years. According to PRODES data (2022), in 2020, 10,800 km^2 was deforested and

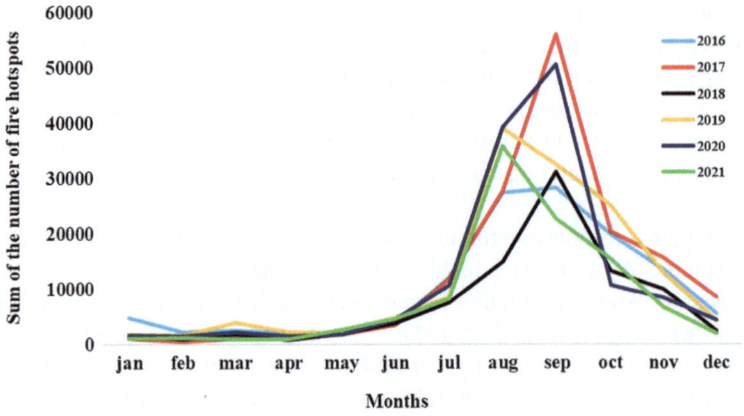

Fig. 5 Sum of the number of fire hotspots in the the Legal Amazonia region between the years 2016 until 2021

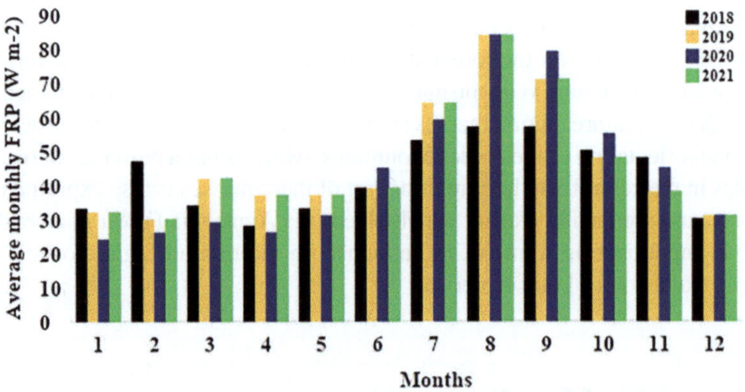

Fig. 6 Average monthly FRP in the Legal Amazonia region between 2016 and 2021

2021 broke the record of the past 15 years, with a loss of 13,000 km^2. The main drivers of deforestation are extensive livestock farming, causing the opening of new areas, and the production of commodities, particularly the growing expansion of soybeans cultivation (Rosana Villar 2022).

Figure 8a to i shows the sum of fire outbreaks per month from 2016 to 2021 for each state in the Legal Amazonia region. Figure 8a shows that in Acre state, the most fire occurs from July to October, with the highest sum of fire hotspots in September 2018. Amapa state had the least sum of fire hotspots throughout the year as shown in Fig. 8b. Most fires occur from September to December, with the highest sum of fire hotspots in November 2016. Amazonas state had most of the fires from July to October as shown in Fig. 8c, with the highest sum of fire hotspots in August 2021. In Maranhao state, as shown in Fig. 8d, most of the fires occur from July to December. The highest sum of fire hotspots of fire hotspots occurred in September 2017. In

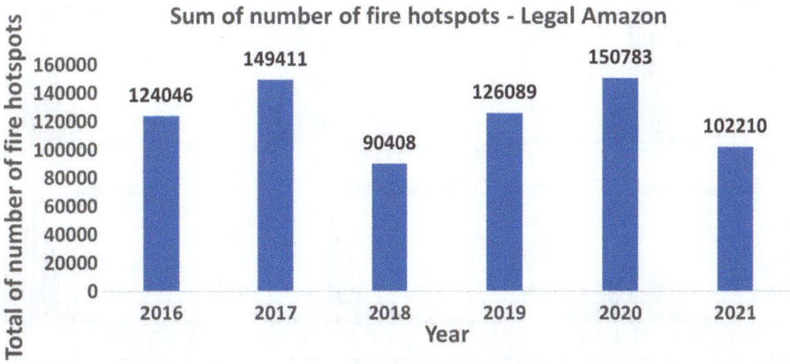

Fig. 7 Annual total of fire hotspots in the Legal Amazonia region between 2016 and 2021

Mato Grosso state as shown in Fig. 8e, most fires occur from August to October, and the highest sum of fire hotspots occurred in September 2020. Para state recorded fire hotspots as the highest sum of fire hotspots in September 2017. Most fires occur from July to December as shown in Fig. 8f. Rondonia state in Fig. 8g showed that most of the fires occur from July to November, and the highest number of fire hotspots occurred in September 2018. Roraima state in Fig. 8h showed the highest number of fires recorded in 2019 from January to April, with March being the month with the highest number of fire hotspots. In Tocantins state as shown in 8i, most fires occur from May to November, while the highest sum of fire hotspots occurred in September 2017. In summary, Fig. 8 shows that September is the month with the highest sum of fire hotspots, as recorded in six of the nine states studied, i.e., Acre, Maranhao, Mato Grosso, Para, Rondonia, Rara, and Tocantins. The states of Roraima, Amazonas, and Amapa had the highest sum of fires in the months of March, August, and October, respectively. Roraima is the only state with the majority of the fire hotspots from January to April, whilst the other states have them mostly from July to November. Amapa state had the least number of fire hotspots, while Mato Grosso/Para had the highest number of fire hotspots among all the states as shown in Fig. 8b and e/f respectively.

Figure 9a to i shows the average FRP per month from 2018 to 2021 for each state in the Legal Amazonia region. Generally, there is an increase in FRP for all the states from June to October, except for Roraima state, which shows an increase in FRP from January to May. These patterns are linked to the region's dry season, coinciding with the highest number of fires. Over the years studied 2018–2021, Amazonas, Mato Grosso, Para, and Rondonia consistently exhibited peak FRP in August, while Roraima's peak occurred in March, Acre and Maranhao's in September, and Amapa's in October.

Figure 10 shows the distribution of the concentration (molecules cm^2) of NO_2 in the atmosphere close to the surface of the states of the Legal Amazonia in Brazil obtained through data from the OMI sensor on the AURA satellite from 2016 to 2021 through monthly seasonal averages. For the time interval analyzed, seasonally, the

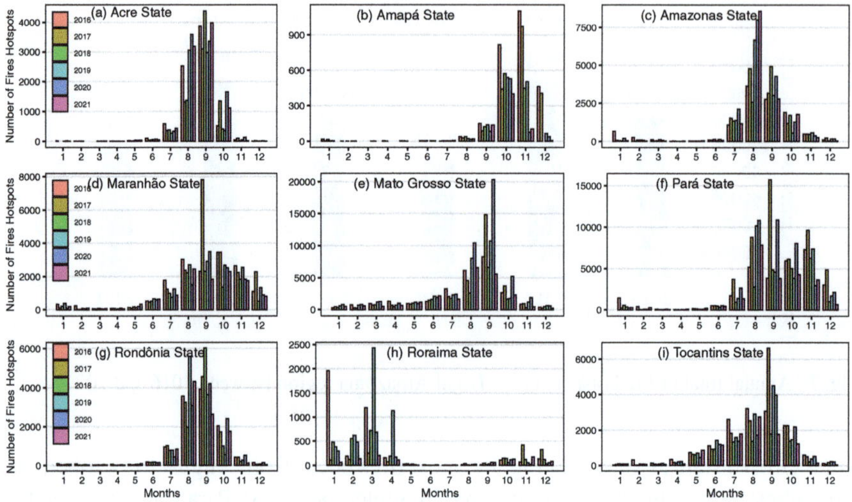

Fig. 8 Sum of the number of monthly fires between 2016 and 2021 in the states: **a** Acre, **b** Amapa, **c** Amazonas, **d** Maranhao, **e** Mato Grosso, **f** Para, **g** Rondonia, **h** Roraima and **i** Tocantins

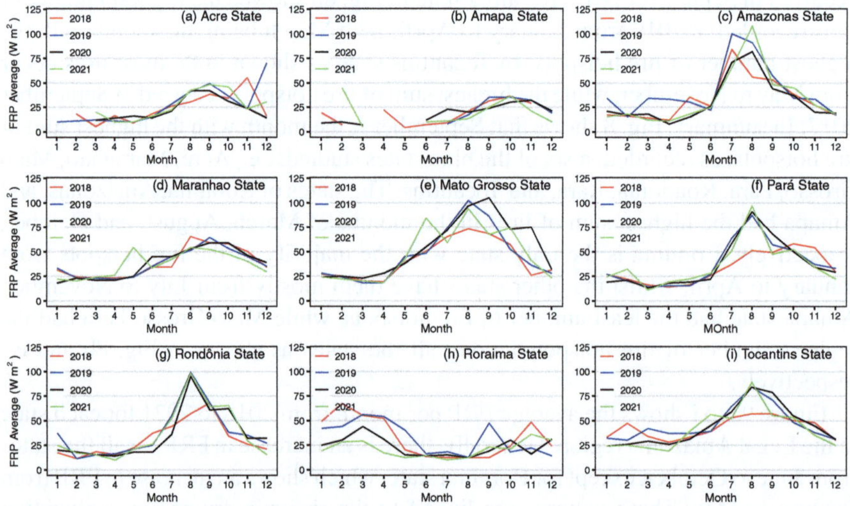

Fig. 9 Average monthly FRP in the states: **a** Acre, **b** Amapa, **c** Amazonas, **d** Maranhao, **e** Mato Grosso, **f** Para, **g** Rondonia, **h** Roraima and **i** Tocantins

lowest concentrations of this pollutant are observed during the period of December-January-February (DJF) to March-April-May (MAM), with emphasis on MAM-2016, presenting the lowest observed concentration. The highest concentration is found in the period August-September-October (ASO), with emphasis on the year 2020, in which there is the highest concentration observed in the interval, occurring mostly in the southern region of Mato Grosso, close to the city from Cuiabá. The region in the extreme north of the Legal Amazonia—States of Amapa, Roraima, northern Para, and northern Amazonas—presents a very small variation in NO_2 concentration on the surface. It is noted that 2018 was the year with the lowest concentration of NO2 during the ASO period. From 2018 onwards, the concentration recorded in the following years are significantly higher than in the previous years.

Figure 11 shows the distribution of CO concentration (ppbv) in the atmosphere near the surface of the study region obtained through MERRA-2 reanalysis data from 2016 to 2021. The lowest CO concentrations were observed during the DJF to MAM periods. From JJA concentrations begin to rise, reaching the highest records in the ASO period. The year in which the highest recorded CO concentration on the surface occurs is 2020, mostly over the States of Rondonia and Mato Grosso in the arc of deforestation. This information agrees with the seasonality graph of the number of fire outbreaks for these States in Fig. 8e and g, since CO emission is directly linked to the burning of biomass as it is emitted in smoke as a trace gas in the burning process.

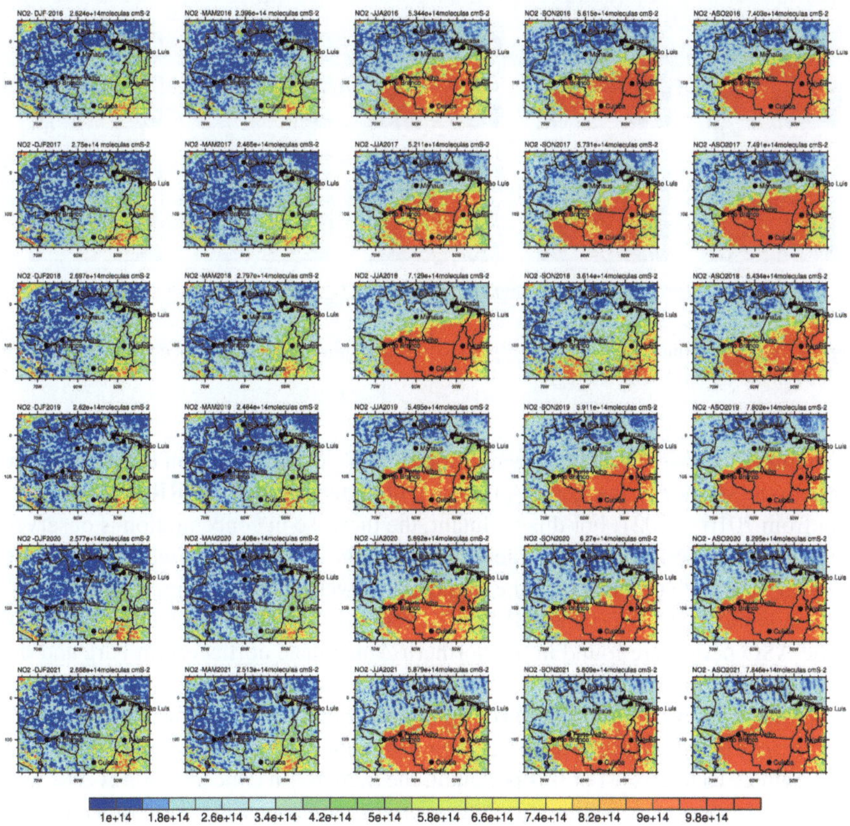

Fig. 10 Seasonal concentration of NO_2 (molecules cm^{-2}) in the atmosphere near the surface for the years 2016 to 2021 in the Legal Amazonia region

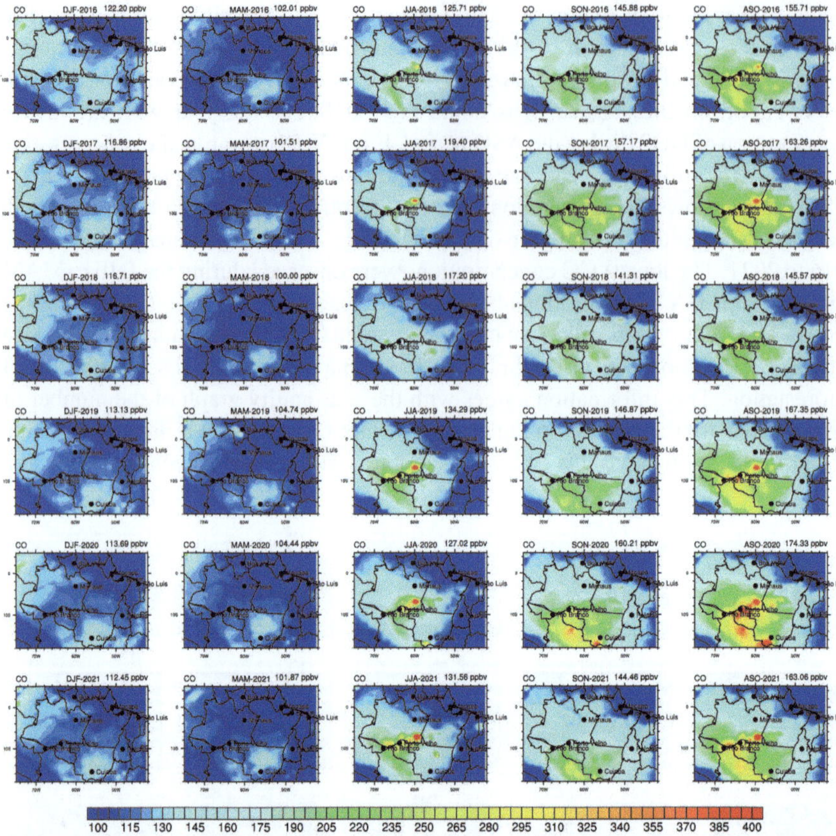

Fig. 11 Seasonal concentration of CO (ppbv) in the atmosphere near the surface for the years 2016 to 2021 in the Legal Amazonia region

Figure 12 shows the SO_2 concentration ($\mu g\ m^{-3}$) distribution in the atmosphere close to the surface of the Legal Amazonia, obtained through MERRA-2 reanalysis data from 2016 to 2021. For this pollutant, the increase in concentration is observed from the JJA period, reaching greater intensity in the ASO period—with emphasis on the year 2020, which presented the highest observed concentration, with emphasis on Rondonia, Mato Grosso and Para. The regions with lower concentrations even in the ASO period are found in the northern Amazonas, northwestern Para and the states of Roraima and Amapa.

Figure 13 shows the distribution of the concentration ($\mu g\ m^{-3}$) of BC in the atmosphere close to the surface of the study region, obtained through MERRA-2 reanalysis data from 2016 to 2021. For this pollutant, the increase in concentration is observed from the JJA period, spreading from the south of Amazonas and north of Rondonia, reaching greater intensity in the ASO period—with emphasis on the year 2020, which presented the highest concentration throughout the state of Rondonia, northwest and

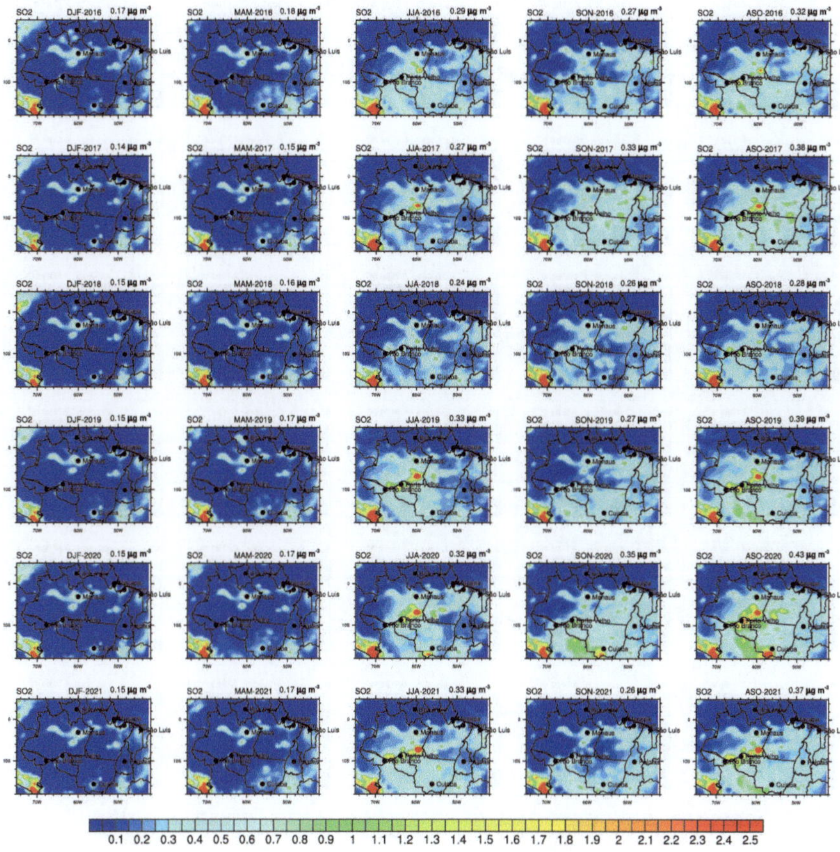

Fig. 12 Seasonal concentration of SO_2 (ppbv) in the atmosphere near the surface for the years 2016 to 2021 in the Legal Amazonia region

southwest parts of Mato Grosso, southern Amazonas and Para. The high concentration of this pollutant in the atmosphere indicates intense burning activity as a cause, considering these regions are poorly industrialized, industrial emissions do not significantly impact this overcome. As for the lowest concentrations, the states Amapa, Roraima, the extreme north of Acre, Amazonas and Para are the regions that present the lowest concentrations of this pollutant throughout the year—with emphasis on Amapa, which is the state that remains in areas with the lowest recorded concentration when compared to other States. As it is a material that characterizes combustion processes due to its long residence time in the atmosphere, it is also an indicator of anthropogenic activity (Ferreira et al. 2005).

Figure 14 shows the distribution of the concentration ($\mu g\ m^{-3}$) of OC in the atmosphere close to the surface of the study region, obtained through MERRA-2 reanalysis data from 2016 to 2021. The DJF and MAM periods have the lowest concentrations, mainly in 2017, being the lowest concentration in the studied years.

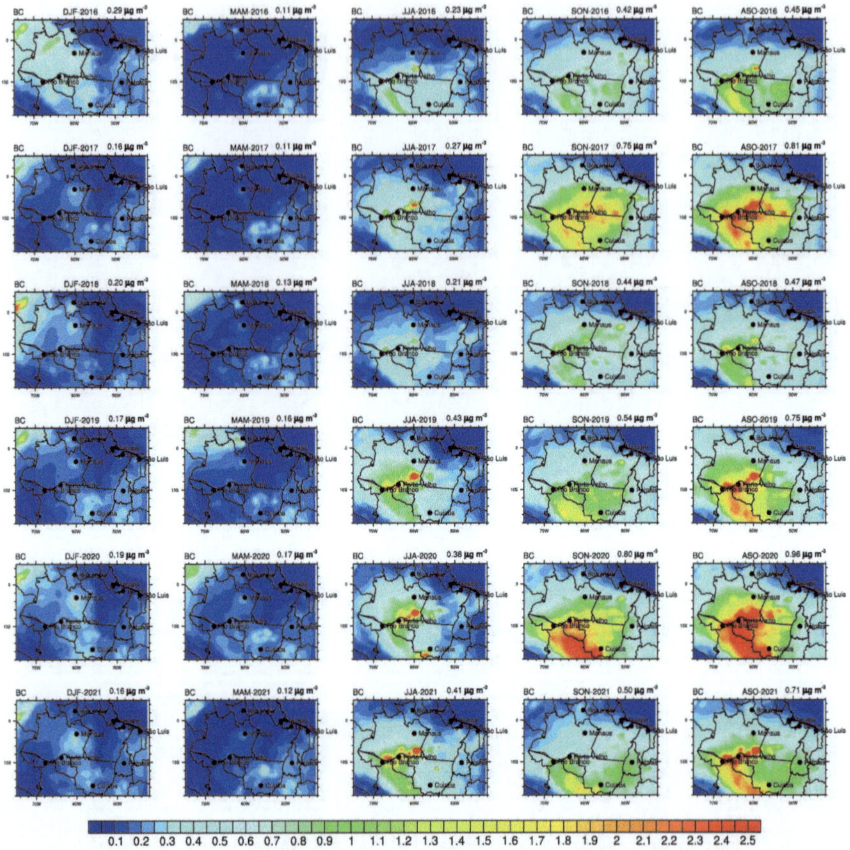

Fig. 13 Seasonal black concentration BC ($\mu g\ m^{-3}$) in the atmosphere near the surface for the years 2016 to 2021 in the Legal Amazonia region

From JJA onwards, it is already possible to notice a significant increase, starting in the north of Rondonia and the south of Amazonas. The highest concentration peaks are observed in the ASO and SON periods, in which concentrations up to approximately four times higher than the concentrations in the first half of the year are recorded. The highest concentration occurs in the ASO period of 2020, in which high concentrations permeate the entire atmospheric surface of the state of Rondonia and run from the northwest of Tocantins and throughout a large part of the territory of the states of Para, Amazonas and Acre. Still in the ASO period, 2017 continues with high concentration in the same regions as mentioned. The behavior of the spatial and temporal distribution of OC in the atmosphere close to the surface of the states of the Legal Amazonia is similar to the concentration distribution of BC, both presents concentration peaks in the same period (i.e., ASO and SON), indicating that the OC emitted at this time of the year is not related to biogenic emissions from trees in the forest but rather OC emissions from biomass burning.

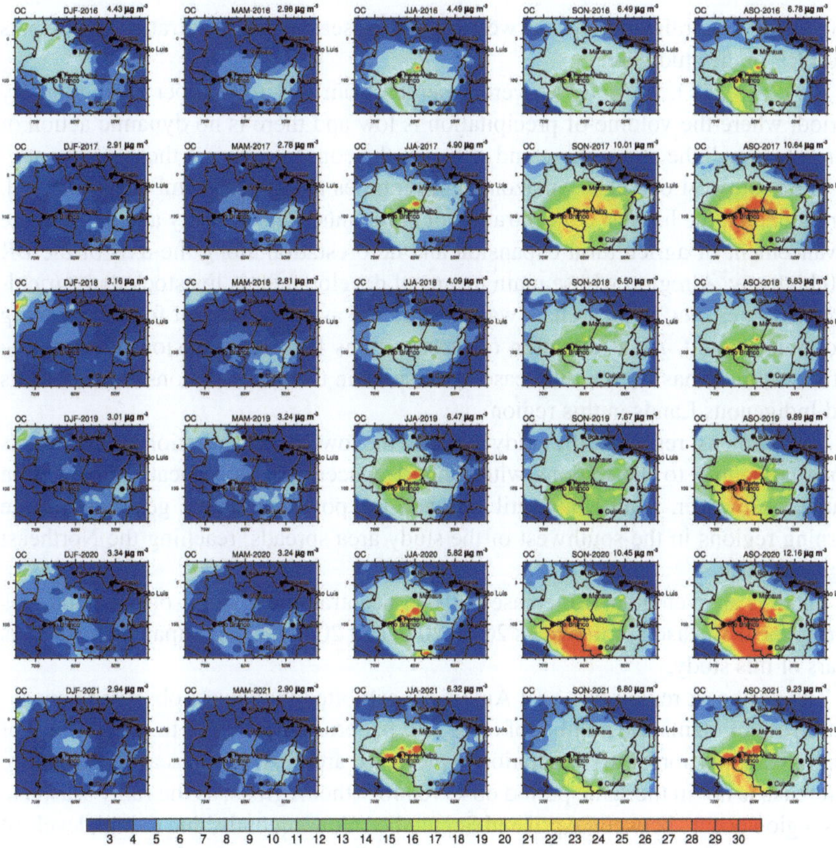

Fig. 14 Seasonal concentration of OC (μg m^{-3}) in the atmosphere near the surface for the years 2016 to 2021 in the Legal Amazonia region

4 Conclusions

This chapter analyzes the concentration of pollutants, i.e., NO_2, CO, SO_2, BC and OC, in the atmosphere near the surface of the states of the Brazilian Legal Amazonia during the period from 2016 to 2021. The study also explores how these concentrations relate to the characteristics of the region.

The seasonality of high concentration of pollutants was identified in the ASO period, with the highest concentration observed mainly in the states of Rondonia, Mato Grosso and Amazonas. Notably, 2020 recorded the highest concentration of the studied pollutants and had the highest number of fires analyzed with 151 thousand registered fires.

For the period from 2016 to 2021, analyses were carried out to identify the concentration of pollutants in the Legal Amazonia region, and their possible relationship with the dry and rainy periods, as well as with the seasonal concentration of hotspots and FRP in the study area.

NO_2, SO_2, CO, BC and OC were observed from July to October during the dry period, where the volume of precipitation is low and there is no dynamic action of rain “cleaning” the atmosphere and reducing the concentration of these pollutants.

The southwest of Legal Amazonia concentrates the largest number of fires and, consequently, the highest concentration of pollutants. This is partly attributed to the advancement of agricultural expansion and deforestation along the axis of the BR 230 highway. A region whose main means of development is livestock and agriculture, activities that, in an archaic way, require fire as an instrument for site cleaning and management. As a reflection of these actions for the expansion of economic activities, there has been an increase in hotspots in the areas of Conservation Units and Indigenous Lands in this region.

The northeast region of the study area had the lowest concentrations of pollutants when compared to the regions with highest concentrations of heat and pollution sources. However, from May until October, the pollution plume generated in the burning regions in the southwest of the study area spreads, reaching the Northeast region as well.

There is an even greater increase in the concentrations of atmospheric pollutants during the dry season in the years 2019, 2020 and 2021 when compared to previous years of this study.

The northeast region of Legal Amazonia exhibited the lowest observed concentration of pollutants and number of fires. This pattern is mostly due to the presence of a significant portion of the vegetation in protected areas. Also, the seasonality of fire outbreaks is not in the same period observed for other regions, as the rainy season in this region occurs between April and September. Consequently, due to high levels of precipitation, fires decrease significantly, and high humidity levels prevent the fire from spreading and rain dilutes the concentration of pollutants in the atmosphere.

The dry season leaves vegetation more susceptible to natural or unintentional fires, but intentional practice for agricultural and extractive purposes, among others, is the main cause of outbreaks, which are also responsible for the significant increase in the concentration of pollutants in the atmosphere. Therefore, anthropogenic modifications are the main causative agents of this scenario observed in the present study. The impacts are alarming for the local and even global environment since these pollutants modify the composition of the atmosphere. Another extremely important impact is on public health, which can cause mild symptoms such as eye or nose irritation, to more serious illnesses in the respiratory and cardiovascular systems. Regarding deforestation and forest fires, the impacts on biodiversity can last for years and may be irreversible. Therefore, the importance of actions to prevent and monitor fire outbreaks with the population and competent authorities is reiterated.

Finally, the research suggests the continuity of studies, as well as the expansion of analyses in relation to meteorological parameters of wind, temperature, relative humidity, and precipitation, to verify whether this increase in concentration of pollutants analyzed in the last three years of the study is related to fires caused by human action or also due to lower precipitation, higher temperature, lower relative humidity during these years.

References

Araújo E, Barreto P, Baima S et al (2017) Most deforested conservation units in the legal Amazon (2012–2015) 92. Imazon, Belém. https://imazon.org.br/en/publicacoes/most-deforested-conservation-units-in-the-legal-amazon-2012-2015/
Brebbia CA, Martin-Duque JF (eds) (2002) Air pollution X. WIT Press; For USA, Canada, and Mexico, Computational Mechanics, Southampton, Boston
Crisostomo A, Alencar A, Mesquita I et al (2015) Indigenous lands in the Brazilian Amazon: carbon stocks and barriers to deforestation. https://ipam.org.br/wp-content/uploads/2015/04/indigenous_lands_in_the_brazilian_amazon-1.pdf
De Oliveira AN, Brito J, Caumo S et al (2015) Biomass burning in the Amazon region: aerosol source apportionment and associated health risk assessment. Atmos Environ 120:277–285. https://doi.org/10.1016/j.atmosenv.2015.08.059
Ferreira LV, Venticinque E, Almeida S (2005) Deforestation in the Amazon and the importance of protected areas. Estud Av 19:157–166. https://doi.org/10.1590/S0103-40142005000100010
Freitas SR, Longo KM, Dias MAFS, Dias PLS (2005) Emissions from fires in South American ecosystems. Estud Av 19:167–185. https://doi.org/10.1590/S0103-40142005000100011
Gewandsznajder F (2021) Biology today: biology: high school, vol 2. Ática, Sao Paulo
Ignotti E, Hacon SDS, Silva AMC et al (2007) Effects of fires in the Amazon: method for selecting municipalities according to health indicators. Rev Bras Epidemiol 10:453–464. https://doi.org/10.1590/S1415-790X2007000400003
IPAM (2020) Everything you wanted to know about fire in the Amazon, but didn't know who to ask. https://ipam.org.br/cartilhas-ipam/tudo-o-que-voce-queria-saber-sobre-fogo-na-amazonia-mas-nao-sabia-para-quem-perguntar/
Jesus LÍ de M, Alvim DS, Guimarães SCP, Gobo JPA (2021) Seasonal analysis of the spatial concentration of atmospheric pollutants in northern Rondônia and southern Amazonas In: Academic, popular and institutional knowledge in climatology: contexts for a socio-environmental agenda, pp 2521–2535
Kampa M, Castanas E (2008) Human health effects of air pollution. Environ Pollut 151:362–367. https://doi.org/10.1016/j.envpol.2007.06.012
Murer BM, Futada SM (2022) Conservation areas. In: socio-environmental institute (ISA). https://uc.socioambiental.org/pt-br/unidadesdeconservacao#sistema-de-unidades-de-conservao-snuc
Oliveira EFCD, Oliveira JFD, Silva JAFD (2020) Legal Amazon, sustainable use and environmental surveillance "systems": historical legacy and future prospects. Rev Bras Ciênc Ambient 56:49–64. https://doi.org/10.5327/Z2176-947820200680
Villar R (2022) The Amazon we need expedition. In: Greenpeace Brazil. https://www.greenpeace.org/international/story/54542/amazon-we-need-expedition/
Veríssimo A, Rolla A, Vedoveto M, Futada S de M (eds) (2011) Protected areas in the Brazilian Amazon: advances and challenges. IMAZON, Belém. https://imazon.org.br/publicacoes/2673-2/

Carbon Stock in Two Tree Species on a Central Avenue in the Municipality of Santarém, Pará State

Darliane Miranda da Rocha, Rodrigo Alex Alves dos Santos, Andressa Jaqueline Viana de Souza, Élcio Meira da Fonseca Junior, and Cristina Aledi Felsemburgh

Abstract Estimates predict that in 2050 the world will have 9.7 billion inhabitants, which will cause an increase in the rates of greenhouse gases (GHG)s in the atmosphere due to human activities, mainly in urban centers. One of the leading gases is carbon dioxide (CO2), and Brazil is now the seventh-emission country. Considering this scenario, governments have created several proposals to reverse this situation, such as urban green spaces. Trees have several functions, such as fixing CO2 during photosynthesis and storing it as carbon in their biomass. Our aim in this study was to estimate the carbon stock and biomass of the aerial part of two tree species, *Handroanthus serratifolius* (Vahl) S. Grose (yellow ipê) and *Mangifera indica* L. (mango), at Avenida Mendonça Furtado, in the Municipality of Santarém, Pará State. The carbon storage can be estimated through allometric equations using the diameter measured at 1.37 m above ground (DBH) and the height of the trees. The findings show that the species *M. indica* L. stored more carbon with 2,273.92 kg and 102 individuals. The species *H. serratifolius* had 138 individuals and stored only 1,589.33 kg of carbon. Despite the smaller number of trees, the greater stocking of the *M. indica* species can be explained by the larger diameter of these trees when compared to the diameter of individuals of *H. serratifolius.* The stock of carbon stored in the form of biomass of the species used in urban afforestation can contribute to the reduction of greenhouse gases. This highlights the ecosystemic importance of urban green areas.

Keywords Urban afforestation · Biomass · Allometric equation · Carbon storage · Environmental services · Santarém Municipality

---

D. M. da Rocha (✉) · R. A. A. dos Santos
Forest Engineering, Federal University of Western Pará, Santarém, Brazil
e-mail: darlirocha03_30@hotmail.com

A. J. V. de Souza
Graduate Program in Natural Resources of the Amazon, Federal University of Western Pará, Santarém, Brazil

É. M. da Fonseca Junior · C. A. Felsemburgh
Institute of Biodiversity and Forests, Federal University of Western Pará, Santarém, Brazil

W. Leal Filho et al. (eds.), *Amazon 2030—Sustainability Issues in the World's Largest Rainforest Region*, World Sustainability Series,
https://doi.org/10.1007/978-3-031-81465-5_23

1 Introduction

The world population, and especially the urban population, has been growing in expressive numbers. Estimates predict that in 2050, the world will have 9.7 billion inhabitants, that is, an additional 2.3 billion people living in cities (United Nations Organization 2017). In addition to the increasing rates of urbanization, in global terms, 75% of the generation of greenhouse gases—GHGs, due to human actions originates in mostly urban activities, such as transport, construction, industry, and energy (Bai et al. 2018).

Studies compiled claim that the concentration of carbon dioxide in the atmosphere rose from about 280 parts per million (ppm) in 1750 to 392 ppm in 2011 (IPCC 2014)). Between 1751 and 2016, Brazil accumulated a total of 13.88 billion tons of carbon dioxide, the main GHG, in the atmosphere (Le Quéré et al. 2018). Brazil is now the seventh world economy in terms of GHG emissions into the atmosphere, in an increasing trend since the mid-twentieth century (Ritchie and Roser 2019).

In recent decades, there has been an exponential increase in gases in the atmosphere, mainly CO_2, which has generated serious changes in the global climate order, being observed mainly in urban centers (Brun 2012).

In this context, among the various alternatives proposed by the government, the creation of green spaces stands out as an important strategy to be addressed by cities, considering the essential ecosystem services provided by trees, such as reducing temperatures, increasing the moisture content of the air, favoring human thermal comfort and conserving biodiversity. In this way, the multiplication of green areas plays a relevant role, as it enhances social well-being, in addition to serving as an indicator of urban quality (Gonçalves 2018).

Considering this scenario, trees are elements that can offset GHG emissions, due to their important functions in reducing the concentrations of these gases in the atmosphere, fixing CO2 during photosynthesis and storing carbon in biomass, thus becoming GHG sinks. In addition, trees provide ecosystem services in the urban perimeter, such as boosting the landscape, delimiting spaces, producing shelter for fauna to feed, purifying the air, reducing CO2 levels, enriching the soil, absorbing noise, participating in the hydrological cycle and influencing in the microclimate (Brianezi et al. 2013; Nowak et al. 2013).

The effects of carbon capture can be quantified by estimating the biomass above and below ground, the stock of wood products and the amount of CO2 absorbed during photosynthesis (Muneroli 2009).

Biomass studies are divided into direct methods (or determination) and indirect methods (or estimates) (Higuchi 1994). The direct method consists of felling and weighing all the trees in a fixed plot, while the indirect method uses estimates from inventories - where the volume of wood is the main variable (Higuchi et al. 1998).

Understanding the contribution of urban trees to mitigating atmospheric carbon is one of the concerns of urban planning managers (Choudhury et al. 2020), increasing the need to study urban green infrastructure, the characterization of tree species in response to storage of carbon, ecological benefits of species and possible future impacts (Chambers et al. 2013). This characterization will help face environmental and social challenges (Davies et al. 2013; Yang et al. 2022).

Individual trees in forests and urban areas are considered carbon sinks (Nowak et al. 2013). Information about individual trees can further improve knowledge and the ability to estimate biomass in urban trees and provide data to support urban ecological research (Duncanson et al. 2014).

Based on the above, the objective of this research was to estimate the carbon stock and biomass of the aerial part of two tree species on a public road located in the Municipality of Santarém, West of Pará State.

2 Material and Methods

2.1 Study Area and Data Collection

This study was carried out in the city of Santarém, western Pará, in the Lower Amazon, on the right bank of the Tapajós River. This municipality has a territorial area of 17,898.389 km2, an estimated population of 306,480 inhabitants, and a demographic density of 12.87 inhabitants/km2 in 2020 (IBGE 2021).

The research was carried out on Avenida Mendonça Furtado, with a median measuring 1.5 m wide and 4.8 km long (Fig. 1).

Fig. 1 Study area—Av. Mendonça Furtado, Santarém, PA—**a** Alameda final—Mapiri and **b** Dom Frederico—Prainha. *Source* Google Earth Pro

Fig. 2 Study area—mango and ipê trees on Av. Mendonça Furtado, Santarém, PA. *Source* Authors

It was used as variables the height and the diameter at breast height (DBH) from the database generated through the course completion work "*Quantitative and qualitative analysis of the afforestation of Avenida Mendonça Furtado, in the municipality of Santarém-PA*", by engineer forestry Rodrigo Alex dos Santos. Data collection took place from March 2017 to April 2018, on adult individuals from the construction sites on Avenida Mendonça Furtado in Santarém (Santos 2018).

The species with the highest frequency present in the database were *Handroanthus serratifolius* (Vahl) S. Grose (yellow ipê) and *Mangifera indica* L. (mango tree) (Fig. 2).

The variables used in the allometric equations were diameter at breast height (DBH) $\geq$ 5.0 cm and total height (Ht).

2.2 Allometric Equation

The biomass and carbon estimates were calculated considering the diameter and height of the species, from tabulated data, and the calculation of the allometric equation for biomass estimation was performed in the Microsoft Office Excel software ® 2010.

To estimate the fresh biomass of the species, the following allometric equations were used according to Higuchi et al. (1998).

To determine the fresh weight of a tree, we used:

$$\text{In PF} = -2.694\,\text{In D} + 0.902\,\text{In H}$$

where:

In Napierian logarithm
FW Fresh Weight (kg)

D Diameter at breast height (cm)
H Total height of the tree (m).

To obtain the amount of dry biomass, the value suggested by Higuchi et al. (1998) was used, considering that 60% of the total weight of a tree corresponds to the dry weight. Where:

$$PS = PF * 0.6$$

where:

FW Fresh Weight
DW Dry Weight
0.6 Percentage corresponding to the dry weight.

According to studies by Higuchi (1994), 50% of the dry weight of a tree corresponds to stored carbon. Thus, we applied the equation suggested by Soares et al. (2006), where the weight is multiplied by 0.5 to estimate the amount of carbon.

$$EC = PS * 0.5$$

where:

CE carbon estimate
0.5 Percentage corresponding to the carbon stock.

3 Results and Discussion

A total of 240 trees of the two species studied, *H. serratifolius* and *M. indica*, were measured.

The species *H. serratifolius* (Vahl) S. Grose belonging to the Bignoniaceae family (Table 1) had 138 individuals. The widespread use of this species in public places is associated with the appreciation of its beauty during flowering by a large part of the population (Mori 2010).

Table 1 Family, scientific name, number of individuals and carbon stock of the species *Mangifera indica* and *Handroanthus serratifolius* registered along Avenida Mendonça Furtado in Santarém, Pará State

Family	Scientific name	Individuals number	Carbon stock (kg)
Anacardiaceae	*Mangifera indica* L	102	2,273.92
Bignoniaceae	*Handroanthus serratifolius* (Valhl) S. Grose	138	1,580.33
	Total	240	3,854.25

M. indica L., Mango, from the Anacardiaceae family, totaled 102 individuals distributed along the avenue (Table 1). The occurrence of this species is mainly due to the fact that it is a fruit tree and provides shade, which encourages spontaneous planting by the population (Almeida and Rondon Neto 2010). Its use in afforestation in the region is important at a sociocultural level for the population, being legally established as a historical heritage in Belém—PA (Porto and Brasil 2013).

The species H. serratifolius had the diameter class with the largest number of trees between 10 and 20 cm, and with 94 trees. No trees with a diameter greater than 40 cm were quantified, and only one tree individual had a diameter between 30 and 40 cm (Fig. 3). The smallest arboreal individual was 1.4 m tall and 14.64 cm in diameter. The tallest individual was 13.20 m tall and 30.94 cm in diameter.

The total amount of carbon stored by the two tree species along the Avenue was 3,854.27 kg. These values varied according to the diameter and height of the individuals found. The species *Mangifera indica*, despite having a smaller number of individuals when compared to the species *Handroanthus serratifolius*, had a higher carbon stock (2,273.92 kg). *M. indica* had the highest DBH and consequently stored more biomass.

Higher values were observed by Santos et al. (2013) in estimating the biomass and carbon stock of tree vegetation in a fragment of the Tucumã urban park, located in Rio Branco, Acre. In this area, the species *M. indica* stocked 11,630 kg for 32 individuals.

The species *H. serratifolius* stored carbon as a function of the average height of its individuals. This estimate was higher than that calculated for *M. indica* (Table 1). This allows us to infer that storage capacity is directly linked to growth in DBH and height and, consequently, in relation to plant age (Sanquetta et al. 2014).

The proportions vary between species, due to intrinsic characteristics, as each species has a growth regime (Ratuchne 2015).

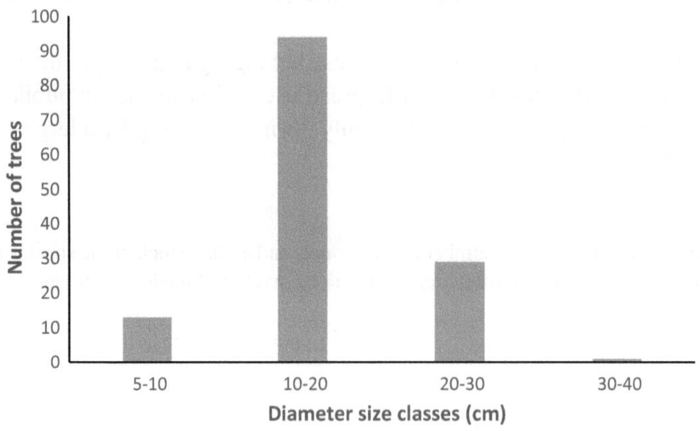

Fig. 3 Number of M. indica trees distributed in diameter classes on Av. Mendonça Furtado, Santarém, PA. *Source* Authors

Carbon storage carried out by individual trees is up to 1,000 times greater in large individuals than in small individuals, with 90 times higher rates of sequestration when healthy (Mcpherson et al. 1994).Thus, when aiming at greater and more efficient carbon storage in urban green areas, it is necessary to properly plan and execute the afforestation of parks and streets. This includes the introduction of medium and fast-growing trees, thus contributing to increasing the efficiency of carbon capture and bringing benefits to the urban environment (Muneroli and Mascaró 2010).

4 Conclusions

Among the functions performed by urban afforestation is the accumulation of carbon in its biomass through photosynthesis. However, urban afforestation in Brazil is still carried out in a disorderly way, including the exacerbated use of exotic species.

Thus, proper planning is necessary with a prior study of the species that will be introduced in urban green spaces. This can occur through beds where the roots can support the densest crowns, and grow in diameter and height according to your needs. The importance of introducing fast-growing species is also highlighted for more excellent carbon storage in the urban environment.

In this study, the findings show that the greater number of individuals of a species does not necessarily imply a greater amount of stored carbon. However, variables such as diameter and height influence the density of carbon sequestration and storage by trees.

The present work contributes to knowledge about the dynamics of species in carbon stocks. However, one must also consider the other species present in these paths and green areas, and consider other attributes such as leaf area, structure, area, and shape of the canopy, in addition to extending the research to larger areas.

References

Almeida DN, Rondon Neto RM (2010) Análise da arborização da arborização urbana de duas cidades da região norte do estado de Mato Grosso. Revista Árvore, Viçosa 34(5):899–906. https://doi.org/10.1590/S0100-76220100005000015

Bai X, Dawson RJ, Ürge-Vorsatz D, Delgado GC, Barau AS, Dhakal S, Dodman D, Leonardsen L, Delmotte-masson V, Roberts DC, Schultz S (2018) Six research priorities for cities and climate change. Nature 555(7694):23–25. https://doi.org/10.1038/d41586-018-02409-z

Brianezi D, Jacovine LAG, Soares CPB, Castro RV, Basso VM (2013) Equações alométrica para estimativa de carbono em árvores de uma área urbana em Viçosa—MG. Revista Árvore, Viçosa 37(6):1073–1081. https://doi.org/10.1590/S0100-67622013000600009

Brun KGF (2012) Avaliação do potencial estoque de carbono por Sibipiruna *(Poincianella Pluoviosa* var *peltophoroides* (Benth.) L.P.Queiróz) na arborização viária de Maringá—PR. Tese (Doutorado em Ciências). Escola Superior de Agricultura "Luiz de Queiroz", Piracicaba. https://doi.org/10.11606/T.11.2012.tde-10122012-112704

Chambers D, Périé C, Casajus N, de Blois S (2013) Challenges in modelling the abundance of 105 tree species in eastern north America using climate, edaphic, and topographic variables. For Ecol Manag 291:20–29. https://doi.org/10.1016/j.foreco.2012.10.046

Choudhury MAM, Marcheggiani E, Despini F, Costanzini S, Rossi P, Galli A, Teggi S (2020) Urban tree species identification and carbon stock mapping for urban green planning and management. Forests 11:1226. https://doi.org/10.3390/f11111226

Davies ZG, Dallimer M, Edmondson JL, Leake JR, Gaston KJ (2013) Identifying potential sources of variability between vegetation carbon storage estimates for urban areas. Environ Pollut 183:133–142. https://doi.org/10.1016/j.envpol.2013.06.005

Duncanson LI, Cook BD, Hurtt GC, Dubayah RO (2014) An efficient, multi-layered crown delineation algorithm for mapping individual tree structure across multiple ecosystems. Remote Sens Environ 154:378–386. https://doi.org/10.1016/j.rse.2013.07.044

Gonçalves MP (2018) Análise da distribuição espacial, funcionalidade e atratividade de áreas verdes pública na cidade de São Carlos, SP. Dissertação (Mestrado em Ciências Ambientais). Universidade Federal de São Carlos, São Carlos. SP, Brasil https://repositorio.ufscar.br/handle/ufscar/10077

Higuchi N, Santos J, Ribeiro RJ, Minetti L, Biot Y (1998) Biomassa da parte aérea da vegetação da floresta tropical úmida de terra-firma da Amazônia Brasileira. Revista Acta Amazonica, Manaus 28(2):153–166. https://doi.org/10.1590/1809-43921998282166

Higuchi N (1994) Fitomassa e conteúdo de carbono de espécies da Amazônia. In: Carvalho Júnior JA (ed) Emissão x Sequestro de CO: Uma nova perspectiva de negócios para o Brasil. CRVD, Rio de Janeiro, pp 125–154. https://repositorio.inpa.gov.br/handle/1/35086

Instituto Brasileiro de Geografia e Estatística (2021) Cidades e Estados

IPCC (2014) Intergovernmental panel on climate change: synthesis report. Contribution of working groups I, II and III to the fifth assessment report of the intergovernmental panel on climate change [Core writing team, Pachauri RK, Meyer LA (eds)]. IPCC, Geneva, Switzerland, 151p

Le Quéré C, Andrew RM, Friedlingstein P, Sitch S, Pongratz J, Manninhg AC et al (2018) Global carbon budget. Earth Syst Sci Data, Geneva 10:405–448. https://doi.org/10.5194/essd-7-349-2015

Mcpherson EG, Nowak, DJ, Rowntree RA (1994) (ed) Chicago's urban forest ecosystem: results of the Chicago urban forest climate project. U.S. Department of Agriculture, Forest Service, Northeastern Forest Experiment Station. California University, Berkeley, 201 p

Mori NT (2010) Variabilidade genética entre e dentro de subpopulações de ipê-roxo *Handroanthus Heptaphyllus* (Vell.) Mattos e seu sistema reprodutivo. Dissertação (Mestrado em Ciência Florestal)—Universidade Estadual Paulista "Júlio Mesquita Filho", Botucatu, SP, Brasil. https://repositorio.unesp.br/bitstreams/42d92142-10af-4f1b-b127-ba3b655942d0/download

Muneroli CC (2009) Arborização urbana: espécies arbóreas nativas e a captura de carbono atmosférico. Dissertação (Mestrado em Engenharia)—Universidade de Passo Fundo, Passo Fundo, Brasil

Muneroli CC, Mascaró JJ (2010) Arborização urbana: uso de espécies arbóreas nativas na captura do carbono atmosférico. Revista da Sociedade Brasileira De Arborização Urbana, Piracicaba, SP 5(1):160–182

Nowak DJ, Greenfield EJ, Hoehn RE, Lapoint E (2013) Carbon storage and sequestration by trees in urban and community areas of the United States. Environ Pollut Amsterdam 178:229–236. https://doi.org/10.1016/j.envpol.2013.03.019

Organização das Nações Unidas (2017) Infográfico das Nações Unidas: sustaintable innovation forum

Porto LMP, Brasil HMS (2013) Manual de orientação técnica de arborização urbana de Belém: guia para planejamento, implantação e manutenção da arborização em logradouros públicos. Belém: EDUFRA

Ratuchne LC (2015) Biomassa e carbono: equações e dinâmica em fragmentos de floresta ombrófila mista no Paraná. Tese (Doutorado em Engenharia Florestal). Universidade Federal do Paraná, Curitiba, PR, Brasil

Ritchie H, Roser M (2019) CO_2 and other greenhouse gas emissions

Santos RAA (2018) Análise quanti-qualitativa da arborização da Avenida Mendonça Furtado, no Município de Santarém-PA. Monografia (Graduação em Engenharia Florestal). Universidade Federal do Oeste do Pará, Santarém, PA, Brasil

Santos LR, Santos EA, Ferreira EJL (2013) Estimativa da capacidade de estoque de biomassa e carbono da vegetação arbórea de um fragmento do parque urbano tucumã em Rio Branco, Acre. Enciclopédia Biosfera. Centro Científico Conhecer, Goiânia 9(17):1305–1321. https://www.conhecer.org.br/enciclop/2013b/CIENCIAS%20AGRARIAS/estimativa%20da%20capacidade.pdf

Sanquetta CR, Corte APD, Mognon F, Mass GCB, Rodrigues AL (2014) Estimativa de carbono individual para Araucaria angustifólia. Pesq Agropec Trop Goiânia 44(1):1–8. https://doi.org/10.1590/S1983-40632014000100006

Soares CPB, Neto FP, Souza AL (2006) Dendrometria e Inventário Florestal, 1st edn. UFV, Viçosa

Yang M, Mou Y, Liu S, Meng Y, Liu Z, Li P, Xiang W, Zhou X, Peng C (2022) Detecting and mapping tree crowns based on convolutional neural network and google earth images. Int J Appl Earth Obs Geoinf 108:10276. https://doi.org/10.1016/j.jag.2022.102764

Challenges in Sustainable Territorial Management in the Brazilian Amazon Region for the XXI Century

Claudio Szlafsztein

Abstract Territorial management in the Amazon region of Brazil presents essential challenges for the coming years, considering the unresolved problems of the central government instrument, the Economic Ecological Zoning of the EEZ, and the changes that are taking place in this territory. The EEZ is an instrument that has been applied at different scales for more than 30 years. Still, it has yet to overcome the problems related to the diverse and profound pressures on the environment and the poor quality of the social and economic development of the population. Climate modifications, deforestation, environmental degradation, expansion of mining activities, water flow alterations, and human occupation and urbanization processes are some of the changing factors contributing to significant uncertainties in the region. In this sense, the article presents, based on its own experiences, government information, and data and interviews with managers who worked in the elaboration of the EEZ, a reflection on the main obstacles resulting from the implementation of the EEZ and the changes in the development processes in the region (increasing demand for energy, economic production based on natural resources, climate change). Finally, possible ways of overcoming these challenges are presented, considering methodological guidelines, governance, and forms of social participation.

Keywords Ecological economic zoning · Amazonia · Landscape zoning · Changes

1 Introduction

The Amazon region in Brazil (Fig. 1) is known due to one of the most extensive and rich tropical forests and the enormous Amazon River Basin, where sustainability is closely associated with biodiversity (Antonelli et al. 2018), deforestation (Barona

C. Szlafsztein (✉)
Núcleo de Altos Estudos Amazônicos (NAEA), Universidade Federal do Pará (UFPA), Campus do Guamá, CEP, 66075-110 Belém, PA, Brasil
e-mail: ioselesz@gmail.com

W. Leal Filho et al. (eds.), *Amazon 2030—Sustainability Issues in the World's Largest Rainforest Region*, World Sustainability Series,
https://doi.org/10.1007/978-3-031-81465-5_24

et al. 2010; Sonter et al. 2017), climate change (Lovejoy and Nobre 2018), and the ways of life of indigenous and traditional communities (Cunha and Almeida 2000). Economic activities are commonly based on the unsustainable exploitation of enormous mineral deposits (Diele-Viegas et al. 2020), hydropower energy resources (Latrubesse et al. 2017), and agriculture and livestock (Maranhão et al. 2019). Exploiting forests for logging and opening new areas for rural activities is a fundamental cause of the high and rising deforestation and forest fire (Brando et al. 2020; Silva et al. 2021). Still, roughly 80% of the region maintains primary forest, particularly in numerous diverse conservation units and indigenous territories (Hansen et al. 2020). These dynamic and growing scenarios of economic expansion, driven by demands from domestic and international markets, have coexisted in this region in the last 60 years with significant socioeconomic inequalities, little awareness of the social organization, low efficiency in the implementation of public policies, and essential evidence of environmental degradation (Gollnow and Lakes 2014).

Governments have taken an interest in the Amazonian territories for geopolitical reasons and to promote sustainable development, considering their natural fragility and socioeconomic potential. The region has implemented territorial zoning over the past 30 years, primarily through Ecological-Economic Zoning (EEZ) on different scales. The EEZ was developed, with few modifications, following a methodology

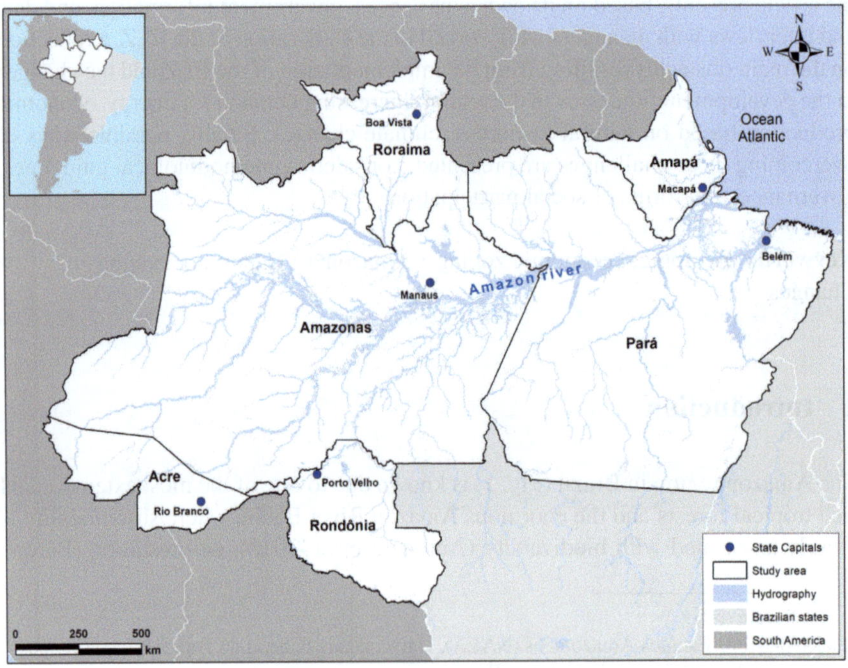

Fig. 1 Location map of the study area

specified in the document "Detailing the Ecological-Economic Execution Methodology Zoning by the States of Legal Amazonia" (Becker and Egler 1996), the norms established in Federal Decree 4297/2002 (Brasil 2002) and the methodological guidelines prepared by the Ministry of the Environment (Brasil 2006).

The planning and management of territorial projects have yielded numerous positive outcomes that aid in the growth of the Amazon region (Szlafsztein et al. 2016). However, the implementation of the project still faces some issues, as noted by Szlafsztein et al. (2016), Brasil (2016), and Lopes et al. (2019). It is essential to grasp the rapid and ongoing changes taking place in the region, as emphasized by Davidson et al. (2012), Phillips et al. (2008), and Timpe and Kaplan (2017). A methodology that reflects the current reality must be adopted to achieve this, rather than relying on outdated Amazonian viewpoints from 20 years ago.

This article examines the difficulties of sustainable planning and management of the Brazilian Amazon in the twenty-first century. The focus is on analyzing the lessons learned from territorial management in the region, comprehending the changing processes happening in the region, and proposals for improving the effectiveness of existing tools. The analyses include data from government institution documents, bibliographic references, and interviews with key actors involved in developing EEZ initiatives. Guided discussions were held using specific questions that considered each interviewee's unique contributions to the project. Some of these findings can be found in the BRASIL (2016) report.

2 Lessons Learned from Territorial Management

In Brazil, territorial management and planning consist of four phases: Planning, Diagnosis, Prognosis, and Implementation (Brasil 2006). The Planning stage involves political-institutional articulations, identifying social demands, and mobilizing financial and human resources. The Diagnosis stage has three components describing the territory's legal-institutional organization, physical-biotic systems, and socioeconomic dynamics. These components need to present integrated results. Physical-biotic information assesses and spatializes "Environmental Vulnerability." Socioeconomic dynamics studies understand land occupation and the use of natural resources, assessing each importance in the description and spatialization of "Social Potentiality." The Prognosis stage describes and maps territorial management subsidies that define sustainable land use, classifying areas as Productive and Critical. Productive areas strengthen human development in occupied territories (Consolidation) and the installation of social and economic activities in sparsely populated regions (Expansion). The Critical Areas are associated with the preservation of environmental heritage, given the high degree of natural vulnerability (Conservation) and the implementation of corrective activities in environments impacted by economic development (Recovery) (Fig. 2).

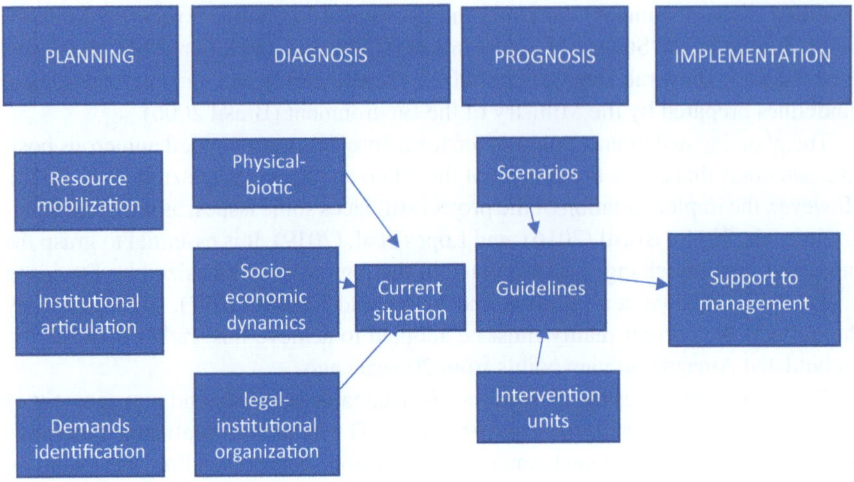

Fig. 2 Diagram of stages and products of the EEZ according to the methodology explained by the Ministry of the Environment (Brasil 2006)

Implementing territorial management and planning in the Amazon has provided governments and society with essential benefits despite the challenges of coordinating various institutions due to inconsistent public policies (D'Almeida et al. 2010). These benefits include advancements in environmental issues within state government management, regional scientific, technical, and leadership capabilities improvements, and establishing and reinforcing data and information collection, storage, and mapping systems. The management and planning projects' outcomes, along with deep scientific, societal, and government considerations about the territory, serve as essential references for various environmental policies in the Amazon region, such as the Action Plan for the Prevention and Control of Deforestation in the Legal Amazon, an instrument of the Climate Change National Policy (Law 12,187/2009), the federal government loans for rural activities system (Banco Central do Brasil 2008), the delimitation of conservation units, biodiversity corridors, and protected areas of rural properties (Brasil 2012), and the environmental licensing processes.

Such a complex and diversified process of designing and implementing the territorial projects in the Amazon also presents trouble spots that become challenges to overcome soon, besides the institutional, financial, and political discontinuities that permeate the long processes of planning and managing in Brazil (Frey 2019; Antunes et al. 2021). Among these problems to be overcome are the support that small and distant municipalities and their populations lack involvement in the process and familiarity with long-term planning culture, hindering their ability to benefit from the instrument results (Marroni and Asmus 2013; Siqueira 2008) and the inaccurate perception that economic growth in the private sector takes a backseat to environmental protection due to territorial planning is a responsibility delegated to state environmental agencies and the Ministry of the Environment. However, it is crucial to note that the EZZ elaboration methods, last updated in 2006, need to consider the changes in the region and the resulting impact on the territories.

3 The Amazon: A Region in Transition and Its Impact on the Territory

Brazil's Amazon region has witnessed several economic cycles that revolved around exploiting its natural resources, including rubber, Brazil nuts, and black pepper. These cycles were relatively simple regarding production, transportation, and marketing infrastructure requirements (Loureiro, 2015). During the latter half of the twentieth century, infrastructure such as highways, ports, and airports were built in the Amazon region. This resulted in new economic opportunities fueled by the increasing availability of energy, minerals, and agricultural products for Brazil and the global market (Silva et al. 2016). However, the increase in development impacted populations, landscapes, biotas, traditional and indigenous communities, and cities (Ometto et al. 2011; Fearnside 2016; Sonter et al. 2017). Amazon's development is influenced by several significant factors that affect the surrounding area. These include population growth and urbanization, an increasing demand for electricity in Brazil and South America, economic production reliant on natural resources, and the impact of climate change and natural disasters. These key factors must be considered for future planning and management of Amazon in the twenty-first century.

Brazil's Amazon region has a population in 2022 of 15.8 million inhabitants (8% of the country's total population) (IBGE 2023). Although the Amazon region has traditionally had low percentages and population density, it is now undergoing a rapid and noticeable process of urbanization and occupation, leading to forest transition (Trindade 2013; Richards and VanWey 2015). The region's population growth rate was the highest in Brazil compared to national values and other areas for 2000/2010 and 2010/2022. In an eminently urban country, the Amazon has one of the highest growth percentages between 1960 and 2010, and Canaã dos Carajas (state of Pará) is the municipality that has the most significant population growth in all the country (IBGE 2023) (Table 1).

South America possesses abundant oil and gas reserves, but their exploration is a contentious issue due to the potential environmental harm it may cause (Finer et al. 2008). Therefore, the region focuses on transitioning to cleaner energy sources by utilizing its strategic mineral resources, such as lithium, and capitalizing on natural conditions like abundant water, ample sun exposure, and strong winds. This strategy helps meet the continent's increasing energy demands, presents opportunities for surplus exports, and promotes regional integration (Sadorsky 2012). Brazil, in particular, has been gradually shifting to clean energy since the 1970s, emphasizing biofuel and hydroelectric power. In the Amazon, producing biofuels made from sugar cane is prohibited. However, there is an increasing production of oils made from fruits, particularly from the dende palm (Walker 2011). The Amazon basin is considered

Table 1 Resident population by regions in Brazil—population growth rate and proportion of the urban population. Data available at the Brazilian Institute of Geography and Statistics—https://censo2010.ibge.gov.br/ and https://cidades.ibge.gov.br/brasil/es/panorama

Region	Population growth rate		Urban population (%)						
	Periods			Years					
	2000/2010	2010/2022		1960	1970	1980	1991	2000	2010
Brazil	1.17	1.03	Total population	45.08	55.98	67.7	75.47	81.23	84.36
Amazon (North)	2.08	1.62	Total urban population	3.25	3.37	4.14	5.35	6.54	7.25
North East	1.06	0.76		24	22.65	21.9	23.23	23.9	24.12
South East	1.04	1		55.67	55.47	53.1	49.74	47.51	46.42
South	0.87	0.96		13.96	14.05	14.82	14.78	14.74	14.45
Center West	1.91	1.59		3.11	4.46	6.04	6.9	7.31	7.76

an "inexhaustible" source of water for hydroelectricity due to the depletion of other Brazilian basins. As of March 2020 (RAISG 2020), there are a total of 833 hydroelectric plants planned or already in existence in the nine countries of the region (Bolivia, Brazil, Colombia, Ecuador, French Guyana, Guyana, Peru, Suriname), with the majority located in Brazil (52%). Of the 181 hydroelectric plants in the Amazon, 137 are small power stations generating less than 30 MW, and the larger plants generate over 2000 MW each.

In Brazil's Amazon region, mining has recently significantly contributed to the country's economy by providing primary products to various industries (Triner 2016). The mining industry in the region has rapidly expanded in recent years, setting records for annual production, profits, and job creation and contributing significantly to Brazil's Gross Domestic Product (Fig. 3). The mining industry's progress in the Amazon region has impacted the socio-economic development of numerous municipalities (Cordani and Juliani 2019).

In recent years, the Amazon region has become a hub for agricultural and livestock activities due to the establishment of large-scale projects. The availability of low-cost land supports these projects, and various plant and animal species are well-suited to the region's unique environmental conditions (Oliveira 2016; Sauer 2018; McManus et al. 2016). The modern growth in the region, which produces more significant deforestation and land use change, is linked to the expansion of consumer markets and the area's increased proximity to them, which has led to a reduction in transport costs. In collaboration with private entities, the government is introducing various programs to build and enhance the highway network (Moran 2016), expand the

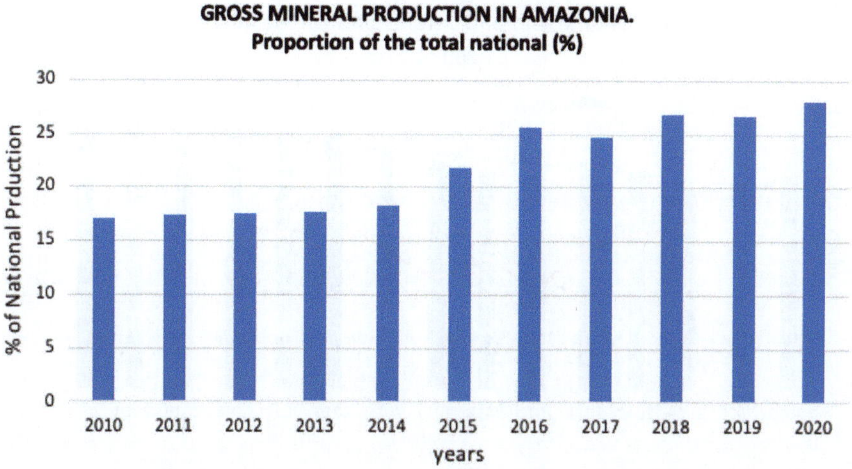

Fig. 3 Gross Mineral production in the Amazon region of Brazil considering the national mining production (2010–2020) *Source* National Mining Agency (*Agencia Nacional de Mineraçao*) (https://www.gov.br/anm/pt-br)

railway system, and overcome any barriers that may hinder the smooth operation of waterways (Fearnside 2015).

The Amazon region's hydroclimatic disasters are often overlooked due to the perception of low population density and a rural lifestyle adapted to natural dynamics. However, several climate scenarios indicate changes in natural conditions, making extreme events more frequent, increasing temperatures, and decreasing precipitation for the coming years in the region (Oliveira et al. 2021).

4 Proposal to Overcome Future Challenges in Territorial Planning for the Amazon

The challenges of territorial planning in the Amazon for the twenty-first century are linked to rapid and growing human activities. The increasing energy demand in Brazil has significant consequences on the hydrologic flows, vegetation cover, biota, protected areas, and populations living in flooded areas of the Amazon (Chaudari and Pokhrel 2022; Cavalcante et al. 2021; Ferrante and Fearnside 2020; Mayer et al. 2021). The primary type of mining extraction, open surface-pit, and placer mining, has significant environmental impacts. It alters vegetation cover and soil characteristics, pollutes rivers, and affects the ability of ecosystems to provide services (Martins et al. 2022). Moreover, mining activities lead to population concentration, urbanization, and transportation logistics that significantly impact the area. Productive agriculture and livestock activities (Santos et al. 2021) are often associated with widespread deforestation and forest fires. These destructive processes significantly

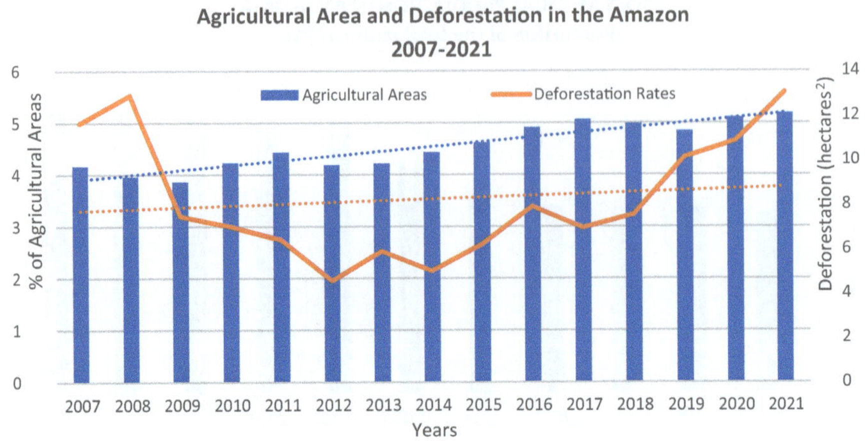

Fig. 4 Compares deforestation rates (square hectares) and the proportion of planted areas over the national total (%) in the Amazon between 2007 and 2021. *Source* IBGE—Levantamento Sistemático da Produção Agrícola (https://www.ibge.gov.br/) e Instituto Nacional de Pesquisas Espaciais (http://www.obt.inpe.br/)

affect vast regions, altering the original and secondary vegetation cover, changing the landscape's use and occupancy, modifying the characteristics of the soil's uppermost layers, and ultimately reducing the ability to provide essential ecosystem services (Fig. 4). In particular, it contributes to changes in climate conditions and disasters that have enormous impacts on defining the territory as risk-prone areas for people and activities (Tanure et al. 2020; Nobre et al. 2019).

It is important to revise the methodological guidelines, governance, and governmental and social participation methods to address the challenges in territorial planning in the Brazilian Amazon. These revisions should take into account the new demands of the territory.

The current methodology for determining the EEZ is the outcome of the continuous development of technical know-how, available data, and political perspectives (Vagula et al. 2021). This methodology was initially developed during the 1990s when the region was not well-known and was partially geographically isolated from the rest of the country. During that period, preservation was closely associated with preventive measures (Vasconcelos et al. 2013; Silva et al. 2017). During the discussion about proposals to enhance the EEZ methodology, various aspects needed improvement. These include the need for more detailed analysis and presentation of results, better diagnosis of the region, more significant consideration of future scenarios, and inclusion of new subjects.

Scales with little detail help recognize and demarcate large areas for planning purposes. However, more detailed scales can assist governments and society in finding answers to their demands for decision-making subsidies. They can also facilitate the identification of potentialities and vulnerabilities in the territories by social actors. Therefore, it is important to promote and qualify the participation of social actors throughout the elaboration and implementation of the EEZ. The process of diagnosing needs to be revised. It currently involves gathering a large amount of data and information, often resulting in high costs and lengthy execution times. This can lead to reports that inaccurately reflect changes in a particular region, making it challenging to conduct critical analysis on managing the territory. Given that the EZZ needs to facilitate the territorial planning of the Amazon region for the future, it is crucial to prioritize using methods that build prospective scenarios and support the predictive stage. This will allow for a stronger focus on future economic and social dynamics trends, emphasizing a forward-looking approach to the EEZ. The region is experiencing dynamic changes that are driving new themes. These include natural disasters due to climate change, population migrations, expansion of hydroelectric plants, growth in agribusiness associated with grain and meat commodities, and increasing deforestation.

However, the primary focus should be on adopting novel approaches to plan and develop the region based on the assessment and recognition of the ecosystem services provided by the territory. These services have the potential to contribute to the region's growth while ensuring the conservation and protection of its natural resources.

Public policies in Brazil face difficulties due to governance issues, hindering implementation, law enforcement, and sustainability (Donadelli 2017). In this regard, collaborating and integrating with other policies and tools, such as the National Policy for Regional Development (PNDR) (Decree No. 9.810/2019), municipal master plans, and the methodological processes for conducting environmental impact assessments is crucial for empowering the EEZ in the future. Improving the governance of Brazil's EEZ is of utmost importance. This can be accomplished by transferring the responsibility of coordinating EEZ programs from the Ministry of the Environment to other ministries, such as the Ministry of Planning, Development, and Management. This change will enhance communication and collaboration between government sectors working within the territory.

Strengthening social participation in territorial planning is crucial to overcoming inequalities in regions like the Amazon. In this sense, it is important to consider the proposed mechanisms.:

1. Promote direct government involvement at the highest levels, ensuring successful implementation.
2. Prepare federal and state managers and technicians to incorporate zoning into their work routines using EEZ-generated results.
3. When building the EEZ, consider the vision and plans of indigenous and conservation unit communities.
4. Ensure that financial resources are available to mobilize actors, allocate time for consultation and technical document explanation, and raise awareness among responsible parties regarding the importance of a broad discussion process for zoning.

5 Conclusion

Despite various attempts, including regional planning, environmental education, satellite-based monitoring systems, and control measures, human activities continue contributing to the deterioration of Amazonian ecosystems. Nevertheless, research indicates that the EEZ remains a crucial tool in Brazil's territorial planning agenda since no other method with a standardized approach and national recognition has been identified.

Although the planning and management proposals for Amazon are important, they must also address illegal activities, territorial discontinuity, and outdated instruments to be effective. It is vital to acknowledge that certain land use practices, such as mining on indigenous territories and deforestation in protected areas, are illegal. These activities pose significant challenges and complexities to territorial planning (Condé et al. 2019; Siqueira-Gay and Sánchez 2021). Despite over 30 years of planning and management in the Amazon region, there are still territorial gaps due to the lack of coverage by the EEZ. This includes areas like the coastal zone of the state of Pará, as well as newly established territories that connect different regions across several states due to the same productive activity, such as AMACRO. The two issues indicate an opportunity to reformulate the instrument's principles, objectives, and methodology, given the particularities of these environments.

The future of the EEZ is associated with the possibility of finding mechanisms that can help with the lowest political, economic, and social cost and as quickly as possible (considering the transformation that affects the territory), the implementation of its results, enhancing its appeal as a reference to other land use planning and management policies.

Acknowledgements This research was supported by CAPES (Brazil) and the Center for Latin American Studies at the University of Pittsburgh.

References

Antonelli A, Zizka A, Carvalho F, Scharn R, Bacon C, Silvestro D, Condamine F (2018) Amazonia is the primary source of neotropical biodiversity. Proc Natl Acad Sci USA 115(23):6034–6039. https://doi.org/10.1073/pnas.1713819115

Antunes W Jr, Borsatto R, Souza-Esquerdo V (2021) Why is it so difficult to promote territorial development through public policies? The obstacles faced by a Brazilian experience. World Dev Perspect 24:100367. https://doi.org/10.1016/j.wdp.2021.100367

Barona E, Ramankutty N, Hyman G, Coomes O (2010) The role of pasture and soybean in deforestation of the Brazilian Amazon. Environ Res Lett 5(2). https://doi.org/10.1088/1748-9326/5/2/024002

Becker B, Egler C (1996) Detalhamento da metodologia para execução do zoneamento ecológico-econômico pelos estados da Amazônia Legal. Secretaria de Assuntos Especiais e Ministério do Meio Ambiente. Rio de Janeiro/Brasília, 48 p

Banco Central do Brasil (2008) Resolução 3545 que estabelece exigências de documentação comprobatória de regularidade ambiental e outras condicionantes, para fins de financiamento agropecuário no Bioma Amazônia. Brasília. https://www.bcb.gov.br/pre/normativos/res/2008/pdf/res_3545_v1_O.pdf

Brando P, Macedo M, Silvério D, Rattis L, Paolucci L, Alencar A, Coe M, Amorim C (2020) Amazon wildfires: scenes from a foreseeable disaster. Flora 268:151609. https://doi.org/10.1016/j.flora.2020.151609

Brasil (2002) Decreto nº 4.297, de 10 de julho de 2002. Regulamenta o art. 9°, inciso II, da Lei no 6.938, de 31 de agosto de 1981, estabelecendo critérios para o Zoneamento Ecológico-Econômico do Brasil—ZEE, e dá outras providências. Brasília. http://www.planalto.gov.br/ccivil_03/decreto/2002/d4297.htm

Brasil (2006) Diretrizes metodológicas diretrizes metodológicas para o zoneamento ecológico-econômico do Brasil. Ministério o Meio Ambiente. Brasília, 126 p

Brasil (2012) Lei Nº 12651 de 25 de maio de 2012. Dispõe sobre a proteção da vegetação nativa. Brasília. http://www.planalto.gov.br/ccivil_03/_ato2011-2014/2012/lei/l12651.htm

Brasil (2016) O Zoneamento Ecológico-Econômico na Amazônia Legal: Trilhando o caminho do futuro. Ministério do Meio Ambiente. Secretaria de Recursos Hídricos e Ambiente Urbano. Departamento de Zoneamento Territorial. Brasília, 100 p

Cavalcante M, Costa G, Silva G, Moret A (2021) Hydroelectric plants and conservation unit in the Amazon. Mercator 20:e20017. https://doi.org/10.4215/rm2021.e20017

Chaudhari S, Pokhrel Y (2022) Alteration of river flow and flood dynamics by existing and planned hydropower dams in the Amazon River basin. Water Resour Res 58(5):e2021WR030555. https://doi.org/10.1029/2021WR030555

Condé T, Higuchi N, Lima A (2019) Illegal selective logging and forest fires in the northern Brazilian Amazon. Forests 10(1):61. https://doi.org/10.3390/f10010061

Cordani U, Juliani C (2019) Potencial mineral de la Amazonia: problemas y desafios. Revista de Estúdios Brasileños 6(11):91–108. https://doi.org/10.14201/reb201961191108

Cunha M, Almeida M (2000) Indigenous people, traditional people, and conservation in the Amazon. Daedalus 129(2):315–338

D'Almeida R, Vaz M, Caldas E (2010) Local development management in Brazil: actors' (dis)articulation, instruments, and territory. Revista De Administração Pública 44(3):559–590. https://doi.org/10.1590/S0034-76122010000300002

Davidson E, Araújo A, Artaxo P, Balch J, Foster Brown I, Bustamante M, Coe M, DeFries R, Keller M, Longo M, Munger J, Schroeder W, Soares-Filho B, Souza Jr C, Wofsy S (2012) The Amazon basin in transition. Nature 481(7381):321–28. https://doi.org/10.1038/nature10717

Diele-Viegas L, Pereira E, Rocha C (2020) The new Brazilian gold rush: Is Amazonia at risk? For Policy Econ 119:102270, 1–5. https://doi.org/10.1016/j.forpol.2020.102270

Donadelli F (2017) Environmental policy integration in Brazil: an analysis of climate and biodiversity policies. Revista De Administração Pública 51(5):734–766. https://doi.org/10.1590/0034-7612153044

Fearnside P (2015) Amazon dams and waterways: Brazil's Tapajós basin plans. Ambio 44(5):426–439. https://doi.org/10.1007/s13280-015-0642-z

Fearnside P (2016) Environmental and social impacts of hydroelectric dams in Brazilian Amazonia: implications for the aluminium industry. World Dev 77:48–65. https://doi.org/10.1016/j.worlddev.2015.08.015

Ferrante L, Fearnside P (2020) The Amazon: biofuels plan will drive deforestation. Nature 577(7789):170. https://doi.org/10.1038/d41586-020-00005-8

Finer M, Jenkins C, Pimm S, Keane B, Ross C (2008) Oil and gas projects in the western Amazon: threats to wilderness, biodiversity, and indigenous peoples. PLoS ONE 3(8). https://doi.org/10.1371/journal.pone.0002932

Frey A (2019) Cash transfers, clientelism, and political enfranchisement: evidence from Brazil. J Public Econ 176:1–17. https://doi.org/10.1016/j.jpubeco.2019.05.002

Gollnow F, Lakes T (2014) Policy change, land use, and agriculture: the case of soy production and cattle ranching in Brazil, 2001–2012. Appl Geogr 55:203–211. https://doi.org/10.1016/j.apgeog.2014.09.003

Hansen M, Wang L, Song X, Tyukavina A, Turubano S, Potapov V, Stehman S (2020) The fate of tropical forest fragments. Sci Adv 6(11):1–10. https://doi.org/10.1126/sciadv.aax8574

IBGE (2023) IBGE Cidades. Instituto Brasiliero de Geografia e Estatística. Brasília. https://cidades.ibge.gov.br/brasil/pa/belem/panorama

Latrubesse E, Arima E, Dunne T, Park E, Baker V, d'Horta F, Wight C, Wittmann F, Zuanon J, Baker P, Ribas C, Norgaard R, Filizola N, Ansar A, Flyvbjerg B, Stevaux J (2017) Damming the rivers of the Amazon basin. Nature 546:363–369. https://doi.org/10.1038/nature22333

Lopes E, Souza J, Albuquerque Filho J, Lourenço R (2019) Caminhos e entraves do Zoneamento Ecológico-Econômico no Brasil. Caminhos de Geografia 20(69): 342–359. https://seer.ufu.br/index.php/caminhosdegeografia/article/view/41305

Loureiro V (2015) História da Amazônia: do período da borracha aos dias atuais. Cultura Brasil, Belém, 335 p. ISBN 978-85-65965-96-5

Lovejoy T, Nobre C (2018) Amazon tipping point. Sci Adv 4:eaat2340. https://doi.org/10.1126/sciadv.aat2340

Maranhão R, Carvalho O Jr, Hermuche P, Gomes R, Pimentel C, Guimarães R (2019) The spatiotemporal dynamics of soybean and cattle production in Brazil. Sustainability 11:2150. https://doi.org/10.3390/su11072150

Marroni E, Asmus M (2013) Historical antecedents and local governance in the process of public policies building for coastal zone of Brazil. Ocean Coast Manag 76:30–37. https://doi.org/10.1016/j.ocecoaman.2013.02.011

Martins W, Rodrigues J, Oliveira V, Ribeiro S, Barros W, Schwartz G (2022) Mining in the Amazon: importance, impacts, and challenges to restore degraded ecosystems. Are we on the right way? Ecol Eng 174:106468. https://doi.org/10.1016/j.ecoleng.2021.106468

Mayer A, Castro-Diaz L, Lopez M, Leturcq G, Moran E (2021) Is hydropower worth it? Exploring Amazonian resettlement, human development, and environmental costs with the Belo Monte project in Brazil. Energy Res Soc Sci 78:102129. https://doi.org/10.1016/j.erss.2021.102129

McManus C, Barcellos J, Formenton B, Hermuche P, Carvalho O Jr, Guimarães R et al (2016) Dynamics of cattle production in Brazil. PLoS ONE 11(1):e0147138. https://doi.org/10.1371/journal.pone.0147138

Moran E (2016) Roads and dams: infrastructure driven transformations in the Brazilian Amazon. Ambiente e Sociedade 19(2):207–220. https://doi.org/10.1590/1809-4422ASOC256V1922016

Nobre C, Marengo J, Soares W (eds) (2019) Climate change risks in Brazil. Springer Cham, 226 p. ISBN 978-3-319-92880-7, ISBN 978-3-319-92881-4 (eBook) https://doi.org/10.1007/978-3-319-92881-4

Oliveira B, Bottino M, Nobre P, Nobre C (2021) Deforestation and climate change are projected to increase heat stress risk in the Brazilian Amazon. Commun Earth Environ 2:207. https://doi.org/10.1038/s43247-021-00275-8

Oliveira G (2016) The geopolitics of Brazilian soybeans. J Peasant Stud 43(2):348–372. https://doi.org/10.1080/03066150.2014.992337

Ometto J, Dutra Aguiar A, Martinelli L (2011) Amazon deforestation in Brazil: effects, drivers and challenges. Carbon Manag 2(5):575–585. https://doi.org/10.4155/cmt.11.48

Phillips O, Lewis S, Baker T, Chao K, Higuchi N (2008) The changing Amazon forest. Philos Trans R Soc b: Biol Sci 363(1498):1819–1827. https://doi.org/10.1098/rstb.2007.0033

RAISG (2020) Amazonía bajo presión. Instituto Socioambiental, São Paulo. 68 p. ISBN 978-65-88037-06-5

Richards P, VanWey L (2015) Where deforestation leads to urbanization: how resource extraction is leading to urban growth in the Brazilian Amazon. Ann Assoc Am Geogr 105(4):806–823. https://doi.org/10.1080/00045608.2015.1052337

Sadorsky P (2012) Energy consumption, output and trade in South America. Energy Econ 34(2):476–488. https://doi.org/10.1016/j.eneco.2011.12.008

Santos A, Silva C, Almeida P Jr, Rudke A, Melo S (2021) Deforestation drivers in the Brazilian Amazon: assessing new spatial predictors. J Environ Manage 294:113020. https://doi.org/10.1016/j.jenvman.2021.113020

Sauer S (2018) Soy expansion into the agricultural frontiers of the Brazilian Amazon: the agribusiness economy and its social and environmental conflicts. Land Use Policy 79:326–338. https://doi.org/10.1016/j.landusepol.2018.08.030

Silva D, Claudino L, Oliveira C, Matel A, Kubo R (2016) Extraction and development in the context of the Brazilian Amazon. Desenvolvimento e Meio Ambiente 38:557–577. https://doi.org/10.5380/dma.v38i0.44455

Silva A, Freitas M, Rodrigues R (2017) Methodological strategy of ecological economic zoning (EEZ) in the municipal scale: an academic exercise of political geography for the management of the territory. L'espace Politique 31. https://doi.org/10.4000/espacepolitique.4221

Silva C, Pessoa A, Carvalho N, Reis J, Anderson L, Aragão L (2021) The Brazilian Amazon deforestation rate in 2020 is the greatest of the decade. Nat Ecol Evol 5(2):144–145. https://doi.org/10.1038/s41559-020-01368-x

Siqueira L (2008) Environmental policies—for whom? Ponto de Vista. Ambiente e Sociedade 11(2). https://doi.org/10.1590/S1414-753X2008000200014

Siqueira-Gay J, Sánchez L (2021) The outbreak of illegal gold mining in the Brazilian Amazon boosts deforestation. Reg Environ Change 21:28. https://doi.org/10.1007/s10113-021-01761-7

Sonter L, Herrera D, Barret D, Galford G, Moran C, Soares Filho B (2017) Mining drives extensive deforestation in the Brazilian Amazon. Nat Commun 8:1013. https://doi.org/10.1038/s41467-017-00557-w

Szlafsztein C, Azevedo A, Alencar A (2016) Amazônia em Pauta No 6: Análise da implementação do Zoneamento Ecológico-Econômico (ZEE) sobre o uso e a ocupação do solo na Amazônia brasileira. IPAM, Brasília

Tanure T, Miyajima D, Magalhães A, Domingues E, Carvalho T (2020) The impacts of climate change on agricultural production, land use and economy of the legal Amazon region between 2030 and 2049. Economia 21(1):73–90. https://doi.org/10.1016/j.econ.2020.04.001

Timpe K, Kaplan D (2017) The changing hydrology of a dammed Amazon. Sci Adv 3(11):1–14. https://doi.org/10.1126/sciadv.1700611

Trindade S (2013) Uma floresta urbanizada? Legado e desdobramentos de uma teoria sobre o significado da cidade e do urbano na amazonia. Espaço Aberto 3(2):89–108

Triner G (2016) Mining and the state in Brazilian development: perspectives in economic and social history. Routledge, New York, 260 p. ISBN-13:978-1-84893-068-1

Vagula P, Uliana M, Arana A (2021) Panorama of ecological-economic zoning in Brazil. Geopauta 5(2):e8431. https://doi.org/10.22481/rg.v5i2.2021.e8431

Vasconcelos V, Hadad R, Martins P Jr (2013) Methodologies for integrated studies of natural resources: a discussion on ecological-economic zoning. Pesquisas Em Geociencias 40(1):21–30

Walker R (2011) The impact of Brazilian biofuel production on Amazonia. Ann Assoc Am Geogr 101(4):929–938. https://doi.org/10.1080/00045608.2011.568885

A Conceptual Alternative for Applying Conservation Banking to Offset Development Impacts in the Brazilian Amazon

Clarice Borges-Matos and Luis Enrique Sánchez

Abstract To compensate the residual impacts of development projects planned according to the Mitigation Hierarchy, biodiversity offsets have been increasingly used. However, they have widely recognized shortfalls; one of the most common is the time-lag between immediate impact and future delivery of offset benefits. Companies committed to offsetting their impacts either buy private land to restore or protect existing vegetation, contribute to the conservation of public protected areas, or acquire biodiversity credits from so-called "conservation bankings". Here, we conceptually explore a different avenue for offsetting in the Brazilian Amazonia through conservation banking, considering the social-environmental context of the region. The goal is to approach a realistic measure to be implemented in the world's largest forest. For this, we accounted for its intrinsic and interconnected characteristics, getting together historical and present features of the region to analyze them qualitatively and in parallel. As results, we came to the suggestion to drop the market-feature of conservation bankings, so that credits would be generated and used by the own developer who offsets. Also, it should be focused on restoration efforts, which should preferably be done by leveraging existing restoration initiatives in the region to be turned into conservation banking for future impacts. Local communities and traditional populations would then be included, opposed to a common credit market, which would likely exclude these people. This approach should enhance offset efficiency and partnerships across the Amazon, and potentialize restoration results and incorporation of social aspect.

Keywords Biodiversity offsets · Ecological compensation · Restoration initiatives · Ecological equivalence · Public policies

C. Borges-Matos (✉) · L. E. Sánchez
Escola Politécnica, Universidade de São Paulo (USP), Av. Prof. Mello Moraes, 2373, São Paulo 05508-900, Brazil
e-mail: clariceboma@ib.usp.br

W. Leal Filho et al. (eds.), *Amazon 2030—Sustainability Issues in the World's Largest Rainforest Region*, World Sustainability Series,
https://doi.org/10.1007/978-3-031-81465-5_25

1 Introduction

Under the Global Biodiversity Framework, established in December 2022 by the 15th Conference of Parties of the Convention on Biological Diversity[1], governments, companies and the civil society organizations faced the challenge known as "30 by 30" (Dinerstein et al. 2019). Countries agreed to join efforts to meet several biodiversity targets, and two of them aim to "ensure that by 2030 at least 30 per cent of areas of degraded terrestrial, inland water, and coastal and marine ecosystems are under effective restoration, in order to enhance biodiversity and ecosystem functions and services, ecological integrity and connectivity" (CBD 2022). Moreover, target 3 aims to "ensure and enable that by 2030 at least 30 per cent of terrestrial, inland water, and of coastal and marine areas, especially areas of particular importance for biodiversity and ecosystem functions and services, are effectively conserved and managed (…)."(CBD, 2022). Protection of remaining biodiversity as well as ecosystem restoration are both necessary if ecosystems are to be conserved and to provide benefits to people.

Therefore, making room for new development projects requires rigorous application of environmental impact assessment (EIA), demonstrating careful adherence to the Mitigation Hierarchy, a framework for project planning that underpins EIA. Genuine and demonstrable efforts should be made to avoid impacts (Bull et al. 2022), particularly on important biodiversity features such as critical habitats (Fernandes et al. 2023) and ecosystems valued by local communities due to the services they provide (Souza et al. 2021). Next, developers are required to minimize the impacts, for example by reducing the project footprint, or relocating structures to areas less important for conservation. The following step is to restore or rehabilitate what is possible on-site, that is, on the same site where the impact has happened. The impacts that still remain after these three steps are called "residual", and should be compensated or offset (BBOP 2012a). Figure 1 shows the steps of the Mitigation Hierarchy.

When the compensation is ecologically equivalent in type and amount to what will be lost due to impacts, it reaches no net loss of biodiversity, whereas a net gain would conceivably be achieved if compensation surpasses the initial biodiversity value lost in the affected areas (Fig. 1). Compensations that reach at least no net loss are called biodiversity offsets (BBOP 2012a). Two principles are non-negotiable to apply offsets to development projects (Souza et al. 2023): (i) offsets should only be considered after clear demonstrations that the other steps of the Mitigation Hierarchy have been rigorously applied, hence offsets are a measure of "last resort"; (ii) there should be acceptance that some potential losses are non-offsetable, i.e. they cannot be counterbalanced neither by equivalent nor by larger gains elsewhere. As a consequence, there should be no-go areas for development.

Although biodiversity offsets have been criticized for either facilitating actual biodiversity losses (Maron et al. 2012; Marshall et al. 2020) or being misused (Maron

[1] https://www.cbd.int/conferences/2021-2022.

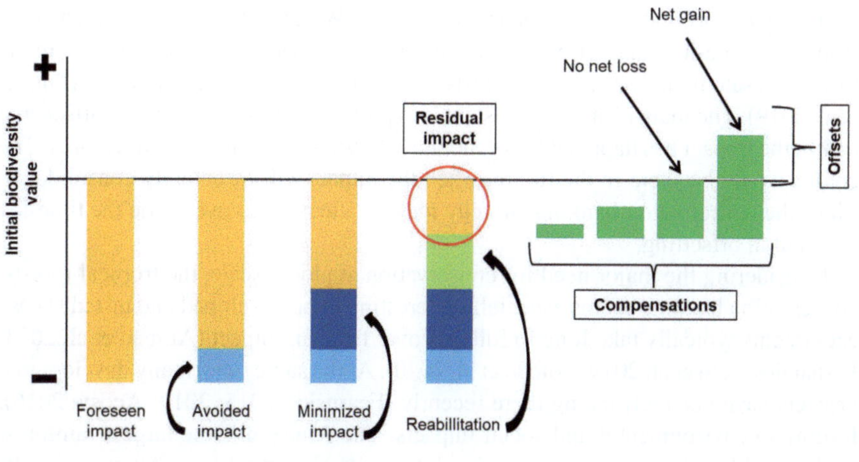

Fig. 1 A scheme presenting the Mitigation Hierarchy steps and the possibilities of compensation for the biodiversity losses due to the impact, including offsets. *Source* Adapted from BBOP (2012b)

et al. 2015), they have been increasingly adopted by legislation in several countries (Bull and Strange 2018; Maron et al. 2016) or are implicit in corporate policies claiming no net loss of biodiversity. For instance, in the Atlantic rainforest of Brazil, offsetting is mandatory (Souza and Sánchez 2018). In the Brazilian Amazon, however, there is no legal requirement for offsetting. Development projects can legally proceed even if they result in biodiversity net loss, except in some protected areas.

Even if implemented as a measure of last resort and respecting no-go areas, biodiversity offsets present several shortfalls that have been largely discussed in the related literature (Apostolopoulou and Adams 2017; Maron et al. 2016; zu Ermgassen et al. 2019, 2020). One of the most widely recognized is the time-lag between immediate impact and future delivery of offset benefits: the biodiversity loss occurs today, but its offset will take longer—or much longer—to be completed (Maron et al. 2012; Souza et al. 2023), especially when it is implemented through restoration (Maron et al. 2012). To implement their offsets, developers either buy private land for conservation purposes (protect biodiversity features or undertake ecological restoration—off-site restoration), contribute to public protected areas conservation, or acquire credits from so-called "conservation banking" (Briggs et al. 2009; USFWS 2011). The latter is the only measure that explicitly addresses the time-lag problem.

Also known as Mitigation Banking in the United States (Theis and Poesch 2022; USFWS 2011), Habitat Banking in England (Briggs et al. 2009), Biodiversity Banking in Australia and Germany (Bekessy et al. 2010; Mann 2015), and Biobanking also in Australia and in China (DECC 2007; Li et al. 2023), the term "Conservation Banking" first arose in 1995 in California, United States, in the context of wetlands protection and mitigation/compensation planning (USFWS 2011). It refers to a system in which biodiversity "credits" are generated by a provider, usually private landowners (Poudel et al. 2019), either by protecting a standing-vegetation area or by restoring the vegetation in an area (USFWS 2011). Then, these credits are made available in a market, where institutions or individuals could buy the credits to compensate their impacts, voluntarily or to comply with legal obligation (Poudel et al. 2019). The main idea of this system is to protect or restore a potential offset area *before* the impact has happened, to generate a "saving" of conservation features. The credits would be spent in the future, when the impact will be actually consolidated. Thus, the conservation banking strategy may be adequate to overcome the time-lag problem in offsetting.

Considering the major need for conservation at global scale, the tropical forests, conservation banking seems especially interesting to be developed and tested. These ecosystems typically take long to fully recover from an impact (Aragão et al. 2023; Fernandes Neto et al. 2019; Toledo et al. 2020). At the same time, many development projects have been advancing there recently (Fearnside 2015, 2019; Acosta 2016), leading to environmental and social impacts. The Amazon is the largest rainforest in the world and its conservation is of international interest (e.g. Amazon Fund[2], since the consequences of its deforestation ultimately affects the whole world. Brazil shelters the largest amount of the Amazon, so what happens there matters for the fate of the forest. Recent news is worrisome: increase in deforestation rates for the past years (PRODES[3]), illegal deforestation, fishing and logging largely unchecked in the region (Siqueira-Gay and Sánchez 2021; Trancoso 2021).

In this chapter, we explore the potential of conservation banking as an offset strategy in the Brazilian Amazon and suggest an approach for its implementation, considering some of the region's intrinsic characteristics.

2 Connecting Concepts with Current Features of the Brazilian Amazon

When trying to approach a solution to compensation of impacts from development projects in the Brazilian Amazon, several features are important to be put together. First, in the region there are constant pressures from human activities that cause environmental impact, including development projects that undergo EIA. The Amazon is the new agricultural frontier in Brazil (Marengo et al. 2022) and has been receiving several large development projects (Fearnside 2007, 2015). Highways, waterways, terminals, hydropower plants, and mines are examples of such projects that may not only have severe direct environmental impacts, but also indirect ones, as they induce other activities (Giljum et al. 2022; Siqueira-Gay et al. 2020). Both legal and illegal activities have led to deforestation and forest degradation. Illegal activities have particularly grown in the last years, since the last Brazilian federal government overlooked—and even encouraged—Amazon deforestation, using the fallacious argument that it was necessary to enhance agriculture and generate more income in the

[2] https://www.fundoamazonia.gov.br/en/home/.

[3] http://terrabrasilis.dpi.inpe.br/.

region (Fearnside 2018; Jakimow et al. 2023; Menezes and Barbosa 2021). Gold-digging and illegal timber extraction have especially increased in Brazil in the last years (Fernandes Neto et al. 2019; Siqueira-Gay and Sánchez 2021). Gold-digging has a particularly hideous consequence besides deforestation, as it severely pollutes the rivers and poisons all the living beings that depend on them—including humans (Pereira et al. 2023; Salomão et al. 2023).

These recent impacts and threats add to the difficulty of supervising environmental irregularities in the region. The Brazilian Amazon has continental dimensions, presents logistical difficulties to move among sites in most cases, and counts on few professionals able to perform supervision activities—proportionally to its size. Moreover, there are imprecisions in the region's land tenure: accordance among landowners lacks, official documentation lacks and there are conflicts related to land ownership, which may become violent (Prioli Duarte et al. 2023). In the last years, the government environmental agencies have suffered great pressure from the federal government, either by a gradual and forceful reduction in staff, or by implicit threatening of the staff inclined to punish environmental violators (Milhorance 2022; Story et al. 2023).

Therefore, applying any type of biodiversity-market system in the region seems unfeasible, since clear land tenure documentation along with good State control over the environmental policy compliance would be essential. The Environmental Reserve Quotas (CRA, Portuguese acronym) is a mechanism from the Brazilian Forest Code that allows landowners with surplus in their compulsory private reserves (a certain percentage of their land area; 80% in the Amazon) to financially negotiate this surplus with other landowners that have deficit in their reserves (Novo Código Florestal 2012; d'Albertas et al. 2021). This is an example of a biodiversity market (Brito 2017) similar to the conservation banking. However, it has not been implemented in the Brazilian Amazon so far (Giannichi et al. 2018; Hissa et al. 2023).

Tropical forests are systems of delicate ecological equilibrium and highly complex (Ali et al. 2019; McDowell 2018), so they often take long to recover after an impact (Fernandes Neto et al. 2019; Toledo et al. 2020). Even though tropical-forest trees may grow fast, ecosystem functions and services take longer to be reestablished, then full recovery is delayed (Aragão et al. 2023; Fernandes Neto et al. 2019; Toledo et al. 2020). This means that achieving NNL would require offset implementation to start earlier, so that regeneration or restoration could advance satisfactorily to actually compensate for the immediate impact. We specifically focus on restoration because in the Amazon the forest loss is already reaching 20% (Gambarini 2020), approaching the 20–25% tipping point, after which the forest could collapse (Amigo 2020; Lovejoy and Nobre 2018). Furthermore, vast areas have also been affected by forest degradation (Qin et al. 2021), hence requiring restoration to recover these areas' ecological functions is crucial. It is clear that only protecting standing-vegetation is not enough; there is a need for additionality in extension of native vegetation in offsetting: the Brazilian Amazon needs to restore the forest and its ecological functions.

The inclusion of the local peoples in conservation efforts, recognized as essential to make them efficient and lasting, is paramount (Mota et al. 2023). First of all,

Amazonian peoples (indigenous, riverside, and afro-descendent communities) have been there for centuries and thus have the ultimate right over the land (Andoke Andoke et al. 2023; Garber et al. 2023). Their historical integration with the forest and their traditional ecological knowledge regarding how to sustainably manage the forest allows them to be its first guardians (Andoke Andoke et al. 2023; Baragwanath et al. 2023; Garber et al. 2023). These communities highly depend on the forest, its products and ecological processes for living (Gavidia and Kemp 2017). They are the most directly affected by negative environmental impacts in the Amazon (Butt et al. 2023; Salomão et al. 2023). Incorporating local Amazonian communities in offsetting is important not only from an ethical and human rights perspective, as also to reach the conservation outcomes and sustainable use desired to the Amazon (Mota et al. 2023).

Despite the recent increase in the deforestation rates, probably due to the consequent need to recompose the forest, there are many restoration initiatives currently happening in the Brazilian Amazon (Gambarini, 2020). The Brazilian multilateral initiative called Alliance for Restoration in the Amazon[4] published a position paper in 2020 that mapped 2,773 restoration initiatives being implemented, summing up to an area of 113.5 thousand hectares (Gambarini, 2020). Despite the numerous initiatives, their total area is still small, when compared to the deforested area. Only 8% of the initiatives are more than 50 ha (Gambarini 2020), which elucidates the need for scaling up restoration in the region. Companies are responsible for only 5.6% of the initiatives, but their small percentage corresponds to 52% of the total restoration area (Gambarini 2020). The majority of the initiatives are implemented by civil society organizations (87.5%) and are agroforestry systems (59%) (Gambarini 2020), showing the region's vocation to engage people and local communities in its environmental processes. All this information can be used to integrate these different stakeholders involved with Amazon restoration to bring scale to this important effort.

3 An Alternative Approach for Biodiversity Offsetting in the Brazilian Amazon

Given the current deforestation rates and further threats to the Brazilian Amazon, the developers worried about compensating their impacts in the biome may be interested in making accurate estimations of the impacts and use it to implement their offsets in advance. The conservation banking strategy could serve well to this purpose, as it allows diminishing the time-lag between current losses and future gains, thus contributing to achieving no net loss of biodiversity in offset trades. This is especially important when a long period of regeneration is needed for the vegetation to recover, as is the case of the Amazon forest (Aragão et al. 2023; Fernandes Neto et al. 2019).

We suggest going beyond the conservation banking concept, though. Since there are numerous restoration initiatives already being implemented in the Brazilian

[4] https://aliancaamazonia.org.br/.

Amazon (Gambarini 2020), these could be leveraged and enhanced by developers, in a system that would count this enhancement as an offset to the developers' impacts. This accounting would be part of a conservation banking strategy, thus made in advance to the impact. The calculation system for accounting offsets should be further detailed in a next step. For now, we want to highlight the possible benefits that could emerge from a partnership between developers and civil society organizations, because this integration has the potential to scale up restoration in the Amazon and contribute to increase the social acceptance of offsets. Besides, considering the time the Amazon forest takes to fully recover after impacts (Aragão et al. 2023; Fernandes Neto et al. 2019), offsets focused on assisted restoration are paramount. Hence, we suggest offsets implemented in the Amazon should preferably be restoration offsets, which would be implemented in a conservation banking system, favoring ongoing restoration initiatives.

Applying bureaucratic systems successfully in the Brazilian Amazon, such as a biodiversity-credit market, appears to be a great challenge, due to the frequent illegal activities, which cause large deforestation, along with the historical difficulty to properly supervise the environment issues in the region (Barlow et al. 2016; Pereira 2022; Siqueira-Gay and Sánchez 2021; Trancoso 2021). If the market were to be implemented in an Amazonian scale, its chances to work would be even lower, since such a large territory would demand good legal compliance and supervision. There is also the problem of ambiguities in the land tenure (Hissa et al. 2023; Prioli Duarte et al. 2023), making it impossible to accurately identify the landowners and land-uses of each property, especially in a large scale. This is one of the most important problems in accomplishing forest restoration (Rakotonarivo et al. 2023), as it makes it hard to clearly define where a restoration project is and who will be responsible for its development. On the other hand, working in small scales in the Amazon may not be sufficient to solve the biome urgent needs; as mentioned above, restoration in the Amazon needs to be scaled-up, not down.

Importantly, Amazonian peoples must be involved in the region's conservation actions, as they are fundamental for these actions success and endurance across time (Nolte et al. 2013). Their involvement is important also because of ethical and human rights reasons (Andoke Andoke et al. 2023). However, a vast conservation market in the whole Amazon would likely exclude local peoples, especially small producers and family farmers, who may not be able to sell credits they would generate. To make them visible and to include them in the system of the biodiversity-credit market would be difficult and unlikely. The need of including Amazonian peoples in the offset equation, plus the land tenure issues, the recent high increase of deforestation and illegal activities, and the difficulty in formally supervising the region, leads us to our last suggestion in conservation banking: it should drop its market feature when applied to the Amazon, especially in Brazil. Thus, there would not exist an open credit market, it would function as a measure to offset own-generated impacts only—in a Single-user bank mode (Poudel et al. 2019).

Therefore, we suggest that a fit-for-purpose conservation banking initiative could be adequate as a tool to facilitate offsetting in the region. It should be exclusively or at least focused on restoration, aiming at recovering the Amazonian ecosystems, and enhancing the provision of ecosystem services. To overcome political difficulties caused by land tenure issues and illegal activities in the region, and also to widely include the Amazonian communities in the offsetting process, the developers should invest in conservation banking to offset their own future impact. To accomplish this goal faster and scale-up restoration, as well as to collaborate with local communities and facilitate social acceptance of offsets, the developers could leverage the restoration initiatives currently being implemented in the region, including the establishment and maintenance of agroforestry systems, and engage with stakeholders. The enhancements due to the collaborations could be accounted for as offsets for the impacts caused by developers' activities.

4 Conclusions

It is argued here that developers seeking to offset the impacts on biodiversity arising from their activities in the Amazon should consider the option of leveraging and scaling up regional ongoing restoration initiatives as part of a fit-for-purpose version of the conservation banking strategy, focused on restoration offsets. This will enable the developers to engage and collaborate with the local communities engaged in these restorations actions. The current credit market model of conservation banks implemented elsewhere should be abandoned. Instead, banks should be used by the developer who creates them to offset the impacts caused by their own projects. In addition to providing an alternative for developers intending to implement corporate biodiversity policies tuned to the current Global Biodiversity Framework, this conceptual approach should bring several benefits, such as: enhancing restoration and expanding restored areas, achieving greater social acceptance for offsets, enhancing partnerships with local communities and traditional peoples, and favoring human–environment integration.

It has been recognized that practical application of this strategy will certainly face many hurdles. Ensuring the long-term permanence of offsets, addressing the fact that many restoration initiatives do not aim at achieving ecological equivalence between biodiversity losses and gains (e.g. agroforestry systems), and developing appropriate offset metrics is some of these hurdles. Therefore, we argue that the proposed format of conservation banking should be tested by simulating its application to actual development projects in the Amazon. For that purpose, the extent and type of biodiversity losses and the potential of restoration initiatives to offset those losses should be carefully measured and adjusted to ensure ecological equivalence and long-term conservation goals.

References

Ali A, Lin S-L, He J-K, Kong F-M, Yu J-H, Jiang H-S (2019) Climate and soils determine aboveground biomass indirectly via species diversity and stand structural complexity in tropical forests. For Ecol Manage 432:823–831. https://doi.org/10.1016/j.foreco.2018.10.024

Amigo I (2020) When will the Amazon hit a tipping point? Nature 578(7796):505–507. https://doi.org/10.1038/d41586-020-00508-4

Andoke Andoke L, Arazi E, Castro Suárez H, Griffiths TF, Gutiérrez Sánchez E (2023) Amazonian visions of Visión Amazonía: indigenous peoples' perspectives on a forest conservation and climate programme in the Colombian Amazon. ORYX 57(3):335–349 (Scopus). https://doi.org/10.1017/S0030605322001636

Apostolopoulou E, Adams WM (2017) Biodiversity offsetting and conservation: reframing nature to save it. ORYX 51(1). https://doi.org/10.1017/S0030605315000782

Aragão JS, Elias F, das Neves EC, Guilherme FAG (2023) Forest recovery by direct seeding on the southern edge of the Brazilian Amazon. Restor Ecol:e14036. https://doi.org/10.1111/rec.14036

Baragwanath K, Bayi E, Shinde N (2023) Collective property rights lead to secondary forest growth in the Brazilian Amazon. Proc Natl Acad Sci US Am 120(22) (Scopus). https://doi.org/10.1073/pnas.2221346120

Barlow J, Lennox GD, Ferreira J, Berenguer E, Lees AC, Nally RM, Thomson JR, Ferraz SFDB, Louzada J, Oliveira VHF, Parry L, Ribeiro de Castro Solar R, Vieira ICG, Aragão LEOC, Begotti RA, Braga RF, Cardoso Jr TM, Souza Jr CMD, Gardner TA et al (2016) Anthropogenic disturbance in tropical forests can double biodiversity loss from deforestation. Nature 535(7610):144–147. https://doi.org/10.1038/nature18326

BBOP (2012a) BBOP glossary. https://www.forest-trends.org/publications/bbop-glossary/

BBOP (2012b) Standard on biodiversity offsets. https://www.forest-trends.org/publications/standard-on-biodiversity-offsets/

Bekessy SA, Wintle BA, Lindenmayer DB, Mccarthy MA, Colyvan M, Burgman MA, Possingham HP (2010) The biodiversity bank cannot be a lending bank. Conserv Lett 3(3):151–158. https://doi.org/10.1111/j.1755-263X.2010.00110.x

Briggs BDJ, Hill DA, Gillespie R (2009) Habitat banking—How it could work in the UK. J Nat Conserv 17(2):112–122. https://doi.org/10.1016/j.jnc.2008.12.006

Brito B (2017) Potential trajectories of the upcoming forest trading mechanism in Pará State, Brazilian Amazon. PLoS ONE 12(4) (Scopus). https://doi.org/10.1371/journal.pone.0174154

Bull JW, Sonter LJ, Gordon A, Maron M, Narain D, Reside AE, Sánchez LE, Shumway N, von Hase A, Quétier F (2022) Quantifying the "avoided" biodiversity impacts associated with economic development. Front Ecol Environ 20(6):370–378. https://doi.org/10.1002/fee.2496

Bull JW, Strange N (2018) The global extent of biodiversity offset implementation under no net loss policies. Nat Sustain 1(12) (Artigo 12). https://doi.org/10.1038/s41893-018-0176-z

Butt EW, Baker JCA, Bezerra FGS, von Randow C, Aguiar APD, Spracklen DV (2023) Amazon deforestation causes strong regional warming. Proc Natl Acad Sci 120(45):e2309123120. https://doi.org/10.1073/pnas.2309123120

CBD (2022) Decision adopted by the conference of the parties to the convention on biological diversity. 15/4. Kunming-Montreal Global Biodiversity Framework. https://www.cbd.int/doc/decisions/cop-15/cop-15-dec-04-en.pdf

d'Albertas F, González-Chaves A, Borges-Matos C, de Almeida Z, Paciello V, Maron M, Metzger JP (2021) Private reserves suffer from the same location biases of public protected areas. Biol Cons 261:109283. https://doi.org/10.1016/j.biocon.2021.109283

DECC (2007) Draft biobanking assessment methodology: under part 7A of the threatened species conservation act 1995. Dept. of Environment and Climate Change

Dinerstein E, Vynne C, Sala E, Joshi AR, Fernando S, Lovejoy TE, Mayorga J, Olson D, Asner GP, Baillie JEM, Burgess ND, Burkart K, Noss RF, Zhang YP, Baccini A, Birch T, Hahn N, Joppa LN, Wikramanayake E (2019) A global deal for nature: guiding principles, milestones, and targets. Sci Adv 5(4):eaaw2869. https://doi.org/10.1126/sciadv.aaw2869

Fearnside P (2015) Hidrelétricas na Amazônia: Impactos ambientais e sociais na tomada de decisões sobre grandes obras (vol 1). Editora INPA. https://repositorio.inpa.gov.br/bitstream/1/4684/1/hidreletricas_na_Amazonia_v1.pdf

Fearnside P (2018) Why Brazil's new president poses an unprecedented threat to the Amazon. Yale Environment 360, Yale Sch Environ 11

Fearnside PM (2007) Brazil's Cuiabá-Santarém (BR-163) highway: the environmental cost of paving a soybean corridor through the Amazon. Environ Manage 39(5):601–614. https://doi.org/10.1007/s00267-006-0149-2

Fearnside PM (2019) Review of: Shoemaker & Robichaud (eds): dead in the water—global lessons from the world bank's model hydropower project in Laos. Biol Conserv 233:239–240. https://doi.org/10.1016/j.biocon.2019.03.008

Fernandes Neto JG, Costa FRC, Williamson GB, Mesquita RCG (2019) Alternative functional trajectories along succession after different land uses in central Amazonia. J Appl Ecol 56(11):2472–2481. https://doi.org/10.1111/1365-2664.13484

Fernandes TN, Dos Santos FMG, Gontijo FD, Da Silva Filho JA, Castilho AF, Sánchez LE (2023) Mainstreaming flora conservation strategies into the mitigation hierarchy to strengthen environmental impact assessment. Environ Manage 71(2):483–493. https://doi.org/10.1007/s00267-022-01756-y

Gambarini A (2020) Panorama e caminhos para a restauração de paisagens florestais na amazônia

Garber PA, Estrada A, Klain V, Bicca-Marques JC (2023) An urgent call-to-action to protect the nonhuman primates and indigenous peoples of the Brazilian Amazon. Am J Primatol (Scopus). https://doi.org/10.1002/ajp.23523

Gavidia M, Kemp D (2017) Company-community relations in the mining context: a relational justice perspective

Giannichi ML, Dallimer M, Baker TR, Mitchell G, Bernasconi P, Ziv G (2018) Divergent landowners' expectations may hinder the uptake of a forest certificate trading scheme: expectations about a conservation scheme. Conserv Lett 11(3):e12409. https://doi.org/10.1111/conl.12409

Giljum S, Maus V, Kuschnig N, Luckeneder S, Tost M, Sonter LJ, Bebbington AJ (2022) A pantropical assessment of deforestation caused by industrial mining. Proc Natl Acad Sci 119(38):e2118273119. https://doi.org/10.1073/pnas.2118273119

Hissa L, Monteiro D, Santos EBD (2023) Código Florestal na Amazônia e regularização fundiária (O Código Florestal nos Biomas: desafios e soluções). Observatório do Código Florestal

Jakimow B, Baumann M, Salomão C, Bendini H, Hostert P (2023) Deforestation and agricultural fires in South-West Pará, Brazil, under political changes from 2014 to 2020. J Land Use Sci 18(1):176–195 (Scopus). https://doi.org/10.1080/1747423X.2023.2195420

Li J, Zhang H, Wang N (2023) Paths and institutional arrangements for the value realization of biodiversity products: experiences and inspirations of foreign bio-banking. Shengtai Xuebao 43(1):198–207 (Scopus). https://doi.org/10.5846/stxb202112293701

Lei da Mata Atlântica, 11428 (2006)

Lei de Proteção da Vegetação Nativa, 12.651, 39 (2012) https://www.planalto.gov.br/ccivil_03/_ato2011-2014/2012/lei/l12651.htm

Lovejoy TE, Nobre C (2018) Amazon tipping point. Sci Adv 4(2) (eaat2340). https://doi.org/10.1126/sciadv.aat2340

Mann C (2015) Strategies for sustainable policy design: Constructive assessment of biodiversity offsets and banking. Ecosyst Serv 16:266–274. https://doi.org/10.1016/j.ecoser.2015.07.001

Marengo JA, Jimenez JC, Espinoza JC, Cunha AP, Aragão LEO (2022) Increased climate pressure on the agricultural frontier in the Eastern Amazonia–Cerrado transition zone. Sci Rep 12(1) (Artigo 1). https://doi.org/10.1038/s41598-021-04241-4

Maron M, Gordon A, Mackey BG, Possingham HP, Watson JEM (2015) Conservation: stop misuse of biodiversity offsets. Nature 523(7561) (Artigo 7561). https://doi.org/10.1038/523401a

Maron M, Hobbs RJ, Moilanen A, Matthews JW, Christie K, Gardner TA, Keith DA, Lindenmayer DB, McAlpine CA (2012) Faustian bargains? Restoration realities in the context of biodiversity offset policies. Biol Cons 155:141–148. https://doi.org/10.1016/j.biocon.2012.06.003
Maron M, Ives CD, Kujala H, Bull JW, Maseyk FJF, Bekessy S, Gordon A, Watson JEM, Lentini PE, Gibbons P, Possingham HP, Hobbs RJ, Keith DA, Wintle BA, Evans MC (2016) Taming a wicked problem: resolving controversies in biodiversity offsetting. Bioscience 66(6):489–498. https://doi.org/10.1093/biosci/biw038
Marshall E, Wintle BA, Southwell D, Kujala H (2020) What are we measuring? A review of metrics used to describe biodiversity in offsets exchanges. Biol Cons 241:108250. https://doi.org/10.1016/j.biocon.2019.108250
McDowell NG (2018) Deriving pattern from complexity in the processes underlying tropical forest drought impacts. New Phytol 219(3):841–844. https://doi.org/10.1111/nph.15341
Menezes RG, Barbosa R Jr (2021) Environmental governance under Bolsonaro: dismantling institutions, curtailing participation, delegitimising opposition. Zeitschrift Für Vergleichende Politikwissenschaft 15(2):229–247. https://doi.org/10.1007/s12286-021-00491-8
Milhorance C (2022) Policy dismantling and democratic regression in Brazil under Bolsonaro: coalition politics, ideas, and underlying discourses. Rev Policy Res 39(6):752–770. https://doi.org/10.1111/ropr.12502
Mota AGSDS, Farias MDL, Silva AAD, Vieira TA, Alves HDS, Silva ADSLD (2023) Management of the Tapajós-Arapiuns extractive reserve: limits and possibilities in the perception of its counsilors. Ambiente & Sociedade 26:e02361. https://doi.org/10.1590/1809-4422asoc20200236r1vu2023l2oa
Nolte C, Agrawal A, Silvius KM, Soares-Filho BS (2013) Governance regime and location influence avoided deforestation success of protected areas in the Brazilian Amazon. Proc Natl Acad Sci 110(13):4956–4961. https://doi.org/10.1073/pnas.1214786110
Pereira A (2022) Populism and dismantling of state capabilities in the Brazilian environmental sector. SSRN Electron J. https://doi.org/10.2139/ssrn.4125350
Pereira WVDS, Ramos SJ, Melo LCA, Dias YN, Martins GC, Ferreira LCG, Fernandes AR (2023) Human and environmental exposure to rare earth elements in gold mining areas in the northeastern Amazon. Chemosphere 340 (Scopus). https://doi.org/10.1016/j.chemosphere.2023.139824
Poudel J, Zhang D, Simon B (2019) Habitat conservation banking trends in the United States. Biodivers Conserv (Scopus). https://doi.org/10.1007/s10531-019-01747-2
Prioli Duarte D, Peres CA, Perdomo EFC, Guizar-Coutiño A, Nelson BW (2023) Reducing natural vegetation loss in Amazonia critically depends on the formal recognition of indigenous lands. Biol Conserv 279 (Scopus). https://doi.org/10.1016/j.biocon.2023.109936
Qin Y, Xiao X, Wigneron JP, Ciais P, Brandt M, Fan L, Li X, Crowell S, Wu X, Doughty R, Zhang Y, Liu F, Sitch S, Moore B (2021) Carbon loss from forest degradation exceeds that from deforestation in the Brazilian Amazon. Nat Clim Change 11(5) (Artigo 5). https://doi.org/10.1038/s41558-021-01026-5
Rakotonarivo OS, Rakotoarisoa M, Rajaonarivelo HM, Raharijaona S, Jones JPG, Hockley N (2023) Resolving land tenure security is essential to deliver forest restoration. Commun Earth Environ 4(1) (Artigo 1). https://doi.org/10.1038/s43247-023-00847-w
Salomão GN, Dall'Agnol R, Sahoo PK, Almeida GSD, Amarante RT, Zeferino LB, Lopes JPN, Souza Filho PWME, Costa NYMD, Guimarães JTF, Silva MSD, Martins GC, Teixeira MFB, Marques ED, Angélica RS, Araújo WEO (2023) Changes in the surface water quality of a tropical watershed in the southeastern amazon due to the environmental impacts of artisanal mining. Environ Pollut 329 (Scopus). https://doi.org/10.1016/j.envpol.2023.121595
Siqueira-Gay J, Sánchez LE (2021) The outbreak of illegal gold mining in the Brazilian Amazon boosts deforestation. Reg Environ Change 21(2):28. https://doi.org/10.1007/s10113-021-01761-7

Siqueira-Gay J, Sonter LJ, Sánchez LE (2020) Exploring potential impacts of mining on forest loss and fragmentation within a biodiverse region of Brazil's northeastern Amazon. Resour Policy 67:101662. https://doi.org/10.1016/j.resourpol.2020.101662
Souza BA, Rosa JCS, Campos PBR, Sánchez LE (2023) Evaluating the potential of biodiversity offsets to achieve net gain. Conserv Biol:e14094. https://doi.org/10.1111/cobi.14094
Souza BA, Rosa JCS, Siqueira-Gay J, Sánchez LE (2021) Mitigating impacts on ecosystem services requires more than biodiversity offsets. Land Use Policy 105:105393. https://doi.org/10.1016/j.landusepol.2021.105393
Souza BA, Sánchez LE (2018) Biodiversity offsets in limestone quarries: Investigation of practices in Brazil. Resour Policy 57:213–223. https://doi.org/10.1016/j.resourpol.2018.03.007
Story J, Lotta G, Tavares GM (2023) (Mis)Led by an outsider: abusive supervision, disengagement, and silence in politicized bureaucracies. J Public Admin Res Theory 33(4):549–562. https://doi.org/10.1093/jopart/muad004
Theis S, Poesch M (2022) Current capacity, bottlenecks, and future projections for offsetting habitat loss using mitigation and conservation banking in the United States. J Nat Conserv 67 (Scopus). https://doi.org/10.1016/j.jnc.2022.126159
Toledo RM, Perring MP, Verheyen K, Martini AMZ, Ferreira MP, Santos RF (2020) Restoring tropical forest composition is more difficult, but recovering tree-cover is faster, when neighbouring forests are young. Landsc Ecol 35(6):1403–1416. https://doi.org/10.1007/s10980-020-01023-7
Trancoso R (2021) Changing Amazon deforestation patterns: urgent need to restore command and control policies and market interventions. Environ Res Lett 16(4):041004. https://doi.org/10.1088/1748-9326/abee4c
USFWS (United States Fish and Wildlife Service) (2011) Conservation banking, incentives for stewardship 2
zu Ermgassen SOSE, Baker J, Griffiths RA, Strange N, Struebig MJ, Bull JW (2019) The ecological outcomes of biodiversity offsets under "no net loss" policies: a global review. Conserv Lett. https://doi.org/10.1111/conl.12664
zu Ermgassen SOSE, Maron M, Corlet Walker CM, Gordon A, Simmonds JS, Strange N, Robertson M, Bull JW (2020) The hidden biodiversity risks of increasing flexibility in biodiversity offset trades. Biol Cons 252:108861. https://doi.org/10.1016/j.biocon.2020.108861

Amazon Riverine Communities: Practice of Family Farming and the Use of Natural Resources in the Tapajós River Basin

Athila Rafael Rego Reis, Arthur Daniel Lopes Frota, Yara Fernanda Ferreira Henn, Raissa Vasconcelos Rego, and Iracenir Andrade dos Santos

Abstract This chapter conducts a diagnosis of how riparian families in the Tapajós region engage in sustainable exploration of the Amazon rainforest. Within this context, there is an identification of plant species used for their sustenance, the types of agricultural management adopted, and the socio-economic significance of agriculture for families belonging to traditional communities in the region.

Keywords Food security · Family farming · Amazon communities · Tapajós river basin

1 Introduction

Riparian communities play a significant socio-environmental role in the Amazon region (Viana et al. 2016). These communities, characterized by their traditional lifestyle and close connection with natural resources (Silva et al. 2019), play a fundamental role in maintaining biodiversity, environmental balance, and preserving

A. R. R. Reis (✉) · A. D. L. Frota · Y. F. F. Henn · R. V. Rego · I. A. dos Santos
Federal University of Western Pará, Santarém, Brazil
e-mail: reisathila@gmail.com

A. D. L. Frota
e-mail: arthurlopesfrota@gmail.com

Y. F. F. Henn
e-mail: henn882.yh@gmail.com

R. V. Rego
e-mail: raissastm20@gmail.com

I. A. dos Santos
e-mail: iracenir.andrade@ufopa.edu.br

W. Leal Filho et al. (eds.), *Amazon 2030—Sustainability Issues in the World's Largest Rainforest Region*, World Sustainability Series,
https://doi.org/10.1007/978-3-031-81465-5_26

ancestral knowledge (Diegues 2019). The practice of family farming ensures subsistence and livelihood for their families while also promoting the conscious use of natural resources (Martins et al. 2012).

Family farming in riparian communities is a system adapted to the unique conditions of the Amazon (Fraxe et al. 2011). Land use involves traditional agricultural practices that combine the cultivation of staple foods such as cassava, corn, and beans with agroforestry techniques that allow the coexistence of agricultural crops and economically and ecologically valuable trees (Gomes 2005). These sustainable practices not only provide food for riparian families (Hecktheuer et al. 2018) but also keep the forest standing and contribute to biodiversity conservation (Costa and Lucas 2022). However, population growth, land pressure, and the expansion of economic activities such as mining and commercial agriculture have led to deforestation and the degradation of forested areas (Laurance et al. 2002). This poses a direct threat to the subsistence systems of riparian communities and the health of local ecosystems (Vasconcelos et al. 2017).

The Tapajós River basin, with an approximate length of about 146 km in the Amazon Region (Santos et al. 2015), extends across the states of Mato Grosso, Pará, and a portion of Amazonas. It is characterized by significant landscape variability that influences, among other factors, the climate, rainfall, and hydrodynamics of the river (Araujo et al. 2015). The diversity in the landscape structure of the Tapajós River basin fosters a high diversity of terrestrial and aquatic fauna and flora, with high numbers of endemic species (Sousa 2013). In addition to environmental diversity, there is social diversity marked by the presence of indigenous peoples, quilombola communities, and riparian communities (Rocha et al. 2022a, b).

Family farmers cultivate a diversity of crops, which ensures a broad and varied supply of food for subsistence and provides greater stability to the production system (Santos et al. 2015), as the fulfillment of basic food needs for families does not depend on commercialization (Santos 2009). The sustainable use of forest resources brings socio-economic improvement to the communities and contributes to the conservation of the forest and managed species (Almeida and Gama 2019). Sustainable approaches such as agroforestry systems and the use of appropriate management techniques can contribute to reconciling agricultural production with the conservation of the Amazon, promoting food security, and environmental protection (Espada and Sobrinho 2015).

In this context, the present study aimed to conduct a diagnosis of family farming in the riparian communities of the Tapajós River Basin. Thus, identifying the main cultivated species, analyzing the various agricultural management methods adopted, and ultimately assessing the socio-economic relevance of agriculture for families belonging to the traditional communities in the region.

2 Materials and Methods

Sampling was conducted in the middle and lower regions of the Tapajós River, in the eastern part of the Amazon basin (Fig. 1), covering an expanse of three municipalities in the western state of Pará, Brazil. The Tapajós River is a clear-water river (Silva et al. 2013), ranking as the fifth-largest tributary of the Amazon, with a drainage area of 493,200 km^2 (Santos et al. 2015; Sousa 2022). Data collection took place in eight communities along the Tapajós River, including the communities of Nazaré and Barreiras in the municipality of Itaituba; Cassuepá, Apacê, and Cametá in the municipality of Aveiro; São Luís do Tapajós in the municipality of Itaituba; and Surucuá and Parauá (in the municipality of Santarém). All data collections were carried out between July 2021 and February 2022.

Interviews were conducted using semi-structured questionnaires, which included both open-ended and closed-ended questions (Manzato and Santos 2012). These interviews followed a predetermined script with the aim of gathering maximum information about the informant's knowledge of the environment and daily life (Silva 2019). The snowball method was employed, involving the identification of individuals directly involved in agriculture (Ferreira 2018). At the end of each interview, the informant was asked to recommend one or more individuals from the community engaged in daily agricultural activities (Oliveira and Buhler 2016).

Exploratory and descriptive statistical analyses were conducted to assess the importance of sets of explanatory variables in understanding the structure of agriculture in the region. Additionally, the analyses aimed to describe how the management process occurred and the roles of farmers in terms of gender, river section, age distribution frequency, ethnicity, educational level, and types of activities conducted. All analyses were performed using the R program (R Development Core Team 2008, Paludo and Costabeber 2012).

3 Results

3.1 Population Traits

In relation to the 141 applied questionnaires, they were used to assess agricultural practices in eight communities in the Middle and Lower Tapajós region. Of these, 34.75% were completed by women, while 65.25% were completed by men. Within the interviewed population, the most frequent age group was individuals between 35 and 54 years old (Fig. 2). Additionally, it was observed that the communities of São Luís do Tapajós and Suruacá had a more elderly population.

A diverse ethnic composition was observed within the communities, with approximately 64.54% identifying themselves as "pardos", which literally means 'brown.' The term refers to a mixed race in Brazil, representing ethnically varied individuals. Indigenous people made up about 12.77% of the sample, while individuals

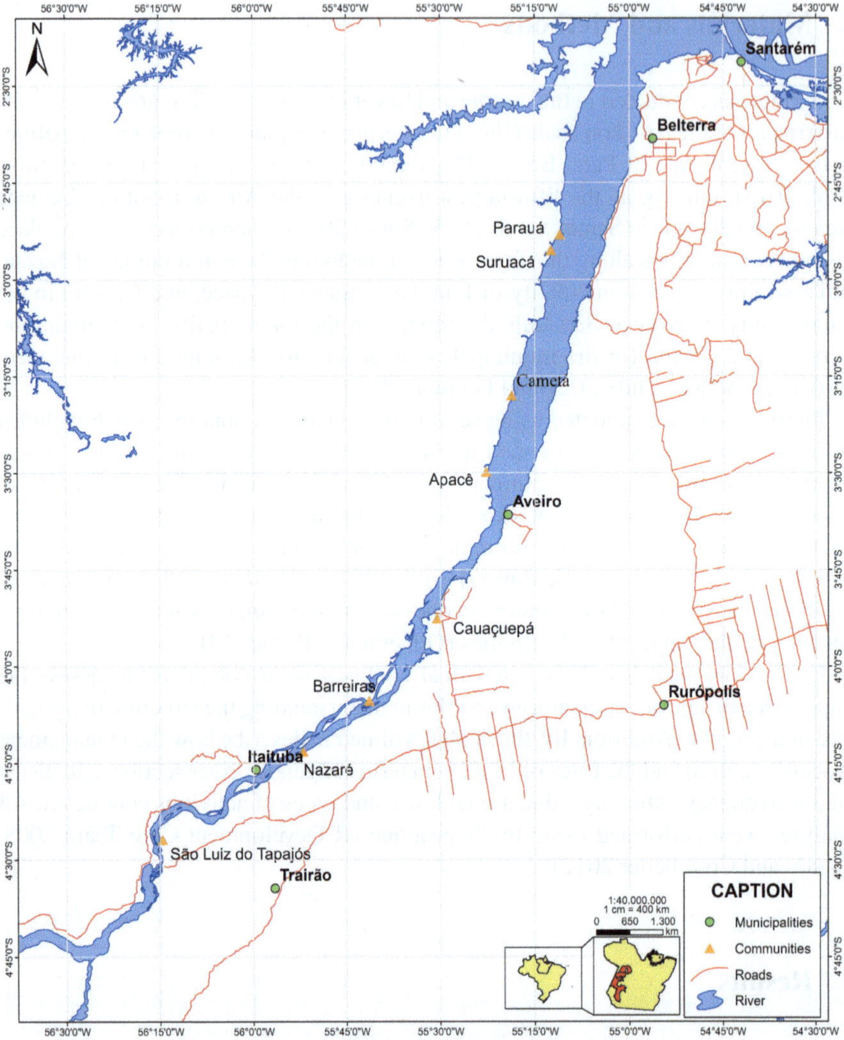

Fig. 1 Location map of data collection: Tapajós River (middle and lower reaches) in the state of Pará, eastern Amazon Basin, Brazil. *Source* Author oneself

who identified as black and white accounted for approximately 11.35% and 6.38%, respectively. A small portion of the interviewed population, approximately 0.71%, identified themselves as 'cabocla,' a term used to describe individuals of mixed indigenous and European descent. Additionally, it is important to note that 4.26% of the interviewed individuals chose not to respond to the question about their ethnic origin.

The population in the communities has a low level of education, with the majority being individuals with only incomplete primary education or secondary education

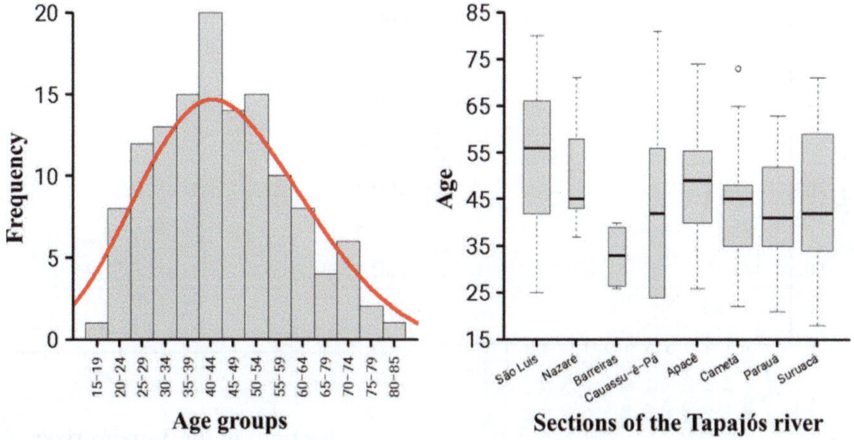

Fig. 2 Age structure **a** of the population distributed by communities **b** in the lower and middle Tapajós Region, Pará, Brazil

(Fig. 3). Despite the existence of a society open to different forms of education, it is still possible to identify a significant portion of the population who are deprived of exercising their rights (Fig. 3).

The source of income in these communities is varied, but for the most part, they have very low income, ranging from \$101.82 to \$202.22 (Fig. 4). Fishing and agriculture stand out as the main sources of income in all communities. In these communities, various types of waste are generated; however, there is no functioning selective waste

Fig. 3 Education level of the population in the middle and lower Tapajós Communities, Pará, Brazil

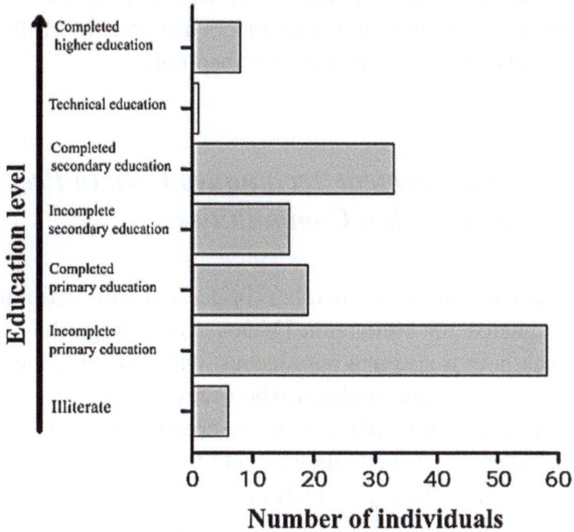

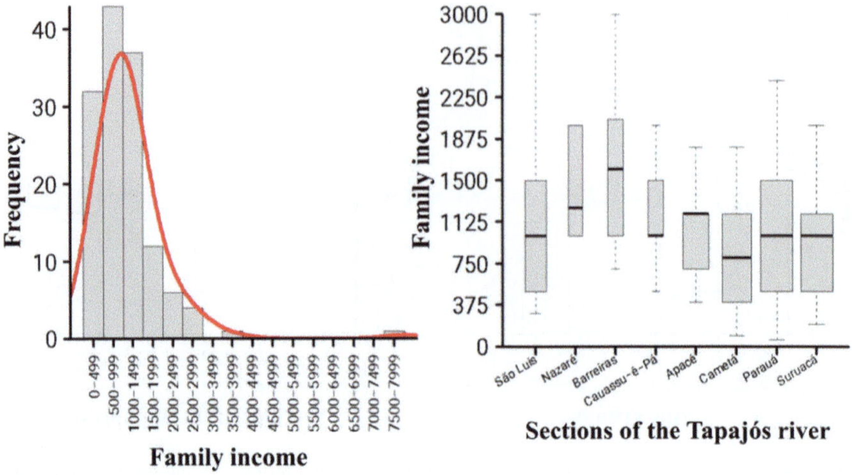

Fig. 4 Income groups of families according to communities in the middle and lower Tapajós Region, Pará, Brazil

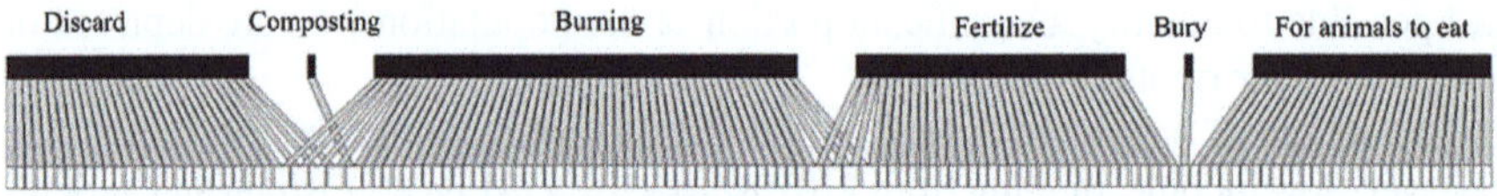

Fig. 5 Methods of waste disposal in the middle and lower Tapajós Communities, Pará, Brazil

collection system in many places, leading people to burn their trash (Fig. 5). Nevertheless, some community members use organic waste for fertilizing plants, feeding animals, and even for improper disposal.

3.2 *Management Techniques Used to Prepare Planting Areas in Riparian Communities*

Riparian community members prefer to use the slash-and-burn method to clear areas designated for cultivation (Matos 2016). Regarding crop management, producers adopt both traditional approaches, which involve the use of various types of pesticides, and organic methods, where they only engage in weeding to reduce competition from regenerative plants with the crop (Hora et al. 2014; Quitério 2022). However, the diversity of methods used and the reports highlight the lack of rural technical assistance in the region (Silva et al. 2014).

The diversity of agricultural methods used by rural communities reflects their local adaptability and traditions (Baiardi and Alencar 2014). However, the lack of agricultural technical assistance poses significant challenges (Sablayrolles and

Azevedo 2022). The absence of specialized guidance limits farmers' ability to optimize their practices, affecting productivity and sustainability (Oliveira and Bühler 2016). Investments in technical assistance are essential to promoting sustainable agricultural practices and improving the quality of life for rural communities (Souza et al. 2016).

Promoting rural technical assistance requires a comprehensive approach, considering not only technical aspects but also social, cultural, and environmental factors (Silva and Caporal 2016). Close collaboration between farmers, research institutions, and government entities is essential to developing solutions tailored to local realities (Pereira and Castro 2020). Training and awareness-building among farmers play a crucial role in adopting sustainable agricultural practices (Abreu et al. 2023). This integrated approach not only drives economic progress but also strengthens the resilience of rural communities in the face of constantly evolving challenges (Grisa 2010).

3.3 Agroforestry Systems

Agroforestry systems are potential techniques for the continuous use of land in the Brazilian Amazon (Paludo and Costabeber 2012). In the communities of the lower and middle Tapajós River, the most commonly used production systems in family farming are mixed-crop swiddens, with analyzed swiddens and home gardens with intercropping. The main crops planted were *Manihot esculenta* (37.28%), *Zea mays* (10.78%), *Cucumis anguria* (6.42%), and *Phaseolus spp.* (4.66%). Agroforestry systems have the potential to serve as a valid alternative for the sustainable use of the Amazon region, particularly in areas with secondary vegetation resulting from the predatory exploitation of native forests (De Andrade 2012) (Fig. 6).

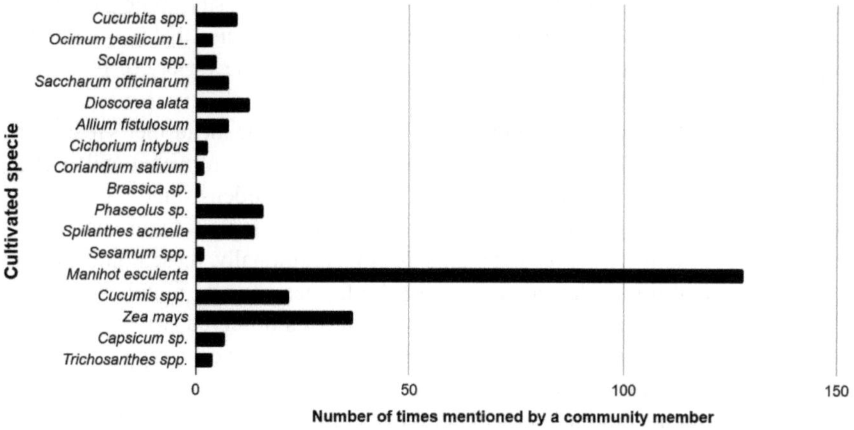

Fig. 6 Diversity of cultivated species in community swidden fields

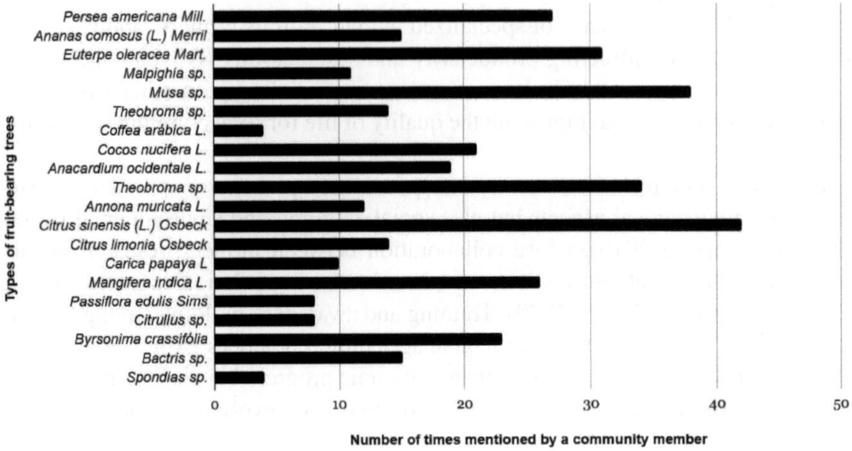

Fig. 7 Diversity of fruit trees produced in backyards and swidden fields in the communities

Cassava flour is a staple food in the lower and middle Tapajós region, as well as in much of the Amazon region (Lima et al. 2020). Its versatility in local cuisine makes it a crucial choice for the subsistence of these communities (Souza et al. 2020). Cassava is transformed into flour, tapioca, and various traditional dishes that form the basis of their diets (Reis 2011). Furthermore, its cultivation is adapted to local conditions, making it a crop of significant economic importance (Castro et al. 2009). The relationship between cassava and riparian communities is an example of how natural resources are sustainably exploited to meet the dietary and cultural needs of these populations (Queiroz and Peralta 2006) (Fig. 7).

Family farming demands alternatives that consider the diversity of local ecosystems and address the economic, environmental, and sociocultural dimensions of sustainability (Paludo and Costabeber 2012). Native species are very common in agroforestry home gardens (Pereira and Neto 2015). Agroforestry home gardens are important as a source of agrobiodiversity conservation, species, habitat, and landscape protection (Kothalawala et al. 2013), obtaining a wide variety of plants in a single area (Duarte and Pasa 2016).

Agroforestry home gardens represent the area surrounding the farmer's residence, where a variety of fruit species, grains, and vegetables are cultivated (Silva et al. 2022). Despite their relatively limited size, these home gardens stand out as Agroforestry Systems (AFS) due to the farmer's ability to dedicate continuous and diligent care to these systems (Almeida and Gama 2014). Additionally, these home gardens play an important role as leisure spaces for the farmer's family members (Siviero et al. 2011). Among the 21 agroforestry home gardens observed, which consist of well-maintained areas near the agricultural producer's residence, a high diversity of plant species cultivation is evident (Fig. 8). These crops are primarily intended for direct consumption by the farmer's own family (Araújo and Ribeiro 2017).

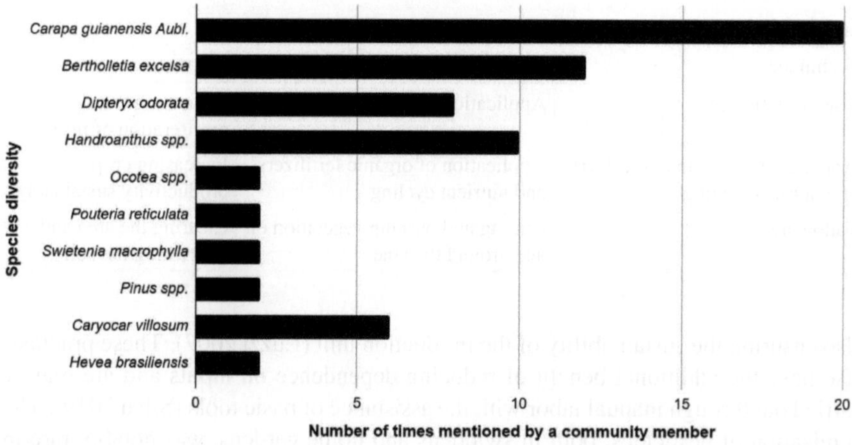

Fig. 8 Tree diversity planted in swiddens in communities of the middle and lower Tapajós, PA, Brazil

Regarding fruit species, the frequent presence of *Musa* spp. (38%), *Citrus* spp. (42%), and *Theobroma grandiflorum* (34%) stands out. These fruits, along with other varieties found in the communities, are grown in mixed cropping systems and play a crucial role in complementing the diets of family farmers (Fraxe et al. 2011). In general, these home gardens serve the function of ensuring access to foods that are sources of essential nutrients, such as proteins, carbohydrates, and minerals, significantly contributing to mitigating the nutritional insecurity faced by these families (Guimarães 2015).

3.4 Management Techniques for Swiddens and Home Gardens

Farmers employ a variety of management techniques in agroforestry systems, aiming to reduce weed propagation and the impact of pests, such as ants (Castro et al. 2009). Through these practices, producers aim to ensure a conducive environment for the development of crops in their swiddens and home gardens, using a variety of management techniques (Table 1).

In swiddens, the predominant technique for preparing cultivation areas is the 'cut and burn' method with fallow (Pedroso et al. 2008). In this process, the first step involves clearing the surrounding vegetation and invasive grass from the region (Moura 2012). Next, this biomass is burned with the purpose of returning nutrients to the soil, allowing it to rest until the beginning of a new cultivation cycle (Gonçalves 2005).

The implementation of techniques involving pruning, weeding, and the production of fertilizers aims primarily to increase crop productivity in swiddens while

Table 1 Main management techniques used by producers

Technique	Method of execution	Purpose
Use of pesticides	Application of pesticides on crops	Preventing the proliferation of insects
Application of organic fertilizers and nutrient cycling	Application of organic fertilizers and nutrient cycling	Increasing crop productivity sustainably
Fallowing	Cutting and burning vegetation on and around the land	Clearing the area and nourishing the soil

also ensuring the sustainability of the production unit (Luzzi 2007). These practices also have the additional benefit of reducing dependence on inputs and are mainly carried out through manual labor with the assistance of basic tools (Silva 2019). The application of pesticides, both in swiddens and home gardens, was notably rare in the examined production units. This highlights the emphasis placed on soil conservation in a way that does not harm the environment or negatively affect the cultivated products (Santos 2009).

The main result of the study highlights the gender dynamics in agricultural management practices within riparian communities along the Tapajós River Basin in the Brazilian Amazon. The focus is on the significant contribution of women in carrying out agricultural tasks, particularly in agroforestry systems. Additionally, the study emphasizes the occasional involvement of young individuals in these tasks, especially when male community members are predominantly engaged in fishing activities throughout the day. These findings underscore the importance of understanding gender dynamics in agricultural practices for a more comprehensive grasp of sustainability and subsistence in these riparian communities.

The main source of income in these communities comes from fishing and agriculture. Sustainable cultivation in agroforestry home gardens plays a crucial role in meeting the dietary and cultural needs of riverine communities, with the use of agroforestry gardens near residences exhibiting a high diversity of plant species. The interconnection between sustainable agricultural practices, food security, and socioeconomic conditions underscores the importance of integrated approaches to promote sustainable development in these riverine communities, including investments in education, technical assistance, and environmental preservation.

4 Conclusion

The results highlight that agriculture plays an extremely important role in riparian communities of the Middle and Lower Tapajós, being fundamental to ensure the food security of these populations. It is through agriculture that families can obtain high-quality, safe, and diverse foods, enriching their diets with essential nutrients. In addition to being the foundation of families' diets, agriculture also significantly contributes to income generation and employment opportunities for family members, playing a crucial role in alleviating hunger in many low-income communities.

Consequently, it is imperative that future initiatives involving natural resource utilization by communities give due consideration to crucial issues such as:

1. Implementing effective rural technical assistance programs with communities to promote sustainable production in the Amazon.
2. Promoting the adoption of agroforestry home gardens and swiddens to conserve the forest, enhance biodiversity, and provide environmental services.
3. Integrating technologies to optimize production, improve health, quality of life, and elevate crop quality.
4. May these productive units continue to play a fundamental role, always correlated with the Sustainable Development Goals (SDGs).

Acknowledgements The authors would like to thank The Nature Conservancy of Brazil (TNC) for their partnership through the project 'Community-Based Conservation in Freshwater Ecosystems: Tapajós River. Thanks for the financial support from the Coordination for the Improvement of Higher Education Personnel (CAPES)—Brazil.

References

Abreu FAK, do Amaral AB, Galvão JR, da Silva RFF, Neves LIA, de Castro VL, dos Santos Galate R et al (2023) Challenges arising from family farming in the municipality of Acará/PA and region. Seven Publisher. https://doi.org/10.56238/futuroeducpesqutrans-041

Almeida LSD, Gama JRV (2014) Agroforestry home gardens: structure, floristic composition, and socio-environmental aspects in a rural settlement area in the Brazilian Amazon. Ciência Florestal 24:1041–1053. https://doi.org/10.1590/1980-509820142404023

Almeida EC, Gama JRV (2019) Socio-ecological system of forest resource use by communities in the Santarém Region, Pará. Ibero-Am J Environ Sci 10(5):231–242. https://doi.org/10.6008/cbpc2179-6858.2019.005.0020

Araujo AM, Ribeiro EM (2017) Vale markets: the destination of productive surpluses in open-air markets in the Jequitinhonha, Minas Gerais. RARR Admin J 7(2):221–244. https://doi.org/10.18227/2237-8057rarr.v7i2.4521

Araujo I, Lima AM, Santos C (2015) Hydrogeomorphological characterization of the landscape components of the Tapajós River basin. Biosph Encycl 11(22). https://doi.org/10.18677/Enciclopedia_Biosfera_2015_046

Baiardi A, Alencar CMMD (2014) Family farming, its academic interest, its constitutive logic, and its resilience in Brazil. J Econ Rural Sociol 52:45–62. https://doi.org/10.1590/S0103-20032014000600003

Castro APD, Fraxe TDJP, Santiago JL, Matos RB, Pinto IC (2009) Agroforestry systems as an alternative for sustainability in floodplain ecosystems in the Amazon. Acta Amazônica 39:279–288. https://doi.org/10.1590/S0044-59672009000200006

Costa G, Lucas FCA (2022) Riparian agroecosystems in the Xingu and their role in biodiversity conservation. Agroecosyst J 14(1):1–20. https://doi.org/10.18542/ragros.v14i1.11602

Da Silva MDSR, Miranda SÁF, Domingos RN, Silva SLR, Santana GP (2013) Classificação dos Rios da Amazônia: uma estratégia para preservação desses recursos. Holos Environ 13(2):163–174. https://doi.org/10.14295/holos.v13i2.7344

De Andrade DBM (2012) Identification of preferential areas for the use of potential forest species in agroforestry systems in the Paraense Green Arch (Doctoral dissertation, Master's Dissertation, Institute of Geosciences, Federal University, Belém, Pará. 78 p). https://ainfo.cnptia.embrapa.br/digital/bitstream/item/77364/1/DissertacaoDaniele.pdf

De Abreu LS, Santos ADS, Watanabe MA (2017) The contribution of family farmers in the southern region of the Brazilian Amazon to the global ecological crisis. http://www.alice.cnptia.embrapa.br/alice/handle/doc/1087729

De Oliveira Silva JR, de Oliveira Silva IC, Coelho MDFB, Camili EC (2019) Plants and their uses in the backyards of Alta Floresta, Mato Grosso. Green J Agroecol Sustain Dev 14(3):420–428. https://doi.org/10.18378/rvads.v14i3.6499

Diegues AC (2019) Knowledge, traditional practices, and the ethnoconservation of nature. Dev Environ 50. https://doi.org/10.5380/DMA.V50I0.66617

Duarte GSD, Pasa MC (2016) Agrobiodiversity and ethnobotany in the São Benedito community, Poconé, Mato Grosso, Brazil. Interactions (Campo Grande) 17:247–256. https://doi.org/10.20435/1984042X2016208

Espada ALV, Sobrinho MV (2015) Manejo comunitário e governança ambiental para o desenvolvimento local: análise de uma experiência de uso sustentável de floresta na Amazônia. Administração Pública e Gestão Soc 7(4):169–177. https://doi.org/10.21118/apgs.v7i4.669

Ferreira ALV, Sobrinho MV (2015) Community management and environmental governance for local development: analysis of a sustainable forest use experience in the Amazon. Public Admin Soc Manag 7(4):169–177. https://doi.org/10.21118/APGS.V7I4.669

Ferreira OMF (2018) The use of productive backyards by family farming in the rural community of João Ferreira in the municipality of Ribeirópolis-SE. https://ri.ufs.br/jspui/handle/riufs/10679

Fraxe TJ, Pereira HS, Witkoski AC (2011) Amazonian riparian communities: ways of life and natural resource use. Reggo, pp 98130–97129.

Gomes I (2005) Sustentabilidade social e ambiental na agricultura familiar. Revista de biologia e ciências da terra 5(1). Redalyc Sustentabilidade social e ambiental na agricultura familiar

Gonçalves JDS (2005) The practice of burning in traditional knowledge and scientific risk conception: study on the use of fire by small rural producers in the Northern State of Minas Gerais. https://locus.ufv.br//handle/123456789/9957

Grisa C (2010) Public policies for family farming in Brazil: an essay based on the cognitive approach. Dev Debate 1(2):83–109. https://revistas.ufrj.br/index.php/dd/article/view/31915/18074

Guimarães AS (2015) Agroforestry home gardens and family nutritional status in the Santa Maria community, Santarém-PA (Doctoral dissertation, Federal University of Western Pará). https://repositorio.ufopa.edu.br/jspui/handle/123456789/140

Hecktheuer PA, Souza CBC, Hecktheuer FR (2018) Development as Freedom in a Riparian Community in the Amazon: an analysis of the effects of the Bolsa Família Program in São Carlos, Porto Velho. Rondônia. Sequence (Florianópolis):119–148. https://doi.org/10.5007/2177-7055.2018v39n78p119

Hora NN, Ferreira Filho HR, Fonseca MDJDCF, Gemaque AMDS (2014) Local knowledge and biodiversity conservation: uses, practices, and experiences of farmers in a community in the municipality of Ananindeua-Pa. https://tede.ufam.edu.br/handle/tede/8895

Kothalawala KIA, Pushpakumara DKNG, Sivananthwerl T (2013) Agro-forestry system to protect both the Sinharaja forest and the peripheral villages (a case study of the southern part of Sin haraja). In: Managing natural resources towards sustainable development, pp 10–11. https://doi.org/10.31357/FESYMPO.V0I0.1666

Laurance WF, Albernaz AK, Costa CD (2002) O desmatamento está se acelerando na Amazônia brasileira? Biota Neotropica 2:1–9.https://doi.org/10.1590/S1676-06032002000100003

Lima RF, da Silva AO, Dias PHB, da Silva BC, do Rosário Guimarães W, dos Santos Vasconcelos E, Silva DAS et al (2020) Cassava production (Manihot esculenta Crantz) in family farming in the Northeast region of Pará: a study from the Jacarequara community, Capanema, Pará. Braz J Anim Environ Res 3(3):1284–1296. https://doi.org/10.34188/bjaerv3n3-047

Luzzi N (2007) The agroecological debate in Brazil: a construction by different social actors. https://tede.ufrrj.br/jspui/handle/tede/693

Manzato AJ, Santos AB (2012) Questionnaire design in quantitative research. Department of Computer Science and Statistics–IBILCE–UNESP, p 17. http://www.inf.ufsc.br/~vera.carmo/Ensino_2012_1/ELABORACAO_QUESTIONARIOS_PESQUISA_QUANTITATIVA.pdf

Martins RK, Silva NS, Fernandes MO, Borsato JMLS (2012) O sistema mandala de produção de alimentos: uma estratégia para o desenvolvimento da agricultura familiar. XXI Encontro Nacional de Geografia Agrária 55. http://www.lagea.ig.ufu.br/xx1enga/anais_enga_2012/eixos/1397_1.pdf

Matos FJRD (2016) Way of life and management of açaí palm groves in the floodplains of the Mazagão River, municipality of Mazagão-AP, Brazil. https://repositorio.ufpa.br/jspui/handle/2011/9668

Moura CMMD (2012) Ethno-ecological knowledge of family farmers in the context of "roça de toco" in Três Riachos–Biguaçu, SC. https://repositorio.ufsc.br/handle/123456789/132623

Oliveira VLD, Bühler ÈA (2016) Technique and nature in the development of "agribusiness." Caderno CRH 29:261–280. https://doi.org/10.1590/S0103-49792016000200005

Paludo R, Costabeber JA (2012) Agroforestry systems as a strategy for rural development in different Brazilian biomes. Revista Brasileira de Agroecologia 7(2):63–76. https://orgprints.org/id/eprint/22937/1/Paludo_Sistemas%20agroflorestais.pdf

Pedroso JNN, Murrieta RSS, Adams C (2008) Slash-and-burn agriculture: a system in transformation. Boletim Do Museu Paraense Emílio Goeldi Ciências Humanas 3:153–174. https://doi.org/10.1590/S0044-59672009000200006

Pereira CN, Castro CND (2020) Technical assistance and rural extension in Brazil: an analysis of the 2017 agricultural census. https://doi.org/10.38116/brua24art12

Pereira PVM, Neto LFF (2015) Conservation of forest species: a study in agroforestry home gardens in the municipality of Cáceres–MT. Revista Eletrônica Em Gestão, Educação e Tecnologia Ambiental:783–793. https://doi.org/10.5902/2236117018364

Queiroz HL, Peralta N (2006) Sustainable development reserve: integrated natural resource management and participatory management. Hum Dimens Biodivers:447–476. https://d1wqtxts1xzle7.cloudfront.net/4225071/DimensoesHumanasBiodiversidade-libre.pdf?

Quitério TDC (2022) Dynamics of invasive plants in cassava crops in Amazon ecosystems in Alto Solimões. https://tede.ufam.edu.br/handle/tede/8895

R Development Core Team (2008) R: a language and environment for statistical computing. R Foundation for Statistical Computing, Vienna, Austria. ISBN 3-900051-07-0. http://www.R-project.org

Reis GT (2011) The foods from inside and outside: eating habits of caboclo-riverine communities of São Francisco, Careiro da Várzea (AM). http://riu.ufam.edu.br/handle/prefix/2031

Rocha BC, Scoles R, Puga BP, Blaser A (2022a) Tapajós under the sun: a dive into the ecological, sociocultural, and economic characteristics of the river basin. International Rivers. https://www.researchgate.net/profile/Bruno-Peregrina-Puga/publication/358854650_Tapajos_sob_o_sol_Mergulho_nas_caracteristicas_ecologicas_socioculturais_e_economicas_da_bacia_hidrografica/links/6218e74f6051a16582f7a6dd/Tapajos-sob-o-sol-Mergulho-nas-caracteristicas-ecologicas-socioculturais-e-economicas-da-bacia-hidrografica.pdf

Rocha BC, Scoles R, Puga BP, Blaser A (2022b) Tapajós sob o sol: Mergulho nas características ecológicas, socioculturais e econômicas da bacia hidrográfica

Sablayrolles PJL, de Azevedo ABA (2022) Public technical assistance for economic organizations of family farming: evidence from Bahia. DRd-Regional Development in Debate 12(spec. ed. 2(Cooperatives Dossier)):96–120. https://doi.org/10.24302/drd.v12ied.esp.2(dossiecooperativismo).3663

Santos C, Barros I, Wanzeler RT, Serrão E, Farias M, Lima A (2015) Hydroclimatological regionalization of the Tapajós River basin. Revista Geográfica Acadêmica 9(1):32–51. https://doi.org/10.18227/1678-7226rga.v9i1.2929

Santos AGD (2009) Technical and socioeconomic viability of agroforestry systems used by family farmers in Roraima. https://www.lume.ufrgs.br/bitstream/handle/10183/18810/000729173.pdf?sequence=1&isAllowed=y

Silva TC, Caporal FR (2016) Law on technical assistance and rural extension in Alagoas, Brazil: contributions or discontinuity. Rural Ext 23(2):7–26. https://doi.org/10.5902/2318179621288

Silva RGDC, Silva VVD, Lima LAP (2019) The new axes of the frontier in western Amazonia. Confins. Franco-Braz J Geogr (43). https://doi.org/10.4000/confins.24950

Silva AM, Pauletto D, da Silva AF, Sousa VS, de Sousa Oliveira TG, da Silva AF, da Silva Santos A et al (2022) Diversidade e multifuncionalidade de quintais agroflorestais em aldeias da Terra Indígena Tupinambá, Santarém, Pará. Biodiversidade 21(4). https://periodicoscientificos.ufmt.br/ojs/index.php/biodiversidade/article/view/14795

Silva SRD (2019) Ethnodevelopment: analysis from traditionally occupied indigenous land. http://repositorio.ufgd.edu.br/jspui/handle/prefix/1993

Siviero A, Delunardo TA, Haverroth M, Oliveira LCD, Mendonça MS (2011) Cultivation of food species in urban backyards in Rio Branco, Acre, Brazil. Acta Botanica Brasilica 25:549–556. https://doi.org/10.1590/S0102-33062011000300006

Sousa MPAD (2013) Study of genetic diversity in stream fish from the Tapajós River basin: a molecular systematic approach using DNA Barcoding (Doctoral dissertation, Universidade Federal do Oeste do Pará). https://repositorio.ufopa.edu.br/jspui/handle/123456789/339

Souza KDB, Leismann EL, Brandalise LT (2016) Documentary analysis of the process of acquiring products from family farming using the food acquisition program. Revista Competitividade e Sustentabilidade 3(1):40–52. https://doi.org/10.5935/2359-5876.20160004

Souza ACF, Soares NRM, Paiva ADC, Souza JF, Mendes JTR (2020) Microscopic and microbiological evaluation of cassava flour, Manihot esculenta Crantz, sold at producer and farmer markets in Macapá, Amapá. BIOFARM J Biol Pharm Agric Manag 16(2):122–136. https://revista.uepb.edu.br/BIOFARM/article/view/2202/1793

Vasconcelos FG, da Silva Braz V, Rodrigues FF (2017) Perception of environmental problems in a riverine community in the Brazilian Amazon region. Anais SNCMA 8(1). https://doi.org/10.22533/at.ed.4351921025

Viana RL, Freitas CMD, Giatti LL (2016) Environmental health and development in the legal Amazon: socioeconomic, environmental, and health indicators, challenges, and prospects. Saúde e Sociedade 25:233–246. https://doi.org/10.1590/s0104-12902016140843

Amazon Energy Cultures and the Transition to Sustainability

Antonio Mauricio Dias da Costa, Colin Nolden, Marcos André Barros Galhardo, Ricardo dos Santos Caçapietra, Wilson Negrão Macêdo, Maria Cecília Rosinski Lima Gomes, Ana Claudeise Silva do Nascimento, Sam Williamson, and Amy Penfield

Abstract Around a quarter of a million people living in remote communities in the Amazon are not connected to Brazil's national electricity grid. For researchers and policymakers, the matter of energy access in such contexts has recently shifted to questions of "energy resilience," which refers to community adaptation to energy absence, volatility, or failure. Using a combination of quantitative and qualitative techniques during fieldwork in forest communities in Ilha das Onças, in the state of Pará, Brazil, our team attempted to gain a better view of their "energy resilience" through everyday engagements with energy sources and policies. In theory, policies such as "Light for All" (*Luz para Todos*) and "More Light for Amazon" (*Mais Luz para a Amazônia*), which attempt to provide universal access to electricity, should serve these communities. In practice, our researchers found that these initiatives fail to fully access these remote settlements which, for various reasons, predominantly use small fossil-fuel-powered generators. In other global contexts, individual systems

A. M. D. da Costa (✉)
História, Universidade Federal Do Pará, R. Augusto Corrêa, 01–Guamá, Belém, PA 66075-110, Brazil
e-mail: macosta@ufpa.br

C. Nolden
Energy Institute, Management School, University of Sheffield, Conduit Road, Sheffield S10 1FL, UK
e-mail: c.nolden@sheffield.ac.uk

M. A. B. Galhardo · R. dos Santos Caçapietra · W. N. Macêdo
Instituto de Tecnologia, Universidade Federal do Pará, R. Augusto Corrêa, 01–Guamá, Belém, PA 66075-110, Brazil
e-mail: galhardo@ufpa.br

R. dos Santos Caçapietra
e-mail: ricardo.cacapietra@itec.ufpa.br

W. N. Macêdo
e-mail: wnmacedo@ufpa.br

M. C. R. L. Gomes
Instituto de Desenvolvimento Sustentável Mamirauá, Estrada Do Bexiga, Tefé, AM 2584, Brazil
e-mail: cecilia@mamiraua.org.br

W. Leal Filho et al. (eds.), *Amazon 2030—Sustainability Issues in the World's Largest Rainforest Region*, World Sustainability Series,
https://doi.org/10.1007/978-3-031-81465-5_27

or microgrids powered by solar photovoltaics have proved to be promising alternatives. To promote resilient energy cultures, we recommend more collaborative energy governance, with the active participation of residents in knowledge production and decision-making, as opposed to the top-down forms of centralized governance that characterize current policy structures for the supply of electricity in remote areas such as the Amazon and beyond.

Keywords Amazon · Energy access · SDG7 · Off-grid · Solar PV · Resilience

1 Introduction

The Amazon rainforest, an area that covers nearly 60% of Brazil's national territory, includes some of the most remote areas in the world. While there exists a legal and administrative delineation as viewed by the Brazilian government ("Amazônia Legal"), the region is vast and as such extremely diverse in terms of the social, political, and economic realities of its inhabitants, both indigenous and non-indigenous.[1] Many inaccessible areas are not integrated into the national electricity grid (National Interconnected System/*Systema Interligado Nacional*—SIN) and suffer from associated socio-economic hardships, despite abundant renewable energy resources ostensibly at their disposal, solar energy in particular.

This is problematic in many ways, as barriers to energy access have knock-on effects in areas of healthcare, education, democratic participation and equitable access to rights and services. Moreover, many of the energy sources available to these communities are carbon-based, polluting, and unsustainable, such as fossil-fuel-powered machinery. In this paper, we develop a model for fostering a deeper understanding of Amazon Forest peoples in their experiences of "energy resilience." Bringing together knowledge from anthropology, engineering, sociology, geography, law, and public policy to develop an analytical and practical approach to everyday

A. C. S. do Nascimento
Universidade Federal do Sul e Sudeste do Pará, Marabá, Brazil
e-mail: ana.claudeise@gmail.com

S. Williamson
School of Electrical, Electronic and Mechanical Engineering, University of Bristol, Bristol BS8 1UB, UK
e-mail: sam.williamson@bristol.ac.uk

A. Penfield
Department of Anthropology and Archaeology, University of Bristol, Bristol BS8 1UU, UK
e-mail: amy.penfield@bristol.ac.uk

[1] The Federal Law 1.806 was created in January 6th 1953, which integrated in the Brazilian Amazon (besides the states of Pará and Amazonas, and the then federal territories of Amapá, Roraima, Acre, and Rondônia) the following states: Maranhão (west of the meridian 44°), Goiás (north of parallel 13° of south latitude, which is presently the state of Tocantins), and Mato Grosso (north of parallel 16° of south latitude).

resilience through the lens of energy access, we shed light on "energy resilience" as a way of life intimately tied to existing social and political frameworks. This analysis suggests that creating more environmentally and socially sustainable energy supplies hinges less on the immediate provision of technology and more on understanding both immediate and long-term social concerns. Combining short-term considerations around security issues with long-term viewpoints on the co-evolution of sociocultural and socio-technical systems in this way provides new avenues for identifying intervention points.

In recent years, the concept of resilience has gained popularity as a means to describe the qualities and capacities that enable communities to recover from catastrophic events or trauma (Sherrieb et al. 2010). The episodic nature of this approach to resilience has tended to overlook quotidian hardships and how communities make sense of and overcome them. Moreover, resilience must be seen as a dynamic process that is contextualized within complex historical, cultural, ecological, political, and economic settings. Underwood et al. have recently described energy resilience as:

> An analytical approach to investigating energy systems which looks specifically at how different actors withstand, adapt and change energy systems and the energy services they supply while experiencing realised risk to their system vulnerabilities (Underwood et al. 2020).

There can be no doubt that the use of, and access to, clean energy is a complex, cross-cutting and at times discordant phenomenon that requires new conceptual frameworks. As such, exploring energy access as a question of resilience can keep those complexities within the frame by considering overlooked factors such as lived experience and daily challenges surrounding affordability, reliability, and sustainability in energy encounters. What this reveals is that energy is more than just a technical provision, but one aspect of a complex terrain of social life and livelihoods.

Our empirical research into energy resilience involved fieldwork around Belém, in the state of Pará, and around Tefé, in the state of Amazonas. In this chapter, we report our findings from the unconnected region of Ilha das Onças, near Belém. The findings of our research reveal that processes of forging everyday forms of resilience by forest inhabitants in response to daily hardships are heightened by the quotidian hardships rooted in the physical and geographical context of the Amazon rainforest, alongside a challenging socio-economic context which includes the (in-)accessibility of grid-connected energy.

This chapter is organized as follows: Sect. 2 introduces the research context, the sustainable energy solutions supported by policy, and localized solutions which are economically and technically viable but culturally unattractive. Section 3 introduces the methodology. Section 4 focuses on the results. Section 5 discusses these results concerning the research context. Section 6 concludes by establishing recommendations for productive and equitable strategies for sustainable and resilient energy provision in the Amazon and other remote regions of the world.

2 Research Context

Despite the long history of Brazilian electricity sector regulation since the process of formulating public policies in the so-called "Vargas Era" between 1930 and 1945 (Corrêa 2005), access to energy by residents of some remote areas in the Amazon, as a basic right, still faces both social and environmental challenges. The centralized electricity generation, transmission, and distribution system of Brazil today is largely hydro-thermo-wind, with a predominance of hydroelectric plants with multiple owners located in sixteen hydrographic basins across the different regions of Brazil. The transmission system consists of four subsystems: South, Southeast/Central-West, Northeast and most of the North region. Thermal plants tend to be located close to the major centers of energy consumption. Their dispatchable nature allows for the management of water stocks in the reservoirs of hydroelectric plants. Distributed and isolated systems received regulatory approval since 2004 (e.g., decrees 5,163/2004, 12,111/2009, and 7,246/2010). These contain rules for the provision of electricity services of systems unconnected to the national transmission and distribution network, although public policies for universal access to electricity remain highly distributed among the Brazilian Federal Government, specifically the Ministry of Mines and Energy (MME) and the National Electric Energy Agency (ANEEL), and to a lesser degree, the distribution network operators. In a cruel irony, some of the communities lacking access to the centralized grid infrastructure have lost access to their land through the flooding of valleys in support of hydroelectric power generation.

Although access to electricity does not expressly appear in the constitutional text as a social right inherent to every citizen,[2] it is implicitly defined as a fundamental right through its numerous policies that aim to extend access to a wider population (Cavalcante 2013: 58). One such policy is the *Luz para Todos* (Light for All) program that was launched by the Brazilian government in 2003 with the aim of providing universal access to electricity by extending the National Interconnected System (SIN), but with a focus specifically on the provision of funds for renewable energy.[3] However, the 2010 census reported that around 716,000 households remained unconnected, representing 1.27% of all Brazilian households. Other studies estimated the number to be as high as 1 million homes (see Valer et al. 2014), but most importantly 99% of those households hailed from the Legal Amazon (Instituto de Energia e Meio Ambiente 2019).

[2] It is worth remembering that PEC 44/2017, which expressly instituted access to electricity as a social right in the constitutional text, was shelved at the end of the legislature in 2022.

[3] The "Light for all" programme implemented through Decree No. 4,873/2003, conceived as an instrument of development and social inclusion, lists as its main objective the universalization of access to electricity in rural areas of the country, initially defined as priorities for attending projects in municipalities with a service rate of less than 80%, rural electrification projects that benefit populations affected by dams, rural electrification projects of public schools, health centres, and water service wells, rural electrification projects aimed at serving rural settlements and electrification projects for the development of family farming.

The government responded by launching the *Mais Luz para a Amazônia* (More Light for the Amazon) program in 2020 to focus on more remote forest communities who tend to present higher social and economic vulnerabilities as a result of their lack of connection.[4] This initiative similarly pays particular attention to renewable energy through the use of photovoltaic solar systems. Nevertheless, while some progress has been made, overall, the programs have fallen short of universality due to the huge scale of Brazilian Amazon and the remoteness of communities therein.

While rainforest terrains such as Amazonia can present major barriers to the expansion of the National Interconnected System electricity grid, solutions in the form of low-carbon technology, such as solar photovoltaic (hereafter PV) technology, might seem—perhaps form the perspective of the external observer or policymaker—like obvious quick-fix provisions, particularly given the abundance of solar energy available in tropical locales. However, our data has shown that this is far from the case, and many households continue to rely on diesel generators and sometimes even clandestine extensions of the official energy grid even though both options are fraught with dangers (electric shocks) and complications (fuel price volatility), as will be outlined in more detail below. This failure to accomplish absolute transition to PV technology is not solely a result of their inaccessibility and unreliable utilities, but rather strongly tied to social norms, expectations and local timeframes. To put it another way, initial versus long-term costs are vital factors for those involved in decision-making.

Despite higher initial investment requirements, renewable energy technologies such as solar PV are actually cheaper from a life-cycle perspective. Higher dependency on on-site resources (solar irradiation) also provides greater opportunities for local embedding (Valer et al. 2017). Diesel generators, on the other hand, while requiring very low initial investment, actually have hidden high life-cycle costs due to fuel dependency and its price volatility (Nerini et al. 2014; Valer et al. 2017). As such, one issue with initiatives such as *Luz para Todos*, according to Nerini et al. (2014), is its short-term-policy target, which fails to allow for long-term embedded knowledge and consolidated supply chains by placing too much emphasis on initial investment costs. Higher life-cycle costs of diesel generators vis-a-vis renewable energy technologies such as solar PV were not taken into account with the result that the former proved difficult to displace (Gomez and Silveira 2012). This policy failure pushed "the energy use pattern in a sub-optimal manner and distorts the energy user's choice" (Nerini et al. 2014: 42).

Thanks to the expansion of the market for solar PV, coupled with falling costs and greater social acceptance, remote areas have witnessed increased uptake of this technology in recent years (International Energy Agency 2022). Unfamiliarity, security concerns, and deeply rooted cultural beliefs nevertheless impact participation

[4] The program "More light for the Amazon" instituted by Decree No. 10,221/2020, with the purpose of providing energy to the Brazilian population residing in remote regions of the legal Amazon, defined as beneficiaries of the program, the families and the respective socioeconomic support units, in addition to the other consumer units located in remote regions of the legal Amazon lacking access to the public electricity service and remote regions of the legal Amazon that have generation of non-renewable energy source.

and adoption of new practices and technologies. Supporting the diffusion of clean technology necessitates additional efforts to foster community buy-in, expanding the direct contribution of stakeholders in building policy solutions, improving security, and sustaining commitments as well as supply chain development. This all points to a need to view energy access and resilience as a socially embedded phenomenon. In the next section, we discuss the methodology and introduce the communities.

3 Methodology

3.1 Data Collection

Given that the objectives of the project are to understand not only energy use but lived experiences of energy and resilience in cases of insecure access, a multi-modal methodology was deployed for research data collection. One way this was achieved was through simultaneous data collection and regular on-site "labs." What this means is that different disciplinary team members in Brazil undertook fieldwork together and frequently discussed findings as they unfolded during the time of data collection. While our interdisciplinary methodology spans anthropology, law, geography, sociology, and engineering, our main methodological approach draws on social analysis, particularly ethnography above all.

Ethnographic methods are usually based on long-term fieldwork, participant observation, informal conversations with research participants, semi-structured interviews and life histories. An advantage of ethnographic approaches is that they allow the gathering of local views and voices that are lacking in purely quantitative approaches (O'Reilly 2012). Energy, for instance, is a phenomenon that spans numerous areas of lived experience, not just knowledge of the workings of motors and solar panels, but matters of political allegiances that provide access to energy, which result in or stem from social dynamics and are intimately tied to moral interpretations of energy supplies.

The purpose of ethnographic research is to embed oneself in the lives of research participants in order to view any specific phenomenon—indeed life itself—from their perspective (Wilson 1977), and as such gain insights into local behaviors, characteristics, motivations, and attitudes in detail (Bernard 2013). Though research questions can be developed in advance, seeking data from the bottom-up in this way results in research priorities ultimately emerging from the lives and concerns of participants. In this way, ethnography is an inherently open and adaptable technique in that it continually modifies the research design during the process of data collection, rather than manipulating variables by external instruments (Tuckman 1999). During data collection for this project, local experiences with and ideas about energy imparted by research participants informed the research trajectory as the project progressed, and the expertise, experience, and understanding of local actors deeply informed the research agenda.

Table 1 Measuring energy access framework quantitative values for electricity access

Attributes		Tier 0	Tier 1	Tier 2	Tier 3	Tier 4	Tier 5
Capacity (W)		<3	3–49	50–199	200–799	800–1999	≥2000
Availability (hrs)	*Day*	<4	Min 4		Min 8	Min 16	≥23
	Evening	<1	Min 1	Min 2	Min 3	Min 4	
Reliability	*Frequency* (per week)	>14				4–14	≤3
	Duration					>2 h (if frequency ≤ 3)	≤2 h
Quality		Yes				No	
Affordability (Standard 365 kWh/year)		≥5% household expenditure			<5% household expenditure		
Formality (Bill payment)		No				Yes	
Health and safety		Yes				No	

Source World Health Organization (WHO) (2021)

We complement this ethnographic approach with the analysis of our survey data using a multidimensional "Measuring Energy Access" framework of the World Bank and the World Health Organization (2021). This approach enables us to assign a tier classification to attributes including capacity, availability, reliability and quality, accessibility, formality, health, and safety. For these attributes, there was a set of questions applied in the interviews, after which each attribute was classified into one of six levels (0–5), resulting in a multilevel table on access to electricity. Level 0 means inadequate access to electricity or very low-quality access and level 5 corresponds to satisfactory access and high quality (see Table 1 for quantification of this). This enables us to assess multiple dimensions of energy supply and security by the approach developed by Sovacool and Mukherjee (2011).

During field research, team members undertook 18 structured quantitative surveys and 11 semi-structured interviews among members of different communities of the island.[5] This work covered aspects related to the energy supply chain, use of different energy sources, energy transition, and experiences with public policies and services. Apart from one individual aged 69 and one 82, all interviewees were between 14 and 49 years old. The youngest person interviewed was a 14-year-old female, who was about to have her first child and form her own family unit. Among this sample, women were commonly heads of families and all interviewees bar one (94.44%) stated that the number of people living in their homes was between 1 and 6.

[5] Here "community" is defined as groupings of extended families located in different points of the island that take as reference the river flowing in front of their houses and the church, Catholic or Evangelical, attended weekly by the residents.

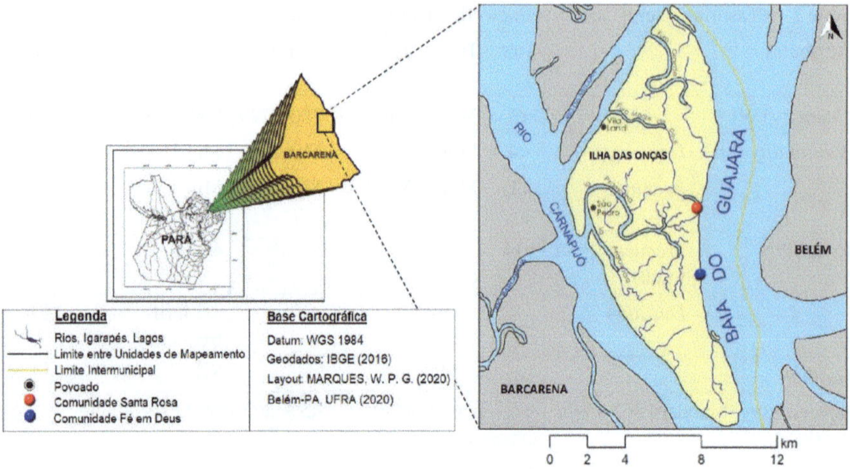

Fig. 1 Map of Ilha das Onças, Barcarena-Pa. *Source* Marques et al. (2020)

3.2 Setting

Our interdisciplinary team undertook this ethnography-oriented research to explore different experiences among Amazon riverside communities known as Ribeirinhos, specifically those existing in the region of Ilha das Onças (Fig. 1),[6] a region accessible only by river transport. This investigation focused on the daily activities of the residents and how local arrangements linked to the use of alternative energy sources are an integral part of the local way of life, directly influencing the cultural, social, familial, and religious context of the islanders.

In January 2023, qualitative data was collected during field research on Ilha das Onças, located in the municipality of Barcarena, which is part of the metropolitan region of Belém, Pará state, in the northern region of Brazil. Ilha das Onças is one of the islands surrounded by the Guajará Bay and is traditionally known for growing and extracting the açaí fruit (*Euterpe oleracea*), as well as the collection of heart of palm and shrimp traded in open markets in Belém and Barcarena. Throughout fieldwork, the research team stayed at the residence of Mr. João Alves, on the banks of the Piramanha River, who assisted the research team in establishing contact with interviewees. Mr. Alves is a key figure in the region, not only as the evangelical pastor for surrounding communities, but also one of the few residents with a private boat and sources of fuel that facilitate the transportation of tradable produce, the purchase of consumer goods in the cities for resale on the island, and the ferrying of passengers and visitors to urban centers such as Belém and Barcarena, as well as

[6] Ribeirinhos are a traditional population in Brazil, descended from migrants from the Northeast of the country. Their settlements, subsistence activities, and travel are associated with rivers which are a central focus of their livelihoods. Due to the geographical location of these rivers, the majority of this population resides in the Amazon.

restaurants located at the entrance of Furo do Nazário that connects the Guajará Bay to the Piramanha River (see Fig. 2). As such, not only was Mr. Alves a key broker for contact with other research participants, but was also a cental player in the energy landscape (access and procurement) for the region so was able to provide insight on that front.

In the absence of electricity provided by the grid, the main sources of electricity in Ilha das Onças are fossil fuel generators and increasingly solar PV. According to our quantitative survey, the riverine communities surveyed on Ilha das Onças use 21 off-grid/isolated systems that constitute a combination of home systems (that serve individual households) and microgrids (solar or diesel-generator powered) that provide energy to a cluster of interconnected households or that work together to improve the reliability of the system as a whole. Of the 21 systems in Ilha das Onças, 11 are PV systems and 10 are diesel or gasoline generators, indicating a range of energy sources in use. These local grids have been part of daily energy systems for

Fig. 2 Map of Furo do Nazário. *Source* Google Earth

many decades, as numerous respondents reported using generator-powered electric lighting since childhood.

4 Findings

4.1 Socio-Economic Arrangements

The main economic activities of Ilha das Onças communities consist of the extraction and trade of natural products such as fish and shrimp, and, above all, açaí, the most profitable resource (see Table 2). These tend to be supplemented by fruits such as taperebá (*Spondias mombin),* andiroba (*Carapa guianensis)* and murumuru (*Astrocaryum murumuru*), which are often sold to cosmetic companies such as Avon, O Boticário, Beraca, and Natura. This extractive work of wild fruit involving harvesting by hand requires mainly human labor and tends to be shared by all members of family groups, regardless of gender and age. In this sense, though these communities have not directly benefitted from either from *Luz para Todos* nor *Mais Luz para a Amazonia* programs (see above), one could argue that this has not negatively impacted their productive activities since they do not require energy-intensive equipment. Nevertheless, access to electricity would improve working conditions and add value by enabling pulping and longer-term storage of raw food and fruit pulps through freezing, thereby increasing both durability and geographical distribution, whilst providing the means to develop new business models around the sale of frozen produce.

Other potential income-generating activities such as the preparation and sale of homemade cakes and sweets according to order, carpentry activities, and welding and sewing services, on the other hand, are dependent on energy-intense electrical

Table 2 Work activities of the interviewees

Product	Count
Açaí berry	15
Fishing	4
Heart of palm	2
Shrimp	3
Machine shop	1
Lumberyard	1
Fuel trade	1
Student transportation	1
Unspecified trade	1
Not applicable	13

Source Field Research

equipment such as ovens and power tools not adequately supported by existing solar PV systems or generators. Consequently, very few residents are able to pursue such activities or other energy-intensive economic pursuits such as selling fuel, ice, or food (in small commercial establishments), or providing transport services (to Belém or Barcarena). This lack of opportunity for transport also means that only a small number of residents have been able to take up public jobs in the Municipality of Barcarena (as, for example, security guards or school caretakers), or jobs in bars and restaurants in nearby towns that cater to visitors on weekends.

Consequently, average incomes are low among the surveyed population; 29% of the respondents receive less than the minimum wage, 47% receive between one and three times the minimum wage, and 24% did not respond. Despite a policy implemented in 1991 that provides financial incentives to deter overfishing during the breeding season, none of the interviewees reported receiving this aid.[7] To supplement household incomes, some interviewees reported receiving supplemental incomes through retirement benefits and the Bolsa Família program, which supports families living in extreme poverty. Some of this has been used to buy solar PV equipment since it is a fixed sum received monthly, not dependent on the harvest or the tide. However, 45% of respondents do not receive any benefits from any of these programs.

Urgent or routine medical treatment is provided by the Basic Health Unit (BHU) of the community of Arapiranga in Ilha das Onças. Traveling to the BHU (from Mr. João Alves's house) takes about 45 min, depending on the level of the river. The BHU Arapiranga serves several nearby communities which leads to demand for care that is not always met. The services it provides hinge on the operation of a generator which receives diesel from the city. However, the engine regularly breaks down and, in such cases, the only way to receive care is to travel to BHUs or alternative healthcare facilities in Belém or Barcarena which are much further away and thus require larger supplies of fuel to travel to. In short, without reliable energy sources, residents cannot access healthcare.

In relation to education, the picture is even more precarious. The schoolhouse, which serves all island communities, operates in a single room unsuitable for the class size of about 30 students which intensifies the heat and in turn creates a challenging learning environment. Despite the supply of ventilators and drinking fountains, these are not enough to meet the demands of students. Moreover, these provisions are powered by a generator which, similar to the BHU of Arapiranga mentioned above, is prone to malfunction. Even when fully operational, the generator can face a number of other problems, such as depletion of fuel and oil, leaving children to attend class in the heat and without water for days. When maintenance complications arise, the school can be left without power for weeks. In the recent past, the community had a wooden bridge that crossed much of the island's edge, facilitating the movement of students to another nearby school without depending on river transport via boats.

[7] A set of federal laws and decrees was created from 1991 on in Brazil to provide resources to fishermen (named "Seguro Defeso"), as a kind of an insurance fund paid monthly during the breeding season of fishes, so that the normal reproduction of specimens could be guaranteed. The legislative pieces are the laws 8.287/1991, 10.779/2003, 11.959/2009, 13.134/2015, and the decrees 8.424/2015, 8.425/2015, 8.967/2017.

However, this bridge has deteriorated over time and is no longer fit for use, which puts further pressure on school children whose families are unable to afford to send them to other schools nearby or in Belém.

4.2 Energy Cultures

It is crucial to place the social setting—the "energy cultures"—at the center of an analysis of energy resilience. The population of Ilha das Onças has traditionally been served by a range of energy sources (see above), which they negotiate on a daily basis. Yet they face many challenges, chief among them safety, cost, and communication.

In terms of safety, many supplies of energy are dangerous to operate and maintain. 7 of the 18 respondents reported accidents related to such energy sources, including fires and burns caused by candles or lamps, injuries when cranking engines, and electric shocks. Many reiterate the need for safer and more reliable options. Since the early 1990s, residents have been acquiring more powerful electricity generation equipment. Current power generators, with a capacity of about 4 kVA, are sufficient to power a wide range of electronic devices (see the surveyed demand profile in Fig. 3). Increasingly, these generators are petrol (gasoline) powered, as opposed to diesel powered, as they are considered more reliable with a reduced need for maintenance which in turn means fewer accidents. Residents also benefit from greater availability of spare parts at lower cost for these new generators. Many interviewees demonstrated an intimate connection with these technologies as many families have been acquiring generators since the 1990s. Almost all community residents know how to use generators and fix common technical issues.

In addition to these new generators, since 2018 residents have started acquiring more photovoltaic (PV) modules, often in combination with one or two batteries and an inverter, equipment that converts direct current to alternating current. 7 of the 10 interviewees now use solar power as the main source of energy and the equipment is usually acquired with the help of social networks between residents (relatives, neighbors, and friends) or external networks, such as traveling salesmen. Initially, residents were uncertain about acquiring this equipment due to cost/benefit, durability, and their perceived unreliability in comparison to generators; they queried whether the system would work, or if it would just be a "waste of money." Following the successful installation of PV systems by innovative families with a higher risk appetite, other residents began admiring their acquisition, contacting known suppliers and acquiring their own PV system (spillover effect).

The current growing interest in acquiring PV modules privately (i.e. personally funded rather than subsidized by the state) is linked to aspirations for improved living conditions, particularly productive business ventures, with 2 of the 18 survey respondents citing this as a key motivation to acquire PV technology. However, high initial costs present the main limiting factor. In addition, there is an expectation that residents who do not own PV systems will get the necessary resources for the investment in the near future, although it is currently impossible for low-income

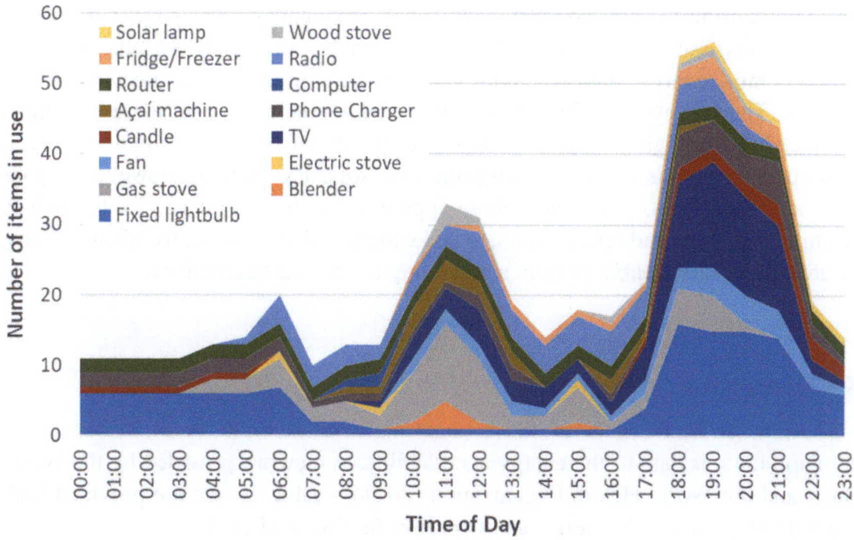

Fig. 3 Energy demand profile on an average day. *Source* Field Research

families to obtain the necessary bank credit for investment.[8] Current funding models and the absence of medium and long-term financing generally restrict the purchase of large solar PV systems, despite the high cost of fuel and maintenance associated with generators.

This precarity also extends into warranties. As it stands, PV salesmen—who are often non-regulated privately contracted personnel rather than state actors—are only responsible for installation while maintenance is usually carried out by residents themselves, or, failing that, relatives and/or acquaintances who reside nearby, despite the lack of appropriate training which can lead to further safety issues and irreparable technology.

Nevertheless, electricity supplied by solar PV has been transformative for these communities, especially during the night. Some activities such as shrimp fishing have been made possible through this availability as they tend to take place at night. Communication, one of their other main concerns, is also enhanced by these new technologies. Mobile phones, for instance, are low-energy appliances that are similarly charged at night while people sleep, increasing opportunities to start a business selling food, ice, or fuel through expanded communication with traders at various times of the day.

When solar PV systems are available and functioning, they do not displace petrol generators but often work alongside them as part of a hybrid system, especially in the context of high-load activities. Solar PV is generally used for lighting, watching

[8] Some of the people interviewed do not have a checking account in banks, in addition to not having mechanisms for control and proof of income, since the trade of açaí and other products is moved with "live" money and face-to-face negotiations.

television during the day, and some low-energy appliances such as blenders, sound systems, fans, and phone chargers. Generators, on the other hand, are generally used to power more energy-intense activities such as water pumping and mechanical açaí pulping. The surveys identified numerous benefits emerging from these combined advanced technologies, including improved quality of light, convenience, as well as reduced accidents and indoor air pollution. Solar PV systems, however, require sufficient capacity to power household appliances with high energy loads, such as washing machines and refrigerators, to make them both economically advantageous, future-proof, and capable of fully supplanting fossil-fuel alternatives.

5 Discussion

By applying our data to the energy accessibility framework provided by the World Bank and the World Health Organization (2021)—Table 1—we are provided with the following picture of energy access in Ilha das Onças (Fig. 4).

These findings suggest that the communities surveyed on Ilha das Onças as part of this research have socio-technical characteristics of energy poverty and insecurity. The cost of acquiring electricity and powering technology consumes a large share of household income. This issue is amplified by the shortage of qualified personnel to install and maintain electrical systems. Furthermore, public policies in support of access to electricity have had little positive impact on the daily lives of these communities. Access to information and communication available through the internet and telephone networks is often a challenge precisely because such communication channels are only available through electricity sources. Consequently, most communication continues to be conducted by direct conversations between people, usually in church meetings, which function as a space for information exchange

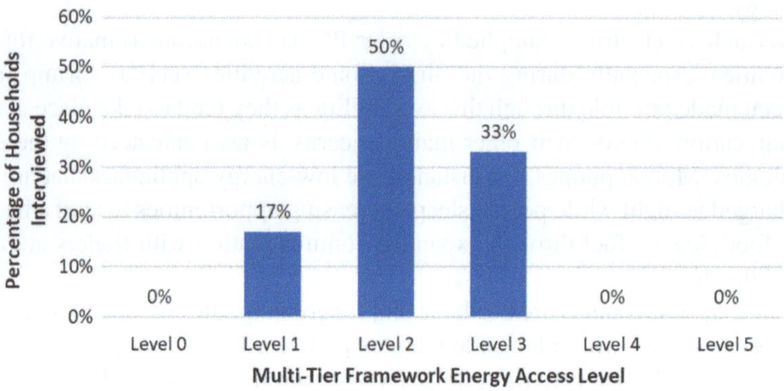

Fig. 4 Energy access levels according to the multi-tier framework in Ilha das Onças. *Source* Fieldwork; WHO (2021)

and leisure. General information, such as news about everyday matters, is obtained through television and radio, which again hinges on the availability of electricity.

According to the interviewees, the efforts of the Municipality of Barcarena to extend the grid electricity supply to the island are announced regularly around election times to secure votes, only to disappear once votes have been cast. Promising meetings involving politicians and residents in Ilha das Onças, Belém and Barcarena in recent years to discuss electricity supply have not delivered any results.[9] There is also apprehension regarding the grid-connected electricity supply due to the need to cut down trees for grid access, which would directly affect the economy of the families that survive on the income of harvesting and trading açaí and other fruits. One of the research participants reported that as part of a proposed grid extension project, it would be necessary to tear down a 6-m swathe of trees, including açaizais that produce the valuable fruit, causing huge economic loss to local families.

As a result, there is a clear trend among the residents to acquire privately owned solar PV systems to mitigate some of the present daily energy challenges. In any case, work activities linked to the residents' main source of income such as açaí extraction and trade rely predominantly on labor power and less on electricity supplied from either solar or combustion energy, further explaining why incentives to connect to the grid are somewhat ambivalent. Thus, while there are many benefits associated with electricity supply, we must view its presence and potential as very much embedded within a complex social terrain which cannot be disentangled from the broader energy landscape.

6 Conclusion

The residents of Ilha das Onças acquire energy from different sources depending on specific activities: fossil fuels for energy-intensive services such as transport, pumping water and pulping açaí, alongside solar PV electricity for less energy-intensive services such as lighting and low-consumption household equipment. While there is a clear trend of expansion in the acquisition of solar PV equipment, it is generally acquired from unregulated sources and the supply chains are not well established. As a result, few working activities hinge on solar power and investment in these new energy sources does not necessarily replace old practices, such as the use of generators to power refrigerators for ice storage or for the açaí pulping machinery.

Data from this research has revealed that energy resilience in Ilha das Onças is limited by unregulated electricity and weak supply chains for clean alternatives, with associated shortfalls in services and economic opportunity. Nevertheless, that is not to say that resilience is entirely absent, as informal structures are at play that provide long-term sustainable resilience in the form of kinship networks, entrepreneurial knowledge, the diversification of energy systems according to shifting needs and

[9] The length of the *Luz para Todos* program was extended to December 31, 2028, by Decree no. 11,628 of August 4th 2023.

expectations, and informal upskilling around solar PV. As mentioned earlier, lack of energy has repercussions in areas of healthcare, education, communication, political inclusion, transportation, and reliable infrastructure. By considering resilience as a mode of enquiry in relation to energy access, we can shift the focus towards local priorities, everyday capacities, and complex systems of investment, to better understand sustainable energy solutions on the ground.

Alongside the practices already incorporated within the culture of Ilha das Onças, residents are diversifying and evolving with new technologies, generating new types of energy access demands and pathways to resilience, pointing towards the need for a policy-mix that supports place-based approaches to sustainable electricity supply. Public policies, such as *Luz para Todos* and *Mais Luz para a Amazônia*, meanwhile, have so far failed to fully serve the communities surveyed in Ilha das Onças. The political terrain is further complexified by the use of electricity as a political asset, as forest and river residents see promises come and go each election cycle. A deeper understanding of governance and public policy—specifically their flaws and incompatibility with social and cultural realities—is required to advance towards universalized access to clean electricity, overcome energy poverty, create desirable low-carbon energy solutions, and support the transition to sustainable living. To improve energy access, improve resilience, and achieve SDG7, a bottom-up approach engaging directly with the local energy cultures would complement and profoundly enhance top-down mandates for energy utilities.

References

Barrios R (2016) Resilience: A commentary from the vantage point of anthropology. Ann Anthropol Pract 40(1):28–38. https://doi.org/10.1111/napa.12085

Bernard HR (1994) Research methods in anthropology: Qualitative and quantitative approaches, 2nd edn. Sage, Thousand Oaks, CA

Bernard HR (2013) Social research methods: Qualitative and quantitative approaches. Sage

Cavalcante HPM (2013) O acesso à energia elétrica no Brasil sob a ótica do desenvolvimento como liberdade. Rev Direito Econômico E Socioambiental 4(2):58–86. https://doi.org/10.7213/rev.dir.econ.socioambienta.04.002.AO04

Corrêa ML (2005) Contribuição para uma história da regulamentação do setor de energia elétrica no Brasil: o Código de Águas de 1934 e o Conselho Nacional de Águas e Energia Elétrica. Política & Sociedade 4(6):255–292. https://periodicos.ufsc.br/index.php/politica/article/view/1955/1706

Gall MD, Gall JP, Bord WR (2005) Applying educational research. Pearson, Boston

Gomez MF, Silveria S (2012) Delivering off-grid electricity systems in the Brazilian Amazon. Energy Sustain Dev 16(2):155–167. https://doi.org/10.1016/j.esd.2012.01.007

Guimarães LAV (2021) Barcarena além da sua emancipação. Available athttp://www.barcarena.pa.gov.br/portal/pagina?id=8&url=hist%C3%B3ria

HdeS S (2000) Uma história de Barcarena. Centro de Filosofia e Ciências Humanas, Universidade Federal do Pará, Belém

Instituto de Energia e Meio Ambiente—IEMA (2019) Um milhão estão sem energia elétrica na Amazônia, mostra IEMA. Available at https://energiaeambiente.org.br/um-milhao-estao-sem-energia-eletrica-na-amazonia-20191125

International Energy Agency (2022) Solar PV global supply chains. International Energy Agency. https://www.iea.org/reports/solar-pv-global-supply-chains
Leal LBB, Alva JCR (2021) Políticas públicas de acesso à energia elétrica, como ferramenta na efetividade dos direitos fundamentais/Public policies of access to electric energy, as a tool to make fundamental rights effective. Braz J Dev 7(8):82796–82823. https://doi.org/10.34117/bjdv7n8-473
Marques WPG, dos Anjos TO, Costa MNRF (2020) Medicinal plants used by riverside communities in the amazon Estuary. Braz J Dev 6(10):74242–74261. https://doi.org/10.34117/bjdv6n10-013
Nerini FF, Howells M, Bazilian M, Gomez MF (2014) Rural electrification options in the Brazilian Amazon: A multi-criteria analysis. Energy Sustain Dev 20:36–48. https://doi.org/10.1016/j.esd.2014.02.005
O'Reilly K (2012) Ethnographic methods, 2nd edn. Routledge, Abingdon, Oxon
Sherrieb K, Norris F, Galea S (2010) Measuring capacity for community resilience. Soc Indic Rec 99:227–247. https://doi.org/10.1007/s11205-010-9576-9
Sovacool B, Mukherjee I (2011) Conceptualizing and measuring energy security: A synthesized approach. Energy 38(8):5343–5355. https://doi.org/10.1016/j.energy.2011.06.043
Tuckman BW (1999) Conducting educational research. Harcourt Brace College Publishers, Orlando
Underwood G, Hill D, Lamichhane S (2020) Earthquakes, blockades and energy crisis: A conceptual framework for energy systems resilience applied to Nepal. Energy Res Soc Sci 69:101609. https://doi.org/10.1016/j.erss.2020.101609
Valer LR, Mocelin A, Zilles R, Moura E, Nascimento ACS (2014) Assessment of socioeconomic impacts of access to electricity in Brazilian Amazon: case study in two communities in Mamirauá Reserve. Energy Sustain Dev 20:58–65
Valer LR, Manito ARA, Selles Ribeiro TB, Zilles R, Pinho JT (2017) Issues in PV systems applied to rural electrification in Brazil. Renew Sustain Energy Rev 78:1033–1043. https://doi.org/10.1016/j.rser.2017.05.016
Wilson S (1977) The use of ethnographic techniques in educational research. Rev Educ Res 47:245265. https://doi.org/10.2307/1170130
World Health Organization (2021) Measuring energy access: a guide to collecting data using the core questions on household energy use. International Bank for Reconstruction and Development. Measuring Energy Access (who.int)

Biosocioeconomic Contributions to Restructuring the Sustainable Growth of the Main Value Chains in the Amazon

Antônio Cordeiro de Santana, Sérgio Castro Gomes, Ádina Lima de Santana, and Éder Silva de Oliveira

Abstract The economic agents in the state of Pará do not consider the value of natural assets to be used in productive activities because they act as price takers of ecosystem services whose adjustment dynamics are influenced by public policies and by the demands for food and raw materials. In this work, the opportunity cost of natural ecosystems was included in the cash flow methodology to estimate the bioeconomic viability of production systems and to reduce the negative environmental externalities generated by current land use systems. The goal of this work was to contribute to the development of value chains with social inclusion and sustainability of Amazonian ecosystems. The conclusion is that agroforestry systems presented higher bioeconomic viability indicators than the other systems analyzed. The agroforestry system can help generate income, employment and reduce deforestation and soil degradation in family farming.

Keywords Bioeconomics · Productive chains · Valuation of ecosystem services · Amazon

A. C. de Santana (✉)
Universidade Federal Rural da Amazônia (UFRA), Belém, Pará, Brazil
e-mail: acsufra@gmail.com

S. C. Gomes
Universidade da Amazônia (UNAMA), Ananindeua, Brazil
e-mail: sergio.gomes@unama.br

Á. L. de Santana
Kansas State University, Manhattan, KS, Estados Unidos, United States

É. S. de Oliveira
Universidade Estadual do Pará, Belém, Brazil

W. Leal Filho et al. (eds.), *Amazon 2030—Sustainability Issues in the World's Largest Rainforest Region*, World Sustainability Series,
https://doi.org/10.1007/978-3-031-81465-5_28

1 Introduction

Land use in the Amazon, due to the favorable conditions of soil, water, climate, and biodiversity of natural ecosystems for the development of agricultural, livestock, mineral, timber, and non-timber forestry value chains, requires as a necessary and sufficient condition the use of more productive, inclusive and sustainable good practices (Santana et al. 2023). Economic and socio-environmental governance must evolve to coordinate vertical and horizontal integration trajectories of production chains. Strategies for regulating the production and commercialization of domestic and global commodities must consider narratives about zero deforestation without anchoring in available knowledge about the channels of cause-and-effect systems that result from agents' decisions about land use and preservation of natural ecosystems (Farley 2008; Droste et al. 2019).

The dynamics of land use in the Amazon are defined by direct relationships between land and agricultural products markets, extractivism, and minerals, gaining momentum from the infrastructure of road, rail, and river modes, large economic projects, and increased urbanization. This dynamic presents complex and differentiated relationships depending on the heterogeneity of business size, technological standard, level of knowledge and organization of agents and institutions, access to credit, information, and the global market for primary and value-added products.

The preservation of ecosystems and mitigation of greenhouse gases (GHG) emissions requires the use of production systems with greater returns, social inclusion, and environmental sustainability. To achieve this, barriers to knowledge, use of technologies, and access to credit and technical assistance must be overcome (Santana et al. 2023). Among the various alternatives, progress can be made in the implementation of the carbon market associated with the payment for ecosystem services for micro and small agents in partnership with large enterprises, for financing regenerative production systems integrated into the links in the value aggregation and commercialization chains in broad markets. One of the quick ways is to estimate the value of the preserved natural assets of rural properties to be used as a guarantee of the credit contribution to be taken by agents to implement systems with high potential for carbon storage and promoting the regeneration of plant, aquatic, and soil microbiological ecosystems. Along the same line, the partnership between large projects and communities of family and extractive producers through Reduced Emissions from Deforestation and forest Degradation (REDD+) Projects so that investments are directed towards the implementation of locally productive and innovative systems can contribute to improving the quality of people's lives and the continued preservation of natural ecosystems (Santana 2020).

In this context, agriculture and extractivism developed on a small scale and with intensive production systems in the use of natural resources, unpaid family labor, and informal labor, together with limited access to credit, technology, and technical assistance (Santana et al. 2023; Olímpio et al. 2022), in addition to causing a strong impact on deforestation requires differentiated treatment for the choice of production systems that generate income, including local communities and reduce GHG emissions, soil degradation, water pollution and poverty and food insecurity. These activities continue to evolve and produce negative socio-environmental externalities with local, national, and international repercussions.

Therefore, which production systems can contribute to simultaneously meeting the different production scales of value chains with the potential for sustainable development in the Amazon? The hypothesis is that agroforestry production systems, silvopastoral systems, and circular integration systems, combining biotechnology with the rotation of activities, along with the use of residual inputs, increase productivity, per capita income, employment of labor, and environmental sustainability. The present study aimed to contribute to the development of bio-socioeconomic viability indicators of the main value chains and enable a new trajectory of economic growth, with social inclusion and environmental sustainability in the Amazon. The justification of this work consists of a methodology that includes the opportunity cost of natural assets and their socio-environmental benefit in the cash flow. With this, viability indicators, when considering private and socio-environmental costs, can attract public and private investments in the requirement of this methodology in the analysis for more productive and sustainable systems in the Amazon. The work also innovates via the estimation of the value of ecosystem services and their inclusion in the methodology for calculating bioeconomic indicators.

2 Methodology

The study area is the state of Pará, as it brings together the historical characteristics of the infrastructure expansion and intensive activities in the use of land and natural ecosystems in the Amazon. Therefore, the application of biosocioeconomic methodologies in the state of Pará serves as a basis for designing projects throughout the Amazon. The government's secondary database (IBGE 2019, 2022a, b, c; INPE 2019) and technical coefficients of productive activities were also used (Santana 2020; Agrianual 2021; Anualpec 2021). The data used in the cash flow structure was organized in Excel spreadsheets to apply the Net Present Value and Internal Rate of Return estimates. SPSS V18 was used to generate the spatial specialization indicators, adopted in the preparation of the map and to estimate the correlation matrix with the respective significance levels.

The biosocioeconomic performance indicators were estimated based on cash flows from a set of agricultural production systems (temporary and permanent crops), beef cattle farming, Agroforestry Systems (AS), and managed logging activities. In the economic dimension, the Uniform Present Value bioeconomic viability indicator was constructed to show the net present value of the annualized system. In the social dimension, the concept of equivalence between the number of man-days per year and formal employment was applied, considering the stability phase of the production systems cycle, in the ratio 280 dh = 1 Employment equivalent (Santana et al. 2023). The environmental dimension is incorporated into the cash flow calculation by including the opportunity cost of the natural assets of soil, water, and forest.

Uniform Present Value (UPV)

The Uniform Present Value (UPV) is an indicator of bioeconomic viability that makes the net present value of projects have a uniform flow per hectare per year. This indicator includes environmental opportunity costs, related to natural assets soil, water, and forestry, through cash flow net biorevenue. For agricultural and livestock activities, the interest rate was used to update flows of 10% per year and 4% for natural assets.

The important points of this indicator are that, on the one hand, it makes the results more understandable for entrepreneurs to support them in decision-making and, on the other hand, it allows the comparison of results between activities with different life cycles, according to Santana et al. (2023). The UPV calculation is given by Eq. 1.

$$\boldsymbol{UPV_t} = \left\{\left[\sum\nolimits_{t=0}^{T}\frac{(PRN_t + NB_t)}{(1+r)^t}\right] \times \left[\frac{r(1+r)^t)}{(1+r)^t - 1}\right]\right\} \tag{1}$$

where: UPV_t = Uniform Present Value, R$/unit/year; PNR_t = Private Net Revenue, R$/unit/year; NBt = Net Biorevenue (includes the costs and benefits of natural assets), R$/unit/year; r = interest rate of 10% p.a.; t = period in years (0, 1, …, T < P ∞).

Socioeconomic Location Quotient

The location quotient was used to identify whether a particular municipality has specialization in a given value chain, using the production value variable associated with the economic dimension and the formal employment variable reflecting the social dimension. In this case, the reference economy was the state of Pará and the economy under study is the municipality that houses each of the value chains.

Equations 2 and 3, respectively, refer to the calculation of the social and economic location quotients.

$$SLQ_i = \left(\frac{E_m^i/E_m}{E_P^i/E_P}\right) \tag{2}$$

$$ESQ_i = \left(\frac{PV_m^i/PV_m}{PV_P^i/PV_P}\right) \tag{3}$$

where: SLQ_i = Social location quotient of activity i; E_{im} = Number of jobs in activity i, in municipality m; E_m = Number of jobs for all activities i, in municipality m; E_{iP} = Number of jobs in activity i, in the state of Pará P; EP = Number of jobs in all productive activities in the state of Pará P; ELQ_i = Economic location quotient of activity i; PV_{im} = Production value of activity i, in municipality m; PV_m = Production value of all activities i, in municipality m; PV_{iP} = Production value of activity i, in the state of Pará; PV_P = Production value of all productive activities in the state of Pará.

For the *ELQ*, the variable Production Value of the production link in the production chains was used and for the *SLQ*, the Formal Employment variable was used for all links in the chains analyzed.

The variable Value of Production of crops, products from plant extractivism and aquaculture were obtained from the government's official statistics body (IBGE 2021a, b, c). The value of livestock production, for the year 2019, was estimated using Eq. 4.

$$LPV = \{[\mathrm{SA} \times (\mathrm{EBo} + \mathrm{EBu} + \mathrm{SGF}/8) \times \mathrm{RC} \times \mathrm{PM} \times \mathrm{PAB}] + \mathrm{VPd}\} \quad (4)$$

where: *VPP* = Livestock Production Value in R$ thousand; *SA* = 20% slaughter rate; *EBo* = Bovine herd in heads; *EBu* = Buffalo herd in heads; *SGF* = Sheep and goat flock in heads; RC = 50% carcass yield; *AW* = Average weight of the herd was considered to be 20 arrobas per head; *PBD* = Price of beef arroba (R$/@); *DPd* = Dairy Derivatives (milk and cheese) Production Value in R$ thousand.

When calculating the production value of poultry (chickens, turkeys, ducks, and quails), Eq. 5 was applied:

$$VPA = [(\mathrm{EA} \times \mathrm{CY} \times \mathrm{WA} \times \mathrm{PPA}) + \mathrm{VEP}] \quad (5)$$

where: *VPA* = Poultry production value in R$; *EA* = Number of birds in beaks; *RC* = 30% carcass yield; WA = Average weight of the flock of 1.15 kg per beak; *PPm* = Price of poultry meat (R$/kg); *VEPo* = Value of egg production in R$ thousand.

To categorize the indicator by degree of specialization and make the reader's visualization and understanding more explicit, the standardization of *QSL* and *QLE* is applied to vary between zero and 1. In this case, a SLQ_i or $ELQ \geq 0.7$ represents high specialization of activity *i* in municipality *m*; $0.4 \leq SLQ_i$ or $ELQ_i < 0.7$ indicates a normal level of specialization; $0 < SLQ_i$ or $ELQ_i < 0.4$ reveals that the municipality is not specialized in the activity under analysis; and SLQ_i or $ELQ_i = 0$ indicates the absence or non-existence of the activity, characterizing the activity, if present, as invisible because it is not included in official statistics. Standardization is based on the results generated by applying Eq. 6.

$$SLQ_{pi} = \left(\frac{SLQ_i - SLQ_{min}}{SLQ_{máx} - SLQ_{min})}\right) \quad (6)$$

where SLQ_{pi} *or* ELQ_{pi} is the standardized location quotient of social or economic specialization of activity *i*, SLQ_{min} *or* ELQ_{min} is the minimum location quotient of social or economic specialization of activity *i*; and $SLQ_{máx}$ *or* $ELQ_{máx}$ is the maximum location quotient of social or economic specialization of activity *i*, in the state of Pará.

3 Results

Agriculture and forestry activities in the state of Pará are still anchored in the use of informal labor and/or unpaid family labor. This, together with obstacles to land ownership rights, limitations on access to credit, use of technology, and remuneration of rural labor in accordance with legislation, has historically affected the productive succession of the main intensive rural activities systems in the use of labor and land.

The problem of land ownership and use is being resolved, as until June 2021, in the state of Pará, 50,162 land title documents were delivered, 47,234 of which were for agrarian reform settlers and 2,924 for occupants of public land (Santana et al. 2023). There is a need to evolve environmental licensing, based on the rural environmental registry, to enable the legal use of land, access to credit, and the broad market. With this, a favorable environment can be created for organized family micro-entrepreneurs to increase income and employment, using good sustainable production practices.

Formal employment in agriculture in the state of Pará represented, in 2019, only 12.8% of the total use of informal labor and/or unpaid family labor. This is a challenge to be resolved to allow the insertion of production generated in rural properties of micro and small family entrepreneurs and extractivists from agribusiness value chains into the broad market.

Agricultural and forestry enterprises are labor intensive, even seasonal, which helps to mitigate rural–urban migration within municipalities, between municipalities in the state, and between states and regions. Table 1 lists the labor employment potential of the main production systems in the state of Pará, estimated in employment equivalent. In this case, the ratio of 280 man-days was used, equivalent to a permanent job. In the area occupied by açaí and AS activities, in general, the employment of labor does not present significant differences between the enterprises. Therefore, the employment-hectare relationship reflects the average potential of these production systems to generate employment.

In effect, the potential of value chains to generate employment and income can be estimated. The permanent and temporary crops analyzed generate a uniform present value per hectare per year between R$663.08 for palm oil and R$8,928.74 for an irrigated AS (Table 1). Extensive beef cattle farming had the lowest economic performance with a value of R$511.65, and logging from state forests with a value of R$956.49. The combination of the two columns in Table 1 estimates the value of labor productivity for each chain.

Some crops are developed by medium and large companies as plantations, given the large area cultivated in monocultures such as soybeans and palm oil, in combination with micro and small family enterprise systems such as açaí, cocoa, AS, cassava, and palm oil produced by small farmers integrated into agro-industries. Much of the production in these chains is distributed through short marketing circuits, generally involving direct transactions with the market and/or with middlemen. The telling point is that such activities, regardless of scale, technology, and market orientation, generate positive net returns.

Table 1 Bioeconomic indicators of employment performance and income from agricultural and forestry activities in the state of Pará, base year: 2021

Agribusiness chains	Area (ha)	ha/Employment*	R$/ha/year
Açaí managed in the floodplain/ irrigated on dry land	188,015	2.8 ha: 1 E	1,505.87
Cocoa (almond)	140,549	4.0 ha: 1 E	1,451.28
Palm oil (bunches of fresh fruit)	164,170	4.5 ha: 1 E	663.08
Cassava (root)	272,905	3.1 ha: 1 E	2,469.02
Soybean (grain)	577,764	75 ha: 1 E	663.08
Irrigated SAF (Açaí, Cocoa, Pepper, Mahogany)	79,864	2.1 ha: 1 E	8,928.74
State forest (logging) (1)	7,102,760	14 ha: 1E	956.49
Beef livestock (average of non-scale systems)	13,703,335	60 ha: 1 E	511.65

Source (*) 280 dh ≅ 1 Job. The IBGE does not separate activities into monocrop, intercropping and AS, so areas with AS can double count the areas with crops that compose them and which are also grown in monocrop such as açaí, cocoa, citrus, and black pepper. (1) Areas of the state forests of Faro, Paru, and Trombetas

Regarding the potential to generate employment, there are differences between intensive systems in the use of informal and/or unpaid family labor, in combination with land, and reduced use of mechanization and external inputs such as cassava, açaí, and cocoa. Other systems are intensive in the use of chemical inputs and mechanization, such as soybeans. In family production systems, the capacity to occupy labor, already converted into employment equivalent, is superior to mechanized grain systems, extensive beef cattle farming, and timber logging. The limitation of these systems is the low scale and organization of producers in cooperatives.

In agroforestry systems and low-carbon agricultural monocultures such as açaí, cocoa, and palm oil, the relationship between cultivated land and labor occupation is much lower than in grains and beef cattle farming. Thus, a trajectory of inclusive growth can quickly evolve, with greater interaction between the economic and social dimensions for the technological management and governance of these agribusiness value chains.

Only the primary production link in these chains, whose labor-land relationship was studied, has the potential to generate the equivalent of 1,008,264 jobs, which represents 8.03 times more than formal employment in 2019, accounted for in all links in the Pará agribusiness value chains (Table 1). Therefore, it is essential to create the conditions for this segment of the economy to evolve into a formal product and factor market. With this, micro and small organized family enterprises tend to break the barriers of scale, environmental degradation, quality, and competitive prices of products, expanding the added value of production, formalizing the employment of labor, and adopting good innovative, inclusive, and sustainable practices. In this way, the configuration of economic growth trajectories changes, evolving from pure cumulative and unequal economic growth to growth with social inclusion and environmental sustainability.

In this evolutionary trajectory of regional development, property rights and access to credit are the potential elements that enable technological transformation and business management. At this point, the value of the natural asset immobilized in rural properties can be used as an important element for access to credit and enable producers to capitalize The estimated value of natural assets immobilized on rural properties can reach the value of R$54,929.66 million per year to form the basis of a receivables fund (Santana 2020), the results of which must be used to enable investments in restoration, recovery and/or enrichment of natural ecosystems and adoption of systems with greater productivity, social inclusion and sustainability.

Thus, the natural asset contributes to entrepreneurs' access to Payment for Environmental Services (PES) programs and policies within the scope of the Forest + Carbon Program, Investment Fund in Agro-industrial Production Chains (FIAGRO), Green Bonds (TV), Reduction of Greenhouse gas emissions from Reducing GHG Emissions, Deforestation and Forest Degradation (REDD+) and, mainly, as a guarantee for access to credit and as capital for negotiation in the global carbon market, defined at the United Nations (UN) Conference for Climate Change (COP26) and COP30 to be held in 2025 in the state of Pará, Brazil. This is one of the points of innovation to leverage investments in the productive restoration of degraded areas, enrichment of natural assets, implementation of agroforestry systems, integrated crop-livestock-forestry systems and their combinations, within the scope of low-carbon and carbon–neutral agriculture programs encouraged by several national and international public and private institutions.

REDD+ can be understood as a PES scheme, structured in multilayers in which liabilities and credits for the reduction of carbon emissions are set by international buyers who emit GHG and remunerate landowners for maintaining forest areas. In this process, there is a set of financial, social, and environmental benefits for stakeholders (Santana 2020). Additionally, the REDD+ policy promotes social benefits, such as improving living conditions for indigenous peoples and forest-dwelling communities and defining land ownership rights (Nhemb et al. 2017). The REDD+ policy seeks to build a governance system that coordinates relationships between stakeholders, indigenous people, communities that depend on the forest, and state and non-state actors. This management arrangement discusses the underlying causes of deforestation, its vectors, and actions to mitigate and repair strategies aimed at engagement and ensuring participation, legitimacy, accountability, and transparency in climate change mitigation and sustainable forest management (Nhemb et al. 2017).

3.1 Biosocioeconomic Connections of Amazon Value Chains

The biggest challenge for Amazon's value chains is to meet national and international demands for social inclusion and the environmental sustainability of production systems. In rural establishments in the Amazon biome, according to the 2017

Agricultural Census, only 36.4% of the area is maintained as legal reserves, permanent preservation areas, reserves, and native forests. In principle, there is a 43.6% imbalance to comply with the Forest Code. This imbalance is due to the use of unsustainable practices in production systems, which continue to destroy the native forest, depleting the natural productivity of the soil, polluting the air and water, and destroying biodiversity.

Given this high degree of environmental liabilities, carried by local productive and non-productive agents, deforestation and forest burning continue without effective control. At this point, the state of Pará is leading the destruction of the natural ecosystems of the Amazon and is putting increasingly at risk the access to the international market achieved by entrepreneurs who use good inclusive, and sustainable practices.

Table 2 presents the correlations between the increase in the area of primary forest deforested in private and public rural areas and the main agricultural, livestock, forestry, and extractive activities. Values with four decimal places and in bold have a statistical significance of less than 10%, and those with two decimal places do not have statistical significance. Correlations in absolute values above 0.6 are considered high, between 0.3 and 0.6 intermediate, between 0.14 and 0.30 low, and less than 0.14, non-significant. Column 1 of the numbers shows the direct correlations between municipalities with an increase in deforested areas and productive activities. The other columns contain cross-correlations between productive activities, which can influence, directly or indirectly, the increase in deforestation.

The results in column 1 of numbers in Table 2 show a high correlation between the increase in deforestation with livestock production and palm oil, a low correlation with logging and cocoa, and a non-significant, or neutral, correlation with soybean and temporary crop. Municipalities, where livestock farming is developed extensively with large herds, maintain high correlations with forest deforestation. The same behavior occurs in municipalities specialized in palm oil. This is a point

Table 2 Correlation matrix between deforestation and the main agribusiness value chains in the state of Pará, 2019

Activity	IADeforested	PV livestock	PV palm oil	PV logging	PV cocoa	PV soybean	PV temporary crop
IADeforested	1						
PV livestock	**0.6143**	1					
PV palm oil	**0.6117**	**0.5272**	1				
PV logging	**0.2347**	–0.070	**0.9032**	1			
PV cocoa	**0.2639**	0.090	**0.3622**	–0.12	1		
PV soybean	–0.090	–0.060	0.000	**0.1576**	**–0.1716**	1	
PV temporary crop	0.060	**0.1494**	**0.5446**	**0.1578**	–0.08	**0.8998**	1

Source Research data. PV = Production value

to be considered to understand the phenomenon and guide actions to mitigate the environmental problem and meet market demands.

Livestock farming in the Amazon is under the global observatory of the consumer market. This positive and significant association with deforestation can worsen the problems of embargoes and the closure of part of the international market for beef cattle products, a fact that is already in advanced consolidation by the European Union. Therefore, change initiatives towards inclusive and sustainable technologies and innovations in production processes and business management must evolve more quickly and transparently. Farms that adopt good practices in municipalities with high deforestation rates suffer the effects produced by those that move in the opposite direction to sustainable development. In effect, it is necessary to have alternative production systems with clear, achievable goals to reduce the deforestation rate, increase the average productivity of soils and herds, recover water sources, preserve biodiversity, expand social inclusion, and implement traceability of livestock value chains and other commodities.

The municipalities that produce palm oil, wood, and cocoa have positive and significant correlations with deforestation, showing that the growth of these activities is associated with high levels of deforestation in the municipality (Table 2). These activities are part of the international observatory, as they lead to deforestation of the Amazon forest, problems with informal labor, and environmental impacts on soil, water, and biodiversity.

Livestock, column 2 of numbers in Table 2, presents a positive and significant correlation with palm oil and temporary crops. In municipalities dedicated to temporary crops and palm oil, there is a positive association with cattle herds. In turn, municipalities with palm oil maintain a strong association with temporary crops and the extraction of wood and non-wood products. This result is important due to the association between agroforestry production systems, in their various combinations, with low-carbon livestock farming.

In general, the results indicate the continuity of deforestation, given that the accommodation of activities replaced by losses in competitiveness and/or expanded by favorable market conditions can be made viable with the inclusion of new areas of land, largely coming from native forests. In addition to the demand for more deforested land, all activities that require energy to process products resort to the use of biomass to produce energy, which reinforces the need for deforestation.

The relationship between agricultural, livestock, and extractive activities with deforestation and wildfire in the Amazon forest is complex. Although livestock farming, soybeans, and palm oil are identified as frontline activities for deforestation in the state of Pará, all rural activities, whose bio-socioeconomic support base is land, influence deforestation. One basis lies in the direct correlation of 0.807 between the increase in population and the deforested area in the period from 2000 to 2021, whose source of needs is met by natural assets of soil, water, and forest.

In the analysis of the environmental dimension, the deforestation variable was used to indicate the degree of sustainability of the agricultural, livestock, timber, and non-timber forestry value chains. The 10 municipalities with deforestation equal to or greater than 20% of the largest deforested area, equal to 575.33 km^2, in 2019, which occurred in Altamira, were classified as having a high level of environmental impact (Fig. 1). The deforested area in these 10 municipalities was 2,896.62 km^2, equivalent to 64.91% of the total. The 114 municipalities that deforested an area greater than zero and less than 20% were considered to have a low level of environmental impact. In this case, deforestation occurred in private areas and forest reserve areas, which requires surveillance to control the destruction of these natural ecosystems. In 2019, there was deforestation in 124 municipalities in Pará and, in another 20 municipalities, no deforestation was recorded according to data from INPE (2021). Municipalities with palm oil, cocoa, and livestock farming showed significant correlations, 0.612, 0.264, and 0.614, respectively, with deforestation.

In Fig. 1, the spatial interaction within the municipalities between livestock farming, temporary crops, and deforestation is qualified. Livestock and temporary crops indicators already combine the economic and social dimensions, through their potential to generate income and employment. In the municipalities with the highest deforestation such as Altamira, São Félix do Xingu, Pacajá, Novo Repartimento, and Novo Progresso, livestock farming appears as the chain most specialized in generating income and employment, followed by the chain of temporary crops at a low level of specialization, that is, there are significant correlations between these activities. In the same line of reasoning, the correlation between livestock farming and temporary crops is verified in the municipalities of Santana do Araguaia, and Santa Maria das Barreiras with a high level of specialization. Highly specialized livestock

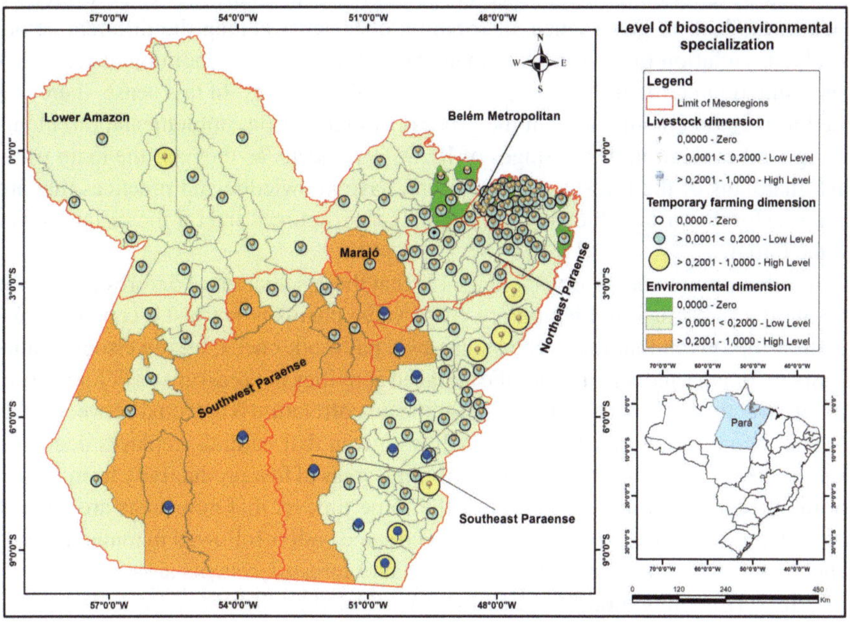

Fig. 1 Spatial location of municipalities specialized and not specialized in livestock farming, temporary crops, and deforestation, state of Pará, 2019. *Source* prepared by the authors

farming is associated with temporary crops with low specialization in the municipalities of Água Azul do Norte, Cumaru do Norte, Xinguara, and Itaituba, both with low levels of deforestation.

Despite the necessary caution in generalizing the results, the livestock and soybean value chains are not meeting the requirements of social inclusion and sustainability. Good bioeconomic practices must be applied and generalized in these production systems to increase productivity, reduce deforestation, recover degraded pasture areas, and restore, enrich, and/or preserve soil fertility, biodiversity, and water sources in natural ecosystems. Failure to meet the social and sustainable requirements of these chains weakens central aspects of REDD+ initiatives such as the right to land ownership, equitable and transparent distribution of benefits, respect for cultural traditions, tacit knowledge, and the preservation and recovery and/or conservation of natural ecosystems.

In the social dimension, there are strong deficiencies due to the high level of use of informal and/or unpaid family labor, limitation in the distribution of added value, the low level of animal comfort and welfare, and respect for the culture and knowledge of local communities. In the municipalities of the Southeast and Southwest mesoregions, there are agrarian reform settlement projects that only received their land titles in June 2021, forest reserves, and indigenous lands where illegal mining is practiced, in addition to land conflicts, limitations on access to credit, product markets, worker rights, and the use of bioeconomic technologies and innovations.

Given this reality, the implementation of REDD+ projects can be seen as an opportunity for rural communities, however, there are some challenges to be overcome such as the lack of governance, distribution of benefits among the different actors involved, violation of territorial, human and cultural rights of communities, which has resulted in conflicts and socio-environmental injustices. In this sense, there is a need to include communities in the process of elaboration, implementation, monitoring, and evaluation of the stages of REDD+ projects, as they are the main users and guardians of the forests, and, therefore, have knowledge and practices that are essential for the sustainable management of these resources and for the sustainability of the initiative.

Another point to be observed in Fig. 1 is that the association of deforestation with the livestock farming and temporary crops value chains is not just a contemporary relationship, given that the soybean, cassava, palm oil, cocoa, and livestock chains compete for area and, in this case, the replaced activities put pressure on deforestation to shift over time. Thus, a positive and significant correlation is detected between palm oil and deforestation in some municipalities, given that palm oil replaces degraded pasture areas from extensive livestock farming, areas of low productivity from family farming (cassava, corn, and beans), at the same time as that new areas are opened for agriculture and livestock production. The same path is followed in municipalities where areas with cocoa and/or agroforestry, and forestry systems are expanding. In fact, the surrounding municipalities are also affected by these activities.

Under these conditions, policies and projects aimed at family micro-entrepreneurs and small and medium-sized capitalist entrepreneurs are a necessary and sufficient condition to achieve an inclusive and sustainable growth trajectory, based on the value chains of meat, soybean, and differentiated products from the Amazonian socio-biodiversity. In the absence of public policies, the scale, product quality, and competitive prices corroborate the expansion of production systems in the form of monocultures, increased concentration of added value, less social inclusion, and greater environmental impact. As a result, carbon–neutral, inclusive, and highly profitable bio-socioeconomic production systems of micro- and small family and capitalist enterprises are further away from the global market.

From this perspective, companies are expanding their corporate social responsibility actions as a way of responding to pressure from consumers and conservation organizations due to environmental damage and climate change. Large companies are adopting the ESG (Environmental, Social, and Governance) concept as a marketing strategy to demonstrate integration with sustainable development objectives and, therefore, reduce the reputational risk in global markets. This transitional shift from a production model that depletes natural resources to processes anchored in social inclusion, reducing deforestation and its effects on the climate, and the creation of sustainable value are adopting REDD+ projects for social inclusion, and environmental sustainability.

The economic, social, and environmental dynamics linked to these value chains depend on the contribution of technical and scientific knowledge in all links of the chains and can be supplied by connection with teaching, research, and extension institutions if engagement is continuous over time. Synergistic and broad actions of this type have not yet been carried out between institutions in the Amazon. Therefore, the shift from current production systems to a trajectory of inclusive and sustainable growth in the Amazonian production chains depends on the joint action of the local institutional arrangement and its cooperative interaction with other knowledge hubs in Brazil and the rest of the world.

4 Conclusions

The heterogeneity in the competitive performance of the production chains in the state of Pará is the result of the historical orientation towards the settlement and development of the Amazon, without the guarantee and respect for the culture and traditions of the local population, in the first place, and without the implementation of regulatory frameworks for the rights of land ownership, worker safety and rights, and productive and management technologies and innovations for the sustainable use of natural ecosystems. These vectors of linear growth have become bottlenecks for the sustainability of enterprises and need to be linked a bio-socioeconomic management of intensification, efficiency, and diversification of the expanding production systems in the municipalities of Pará.

The expansion of commodities, settlements, extractive reserves, and population growth has had a strong impact on deforestation and a trajectory of non-inclusive and unsustainable local development. The implementation of local low-carbon production and innovation systems requires a large contribution of resources for investment and the formation of human and social capital. In the global context of the demand for sustainable products, using the value of natural assets of rural properties as a guarantee for access to credit is a viable alternative for the widespread implementation of more productive, inclusive, and sustainable good bio-socioeconomic practices in rural establishments.

By valuing natural assets and ecosystem services, REDD+, PSA, and other investment modalities can be made viable to reduce deforestation, and degradation, and restore and/or enrich degraded natural ecosystems. It is a way of reducing inequality between family entrepreneurs and capitalists, through productive and integrative management of economic, social, and environmental dimensions.

For this, we rely on the institutional arrangement formed by universities, and governmental and non-governmental organizations to develop innovative projects of good sustainable practices to restructure the growth of agribusiness. Integrated cooperation between the agents of this institutional arrangement can overcome the technological and innovative limitations of production systems via increased productivity, bioeconomic efficiency, distribution of added value, social inclusion, environmental sustainability, and connection with markets.

The main limitation of the work is the difficulty in spreading and implementing this methodology in family production systems, due to lack of information, technical assistance and rural credit.

References

Agrianual (2021) Anuário da agricultura brasileira. São Paulo

Alencar A, Pereira C, Castro I, Cardoso A, Souza L, Costa R, Bentes A J, Stella O, Azevedo A, Gomes J, Novaes R (2016) Desmatamento nos assentamentos da Amazônia: histórico, tendências e oportunidades. IPAM Brasília

Amorim IA, Mello AH, Homma AKO, Pontes AN (2020) Dinâmica do desmatamento em área de assentamento de reforma agrária no sudeste paraense. Rev Caminhos Geogr, 21(78):21–35.https://doi.org/10.14393/RCG217851483

Anualpec (2021) Anuário da Pecuária Brasileira. São Paulo

Droste N, Farley J, Ring I, May PH, Ricketts TH (2019) Designing a global mechanism for intergovernmental biodiversity financing. Conserv Lett e12670:1–8 https://doi.org/10.1111/conl.12670

Farley J (2008) The role of prices in conserving critical natural capital. Conserv Biol 22(6):1399–1408. https://doi.org/10.1111/j.1523-1739.2008.01090.x

Hajjar R, Enbringe G, Kornhauser K (2021) The impacts of REDD+ on the social-ecological resilience of community forests. Environ Res Lett 16(20):1–10. https://doi.org/10.1088/1748-9326/abd7ac

IBGE—Instituto Brasileiro de Geografia e Estatística (2019) Censo Agropecuário. Rio de Janeiro. http://www.ibge.gov.be

IBGE—Instituto Brasileiro de Geografia e Estatística (2022b) Produção pecuária municipal, 2019. http://www.ibge.gov.be

IBGE – Instituto Brasileiro de Geografia e Estatística (2022a) Produção agrícola municipal, 2019. http://www.ibge.gov.be

IBGE—Instituto Brasileiro de Geografia e Estatística (2022c) Produção da Extração vegetal e da silvicultura. http://www.ibge.gov.be

INPE—Instituto Nacional de Pesquisas Espaciais (2019) Desmatamento dos municípios da Amazônia. http://www.obt.inpe.br/OBT/assuntos/programas/amazonia/prodes

Nhemb S, Lee YJ, Phin S (2017) Sustainable management of forest in view of media attention to REDD+ policy, opportunity and impact in Cambodia. Forest Policy Econ 85:10–21. https://doi.org/10.1016/j.forpol.2017.08.011

Olimpio SCM, Gomes SC, Santana AC (2022) Patterns of production and sustainability of cattle ranching in the state of Pará—Brazilian Amazon. Semin: Cienc Agrar 43:541–560. https://doi.org/10.5433/1679-0359.2022v43n2p541

Oliveira GMTS, Santana AC, Oliveira ES, Silva RJ, Santos WAS, Santana AL, Costa VCN (2020) The value of agroforestry ecosystem services provided in rural communities in the Eastern Amazon (Tomé-Açu—PA, Brazil). J Agric Stud 8(4):203–216

Santana AC, Santana Ádina L, Santana Ádamo L, Oliveira GMTS, Santos MAS (2023) Bioeconomic evaluation of an agroforestry system and the potential to recover degraded areas and capitalize producers in the state of Pará, Brazilian Amazon. Desenvolvimento e Meio Ambiente. https://doi.org/10.5380/dma.v61i0.80516

Santana AC (2020) Bioeconomia aplicada ao agronegócio: mercado, externalidades e ativos naturais. Piracanjuba: Editora Conhecimento Livre. https://doi.org/10.37423/2020.edcl190

Environmental, Organizational and Productive Perceptions of Artisanal Fisherwomen and Family Farmers: A Case Study in Santarém, Pará, Brazil

Wandicleia Lopes de Sousa, Samela C. Bonfim, Elizabete de Matos Serrão, and Thiago Almeida Vieira

Abstract This study examined the environmental perception, organization, and production of artisanal fisherwomen and family female farmers. This is a case study conducted with 11 artisanal fisherwomen associated with the Z-20 Fishermen's Colony in Santarém Pará and 07 family female farmers associated with the Union of Rural Workers of Santarém (Pará), who live in the neighborhoods Maicá and Pérola do Maicá in Santarém, Maicá Lake. This is an area in the process of struggle and resistance due to the imminence of the implementation of bulk ports. The results indicate that the living space in the neighborhood is of high importance in their lives. The participating women were organized into class entities representing their categories (fisherwomen and female farmers). In this way, the environmental perception and the relationship with the community organization influence the productive process of these women, especially in the context of resistance to agribusiness, as a way of conserving natural resources and their ways of life.

Keywords Family agricultural · Artisanal fishing · Environment perception · Amazonia

W. L. de Sousa (✉) · S. C. Bonfim · E. de Matos Serrão
Environmental Sciences at the Graduate Program in Society, Nature and Development (PPGSND/UFOPA), Santarém, Pará, Brazil
e-mail: wandicleia@ufopa.edu.br

S. C. Bonfim
e-mail: samelabonfim@gmail.com

E. de Matos Serrão
e-mail: elizabete.serrao@tnc.org

T. A. Vieira
Federal University of Western Pará (UFOPA), Santarém, Pará, Brazil
e-mail: thiago.vieira@ufopa.edu.br

W. Leal Filho et al. (eds.), *Amazon 2030—Sustainability Issues in the World's Largest Rainforest Region*, World Sustainability Series,
https://doi.org/10.1007/978-3-031-81465-5_29

1 Introduction

This study was carried out involving fishing and agriculture in peri-urban areas of Amazonian cities and reveals how these activities are part of the daily lives of people who migrate from their communities to live in other environments. Thus, this research was implemented with women farmers and fisherwomen in two neighborhoods located on the banks of the Maicá River, in the municipality of Santarém, Pará. In this way, the results highlight the limits and challenges, contributing to the proposal of public policies aimed at organizing the use of natural resources as a means of subsistence.

This reality is beginning to change as a result of effective participation in social and trade union struggles and, more recently, involvement in political decisions across the country (Leone and Teixeira 2010). As a result of this process, the self-affirmation of women's presence in the family, social, political and economic environment began, culminating in the recognition of their productive and reproductive work. This process became evident after industrialization (Hoffmann and Leone 2004).

The activities of artisanal fishing and family farming have been carried out by families since primitive times, using natural resources to guarantee the livelihood and quality of life of their family units. Artisanal fishing is characterized as the action of extracting, harvesting or capturing fish from the environment. It is an important source of protein and in the Amazon region, it is consumed at approximately 370 g/day, corresponding to 135 kg/year, an amount considered high in terms of per capita fish consumption worldwide (Barthe and Fabré 2004; Borcem et al. 2011; Serrão 2018). Family farming is developed using rudimentary planting techniques, considered to have a low impact on the environment and important for the balanced use of existing natural resources, involving cultural, social, and economic diversity, and can change from the traditional peasant way to modernized small-scale production (Wanderley 2003).

Women have a strong presence in the development of artisanal fishing and family farming, but this female effort began to be evidenced in the 1970s (Bruschini 1994). The lack of scientific publications on the presence of women in these activities has contributed to the invisibility of their work in generating family income (Santos et al. 2017).

Brazil has five regions with different ecosystems, including the North, where the Amazon is located, known for its ethnic, cultural, social, environmental and economic diversity, inherited from the traditional populations (indigenous people, rubber tappers, extractivists, Quilombolas,[1] family farmers, artisanal fishermen and others) who live and practice subsistence economic activities in this territory (Lima and Pereira 2007).

[1] "They consist of groups that have developed daily practices of resistance in the maintenance and reproduction of their characteristic ways of life and in the consolidation of a territory of their own" (O'Dwyer 2002).

In this context, the study was carried out in Santarém, Pará, located on the banks of the Tapajós and Amazonas rivers, which serves as an interchange between the two capitals Manaus (AM) and Belém (PA). The municipality uses three modes of transport, the main ones being waterway and road, and is considered a "polarizing center because it offers better economic and social infrastructure (schools, hospitals, universities, roads, ports, airport, communications, industry and commerce, etc.) and has a more developed service sector" (Oliveira et al. 2010; Prefeitura Municipal de Santarém 2013).

The economy is diversified, with artisanal fishing and family farming playing key roles in generating employment and income. Both are responsible for ensuring food security for the majority of families. This article aims to highlight the environmental perception, organization and production of women artisanal fishermen and family farmers.

2 Methodology

This research is characterized as quantitative and qualitative because, according to Oliveira (2008), the use of instruments to collect data through questionnaires, interviews, observations and the use of statistical techniques helps to describe the socioeconomic profile of the artisanal fisherwomen and family farmers participating in the study. Qualitative research helped in the analysis of the information collected from the research group, enabling a detailed interpretation of phenomena in "social processes and the interpretation of the particularities of individuals' behaviors or attitudes" (Prodanov and Freitas 2013, p. 70).

The techniques used make up the characteristics of the case study, which helped to carry out a detailed examination of the facts. According to Yin (2005, p. 34), "the case study allows an investigation to preserve the holistic and significant characteristics of real-life events—such as individual life cycles, organizational and administrative processes, changes occurring in urban regions". Following Mattos' (2011, p. 51) assumptions, it is through this research method that the daily, community or individual experience of the researched public is perceived.

The study is made up of primary and secondary data. The primary data was obtained through in-loco research with artisanal fisherwomen and family farmers, using semi-structured interviews. Secondary data were collected through documentary research in the CAPES theses database.

2.1 Context of the Geographical Space of the Research

The neighborhoods where the research was carried out are located on the shores of Lake Maicá, "a complex ecosystem made up of islands and a part of dry land, which suffer minor flooding from the flooding of the streams that exist there, and the lake that surrounds it" (Vieira 2013, n/p), where approximately 604 (six hundred and four) families live, around 2,980 inhabitants, including traditional populations (artisanal fishermen, family farmers and Quilombolas) and other people (Prefeitura Municipal de Santarém 2013). The site is considered a natural nursery for various species of fish, turtles, reptiles, birds, as well as flora, which includes a biodiversity of aquatic plants typical of the Amazon region that are environmentally important to the Amazon ecosystem (Vaz 2016). A significant part of the residents of these neighborhoods survives on primary activities with rural characteristics linked to artisanal fishing, vegetable production, small and large animal husbandry, with production destined for consumption and supply from the local market (Fig. 1).

Maicá is an area that has undergone recent transformations that can cause changes to the landscape and ecosystem, as well as damage the way of life of the traditional Quilombola, indigenous, fishermen and family farmers who live in this region (Prefeitura Municipal de Santarém 2013). These can be seen in the changes to the landscape resulting from investments in recent years, with these economic activities causing damage to the region's ecosystems and accelerating processes such as soil erosion, contamination of watercourses, destruction or fragmentation of natural habitats and vegetation, migrations which, when located in densely populated areas, can result in expropriation for the management of economic activities (Silva 2014).

2.2 Data Collection

Data collection took place in 2018, following the requirements of the Resolution of the Brazilian National Health Council—CNS No. 4.666/2012, which guides research involving human beings, respecting the principles of autonomy, beneficence, non-malice, justice and equity. In addition, we spoke to the coordinators of the Santarém Rural Workers' Union and the Z-20 Fishermen's Colony to ask for authorization.

With the authorization of the Z-20 Fishermen's Colony and the Sindicato dos Trabalhadores e Trabalhadoras Rurais de Santarém (STTR), and after obtaining the Consubstantiated Opinion of the Research Ethics Committee, number 2.452.021, from the Instituto Esperança de Ensino Superior (CEP- IESPES), the research began in the Maicá and Perola do Maicá neighborhoods. We followed the meetings of the organizations linked to the STTR and the Z-20 Fishermen's Colony. The purpose of attending these meetings was to get to know the organizational routine of artisanal fisherwomen and family farmers.

Each participant had access to the Free and Informed Consent Form (TCLE), a tool through which artisanal fisherwomen and family farmers expressed their consent to participate. According to the coordination of the Pérola do Maicá base group linked to the Z-20 Fishermen's Colony, there were 19 artisanal fisherwomen linked to the organization, but only 57.9% of the fisherwomen (11 women) accepted to participate in the study. As for the total universe of family farmers, STTR coordinators have a list of 20 female members who work in these neighborhoods. The study was carried out

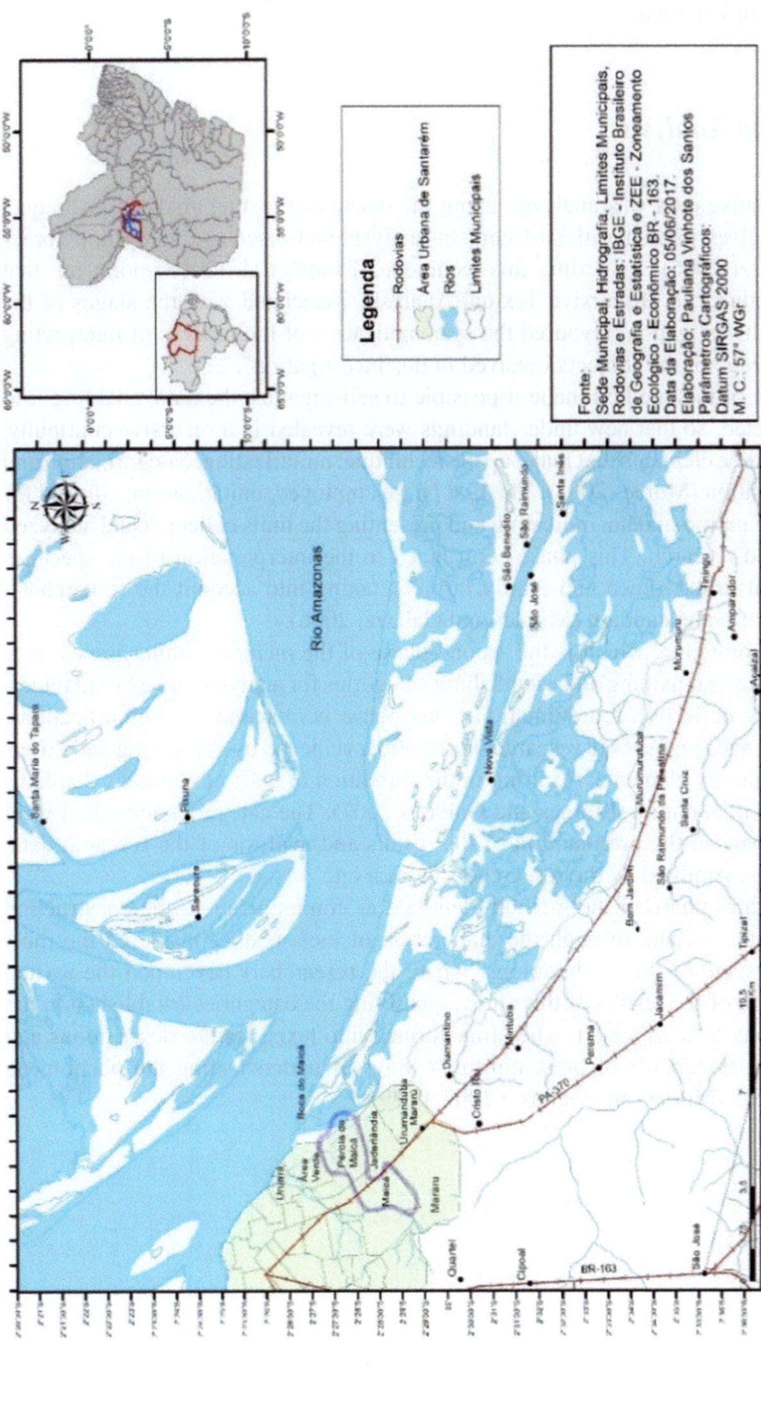

Fig. 1 Maicá region, Santarém, Pará, Brazil. *Source* Prepared by the authors

with 35% of female family farmers who met the pre-established inclusion criteria, representing 7 in total.

2.3 Data Analysis

The qualitative data was analyzed using the discursive textual analysis technique, which involves particularities of content analysis and discourse analysis (Moraes and Galiazzi 2006). Regarding this technique, Torres et al. (2008) point out that "the procedures of Discursive Textual Analysis, associated with the stages of the Thematic Investigation, favoured the systematization of the process of interpreting the knowledge of the subjects involved in this investigation".

The use of this method made it possible to self-organize the data, enabling it to be interpreted, so that new understandings were revealed in a recursive continuity, using the three elements that make up the technique: unitarization, categorization and communication (Moraes 2003). The first phase employed, unitarization, allowed for the fragmentation of data, indicating and presenting the units collected and analyzed during field research. This phase contributed to the interpretation of the speeches of artisanal fisherwomen and family farmers, taking into account the researcher's condition of understanding (Moraes and Galiazzi 2006).

In categorization, which is the second phase of the method, similar issues were brought together, naming and establishing categories for analysis and a broad understanding of collective representation. This phase corresponds to simplifications, reductions and syntheses of research information, achieved by comparing and differentiating unitary elements, resulting in the formation of a set of elements that have something in common (Moraes and Galiazzi 2007). The categories identified were used to organize the presentation of the results and analysis of the research data, using phrases emitted by the researchers in each one.

In the third phase of the method identified as communication, the construction of the content began, through the elaboration of meta-texts, considered the most important stage of the methodology, where the researchers developed the textual construction of the results of this study, qualifying the categories established in the previous phase "[…] which, when transformed into texts, lead to descriptions and interpretations capable of presenting new ways of understanding the phenomena investigated" (Moraes and Galiazzi 2007, p. 89).

3 Discussion and Results

3.1 *Environmental Perception of Artisanal Fisherwomen and Family Farmers*

Perception is a way of dialoguing, understanding and valuing the relationship between the body and the environment, so works that seek to understand perception must take into account social, cultural, environmental, economic and historical aspects (Chauí 2002). Costa and Colesanti (2011) state that research into environmental perception is essential to help understand how human beings perceive, analyze, live with and adapt to the environment, and how environmental issues influence people's daily lives collectively or individually. Understanding environmental perception requires bringing together elements from various fields of knowledge. It is essential to have an interdisciplinary vision because in order to carry out a good analysis of biological data, it is necessary to add socioeconomic information (Bresolin et al. 2010).

According to Serrão (2018, p. 61), "the environment is perceived by people in various ways and some factors influence the relationship between the perception of different communities, which can generate different responses between individuals [...]". The perceptions of the participants in the survey were at times different and at others similar in terms of the topics covered by the survey form. The environmental perception was considered to be the value judgment made by the interviewed women in relation to their homes (Fig. 2).

Analyzing the perception of fauna and flora, it was noted that these elements are considered important or highly important by the respondents. The interviewees consider Maicá Lake to be an environment of high importance (73% of the fisherwomen and 71% of the farmers) or important (27% of the fisherwomen and 29% of the farmers). However, the environment is used differently by them: 73% of the artisanal fisherwomen use the lake to carry out their economic activity, 36% use it as a leisure space and 9% to carry out tourism-related activities. As for the family farmers, 57% use the lake for leisure activities, 29% do not use it for any activity and 14% use it for domestic activities.

This is a reality affirmed by Diegues (2008, p. 63) in "O Mito Moderno da Natureza Intocada" (The Modern Myth of Untouched Nature), in which he points out that traditional populations (family farmers, artisanal fishermen, Quilombolas, indigenous people, extractivists) have a mystical relationship of dependence and appreciation of natural resources, so the strategies used to use the resources existing in nature follow the cycles of nature. The author also states that "[...] there is a symbiosis between man and nature, both in the field of doing, techniques and production, and in the symbolic field [...] the time for fishing, hunting and planting is marked by ancestral myths [...]".

The Maicá area has been undergoing constant transformations caused by various factors: disorderly urbanization, monoculture plantations, industrialization, population growth, implementation of major projects, among others. These man-made

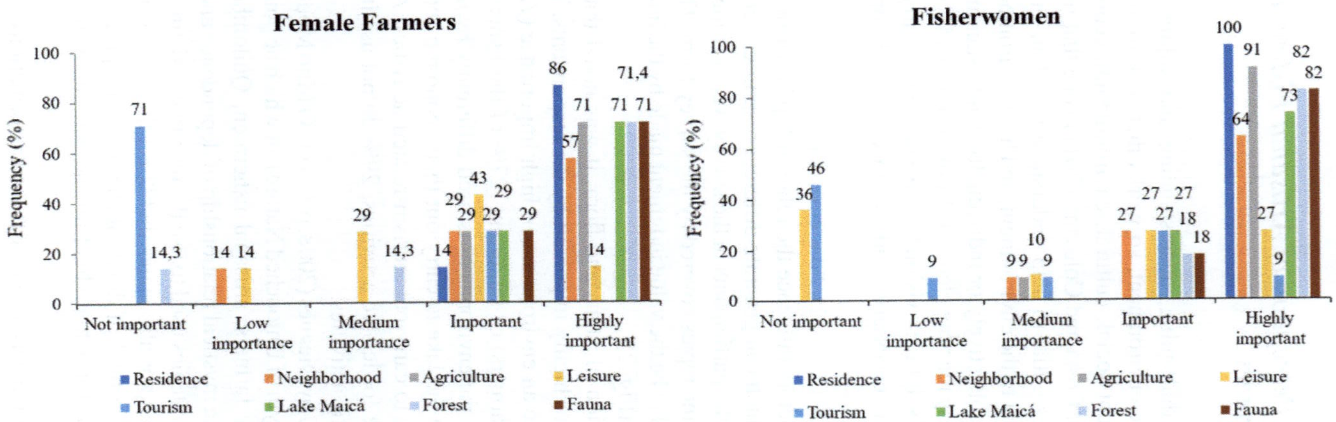

Fig. 2 Degree of importance of the different places of use of the interviewees. *Source* Research data

activities have been blamed for the environmental disasters that have damaged the dynamics of ecosystems, especially in the Amazon region (Krzysczak 2016). According to Tuan (1980, p. 01), "[…] without self-understanding we cannot hope for lasting solutions to environmental problems, which are fundamentally human problems […]", so in order to understand environmental problems it is essential to know one's self in order to find solutions to the problems caused by human action.

The survey identified the main environmental problems based on their perceptions and the most cited among the fisherwomen was the improper disposal of waste (garbage) (Table 1), which they believe causes an increase in venomous animals and attracts diseases. In the opinion of family farmers, the problem most emphasized was the decrease in hunting (Fig. 3). According to Machado (2016, p. 86), the two most common problems pointed out by the interviewees are caused by the disorderly increase in urbanization in outlying districts, a factor that "[…] has damaged ecosystems in countless ways and on a global scale, permeated by a lack of understanding of the effect of growth and development on the functionality of socio-environmental systems […]".

Another issue of concern was the breeding of buffalo (*Bubalos* sp.) in the area surveyed. This problem was identified by 82% of the fisherwomen and 71% of the family farmers, and this issue is related to the other problems reported, such as the reduction in the number of aningais (*Montrichardia* sp.), the supply of fish, water pollution and deforestation. This has led to other social problems, such as conflicts between buffalo farmers and users of the area since the animals tear up fishermen's fishing nets and invade the land of people who live near Lago do Maicá.

This reality is similar to the problems identified by research carried out in the Pixuna do Tapará region (Várzea region, Santarém) by O'Dwyer (2005), where buffalo farming poses a threat to fishing reproduction and aquatic vegetation, such as aningais "[…] the buffalo is considered to be one of the biggest predators of fish barns, as it eats away at the aquatic vegetation and thus scares away the species […]".

Identifying the existence of predatory fishing (trawling, use of pumps, etc.) in Maicá Lake also drew attention, as it is a daily concern for 55% of artisanal fisherwomen and 43% of family farmers. This practice jeopardizes the natural reproduction of fish and the subsistence of families who consume fish as their staple diet. This is close to what has been happening in aquatic ecosystems (rivers and lakes) in the state of Rondônia "[…] predatory and clandestine fishing practiced in the rivers and lakes of Rondônia is exterminating several species of fish native to the Guaporé, Mamoré and Madeira valleys" (Oliveira 2008, p. 1).

According to Cerdeira (2009), the intensification of predatory fishing in Lago do Maicá is considered a step backwards within the policy of community fisheries co-management, since the region was the first in the municipality to build a fishing agreement as a planning instrument. In this context, the community initiative aims to guarantee control of the common use of these resources, based on traditional knowledge rooted in the communities of the Amazon floodplain (Isaac et al., 1993; Mcgrath et al., 2008).

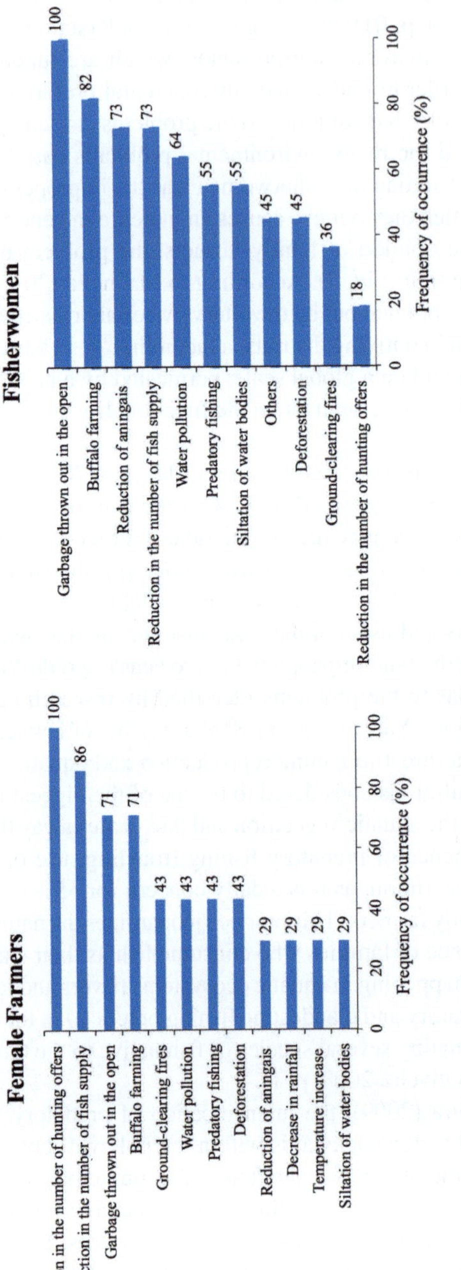

Fig. 3 Frequency of occurrence of the main environmental problems cited by interviewees. *Source* Research data

3.2 Organizational and Productive Overview of Artisanal Fisherwomen and Family Farmers

As for women's participation in some kind of social organization, those who make a living from artisanal fishing and family farming are organized into class entities that represent their categories. The participants in the research are linked to the grassroots nucleus of the Z-20 Fishermen's Colony, which operates in the Pérola do Maicá neighbourhood. The organization meets monthly on the third Sunday of each month to discuss the demands of the category and the deliberations of the Fishing Council, as well as to plan actions for the planning and use of Maicá Lake.

The female family farmers are organized into two entities that bring the category together. Of those surveyed, 14.3% are members of the Santarém Rural Producers' Association (APRUSAN) and 85.7% of the Santarém Rural Workers' Union (STTR). The region's police station is located in the Pérola do Maicá neighbourhood, and the members meet every second Saturday of the month to discuss and deliberate on class and union organization issues.

These union organizations have branches in the rural regions of the municipality, so this link guarantees them access to public labor, social security and welfare policies, among others. According to Leone and Teixeira (2010), women's involvement in social and trade union organizations has helped to develop and strengthen the female workforce in their work, as well as the organizations to which they belong.

Navarro (1996), Leone and Teixeira (2010) point out that women's right to join trade union organizations was won recently, from the 1980s onwards, a period in which they slowly began to occupy spaces in the countryside and cities, gaining political and social visibility. According to Deere (2004), the inclusion of women in trade union struggles is the result of the training process promoted by the Basic Ecclesial Communities (CEBs) and the Pastoral Land Commission (CPT), which encouraged them to question the social condition in which they lived, because women have specific problems and need differentiated public policies.

The women in the area surveyed reported that the purpose of their productive efforts is for self-consumption and commercialization, demonstrating how important they are to the local economic dynamic. In this sense, the survey identified that all the artisanal fisherwomen and family farmers interviewed have been carrying out their productive activities for more than ten years.

It is worth noting that 9.1% of the fisherwomen dedicate two days a week to their work, 36.4% dedicate three days and 54.5% six days a week. An important aspect detected concerns the model of boat used to carry out their activity, in this case, the use of the canoe was highlighted by 73%, in second place comes the rabeta with 27% and the bajara with 18%. The boat identified as the most used shows that they don't travel long distances to carry out their activities, so they use the one that requires manual effort.

The fish caught is marketed 91% live, sold as "cambada" and 9% chilled, at the port and inside the Pérola do Maicá neighborhood, or Porto do Urumanduba. The average weekly value acquired from the sale of the fish was R$112.72, where the cambada is equivalent to an average of 2 kg of fish, and the cambada varies according to the species caught. Adding up the weekly amounts received with the earnings from the productive effort of selling the fish gives an average monthly value of approximately R$450.9, which corresponds to 47.2% of the minimum wage in 2018.

The main consumers are the people who live in the neighborhoods where the research was carried out and near Lago do Maicá. This reality has already been described by Vaz et al. (2017), who state that "[...] marketing takes place mainly in the form of chilled and "in natura" fish, usually in front of houses [...] or on the streets by hawkers, or on the lake shore directly from the boats in the "Porto do Maicá" [...]". The strategy of trading fish is part of the daily life of the residents of the neighborhoods on the shores of the lake, and according to the interviewees there are no difficulties in selling their product.

The analysis of the variety of fish showed that the interviewees caught 33 species (Fig. 4), of which the five most cited were pacu (Subfamília Myleinae incluído *Myleus*, *Metynnis*, *Mylossoma aureum* (Agassiz 1829), *Mylossom aalbiscopum* (Cuvier, 1818) entre outras) with 91%, followed by aracu (*Leporinus* spp., *Schizodon* spp.) with 82%, branquinha (Família Curimatidae incluindo *Potamorhina* spp., *Psectrogaster amazônica* (Eigenmann and Eigenmann 1889), *Curimata inornata* (Vari, 1989) entre outras), with 64% and with the same percentage surubim (*Pseudoplatystoma punctifer* (Castelnau, 1855)) and tambaqui (*Colossoma macropomum* (Cuvier, 1818).

The five most cited species are among the ten most caught in the municipality of Santarém, as described by Martins (2009, p. 34); Barthem and Fabré (2004) and Cerdeira (2009) state that there is a significant diversity of fish species in Lago do Maicá, demonstrating the lake's productive potential, similar to communities in the Amazon floodplain region, where the abundance of nutrients is conducive to feeding fish.

The soil in the region studied is considered fertile, which favors the development of activities such as planting vegetables and fruit trees, and raising small animals. The purpose of this production is to ensure the subsistence of the families and supply the local market, fairs, shops, among others, guaranteeing food security and generating income for the families (Vieira 2013, pp. 4–5).

In this way, the research identified that female family farmers reconcile their productive activities with their domestic ones. It was observed that 14.3% dedicate five days a week to farming and 85.7% seven days to their productive activity. Understanding the productive dynamics of the women surveyed a little more, it was clear that a large part of their production is destined for self-consumption and the rest is sold, which is in line with what was mentioned by Menezes and Homma (2001) when they studied agricultural establishments in Nova Ipixuna, Pará.

The family farmers surveyed reported that they raise crops for consumption and sale, of which 86% raise chickens (*Gallus gallus domesticus*) and eggs, 29% duck (*Cairina moschata*), 14% goose and 14% cattle. Although family farmers grow a variety of products, 43% of those interviewed said that vegetables such as chives (*Allium schoenoprasum*), coriander (*Coriandrum sativum*) and chicory (*Cichorium intybus*) were the most commonly grown. Other varieties of vegetables can also

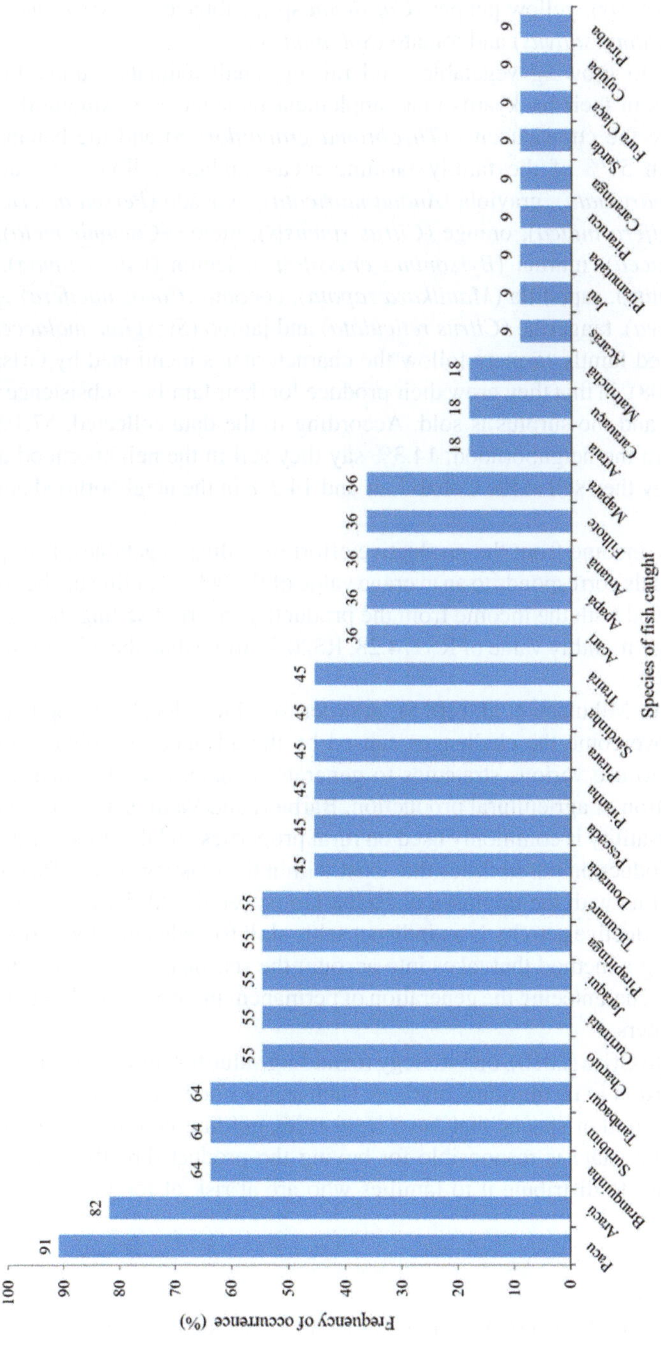

Fig. 4 Fish caught by the interviewed artisanal fisherwomen. *Source* Research data

be found on the farms, such as: jerimum (*Cucurbita* spp), corn (*Zea mays*), beans (*Phaseolus vulgaris*), yellow pepper (*Capsicum* sp.), cabbage (*Brassica oleracea*), cucumber (*Cucumis sativus*) and tomato (*Solanum* sp.).

In addition to growing vegetables and raising small animals, family farmers have fruit trees in their backyards that supplement their income. Among the varieties found are the cupuaçuzeiro (*Theobroma grandiflorum*) and the banana tree (*Musa* spp.) in 57% of the family farming areas studied, followed by acerola (*Malpighia emarginata*), graviola (*Anona muricata*), avocado (*Persea americana*), mango (*Mangifera indica*), orange (*Citrus sinensis*), melon (*Cucumis melo*), açaí (*Euterpe oleracea*), muruci (*Byrsonima crassifolia*), lemon (*Citrus limon*), abiu (*Pouteria caimito*), sapodilla (*Manilkara zapota*), coconut (*Cocos nucifera*), guava (*Pidium guajava*), tangerine (*Citrus reticulata*) and jambo (*Syzygium malaccense*).

The surveyed family farmers follow the characteristics mentioned by Grisa and Schneider (2008), in that they grow their produce for their family's subsistence (self-consumption) and the surplus is sold. According to the data collected, 57.1% sell their produce in the neighborhood, 14.3% say they sell in the neighborhood and at fairs, 14.3% say they sell at the Cohab Fair and 14.3% in the neighborhood and to a supermarket.

The weekly income from the productive effort of selling vegetables, fruit plants and small animals corresponds to an average value of R$243.6. Adding up the weekly amounts received with the income from the productive effort of selling the products gives an average monthly value of R$974.28, R$20.28 more than the minimum wage in force in 2018.

According to Mikulcak et al. (2015), in order to achieve local development it is necessary to overcome the challenges caused by the advance of globalization, so it is essential to use various strategies to generate income, one of which may be the diversification of agricultural production. Barbieri and Valdivia (2010) state that productive versatility is commonly used on rural properties and that this is a strategy to value the production possibilities that exist within the existing space. This understanding is in line with the premises of Padua-Gomes et al. (2015), who states that variation in productive activity is an indispensable skill for balancing the ecosystem, as well as being a method that takes into account the seasonal periods of each crop produced, thus guaranteeing the generation of permanent income throughout the year for family farmers.

According to Grisa (2010), one strategy for adding value to family farm production is to seek insertion in institutional markets, such as the Food Acquisition Program,[2] which are negotiation spaces that have legal rules guiding commercialization. In these spaces, the state is responsible for buying the product directly from family farmers in order to distribute it to families who are at risk of food and nutritional problems.

[2] The Food Acquisition Program (PAA), created by art. 19 of Law No. 10,696, of July 2, 2003, has two basic purposes: promoting access to food and promoting family farming. https://www.gov.br/mds/pt-br/acoes-e-programas/inclusao-produtiva-rural/paa.

4 Considerations

The study aimed to highlight the environmental perception, organization and production of women artisanal fishermen and family farmers who live in the neighbourhoods involved in the research Maicá and Perola do Maicá. It was found that the participants survive on primary activities with rural characteristics: artisanal fishing, growing vegetables and raising animals. For this reason, they consider the space where they live to be of great importance in their lives, with a symbolic relationship of dependence and respect for the sustainable use of natural resources.

Concerning the marketing of their produce, it was noted that the participants in the research have significant production. However, they sell their products in the neighborhoods where they live and at fairs and markets in the surrounding neighborhoods. This reveals the need for them to access public institutional marketing policies, a space where they can negotiate their products at a fair price.

The surveyed women are members of trade unions that represent the category of artisanal fisherwomen and family farmers, which is a way of guaranteeing their social security rights and access to public policies. The systematization of academic work is important to help discuss and implement public policies that overcome the difficulties faced by the municipality's artisanal fisherwomen and family farmers daily.

These results can serve as theoretical input for research institutions, trade unions and the participants in the study to debate the development model that guarantees the survival of these women in their family units, especially in fragile environments and ecosystems, such as peri-urban neighborhoods and rural areas.

Environmental perception and the relationship with community organizations influence the production process of these women, especially in the context of resistance to agribusiness, as a way of conserving natural resources and their ways of life, since artisanal fishing and family farming are important in the lives of the surveyed women, being an element that goes beyond economic and social value, reaching a cultural and symbolic value.

References

Alves JED (2016) Desafios da equidade de Gênero no Século XXI. Rev Estud Fem 24(2):629–638.https://doi.org/10.1590/1805-9584-2016v24n2p629

Andrade FV (2009) Metais pesados em solos de área de mineração e metalurgia de chumbo. II—formas e Disponibilidade para plantas. Revista Brasileira Ciência Do Solo 33:1889–1897. https://doi.org/10.1590/S0100-06832009000600038

Barbieri C, Valdivia C (2010) Recreation and agroforestry: examining new dimensions of multifunctionality in family farms. J Rural Stud 26(4):465–473. https://doi.org/10.1016/j.jrurstud.2010.07.001

Barthem RB, Fabré NN (2004) Biologia e diversidade dos recursos pesqueiros da Amazônia Brasileira. In: Ruffino ML (eds.) A pesca e os recursos pesqueiros na Amazônia brasileira. Projeto Manejo dos Recursos Naturais da Várzea. Manaus: Ibama/ProVárzea, pp 17–51

Borcem ER, Furtado Júnior I, Almeida IC, Palheta MKS, Pinto IA (2011) A atividade pesqueira no município de Marapanim-Pará. Brasil. Revista De Ciências Agrárias 54(3):189–201. https://doi.org/10.4322/rca.2012.014

Bresolin AJ, Zakrzevski SBB, Marinho JR (2010) Percepção, comunicação e educação ambiental em unidades de conservação: um estudo no Parque Estadual de Espigão Alto—Barracão/RS—Brasil. Perspectiva, 34(128):103–114. https://www.uricer.edu.br/site/pdfs/perspectiva/128_141.pdf

Bruschini C (1994) Trabalho feminino: trajetória para um tema, perspectiva para o futuro. Estud Fem, 1:17–32. https://periodicos.ufsc.br/index.php/ref/article/view/15798/14291

Cerdeira RGP (2009) Acordo de pesca como Instrumento de Gestão Participativa na Amazônia. Dissertação apresentada ao Programa de Pós-Graduação em Direito Ambiental da Universidade do Estado do Amazonas. Manaus, Amazonas. p 134. http://repositorioinstitucional.uea.edu.br//handle/riuea/1953

Chauí M (2002) Experiência do pensamento. São Paulo, Martins Fontes, p 326

Chayanov AV (1966) On the theory of non-capitalist economic systems. In: Thorner D (compil). The theory of peasant economy, Kerblay y Smith. Illinois

Costa RGS, Colesanti MM (2011) A contribuição da percepção ambiental nos estudos de áreas verdes. O Espaço Geográfico Em Análise 22:238–251. https://doi.org/10.5380/raega.v22i0.21774

Deere CD (2004) Os direitos da mulher à terra e os movimentos sociais rurais na Reforma Agrária Brasileira. Rev Estud Fem 12(1):175–204. https://doi.org/10.1590/S0104-026X2004000100010

Diegues ACS (2008) O Mito Moderno da Natureza Intocada. 6. Ed. Revista e Ampliada. Editora Hucitec. São Paulo: NUPAUB—Núcleo de Apoio à Pesquisa sobre Populações Humanas e Áreas Úmidas Brasileiras—USP/Hucitec. https://nupaub.fflch.usp.br/sites/nupaub.fflch.usp.br/files/O%20mito%20moderno.compressed.pdf

Eigenmann, Eigenmann (1889) From upper Rio Tapajós, Mato Grossostate, Brazil (Pisces: Ostariophysi: Loricariidae). Braz J Biol 67(3):413–420

Flick U (2004) Introdução à pesquisa qualitative, 2edn. Bookman, Porto Alegre, p 312. https://www2.fct.unesp.br/docentes/geo/necio_turra/PPGG%20-%20PESQUISA%20QUALI%20PARA%20GEOGRAFIA/flick%20-%20introducao%20a%20pesq%20quali.pdf

Gil AC (2010) Como elaborar projetos de pesquisa, 5edn. Atlas, São Paulo, p 176. https://files.cercomp.ufg.br/weby/up/150/o/Anexo_C1_como_elaborar_projeto_de_pesquisa_-_antonio_carlos_gil.pdf

Grisa C, Schneider S (2008) "Plantar pro gasto": A importância do autoconsumo entre famílias de agricultores do Rio Grande do Sul. Rev Econ Sociol Rural 46(2):481–515. https://doi.org/10.1590/S0103-20032008000200008

Grisa C (2010) As redes e as instituições do Programa de Aquisição de Alimentos (PAA). Rev Bras Gestão E Desenvolv RegNal, 6(2):97–129. https://doi.org/10.54399/rbgdr.v6i2.274

Hoffmann R, Leone ET (2004) Participação da mulher no mercado de trabalho e desigualdade da renda domiciliar per capita: 1981–2002. Rev Nova Econ, Belo Horize, 14(2):35–58. https://core.ac.uk/download/pdf/7057585.pdf

Isaac VJ, Rocha VLC, Mota S (1993) Considerações sobre a legislação da "piracema" e outras restrições da pesca da região do Médio Amazonas. In: Furtado LG, Mello AF, Leitão W (eds) Povos das águas: realidade e perspectiva na Amazônia. Belém, Museu Paraense Emilio Goeldi, pp 187–211. http://repositorio.museu-goeldi.br:8080/bitstream/mgoeldi/777/1/POVOS%20DAS%20AGUAS%201993%20ALENCAR.pdf

Krzysczak FR (2016) As diferentes concepções de meio ambiente e suas visões. Rev Educ Do Inst Desenvolv Educ Do Alto Uruguai, 11(23):1–18. https://www.bage.ideau.com.br/wp-content/files_mf/9c9c1925f63120720408c5260bb0080d355_1.pdf

Leão RFC, Oliveira JMGC (2011) O Plano Diretor e a cidade de fato: o caso de Santarém-Pa. Revista Geográfica de América Central, Número Especial EGAL, pp 1–15. file:///C:/Users/PC%20Mopebam/Downloads/2646-Texto%20del%20art%C3%ADculo-5980–1-10–20111212%20(1).pdf

Leone ET, Teixeira MO (2010) As mulheres no mercado de trabalho e na organização Sindical. In: Anais XVII Encontro Nacional de Estudos Populacionais, Caxambu, Minas Gerais, Brasil, pp 1–21

Lima MGM, Pereira EMB (2007) Populações tradicionais e conflitos territoriais na Amazônia. Geogr, Belo Horiz, 3(1):107–119. file:///C:/Users/PC%20Mopebam/Downloads/wassis,+geografias-v3-n1-2007_107.pdf

Machado GS (2016) A qualidade de vida é sustentável? um estudo de caso em Florianópolis/Sc. Tese de Doutorado em Psicologia, Programa de Pós-Graduação em Psicologia, Centro de Filosofia e Ciências Humanas da Universidade Federal de Santa Catarina. Florianópolis. pp 1–247. https://repositorio.ufsc.br/xmlui/handle/123456789/168051

Martins EV (2009) Dinâmica da economia e das relações do trabalho da pesca artesanal no município de Santarém. Dissertação (Mestrado)—Universidade Federal do Pará, Núcleo de Altos Estudos Amazônicos, Programa de Pós-Graduação em Desenvolvimento Sustentável do Trópico Úmido, Belém, pp 1–106. https://www.ppgdstu.propesp.ufpa.br/ARQUIVOS/dissertacoes/2009/EVERALDO%20MARTINS.pdf

Mattos CLG (2011) A abordagem etnográfica na investigação científica. In: Mattos CLG, Castro PA (orgs) Etnografia e educação: conceitos e usos. EDUEPB, Campina Grande, pp 49–83. https://books.scielo.org/id/8fcfr/pdf/mattos-9788578791902-03.pdf

McGrath DG, Cardoso A, Almeida OT, Pezzuti J (2008) Constructing a policy and institutional framework for an ecosystem-based approach to managing the Lower Amazon floodplain. Environ, Dev Sustain, 10(1):677–695. https://doi.org/10.1007/s10668-008-9154-3. https://ideas.repec.org/a/spr/endesu/v10y2008i5p677-695.html

Menezes AJEA, Homma AKO (2001) Participação da "Produção Invisível" nos Estabelecimentos Agrícolas Familiares, no Município de Nova Ipixuna, Pará. In: Congresso Brasileiro de Economia e Sociologia Rural, Recife. Anais. SOBER, p 39. CD-ROM

Mikulcak F, Haider JL, Abson DJ, Newig J, Fischer J (2015) Applying a capitals approach to understand rural development traps: A case study from post-socialist Romania. Land Use Policy 43:248–258. https://doi.org/10.1016/j.landusepol.2014.10.024

Moraes R (2003) Uma tempestade de luz: a compreensão possibilitada pela análise textual discursiva. Ciência Educação: Bauru SP 9(2):191–210

Moraes R, Galiazzi MC (2006) Análise textual discursiva: processo reconstrutivo de múltiplas faces. Ciência & Educ, 12(1):117–128. http://www.scielo.br/pdf/ciedu/v12n1/08.pdf

Moraes R, Galiazzi MC (2007) Análise textual discursiva. Editora UNIJUÍ, Ijuí

Navarro Z (1996) "Democracia, cidadania e representação: os movimentos sociais rurais no Estado do Rio Grande do Sul, 1978–1990. In: Navarro Z (org) Política, protesto e cidadania no campo. Editora da Universidade, Porto Alegre

O'dwyer EC (2002) Quilombos—identidade étnica e territorialidade. Rio de Janeiro: Ed. FGV

Oliveira MM (2008) Como fazer pesquisa qualitative, 2edn. Vozes, Petrópolis, RJ

Oliveira JMGC et al (2010) Espaços híbridos e espacialidades rural-urbanas na amazônia: o caso de uma periferia urbana em Santarém-PA. Anais XVI. Encontro Nacional dos Geógrafos. Crises, práxis e autonomia: espaço de resistência e esperanças. Espaço de Diálogo e Práticas. Porto Alegre—RS, p 12

Padua-Gomes JB et al (2015) Produção orgânica no Assentamento Itamarati, em Ponta Porã, estado de Mato Grosso do Sul, Brasil. Cadernos de Agroecologia, [S.l.], 9(4). https://revistas.aba-agroecologia.org.br/cad/article/view/16559

Pereira JCM (2004) Importância e significado das cidades médias na amazônia: uma abordagem a partir de santarém (PA). Dissertação (Mestrado em Planejamento do Desenvolvimento)—Universidade Federal do Pará. https://www.gov.br/saude/pt-br#%20search=%22%22Sposito%22%20%22cidades%20m%C3%A9dias%20*%20os%22%22

PMS—Prefeitura Municipal de Santarém. Secretaria Municipal de Meio Ambiente. Centro Municipal de Informações Ambientais—CIAM. In: Informações Municipais de Santarém Semma—Ciam. Publicado em 2013. Disponível em: http://www.santarem.pa.gov.br/arquivosdb/basico1/0.66876400135758053 2informacoes_2.pdf. Acesso em: 24 Mar 2016

Prodanov CC, Freitas EC (2013) Metodologia do trabalho científico: métodos e técnicas da pesquisa e do trabalho acadêmico. Feevale, Novo Hamburgo, RS, p 276. https://www.feevale.br/Comum/midias/0163c988-1f5d-496f-b118-a6e009a7a2f9/E-book%20Metodologia%20do%20Trabalho%20Cientifico.pdf

Santos ÁO et al (2017) A (In)visibilidade das mulheres agricultoras familiares do município de santarém-pará nas políticas públicas. 1° Simpósio SOBER NORTE: "A crise econômica e o futuro da agropecuária no Norte: desafios e potencialidades. Belém, Pará. p 4

Serrão EM (2018) A pesca e o conhecimento tradicional dos pescadores de um lago de inundação no baixo amazonas: sugestões para manejo e conservação. Dissertação em recursos aquáticos continentais amazônicos, universidade federal do oeste do Pará, p. 106. https://repositorio.ufopa.edu.br/jspui/bitstream/123456789/210/1/Disserta%C3%A7%C3%A3o_ApescaeoConhecimentoTradicional.pdf

Silva YAR (2014) Agricultura mecanizada e expansão urbana em Santarém-PA: Padrões e Processos Espaciais. Trabalho de Conclusão de Curso de Graduação em Licenciatura Plena em Geografia, no Programa de Humanas da Universidade Federal do Oeste do Pará, Santarém-Pará, p 73

Simonian LTL (2001) Mulheres, cultura e mudanças nos castanhais do sul do Amapá. In: Mulheres da Amazônia Brasileira: entre o trabalho e a cultura. UFPA/NAEA, Belém

Soares SM, Scherer E (2013) Pescadoras mesmo: seguro defeso e reconhecimento em comunidades pesqueiras do Lago do Rei no Amazonas. In: Scherer E (orgs) Trabalhadores e Trabalhadoras na pesca: ambiente e reconhecimento. Garamond, Rio Janeiro, pp 79–105

Torres JR, Gehlen ST, Muenchen C, Gonçalves FP, Lindemann RH, Gonçalves FJF (2008) Ressignificação curricular: contribuições da Investigação Temática e da Análise Textual Discursiva. Revista Brasileira de Pesquisa em Educação em Ciências 8(2). ISSN 1234-5678

Tuan YF (1980) Topofilia: um estudo da percepção, atitudes e valores do meio ambiente. Tradução por Lívia de Oliveira. São Paulo/Rio de Janeiro. Disponível em: https://public.e-livros.download/Tuan%2C%20Yi-fu%20Topofilia.pdf. Acesso em: 29 jul. 2018

Vaz EM, Rabelo YGS, Corrêa JMS, Zacardi DM (2017) A pesca artesanal no lago Maicá: aspectos socioeconômicos e estrutura operacional. Biota Amazônia. Macapá, 7(4):6–12. https://doi.org/10.18561/2179-5746/biotaamazonia.v7n4p6-12

Vaz EM (2016) Caracterização da atividade pesqueira praticada no Lago Maicá, Município de Santarém, Pará. Santarém—Pará. 49f. Trabalho de Conclusão de Curso (Bacharelado de Engenharia de Pesca), Universidade Federal do Oeste do Pará.

Vieira JC (2013) Quilombolas na cidade: Dilemas do planejamento urbano frente ao território do quilombo do maicá em santarém-Pará. In: XXII Encontro Nacional do Conselho Nacional de Pesquisa e Pós-Graduação em Direito, Curitiba-PR

Wanderley MNB (2003) Agricultura familiar e campesinato: rupturas e continuidade. Estud Soc E Agric, Rio J, 11(2):42–61. https://revistaesa.com/ojs/index.php/esa/article/view/238/234

Yin RK (2005) Estudo de Caso: planejamento e métodos, 3 ed. Bookman, Porto Alegre, p. 212. https://edisciplinas.usp.br/pluginfile.php/6598416/mod_resource/content/1/Livro%20Robert%20Yin.pdf

Zhao M, Tyzack M, Anderson R (2013) Estera Onoakpovike. Women as visible and invisible workers in fisheries: A case study of Northern England. Mar Policy 37:69–76. https://doi.org/10.1016/j.marpol

Fisheries Co-management in the Lower Amazon: The Political Re-establishment of the Fishing Agreements in *Pará*, Brazil

Neriane Nascimento da Hora, Antônio José Mota Bentes, Wandicleia Lopes de Sousa, Poliane Batista da Silva, and David Gibbs McGrath

Abstract Fishing agreements are community and inter-community initiatives designed to regulate fisheries in floodplain (*várzea*) lakes in the Amazon region. Initially, lacking legal recognition, these agreements were officially acknowledged by the Brazilian Government in 2002 as a fisheries management instrument. Their regulation was later transferred to the states under Complementary Law 140/2001, and in *Pará* specific guidelines were established with the publication of Decree No. 1686 in 2021. This text explores the process of revitalizing fishing agreements as a tool for fisheries co-management and environmental conservation in the lower Amazon, focusing on the cases of *Arapixuna* and *Lago Grande*. These territories face significant fishing pressure and other socio-environmental threats. The fishing agreements for those regions, originally approved in 2004 and 2005, were considered outdated by local fishing communities. In response, the Regional Sustainable Development Councils of *Lago Grande* and *Arapixuna* reviewed these agreements in 2018 and submitted revision proposals to the environmental agency between 2018 and 2019. After the publication of the aforementioned decree in 2021, the evaluation process of these agreements started, leading to the approval of three fishing agreements in the lower Amazon, including that of the *Lago Grande*. This legal framework recovered the foundations of the artisanal fisheries co-management policy in *Pará*, offering a significant contribution to the socio-environmental sustainability of the Amazon floodplain.

N. N. da Hora (✉)
Sociedade para Pesquisa e Proteção do Meio Ambiente—Sapopema, Av. Rui Barbosa, 3040 Santarém-PA, Brazil
e-mail: neriane.dahora@ird.fr

A. J. M. Bentes · P. B. da Silva
Sociedade para Pesquisa e Proteção do Meio Ambiente—Sapopema, Santarém, Pará, Brazil

W. L. de Sousa
Universidade Federal do Oeste do Pará—UFOPA e Sociedade para Pesquisa e Proteção do Meio Ambiente, Santarém, Pará, Brazil

D. G. McGrath
Earth Innovation Institute, Berkeley, USA

W. Leal Filho et al. (eds.), *Amazon 2030—Sustainability Issues in the World's Largest Rainforest Region*, World Sustainability Series,
https://doi.org/10.1007/978-3-031-81465-5_30

Keywords Artisanal fisheries · Colaborative management · Floodplain conservation

1 Introduction

The *várzea* consists of the floodplains along the banks of whitewater rivers, particularly the Amazon River, which are submerged annually for approximately six months (Junk et al. 1989). These floodplains have long been inhabited by human populations whose way of life is closely adapted to the seasonal river cycle. The nutrient-rich soils of floodplains support productive agriculture, while its lakes provide abundant fish stocks. During the low-water season, the natural grasslands become grazing areas for cattle. Given the opportunities and restrictions of this dynamic environment, floodplain inhabitants adopt diversified livelihood strategies, making use of the various resources available across the floodplain landscape (Furtado 1993; McGrath et al. 1993).

Fishing plays a central role in the economy of floodplain households, serving both for income generation and subsistence. Fish is the main animal protein for riverside communities (Almeida et al. 2001; Isaac and Almeida 2011). However, in the late 20th century, fishing underwent significant transformations due to the expansion and intensification of commercial fisheries (McGrath et al. 2004). This shift led to growing conflicts between local fishers and outsiders (McGrath et al. 1993), as well as increased pressure on fish stocks, particularly in floodplain lakes (Castello et al. 2013).

In the Lower Amazon, fishers report a decline in fish stocks over the past few decades, directly impacting their livelihoods (McGrath et al. 1993). Studies have highlighted the overexploitation of high-commercial valuable species in the Amazon region, such as pirarucu (*Arapaima* spp.) and tambaqui (*Colossoma macropomum*) (Castello et al. 2013, 2015). These declines have been linked to overfishing and environmental degradation driven by deforestation and extensive cattle ranching, with their effects further exacerbated by climate change (Castello and Macedo 2016; Isaac et al. 2016).

In this scenario, since the 1970s, communities have sought mechanisms to mitigate the impacts on fishing resources while ensuring their quality of life. The main strategy adopted by community members to regulate access to and exploitation of fish has been the fishing agreements. Initially informal and documented in meeting minutes, these agreements later gained official recognition through Ibama, the federal environmental agency, following Normative Instruction 29/2002 (Castro and McGrath 2001; McGrath et al. 2008).

The Lower Amazon floodplain is federal public land (Benatti 2005). To regularize the occupation of this territory, the Government established 41 agro-extractivist settlement projects (PAEs) in the region starting in 2006, through a technical cooperation agreement between the Federal Property Office (SPU) and the National Institute for Colonization and Agrarian Reform (INCRA) (McGrath et al. 2011). As part of this process, Utilization Plans (PUs) for the PAEs were developed through participatory approaches. These plans incorporated the rules of the pre-existing fishing agreements and the Conduct Adjustment Terms (TACs), which regulate cattle and buffalo farming in floodplains. Moreover, the PUs established guidelines for land use, forestry, animal husbandry, and agricultural activities, among other aspect. By proposing an integrated management approach to the floodplain's natural resources, the PUs represented a significant advancement, laying the foundations for a co-management system of the Lower Amazon floodplain (McGrath et al. 2008).

In 2011, the Federal Government transferred the authority to regulate fishing agreements to the states through Complementary Law 140/2011. In *Pará*, this responsibility was assigned to the State Secretariat for the Environment and Sustainability (SEMAS) and the Pará State Institute for Forestry and Biodiversity Development (IDEFLOR-Bio). However, between 2011 and 2021, no legal instrument was enacted to regulate fishing agreements in the state (Lima et al. 2023). During this period, fishing communities called for a review of the fishing agreements. Yet, in the absence of specific regulations, the environmental agencies were unable to evaluate and approve new proposals. As a result, processes for new fishing agreements filed during that time came to a standstill, undermining the foundations of the co-management system established between the mid-1990s and the early 2000s.

Fisheries co-management in *Pará* state began to regain momentum only after the publication of Decree No. 1686 on June 29, 2021, which defined guidelines of fishing agreements. Based on cases from fishing regions in the Lower Amazon—where communities have made significant efforts to manage their local fisheries, we analyze the context of fishing agreements in *Pará* before and after the legal framework introduced by Decree 1686/2021. Additionally, we examine how fishers have mobilized and adapted to this new regulatory phase. The recent process of recognizing fishing agreements, led by SEMAS as a fisheries management tool, has yet to be explored in the scientific literature. This analysis is crucial to identifying key aspects that can be improved in the implementation of this policy.

The description and analysis presented here are based on the authors' direct involvement in the development, review, and implementation of fishing agreements in the Lower Amazon region. The authors participated in both community and inter-community meetings on fishing agreements and examined documents submitted by fishing communities, as well as those issued by the environmental agency. These activities are part of projects carried out by the authors within the framework of a Non-Governmental Organization based in *Santarém-PA*, which has a technical cooperation agreement with SEMAS-PA to support the drafting of fishing agreements for fishing communities.

This article is structured into five sections: (1) the first section provides an overview of the evolution of fisheries management in the lower Amazon; (2) the second discusses the social organization of fishers through Fishing Unions and their contribution to fisheries management in the region; (3) the third examines the fishing agreements approved under the management of Ibama/MMA, the first agreements sanctioned by SEMAS in the Lower Amazon, and the ongoing approval process for the Arapixuna region; (4) the fourth explores the progress and challenges following the publication of the state decree regulating fisheries management in floodplain areas; and (5) the final section presents key considerations for strengthening fisheries management in the Lower Amazon floodplain.

2 Social and Political Organization of Fishermen and Fisherwomen in the Context of Fishing Agreements in the Lower Amazon Region of *Pará*

During the 1960s and 1970s, various legal instruments—including laws, decrees, normative instructions, and ordinances were issued to advance the organization of artisanal fisheries in Brazil. However, despite the range of regulations enacted, many failed to consider the diverse realities of artisanal fisheries across different regions, as they were developed without input from fishers or their representative organizations (Leitão 1995). The first legal instrument was the Decree Law No. 221/1967, which established the General Register of Fishing Activities (RGP). Following its creation, the first official registry of artisanal fishers was conducted by the Superintendence of Fisheries Development—SUDEPE, founded in 1962, granting legal recognition to the profession (Brasil 1967).

Since the organization of fishers was placed under the oversight of the Brazilian Ministry of Agriculture (Brazil 1933), a new statute for the fishermen's unions was introduced through Ordinance No. 471 of 26 December 1973. There changes were minimal, as the unions retained their classification as civil societies, but remained subjected to the oversight of the State, the Federations, and the National Fishers' Confederation. This statutory framework continues to be infect to this day in some Fishers' Unions of Brazil.

With the enactment of the 1988 Constitution, significant advances was made in the organization of artisanal fishers: the unions were granted the same social rights as the rural workers' unions. This shift allowed Fishers' Unions to draft their own statutes, tailoring them to the specific contexts of the municipalities where they operate.

During this same period, the Fishers Pastoral Council (CPP) was established in *Pernambuco*, introducing a new organizational framework for artisanal fishers centered on the empowerment of class-based organizations. The creation of CPP marked a turning point in the history of the fishers' unions, as its training initiatives led to transformative changes in the leadership of fishers unions across the country. Previously, these fishers unions had been dominated by military officials, municipal politicians, and business leaders (Moraes 2000; Leitão 1995). The initial changes emerged in the North and Northeast regions, gaining momentum as part of the broader resistance and struggle against the military dictatorship. Throughout this process, the CPP, alongside organizations such as the Basic Education Movement (MEB) and the Federation of Social and Educational Assistance Organizations (FASE) played a key part in driving the success of these efforts (Leroy 1988).

There are approximately 1,000 fishers unions across the country, organized into 27 State Federations and a National Confederation of Fishers and Aquaculturists (CNPA). These entities collectively represent around 1 million artisanal fishers (Amorim 2014). In *Pará*, there are approximately 65 (sixty-five) Fishers Unions affiliated with the Federation of Fishers of *Pará* State (FEPA) and the Fishers' Movement of the State of *Pará* (MOPEPA). At the national level, these organizations maintain a relationship is with the Confederation of Fishers of Brazil (Santos 2005).

In the Lower Amazon and Tapajós region, there are 14 (fourteen) fishing unions that unite approximately 30,000 artisanal fishers. These organizations have a rich history of struggle, resistance, organization, and the development of participatory management tools (Mopebam 2021). This legacy was further reinforce with the establishment of the Association of the Fishers' Movement of the Lower Amazon (MOPEBAM) in 1996, an organization dedicated to coordinating, training, and strengthening the efforts of artisanal fishers' unions in this region (Cartilha Colônia de Pescadores 2004).

When examining the social organization of fishers within the context of fishing agreements, we encounter a diverse array of organizations that have served as the foundation for discussing and formulating community rules in the Lower Amazon. This organizational process among artisanal fishers have been crucial not only in establishing fishing agreements but also in ensuring compliance with the rules. Although less prominent than the Forest Peoples' movement, the mobilization and social organization along the Amazon River floodplain similarly emerged from the concerns of traditional communities about threats to their ways of life posed by development in the Amazon (McGrath et al. 2004).

The fishers' movement for the preservation of floodplain lakes, known as the Fishing Agreements, emerged in response to the intensification of commercial fisheries in the region. Concerned about the decline in fish stocks, floodplain communities organized themselves to take control of lakes by prohibiting access to external commercial fishers and establishing measures to regulate fishing activities within their territories. This process of social mobilization of traditional communities in the Lower Amazon floodplain sparked a transformative process that reshaped fisheries management. It ultimately led to the development of a fisheries co-management policy, achieved through collaboration between government agencies and organized civil society.

The municipality of Santarém has been one of the pilot regions in this process. Throughout the 1990s, a municipal fisheries co-management system was established, encompassing seven major floodplain lake systems. This area spans 2,700 km^2 and supporting over 150 communities, with and estimated population of 35,000 to 40,000 families. Regional Fishing Councils were created for each lake system, ensuring representation from all adjacent communities. The organization of fishers into these Regional Councils was a strategic initiative led by the Fishers Union to strengthen their efforts in defending fishing territories. These Councils had a essential role in formulating proposals for fishing agreements, which were submitted in the early 2000s for regulation by IBAMA and subsequently formalized as Normative Instructions. Initially legislated by IBAMA and more recently by SEMAS, these fishing agreements have been critical in strengthening community-based social organization, promoting the conservation of fish stocks (Almeida et al. 2006), and consolidating territorial governance. This governance model is structured around the regional fishing councils, with support of the Fishers Unions, government regulatory agencies, and civil society organizations operating in the region.

Therefore, the organization of fishermen has been fundamental in ensuring both the participatory development and effective implementation of fishing agreements. Without a structured organizational foundation to support territorial governance, the establishment of conservation rules for fishery resources and their enforcement would be not feasible.

3 Fishing Agreements on the Lower Amazon

Between 1999 and 2005, a total of 14 fishing agreements in the Lower Amazon were formalized as ordinances or normative instructions by IBAMA and the Ministry of the Environment (MMA), as presented in Table 1. However, between 2006 and 2021, no new fishing agreements were officially published in the region. Although several fishing agreements were signed or revised during this period, they were not formally recognized through official regulatory mechanisms (Santos 2021).

The formalization of new fishing agreements took place approximately one year following the enactment of Decree No. 1,686, dated June 29, 2021. The first agreement to receive approval from SEMAS was the Tapajós River Fishing Agreement, formalized through SEMAS Ordinance No. 2816 on November 4, 2022. This agreement delineates a specific area within Tapajós River channel, situated between the territories of two conservation units, Tapajós National Forest (FLONA Tapajós) and the Tapajós-Arapiuns Extractive Reserve (RESEX Tapajós-Arapiuns). Notably this agreement represents a pioneering initiative, as it is the first to establish regulatory frameworks for fishing activities within the main channel of a major river in the region.

The necessity for the Tapajós River Fishing Agreement emerged due to the significant pressure exerted by large-scale commercial fisheries, commonly referred to as "glaciers" fishing, on local stocks. This pressure was adversely affecting the livelihoods of over 100 traditional communities that rely on the area for subsistence and cultural practices. Consequently, the Agreement introduced a series of regulatory measures, including restrictions on the number and size of fishing vessels, as well as limits on the storage capacity of "glaciers" fisheries. Furthermore, the Agreement established specific regulations regarding the gear size, namely the gillnet and longline. The development of the Tapajós River Fishing agreement was significantly supported by the Chico Mendes Institute for Biodiversity Conservation—ICMBio, which has also contributed in the enforcement and monitoring of the agreement following its formal approval.

Subsequently, through SEMAS Ordinance No. 1,129, dated June 15, 2023, the Lago Grande do Curuai Fishing Agreement was formally approved. This agreement

Table 1 List of fishing agreements approved by Ibama/MMA in ordinances or normative instructions between 1999 and 2005

Normative	Fishing agreement area	Municipalities	Legal situation
Ibama/PA Ordinance N° 16, de 31 August 1999	Maicá region	Santarém	Revoked
Ibama Ordinance no 20-N, de 21 September 1999	Ituqui region	Santarém	Revoked
Ordinance no 21-N, de 27 September 1999	Lago grande region	Juruti, Óbidos e Santarém	Revoked
Ibama Ordinance N° 03, de 10 March 2000	Urucurituba region	Santarém	In effect
IBAMA Ordinance no 189, de 28 December 2001	Arapixuna region	Santarém	Revoked
Ibama Ordinance n° 41, de 15 March 2001	Aritapera region	Santarém	Revoked
Normative Instruction nº 19, 14 October 2004	Ituqui region	Santarém	In effect
Normative Instruction MMA nº 21, October 2004	Maicá region	Santarém	In effect
Normative instruction nº 11, 14 October 2004	Aritapera region	Santarém	In effect
Normative instruction MMA n° 30, 06 December 2004	Lago grande region	Juruti, Óbidos e Santarém	Revoked
Normative instruction nº 13, 14 October 2004	Tapará region	Santarém	In effect
Normative instruction nº 19, 24 June 2005	Arapixuna region	Santarém	In effect
Normative instruction MMA nº 33, 22 September 2005	Lago mamauru region	Juruti e Óbidos	In effect
Normative instruction nº 50, 31 September 2005	Lago Maria Pixi region	Terra Santa e Oriximiná	In effect

Source Elaborated by the authors

encompasses a portion of the territories within the municipalities of Juruti, Óbidos and Santarém. The region had previously been governed by an Agreement established under Normative Instruction (NI) from MMA No. 30/2004 (Table 1), which, at the time, contributed to mitigate local conflicts and alleviating pressure on fish stocks in the area. However, over time, certain provisions in NI 30 became outdated and inconsistent with contemporary fishing practices and dynamics. Consequently, in 2017, local fishers initiate efforts to update the Fishing Agreement, supported primarily by a non-governmental organization and, to a lesser extent, by the municipal environmental departments of Juruti and Óbidos. These efforts culminated in the submission of a formal request for the agreement's ratification in 2018, predating the publication of Decree No. 1,686, which later established the formal criteria for the institutionalization of Fishing Agreements.

Finally, SEMAS recently published Ordinance No. 1,580, dated July 31, 2023, ratifying the fishing agreement established by communities in the municipality of Monte Alegre. This agreement reintroduces the mechanism involving volunteer environmental agents, who are directly linked to the signatory communities. These agents are tasked with collaborating with environmental agencies to monitor and enforce compliance with the regulations outlined in the fishing agreement.

Additionally, several other fishing agreements in the lower Amazon are currently undergoing the ratification process. These include agreements for the Tapará and Arapixuna regions, both located in the municipality of Santarém-PA. Among these, the Arapixuna fishing agreement is at the most advanced in terms of analysis and is expected to be formally published in the near future. The Arapixuna region initially formalized its first fishing agreement in 1999, which was subsequently revised, leading to the approval in 2005 (Table 1).

However, beginning in 2017, the communities mobilized efforts to revise the existing agreement, engaging in extensive discussions especially within the assemblies of the Sustainable Development Council of the Arapixuna region (formerly known as the Fishing Council of the Arapixuna Region). Between 2017 and 2019, numerous community and inter-community meetings were conducted, culminating in the approval of a new fishing agreement proposal by the communities. Afterwards, this proposal was submitted to SEMAS in November 2019.

From late 2019 to 2021, the evaluation process of the draft agreement remained stalled at SEMAS. In December 2021, five months after the publication of the state decree, SEMAS issued a notification regarding the Arapixuna fishing agreement, recommending that it be revised to align with the provisions of the decree. In response, fishing communities mobilized once again to deliberate on the points raised in the notification. By April 2022, the requested adjustments to the Arapixuna fishing agreement were finalized and resubmitted to SEMAS. In February 2023, the Arapixuna Sustainable Development Council received a second notification from SEMAS. In May 2023, the fishing communities convened to discuss this second notification, with the primary focus of deliberation being the environmental agency's recommendation to remove the article prohibiting the entry of large commercial fishing vessels, the "glaciers", into the lakes of the Arapixuna region.

The community members rejected SEMAS's recommendation and decided to keep the article prohibiting the entry of glaciers and to send the fishing agreement to the Federal Public Prosecutor's Office for evaluation. In July 2023, the MPF granted a favorable opinion to the Arapixuna Fishing Council for the maintenance of the article banning the entry of glaciers into the region's lakes. As a result, the fishers approved at an inter-community assembly in August 2023 the re-submission of the text of the fishing agreement, maintaining the ban on glaciers, back to SEMAS.

4 Discussion

In the Lower Amazon region, the fishing communities within the municipality of Santarém have historically had the highest number of informal agreements approved between 1999 and 2005. Since the publication of the state decree, the fishing communities of Santarém have secured approval for three agreements, followed by the municipalities of Juruti and Óbidos, which jointly approved one agreement with Santarém, and Monte Alegre, which also ratified one agreement. Consequently, when considering the total number of municipalities in the Lower Amazon (15), both during the period under Ibama/MMA management and currently under state management, approximately 10 municipalities in the Lower Amazon remain without any formalized fishing agreements.

The majority of the fishing agreements ratified between 1999 and 2005 encompass areas within Agro-Extractivist Settlement Projects (PAEs). The Lower Amazon region comprises 41 floodplain PAEs, whose Utilization Plans include specific regulations for fishing activities. However, these PUs are not systematically considered for inspection by environmental control agencies. Among the agreements approved by SEMAS to date, only two include PAEs. Therefore, significant progress remains to be made in formalizing fishing agreements within the floodplain PAEs of the Lower Amazon.

Unlike Conversation Units (UCs), which may receive support from environmental agencies such as ICMbio (for federal UCs) and IDEFLOR (for state UCs) for the supervision and monitoring of fishing activities—as exemplified the Tapajós River fishing agreement—the PAEs rely primarily depend on municipal environmental secretariats for oversight. Given that many agreements remain informal, environmental agencies often resort to other regulator a frameworks, such as closed season regulations, to monitor fishing activities. The formalization fishing agreements is therefore crucial to enable environmental control bodies to effectively enforce and monitor the rules established within these fishing agreements.

However, even in a highly favorable scenario of ratified agreements, environmental agencies in Brazil have historically demonstrated deficiencies in enforcement actions (Kalikoski 2009). Pará state is no exception to this trend (Faria Jr et al. 2018). While efforts to combat deforestation tend to receive greater attention from environmental agencies, enforcement actions targeting the fishing sector remain insufficient relative to the substantial demand in the Amazon region (Pantaleão et al. 2020). This enforcement gap has led some user groups to mobilize and take on fishing enforcement responsibilities themselves. However, such initiatives often entail high costs and may pose significant safety to the individuals involved (Hora 2020; Oviedo and Bursztyn 2016).

Collaborative management of natural resources inherently requires shared responsibility between the state and stakeholders. The effective fulfillment of roles by all parties involved is a critical condition for the success of such collaborative management systems. When one part fails to act, the burden often falls disproportionately on others, who may lack the capacity or authority to address the gaps adequately. This imbalance can undermine the division of responsibilities and, as a result, the sustainable management of fishery resources (Trimble and Berkes 2013).

Fishermen and fisherwomen in regions with approved agreements, represented by their respective Regional Sustainable Development Councils, have been actively mobilizing to participate in the development and implementation of more sustainable fishing practices in the Lower Amazon region. Additionally, non-governmental organizations have also been fundamental in this process, providing technical support to fishing communities. On the other hand, public authorities have been notably less engaged in these efforts (Santos 2021). The Tapajós River agreement stands an exception, with more active government participation, through ICMBio, alongside some involvement from municipal environmental departments in the case of the Lago Grande agreement. Following the publication of the state decree, SEMAS has assumed a slightly more prominent role in fisheries management. However, co-management does not conclude with the legitimization of users' proposals; it must also ensure the effective implementation of these proposals, particularly through its enforcement. Even with the contributions of local fishers, environmental enforcement remains fundamentally the responsibility of the state (Mansbridge 2014).

Fishing agreements are not static structures, rather they are dynamic treaties that require periodic adjustment to remain effective, particularly given the complexity and ever-changing nature of artisanal fishing in the Amazon floodplain. As such, fishing agreements are integral to an adaptive management strategy, evolving is response to shifts and transformations within the socio-ecological system (Berkes et al. 2001). In this context, between 1999 and 2005—a span of seven years several fishing regions saw the approvals of two successive proposals, with the earlier agreements being revoked. Between 2011 and 2021, a period during which Pará lacked a legal framework for fishing agreements in Pará, the existing formalized agreements likely became outdated and necessitated updates, as exemplified by the case of Lago Grande. Social mobilization to revise the Lago Grande fishing agreement began in 2017, yet it was not until 2022, five years later, that the agreement was officially ratified. Without the presence of active leaders, the social mobilization of fishers could have been hindered during this prolonged delay in SEMAS' response. Therefore, establishing an efficient and responsive process for ratifying agreements, on that accounts for their inherent need to adapt to changing conditions is essential for strengthening adaptive fisheries management.

Another highlight in the process of resuming of fishing agreements in the state of Pará is the revival of discussions about exclusivity of fishing resources for local users, as exemplified by the *Arapixuna* agreement. The fishing agreement movement of the 1990s sought the exclusivity of the fishing resources of the floodplain lakes to local fishers, as they are seen as communally owned areas. However, due to a legal interpretation of the Water Code, which considers water resources to be public, the proposals for exclusive use of the lakes by local fishing communities were not included in the agreements made official by IBAMA and MMA at the beginning of the century (McGrath et al. 2004). Carrillo et al. (2019) also point out that fishers of Costa Rica claimed for the exclusivity of resource use as a key factor for fisheries sustainable management. Castro and McGrath (2001) showed that, in the lower Amazon, fishing agreements are considered by users to be an instrument that enforce access rules. Without a legal guarantee of a common property regime, local fishers' engagement to participate in co-management become weaker.

With the MPF's issuing a favorable decision, the *Arapixuna fishing agreement* proposal establishes a precedent for future agreements to also address the issue of exclusivity over fishing resources. Should SEMAS accept the MPF's recommendation, this would mark a significant milestone for the artisanal fishing sector, representing a hard-won achievement after more than two decades of persistent struggle and resistance against the encroachment of large-scale commercial fishing operations.

5 Conclusion

In Brazil, the northern region stands out as the area with the highest number of fisheries co-management initiatives compared to other regions of the country, with the states of Pará and Amazonas hosting the majority of these initiatives (Kalikoski 2009). Unlike the state of Amazonas, which has made significant advancements in environmental policy, including the regulation of fishing activities within its territory, the state of Pará has considerable progress to make in this regard (McGrath et al. 2015). Nevertheless, a notable step forward has been taken with the publication of the state decree formalizing fishing agreements in Pará. This development has reinvigorated efforts among fishers to engage in discussions on strategies and rules aimed at promoting fisheries sustainability and floodplain conservation through the establishment of fishing agreements. In this context, Pará is once again laying the groundwork for fisheries co-management, building on the foundations previously established by the federal government.

The co-management approach currently underway in the Lower Amazon demonstrate a growing consideration for the perspectives and demands of local users, largely due to the resistance and institutional articulation of fishers with governmental and non-governmental organizations. For instance, SEMAS's position has not been passively accepted by fishers, who have seized this opportunity to reaffirm a long-standing demand: the exclusivity of fishing activities for local users in floodplain lakes. This resistance movement among fishers in the Lower Amazon is rooted in historical mobilization efforts, organized through Fishing Councils and Fishing Unions. To achieve broader social environmental impacts and influence the political direction of Pará state is essential to strengthen the social movements of fishers and other community groups in other sub-regions of the state. This is particularly critical given the mounting pressures from several sectors, agribusiness, mining, logging, cattle ranching, hydroelectric dams, and other activities.

In summary, the ratification of fishing agreements represents a fundamental step towards the co-management of fisheries in the Amazon region. However, further progress is necessary to ensure the effective implementation of these agreements post-approval, especially in terms of environmental monitoring and enforcement. Recognize the exclusive use of fishing resources for local fishers would provide significant incentives for these communities to actively collaborate in the sustainable management of fisheries, thereby supporting both their traditional way of life and the environmental sustainability of the floodplain ecosystem. Such recognition could also serve as a catalyst for other fishing communities in Pará state, many of which are currently struggling to manage their fisheries resources, to mobilize and seek formal recognition of their own fishing agreements.

The decree formalizing fishing agreement is a relatively recent policy tool adopted by the state government. As such, the analysis presented in this study is constrained by the limited time frame since its implementation. To fully assess the broader contributions of this policy to fisheries co-management and floodplain conservation in the Lower Amazon, future research will be essential. Such studies should aim to evaluate the long-term impacts and effectiveness of these agreements in achieving their socio-environmental and conservation objectives.

References

Almeida OT, McGrath DG, Ruffino M (2001) The commercial fisheries of the lower Amazon: an economic analysis. Fish Manage Ecol 8:253–269

Almeida O (Org) (2006) Manejo De Pesca Na Amazônia. Peirópolis, São Paulo

Amorim C (2014) História das Colônias de Pescadores no Brasil. Jornal Martim-Pescador 125

Benatti JH (2005) Aspectos jurídicos e fundiários da várzea: uma proposta de regularização e gestão dos recursos naturais. In: Benatti JH, Surgik ACS, Treccani GD, McGrath DG, Gama A (ed) SP A questão fundiária e o manejo dos recursos naturais para a elaboração de novos modelos jurídicos. Edições Ibama/Provázea, Manaus

Berkes F, Mahon R, Mcconney P, Pollnac R, Pomeroy R (2001) Managing small-scale fisheries: Alternative directions and methods. International Development Research Centre, Ottawa, ON, Canada

Brasil (1933) Decreto nº 23.134, de 9 de setembro de 1933. Legislação Informatizada—Transfere os Serviços da Pesca e Saneamento do Litoral do Ministério da Marinha para o Ministério da Agricultura. Rio de Janeiro, 9 de setembro de 1933, 112º da Independência e 45º da República

Brasil (1967) Decreto-Lei nº 221, de 28 de fevereiro de 1967. Dispõe sobre a proteção e estímulos à pesca e dá outras providências

Brasil (1973) Portaria nº 471 de 26 de dezembro de 1973. Institui o novo estatuto para as colônias de pescadores. Ministério da Agricultura. Rio de Janeiro, 26 de dezembro de 1973

Brasil (1988) Constituição da República Federativa do Brasil de 1988. Constituição da República Federativa do Brasil. Publicação do Diário Oficial da União—Seção 1—Brasília, 5 de outubro de 1988. Página 1 (Publicação Original)

Campos AJT (1993) Movimentos sociais de pescadores da Amazônia. In: Furtado LG, Leitão W, Melo ABF (Orgs) Povos das águas: realidade e perspectiva na Amazônia. MPEG, Belém

Carrillo IIC, Stefan P, Madrigal-Ballestero R, Schlüter A, Gutierrez-Montes I (2019) Do responsible fishing areas work? Comparing collective action challenges in three small-scale fisheries in Costa Rica. Int J Commons 13(1):705–746. https://doi.org/10.18352/ijc.923

Cartilha Colônia De Pescadores. Colônias de pescadores. Santarém Baixo Amazonas, Pará, abril. 2004. Disponível em: http://documentslide.com/Documents/cartilha-colonia-de-pescadores-55a0fb3dab4de.html

Castello L, Macedo MM (2016) Large-scale degradation of Amazonian freshwater ecosystems. Glob Change Biol 22:990–1007

Castello L, McGrath DG, Hess L, Coe MT, Lefebvre PA, Petry P, Macedo MN, Renó VF, Arantes CC (2013) The vulnerability of Amazon freshwater ecosystems. Conserv Lett 6:217–229

Castello L, Arantes CC, McGrath DG, Stewart DJ, Sousa FS (2015) Understanding fishing-induced extinctions in the Amazon. Aquat Conserv Mar Freshwat Ecosyst 25:587–598

Castro F, McGrath D (2001) O manejo comunitário de lagos na Amazônia. Parcerias Estratégicas. 12:112–126

Diegues ACS (1983) Pescadores, camponeses e trabalhadores do mar. Ática, São Paulo

Faria CH Jr, Souza JS, Pereira TM (2018) Diagnostico da ação do IBAMA na explotação dos recursos pesqueiros no Baixo Amazonas (PA). Revista Ibero-Americana De Ciências Ambientais 9:273–282. https://doi.org/10.6008/CBPC2179-6858.2018.006.0026

Furtado LG (1993) "Reservas pesqueiras", uma alternativa de subsistência e de preservação ambiental: reflexões a partir de uma proposta de pescadores do Médio Amazonas. In: Furtado L, Leitão W, Mello AF (Orgs.) Povos das Águas: Realidade e Perspectiva na Amazônia. Coleção Eduardo Galvão, PR/MCT/CNPq, MPEG, Belém

Furtado LG (1997) Problemas ambientais e pesca tradicional na qualidade de vida da Amazônia. In: Furtado LG (Org.) Amazônia: desenvolvimento, sócio-diversidade e qualidade de vida. UFPA/NUMA, Belém

Hora NN (2020) Sistemas de co-manejo pesqueiro com enfoque no pirarucu (Arapaima spp.) em comunidades de várzea do Baixo Amazonas: análise institucional e custos de transação. 242 f. Tese (Doutorado em Sociedade, Natureza e Desenvolvimento)—Programa de Pós-graduação em Sociedade, Natureza e Desenvolvimento, Universidade Federal do Oeste do Pará, Santarém. https://repositorio.ufopa.edu.br/jspui/handle/123456789/351

Isaac VJ, Almeida M (2011) El consumo de pescado en la Amazonia Brasileña. Food and Agriculture Organization (FAO). COPESCAL 13:1–43

Isaac VJ, Castello L, Santos PRB, Ruffino ML (2016) Seasonal and interannual dynamics of river-floodplain multispecies fisheries in relation to flood pulses in the Lower Amazon. Fish Res 183:352–359

Junk WJ, Bayley PB, Sparks RE (1989) The flood pulse concept in river-floodplain systems. In Dodge DP (ed) Proceedings of the international large river symposium. Can Spec Publ Fish Aquat Sci 106:110–127

Kalikoski D et al (2009) Gestão compartilhada do uso sustentável de recursos pesqueiros: sefletir para agir. IBAMA, Brasília

Leitão W (1995) Pesca e políticas públicas. Bol. Mus. Para. Emílio Goeldi, Antropol. Belém 11:185–198

Leroy JP (1988) O bote contra a geleira. Os pescadores de Santarém em busca da sua afirmação. Revista Proposta. Entre o passado e o futuro. Pescadores Artesanais Rio J 38:37–50

Lima MW, Praxedes TN, Santos VA, Alves MAMS, Bastos RZ, Henriques FM, Palheta MKS, Silva IGV (2023) Os desafios e as perspectivas dos acordos de pesca após a regulação do decreto estadual 1.686/2021 na Amazônia Oriental. In: Congresso brasileiro de engenharia de pesca—XXII CONBEP, Porto de Galinhas-PE

Mansbridge J (2014) The role of the state in governing the commons. Environ Sci Policy 36:8–10

McGrath DG, Cardoso A, Almeida OT, Pezzuti J (2008) Constructing a policy and institutional framework for an ecosystem-based approach to managing the Lower Amazon floodplain. Environ Dev Sustain 10:677–695
McGrath D, Castello L, Almeida OT, Estupiñán GMB (2015) Market formalization, governance and the sustainable management of community fisheries in the Brazilian Amazon. Soc Nat Resour 28:513–529. https://doi.org/10.1080/08941920.2015.1014607
McGrath DG, Castro F, Futemma BDA, Calabria J (1993) Fisheries and the evolution of resource management on the lower Amazon Floodplain. Human Ecol 21
McGrath DG, Cardoso A, Sá EP (2004) Community fisheries and co-management on the lower Amazon floodplain of Brazil. Proceedings of the second international symposium on the management of large rivers. Phnom Penh, FAO:207–221
McGrath DG, Gama SP, Cardoso A, Almeida OT, Benatti JH (2011) Integrating comanagement and land tenure policies for the sustainable management of the Lower Amazon floodplain. In: Pinedo-Vasquez M, Ruffino ML, Padoch C, Brondízio ES (eds) The Amazon várzea: the decade past and the decade ahead. Springer, London, New York
Mopebam (2021) Cartilha gestão de colônias. Santarém, Pará
Moraes SC (2000) Colônias de pescadores e a luta pela cidadania. Campus de Bragança—RN. Mestrando (Educação)—Universidade Federal do Rio Grande do Norte—UFR
Oviedo AFP, Bursztyn M (2016) The fortune of the commons: Participatory evaluation of small-scale fisheries in the Brazilian Amazon. Environ Manage 57:1009–1023. https://doi.org/10.1007/s00267-016-0660-z
Pantaleão GSL, Cintra IHA, Silva KCA, Santos MAS, Bezerra AS (2020) Environmental infractions and fines imposed for illegal fishing in the State of Pará, Amazon, Brazil. J Agric Stud 8:251–264. https://doi.org/10.5296/jas.v8i2.15437
Santos MAS (2005) A cadeia produtiva da pesca artesanal no estado do Pará: estudo de caso no nordeste paraense. In: Amazônia: Cia and Desenvolvimento, Belém 1.
Santos AO (2021) Mapeamento, diagnóstico e análise dos acordos de pesca do Baixo Amazonas, Sapopema, Santarém-PA
Trimble M, Berkes F (2013) Participatory research towards co-management: lessons from artisanal fisheries in coastal Uruguay. J Environ Manage 128:768–778

Floristic and Phytosociological Survey of Fabaceae and Potential Use of Species for Soil Recovery

Jane Eire Souza da Cunha, Ana Sofia Sousa de Holanda, Cauan Ferreira Araújo, and Dávia Marciana Talgatti

Abstract Concerns about anthropogenic disturbances to soils have led to concerns about reversing degraded soils. This has led to studies on the floristic composition and phytosociological structure of species that can be used to recover degraded soils, such as species from the Fabaceae family. This study aimed to survey the biodiversity of Fabaceae in a floodplain area in Oriximiná-Pará and the potential use of species for recovering degraded soils. Important phytosociological values were observed, demonstrating the diversity of Fabaceae in the study area. Nine species were found in the survey, but in the literature only *Cassia leiandra* Benth., *Campsiandra laurifolia* Benth. and *Bergeronia Sericea* Micheli are mentioned for use in recovering degraded areas. However, the species were predominant *Zygia latifolia* var. *communis* and *Albizia subdimidiata* Barneby & J. W. Grimes in the area studied, emphasising the importance of future studies into the importance of these species in soil recovery. In the granulometric analysis, plots 2 and 4 showed a predominantly sedimentary nature in floodplain soil, while plots 1 and 3 did not follow the pattern, which can be explained by anthropogenic action. In the analysis of pH and exchangeable aluminium, in the 4 study plots, it was possible to identify that the pH result stood out as soil with high acidity and exchangeable aluminium (Al^{+3}) at a high level (acid). The soil in the study area is considered to be of low fertility, and the pH and Al^{+3} values are harmful to plant growth. Finally, there was a shortage of floristic surveys of Fabaceae carried out in the Oriximiná-PA region, so more studies should be carried out in order to reduce this degree of data scarcity, to increase publications of floristic surveys in the region in question and serve to draw up a list of regional species in a given area that can contribute to restoration strategies.

Keywords Degradation · Inventory · Legumes · Amazon river

J. E. S. da Cunha (✉)
Master's student in the Postgraduate Program in Biodiversity (PPGBEES), Federal University of Western Pará, Santarém, Pará, Brazil
e-mail: jane.eire.orixi@gmail.com

A. S. S. de Holanda · C. F. Araújo · D. M. Talgatti
Federal University of Western Pará, Oriximiná Campus, Oriximiná, Pará, Brazil

W. Leal Filho et al. (eds.), *Amazon 2030—Sustainability Issues in the World's Largest Rainforest Region*, World Sustainability Series,
https://doi.org/10.1007/978-3-031-81465-5_31

1 Introduction

Forest inventories provide information on floristic composition and forest structure, which are the basis for understanding ecosystems. It makes it possible to characterise a given area and assess the qualitative and quantitative variables of the forest and their interrelationships, as well as the dynamics of forest growth and succession, relating them to ecological and floristic components (Vibrans et al. 2010). From forest and floristic inventories, it is possible to check which families are most representative in the study area, measure the degree of importance of the species sampled, demonstrate species diversity and richness, and determine whether any species is important in the area studied, as well as finding possible rare species (Chaves et al. 2013; Santos et al. 2016). Thus, the knowledge generated from these studies can support decision-making processes about the use of natural vegetation resources, enabling rational exploitation combined with the conservation and management of biodiversity in a given area (Chavan and Dattaji 2013).

Phytosociology is a branch of ecology that studies the composition, structure and dynamics of plant communities and is a tool with numerous interfaces. Based on the principles of floristics, phytosociology combines qualitative and quantitative methods to describe and analyse plant species, providing valuable information that can be used for the planning and sustainable management of vegetation and forestry, and is also fundamental for projects to recover degraded areas and conserve plant diversity (Rodrigues et al. 2009; Oliveira 2011; Chaves et al. 2013; Dufrêne and Legendre 1997).

Floodplain areas have soils with limiting physical characteristics and are susceptible to degradation (Cravo et al. 2002). They can suffer disturbances caused by natural actions where the loss of organic matter is low, but degradation can occur at different levels and reach its most advanced stages, with other different causes such as removal of the native layer, anthropogenic intervention through intensive cattle farming and burning (Franco et al. 2003). These disturbances can cause chemical changes in the soil related to pH, aluminium, carbon, phosphorus, potassium, magnesium and iron, among other elements (Magalhães and Gomes 2013). Furthermore, due to the mineralogy and constitution dynamics of floodplain soils, nitrogen is the main limiting nutrient, even in conserved soil conditions (Fajardo et al. 2009).

A sustainable and affordable alternative for family farmers to recover degraded soil, enabling a constant supply of nitrogen, is to plant species from the Fabaceae family (APG 2023), also known as the legume family (Brienza 2012). This is also a recommended practice for recovering degraded areas, as legumes use the vegetation itself to protect the soil from erosion (Bertoni and Lombardi Neto 2008).

The Amazon is suffering from major impacts resulting from deforestation and other forms of land use change that have resulted in the significant degradation of many of its ecosystems (Albert et al. 2023; Nobre et al. 2016). The floodplain forests of the Oriximiná region have also suffered from the impact of anthropogenic intervention associated with intensive cattle farming, removal of native vegetation and burning. Knowledge of the diversity of floodplain forests is fundamental for understanding and quantifying the impacts of these actions on this ecosystem and an important tool is to carry out floristic inventories and surveys, where information can be obtained on important species, such as those of the Fabaceae, in order to propose the use of these species in places with disturbances resulting from anthropogenic actions and/or natural actions, ensuring the persistence of plant communities and healthy soil over time.

Along with the species survey, analysing the soil in the studied area provides important data that can be used to correlate the composition and distribution of Fabaceae with the characteristics of the soil, and laboratory methods of soil analysis are used because they provide reliable and accurate results (Niero et al. 2010). This relationship between surveying species and analysing the soil is important for gaining knowledge about a particular area that has not yet been studied in the literature, making it possible to gather important data. Given the above, this study aimed to characterise the soil and survey the diversity of Fabaceae in a floodplain area in Oriximiná-Pará and their potential use as species for recovering degraded soils.

2 Material and Methods

2.1 Study Area

The research was carried out in the floodplain area located on the Nhamundá Igarapé, in the Nossa Senhora da Conceição community, 14 km from the municipality of Oriximiná-PA, 1°50′22.06"S–55°57′48.85"O (Fig. 1).

2.2 Sample Collection and Floristic Survey

The floristic survey took into account the individuals sampled in the phytosociological survey. Fertile samples of the representatives were collected and processed by the techniques proposed by Mori (2011). Identification of the botanical material was based on consultation with specialised literature, comparison with the SpeciesLink database (CRIA 2021), as well as input from expert botanists. The samples will be deposited in the HSTM (Herbarium of the Federal University of Western Pará at the Santarém Campus). The scientific names, botanical synonyms, spellings and abbreviations, and habit information were checked against the List of Species of the Flora of Brazil (Flora do Brasil 2020). The species were classified according to the APG IV system (Angiosperm Phylogeny Group [APG] 2016).

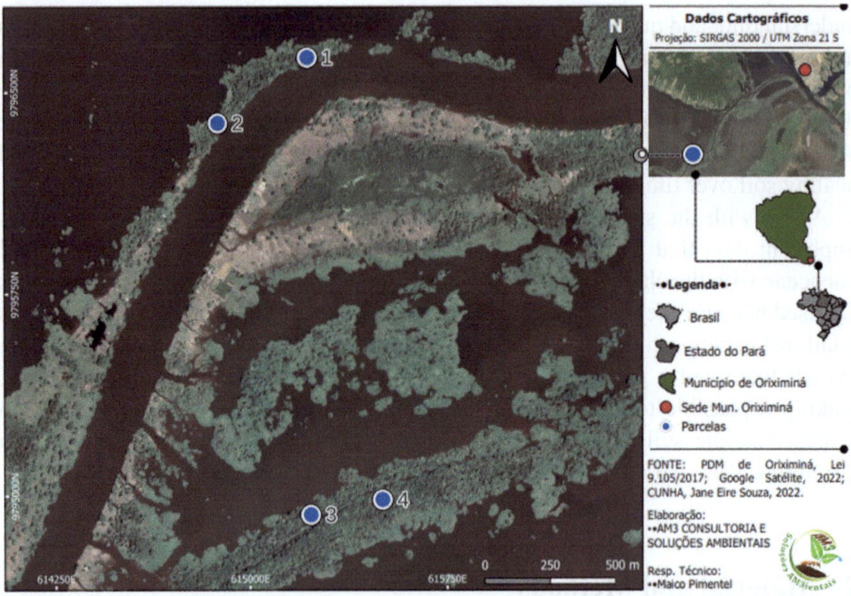

Fig. 1 Location of the collection area and plots (points in blue) where the survey was carried out; PDM of Oriximiná, Law 9.105/2017; Google Satellite (2023). *Source* Prepared by the authors (2023)

2.3 Phytosociological Survey

Four 125 m long by 20 m wide plots were delimited and georeferenced in a floodplain area, totalling one hectare sampled, where the species of flora that occur in this biome were identified and the state of conservation of the vegetation was characterised. A tape measure was used to measure all trees with a circumference greater than or equal to 31.4 cm (DBH $\geq$ 10 cm) at 1.30 m from the ground. For bifurcated individuals, only those in which at least one stem reached the size class determined at breast height were measured; in these cases, all the stems were measured for later totalling.

To identify the material, a fertile or non-fertile branch was collected and photographed. The collections were duly herborised following the techniques of Mori (2011), and dissected in an oven for later determination in the laboratory, aided by the photographs (Fig. 2).

Taxon identification was carried out primarily in the field by local collaborators who knew the area, such as Mr Raimundo Rosi Dos Passos Costa, a local resident. Scientific identification, classification and check-listing were carried out at the UFOPA-CORI Laboratory of Amazonian Plants and Algae (LAPAM), using identification keys available in specialised literature and comparisons with botanical herbarium collections available in the SpeciesLink database (CRIA 2021).

The classification of family, genus, species, author, popular name, occurrence, endemism and origin was checked, while in the virtual herbarium, a query was made

Fig. 2 **a** Measurement of the study plots; **b** Measurement of the circumference at breast height (CBH); **c** Collection of the botanical material (branch) for identification in the laboratory. *Source* Prepared by the authors (2023)

that allowed the samples to be re-identified and typified, as well as other functionalities. The classification system used was APG IV.[1] The checklist and spelling were based on information on the REFLORA website[2] and Tropicos website.[3]

2.4 Soil Collection and Sampling

Soil block samples were collected at a depth of 0–30 cm for characterisation analyses. 10 sub-samples were taken from each plot to obtain an average of the area sampled. The collection area for the soil blocks was the same as for the Fabaceae survey (125 m long × 20 m wide) and the samples were collected in a zigzag pattern in the 4 plots.

Still, at the collection site, the soil was transferred from the auger to the bucket, where it was mixed with the other samples from the plot under study, which were then called composite samples (ESALQ 2023). The sample was then labelled with the name of the area, plot and date (Fig. 3).

The composite samples were taken to the laboratory where they were partially dried, destressed, weighed and visually sieved for soil properties.

The structural quality of the aggregates was observed; the size of the aggregates; and physical characterisation, usually carried out using simple routine tests such as granulometric analysis, compaction tests, determination of consistency limits and

[1] Available at: http://mobot.org/.

[2] Available at: https://reflora.jbrj.gov.br.

[3] Available at: https://www.tropicos.org.

Fig. 3 **a** Removing the soil block with an auger; **b** Bucket with soil block. *Source* Prepared by the authors (2023)

natural water content, determination of density through representative samples and determination of solid particle density.

2.5 *Granulometry*

Analyses to determine granulometry followed the methodology of dispersion in hexametaphosphate and sedimentation of silt, according to the Embrapa protocol (2017).

Firstly, 15 g of TFSA (fine air-dried soil) was weighed into a pre-weighed 250 mL beaker and 50 ml of 3% sodium hexametaphosphate solution was added, after which the solution was left to react for 1 h. After this, it was stirred on a magnetic stirrer for 30 min at 600 rpm and then sieved through a 0053-mm sieve to collect the sand fraction. The sub-sample was then transferred to a pre-weighed beaker and left to dry in an oven at 105 °C for 1 h.

After the above procedures, the container with the sub-sample was weighed and subtracted to obtain the sand fraction. The remaining sub-sample (silt + clay) was then manually stirred and left to stand for 2 h to decant the silt.

The suspended clay solution was carefully discarded using a 10 ml syringe, then the beaker with the silt was placed in an oven at 105 °C for 1 h, and finally, the beaker with the decanted silt was weighed and subtracted to obtain the silt fraction.

Calculation formulas:

% Sand–(weight sand/15 g) × 100
% Silt–(weight silt/15 g) × 100
% Clay–100–(Sand + Silt).

2.6 *pH*

For the pH analysis, 10 g of fine air-dried soil was added to the already identified containers, 1 with standard soil and the other with the soils from the 4 plots under study, 25 ml of distilled water was added to each sample, where it was stirred with a stick for approximately 30 timed seconds, then left to rest for 40 min, after which the samples were stirred again with the glass stick, then the multiparameter device was used to measure the pH of each of the samples.

2.7 *Exchangeable Aluminium (Al^{+3})*

Exchangeable aluminium was measured by extraction with a 1N KCL solution and volumetric determination with a diluted NaOH solution. The solutions were prepared for the analyses.

1N KCl extraction solution: 74.55 g of KCL was added to a 300 ml beaker, then distilled water was added and stirred with a glass rod until all the substance was diluted, then transferred to a 1000 L volumetric flask, where it was identified with the solution already prepared.

1N NaOH solution: 4 g of NaOH was weighed out in a beaker, topped up with distilled water and stirred with a glass rod. When it reached room temperature, it was transferred to a 100 ml volumetric flask.

0.1 N NaOH solution: From the 1N NaOH solution, 10 ml was taken and transferred to a 100 ml volumetric flask and topped up with distilled water.

0.025 NaOH determination solution: 25 ml was taken from the 1N NaOH solution, transferred to a 1000 ml volumetric flask and topped up with distilled water.

Solutions, separated and weighed from the soil samples: 5 g of soil was weighed for each sample and 50 ml of KCL was added, then the ballerinas were capped with watch caps and taken to the magnetic stirrer for 4 min at 600 rpm. After the time had elapsed, each sample was poured separately into 50 ml beakers using a funnel and a coffee strainer for the dripping process.

Bromothymol blue indicator solution 1 g/L: 0.1 g was weighed and deposited on a porcelain grating and mixed with 3 drops of 0.1 N NaOH solution, then macerated for approximately 2 h until it reached a greenish blue colour, during this maceration stage 4 drops of 0.1 N NaOH solution were added every time it was dry, after a running time of approximately 2 h, a greater quantity of 0.1 N NaOH was added until the porcelain grating was abundant.

Measurement of Al^{+3}: First, 25 ml of KCL was separated in a beaker, this being the standard sample (white sample). For the other samples, 25 ml of KCL was also weighed and 3 drops of Bromothymol Blue Indicator Solution were added, then poured into a burette already containing the 0.025 N NaOH solution, then the colour was observed, where blue indicated zero aluminium. The same process was carried out on the 4 samples from the plots, where the results were calculated using the formula:

$$Al^{+3}\left(\text{cmol} \cdot \text{kg}^{1}\right) = (\text{a} - \text{b})$$

where:

a volume spent titrating the sample;
b volume spent titrating the blank.

2.8 *Analysing the Data*

The Shannon–Weaver diversity index (H') and the Pielou equability index (J) were used to interpret diversity.

The following parameters were assessed Absolute Density (AbsDe) and Relative Density (RelDe), Absolute Frequency (AbsFr) and Relative Frequency (RelFr), Absolute Dominance (AbsDo) and Relative Dominance (RelDo), as well as the Importance Value Index (IVI) and Cover Value Index (IVC).

The average was used to analyse the Diameter at Breast Height (DBH), which was calculated based on the minimum and maximum values of the diameters sampled for each species. For height, the average was used based on the minimum and maximum values sampled.

The data was analysed and all the phytosociological parameters were assessed using the Fitopac 2.1 programme (Sheppard 2010). The CAP and height graphs were made in the GraphPad Prism 8.0.1 programme (GraphPad Software Inc., San Diego, CA, USA), based on the averages made in the Fitopac 2.1 programme.

3 Results and Discussion

3.1 *Floristic and Phytosociological Survey*

Eighty-two individuals were recorded, corresponding to nine species, distributed in eight genera, as shown in Table 1.

Table 1 Fabaceae species sampled in the phytosociological survey of the Nossa Senhora da Conceição community, Oriximiná-PA, with a description of their popular name, life form and origin

Species	Popular name	Way of life	Origin
Albizia subdimidiata Barneby and J. W. Grimes	paricá-do-lago	Tree	Native
Bauhinia rutilans Barneby and J. W. Grimes	escada-de-jabuti	Tree	Native
Bergeronia sericea Micheli	buteira	Tree	Native
Campsiandra laurifolia Benth	manaiara	Tree	Native
Cassia leiandra Benth	mari-mari	Tree	Native
Ormosia macrocalyx Duke	tento	Tree	Native
Senegalia loretensis (J. F. Macbr.) Seigler	paricá-de-espinho	Tree	Native
Zygia latifolia (L.) Fawc. and Rendle	ingá-do-baixo	Tree	Native
Zygia latifolia var. *communis* Barneby and J. W. Grimes	ingá-de-sapo	Tree	Native

Source Prepared by the authors (2023)

As for species diversity, the values found for the Shannon index (H') and equability were 1.702 and 0.775 respectively. These results show that there was uniformity in the diversity and distribution of individuals in the species.

Table 2 shows the phytosociological parameters with the species listed in descending order of importance.

The species *Z. latifolia* var. *communis* is widely distributed and stands out from the other species, showing the highest Absolute and Relative Density (AbsDe and RelDe), Absolute and Relative Dominance (AbsDo and RelDo), as well as the highest Importance Value and Coverage Index (IVI and IVC). This is a species restricted to South America, which leads to a bias of concern in conserving this species. In Brazil, it is found in almost all the states of the Brazilian Amazon and Goiás. It generally occurs in floodplain forests and campinarana (white sand forests), igapó forest, and occasionally in terra firme forest with clay soil. It flowers from March to December and bears fruit from August to May (Garcia et al. 2022).

The second most commonly found species was *A. subdimidiata* (paricá-do-lago), which occurred in all four survey plots and had the second-best AbsDe, RelDe, AbsDo, RelDo, IVI and IVC values. It is also important to note that this species had the highest Absolute and Relative Frequency values (AbsFe and RelFe), as shown in Table 2. This species occurs in the northern region, in the states of Acre, Amazonas, Amapá, Pará, Rondônia and Roraima. It occurs in floodplain fields, igapó forest, floodplain forest and terra firme forest (Scheidegger and Rando 2022).

The third species with the highest number of individuals was *C. leiandra* (mari-mari), with the third highest values for AbsDe, RelDe, AbsDo, RelDo, as well as the third highest IVI and IVC in Brazil. It is found in the northern region (Acre, Amazonas, Amapá, Pará, Rondônia and Roraima), occurring in riparian or gallery forests, floodplain forests and terra firme forests, and is adaptable to dry, humid or flooded soils: 10 to 15 m, with flowering from December to January and fruiting

Table 2 List of species sampled in the 4 survey plots and their respective phytosociological parameters

Species	Nind	Nam	AbsDe	AbsDo	AbsFr	RelDe	RelDo	RelFr	IVI	IVC
Zygia latifolia var. *communis*	31	2	28.7	1.26	50.00	37.80	47.82	11.76	97.39	85.63
Albizia subdimidiata	23	4	21.3	0.57	100.00	28.05	21.73	23.53	73.31	49.78
Cassia leiandra	10	2	9.3	0.33	50.00	12.20	12.68	11.76	36.64	24.88
Campsiandra laurifolia	6	2	5.6	0.06	50.00	7.32	2.24	11.76	21.32	9.56
Ormosia macrocalyx	4	1	3.7	0.12	25.00	4.88	4.48	5.88	15.24	9.36
Zygia latifolia	3	1	2.8	0.22	25.00	3.66	8.48	5.88	18.02	12.13
Bauhinia rutilans	2	2	1.9	0.03	50.00	2.44	1.10	11.76	15.31	3.54
Senegalia loretensis	1	1	0.9	0.03	25.00	1.22	0.99	5.88	8.09	2.21
Bergeronia sericea	1	1	0.9	0.00	25.00	1.22	0.00	5.88	7.10	1.22

Where: **NInd**—Number of Individuals; **NAm**—Number of Samples; **AbsDe**—Absolute Density; **AbsDo**—Absolute Dominance; **AbsFr**—Absolute Frequency; **RelDe**—Relative Density; **RelDo**—Relative Dominance; **RelFr**—Relative Frequency; **IVC**—Cover Value Index; **IVI**—Importance Value Index
Source Prepared by the authors (2023)

from February to March, cited in the literature as a species used to recover degraded areas, as it aims to improve soil fertility as a nitrogen fixer, the limiting nutrient in flood plains, and its fruit serve as food for fauna (Chagas and Dutra 2022; FAO 1986; Paisagem Nativa 2023; Rioterra 2021). The species *C. laurifolia* (manaiara) and *B. rutilans* (escada-de-jabuti) are found in Brazil in the northern region, in the states of Amazonas, Amapá, Pará, Rondônia and Roraima. The species *C. laurifolia* occurs in igapó forest and ombrophilous forest (rainforest) and is a nitrogen-fixing species (Sprent and Parsons 2008) and recommended for recovering degraded areas (Vieira et al. 2017) and *B. rutilans* is found in the Amazon in terra firme forest, which differs from the survey in this study in which it was found in floodplain forest (Cota 2022; Kuntz and Vaz 2022).

Zygia latifolia is a species that, although not endemic to Brazil, is found in the following regions: North (Acre, Amazonas, Amapá, Pará, Rondônia, Roraima), Northeast (Bahia, Pernambuco), Centre-West (Goiás, Mato Grosso), Southeast (Espírito Santo, Minas Gerais, Rio de Janeiro, São Paulo), and South (Paraná). It occurs in the vegetation types of anthropogenic area, campinarana, igapó forest, terra firme forest, várzea forest, deciduous seasonal forest, semi-deciduous seasonal forest and ombrophilous forest (rain forest) (Garcia et al. 2022). Unlike the species *Z. latifolia* var. *communis*, which belongs to the same genus and was the most commonly found, this species is restricted to South America, found in almost all the states of the Brazilian Amazon and Goiás, but occurs in vegetation very similar to *Z. latifolia* (Garcia et al. 2022).

The species *O. macrocalyx* (Tento) and *S. loretensis* (paricá-de-espinho) have been confirmed in Brazil in the northern region (Acre, Amazonas, Amapá, Pará, Rondônia and Roraima). O. coccinea occurs in the Amazon in campinarana, ombrophilous forest (rainforest) and terra firme forest (Cardoso et al. 2022). The species *S. loretensis* occurs in anthropogenic areas, terra firme forests, semideciduous seasonal forests (rain forests), floodplain forests and deciduous seasonal forests (Terra and Morin 2022).

The species *B. sericea* (buteira) colonises roadsides and has the potential to recover degraded areas. It occurs abundantly in flooded river riparian forests, and sometimes in drier forests on clay soils (Pott and Pott 1994).

The highest Number of Individuals (NInd) and highest Number of Species (NSpp) occurred in plot 2, with 36 individuals and 6 species, also having the highest Absolute Density. This was followed by plots 1 and 4, respectively, with 31 and 10 individuals and 5 species in both, with plot 1 having the highest basal area (Table 3). One observation is that only 1 species occurred in plot 3, *A. subdimidiata*, noting that this was the only species that occurred in all 4 plots, suggesting that it may be a widely distributed species in the study region.

Figure 4 shows that the species with the highest average DBH was *B. sericea*, with an average of 57.3 cm, followed by the others, which maintained an average between 35.1 and 18.9, as can be seen in the figure.

Figure 5 shows that the species with the highest average height was *O. macrocalyx* at 11.3 m, followed by *C. laurifolia* and *S. loretensis.*

According to the authors Carim et al. (2007), Junior et al. (2015) and Silva (2018a, b), the Fabaceae family plays an important role in the constitution of secondary forests because it has an abundance of individuals with different habits. In addition to playing a fundamental role in recovering areas that have been damaged or degraded, it performs ecological functions of great importance in a forest, such as the accumulation of biomass and the ability to recover the soil through green manure,

Table 3 Phytosociological pattern values for the 4 survey plots

Samples	NInd	NSpp	AreaBas	AbsDe	AbsDo	RelDe
Parcel 1	31	5	1,47	114,8	5,44	37,80
Parcel 2	36	6	0,97	133,3	3,61	43,90
Parcel 3	5	1	0,13	18,5	0,50	6,10
Parcel 4	10	5	0,26	37,0	0,98	12,20

Where: **NInd**—Number of Individuals; **NSpp**—Number of Species; **AreaBas**—Basal Area; **AbsDe**—Absolute Density; **AbsDo**—Absolute Dominance; **RelDe**—Relative Density
Source Prepared by the authors (2023)

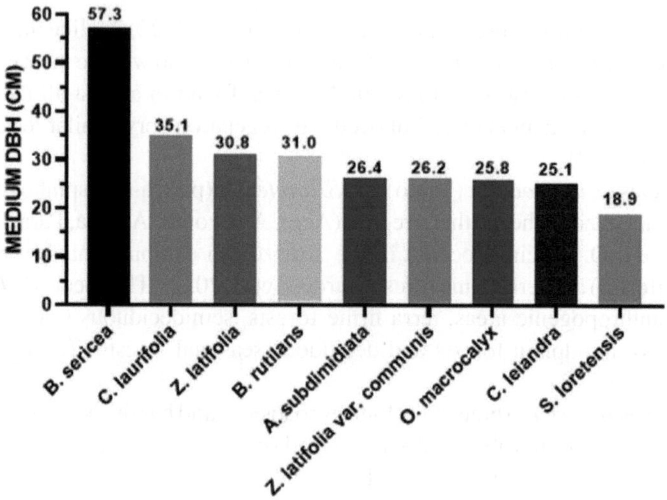

Fig. 4 Average diameter at breast height (DBH) of the species. *Source* Prepared by the authors (2023)

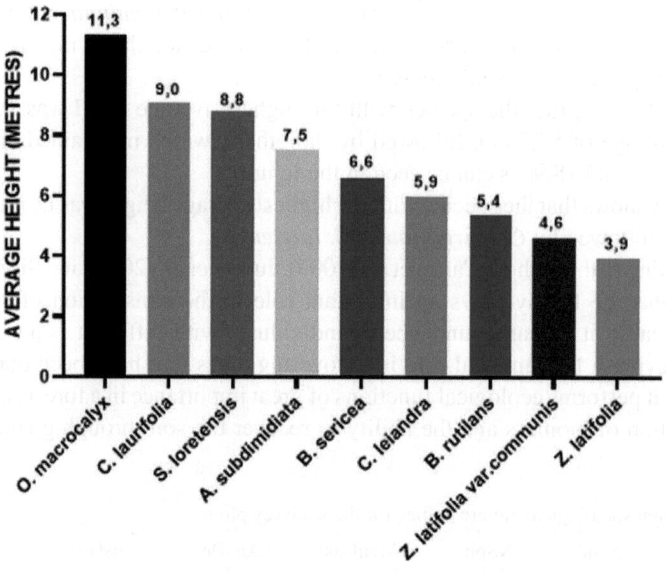

Fig. 5 Average species height (metres). *Source* Prepared by the authors (2023)

which provides potassium, phosphorus, calcium and especially nitrogen, the latter being assimilated into the soil through symbiotic relationships between bacteria and roots.

In order to select species for the restoration of degraded areas, a strategy should be used of using species with a diversity and structure similar to those of a reference ecosystem, so that the species evolve properly since they need to be well adapted and with the presence of pollinators and seed dispersers, so it is important to choose the right species, in which case this survey contributes to the selection of species for floodplain areas (Kageyama and Gandara 2000; Suganuma et al. 2013).

Reinforcing the above, Rodrigues et al. (2007) emphasise that in Brazil, native species of regional occurrence are preferably used, looking at the functions that each species performs, as well as the activities of the local fauna, in isolation or together.

Survey studies are helpful and prevent the use of exotic species, as they can show invasive behaviour and interfere with the process, although they have legal instruments for use in degraded areas (Costa and Durigan 2010; Assis et al. 2013).

Espirito-Santo et al. (2005) carried out a floristic inventory of 8.75 ha in the primary forest in the Tapajós National Forest in the state of Pará and found 3877 individuals with a DBH greater than 10 cm, distributed in 42 species, with the Fabaceae family having the highest species richness. Another study carried out by Almeida et al. (2012) in Santarém do Pará analysed a terra firme forest in diversified plots, with Fabaceae being the family with the highest number of species found, a total of 45 species. Other studies carried out in Pará also showed Fabaceae as the most diverse botanical family Andrade et al. (2015) and the most abundant Valente (2019), showing the importance of this family in the floristic composition and structure of vegetation in terra firme forests in the Amazon.

Borges (2019) also carried out a survey in dryland forests, ombrophilous forests, igapó forests and floodplain forests, which aimed to systematise information on the native plant species of the Guamá Microregion in the state of Pará, and the Fabaceae family showed the greatest representation, with 119 species of interest for restoration, being species with the Amazon Phytogeographic Domain and endemism in the region, with *Zygia latifolia* var. *communis* and *Campsiandra laurifolia*, which were also found in this study and have potential for use in degraded areas. These studies corroborate the fact that Fabaceae has a very high species richness in the Amazon, as in the state of Pará.

3.2 Soil Granulometry and Chemical Analysis

Particle size analysis identifies the levels of silt, clay and sand in the soil sample and is an important tool capable of providing a scientific basis for understanding how components affect water retention, porosity and nutrients (Silva 2018a, b).

When analysing the soil granulometry of the 4 study plots, it was possible to determine the silt, argyl and sand values (Table 4).

Soil chemical analyses focus mainly on determining pH (hydrogenionic potential), which represents the hydrogen in the soil solution [H+] and its concentration, which is also known as measuring the active acidity of the soil. Another is aluminium

Table 4 Values in % of sand, clay and silt defined in soil samples through granulometry analysis

	Silt %	Argyl %	Sand %
Parcel 1	36.8	23.8	39.3
Parcel 2	54.3	37.2	8.4
Parcel 3	32.3	33.9	33.8
Parcel 4	37.0	46.8	16.2

Source Prepared by the authors (2023)

(Al^{+3}), which is considered a toxic element for plants and is associated with acidity (EPAMIG 2012).

In the analysis of the pH and exchangeable aluminium (Al^{+3}) values of the 4 study plots, it was possible to identify the values described in Table 5. For all the plots, the pH result highlighted that the soil has high acidity and exchangeable aluminium is at a high level.

About pH values, according to the chemical acidity classification of Alvarez et al. (1999), the soil in the 4 study plots stood out for its high acidity, which lies between the 4.5–5.0 range. The results for exchangeable aluminium were high (>1.0) according to the chemical classification of the same authors (Table 6).

According to the distribution of the main soil classes contained in the different soil mapping units in the pedoenvironments of the Amazon region, the floodplain

Table 5 pH and exchangeable aluminium values of the 4 study plots, defined by soil analysis

	pH	Exchangeable aluminium (mg)
Parcel 1	4.97	1.0
Parcel 2	4.91	1.0
Parcel 3	4.78	3.0
Parcel 4	4.53	3.0

Source Prepared by the authors (2023)

Table 6 Chemical classification of pH and exchangeable aluminium according to Alvarez et al. (1999)

Chemical classification						
pH						
Very high acidity	High acidity	Medium acidity	Low acidity	Neutral	Low alkalinity	High alkalinity
<4.5	4.5–5.0	5.1–6.0	6.1–6.9	7.0	7.1–7.8	>7.8
Exchangeable aluminium						
Low		Medium		High		
<0.3		0.3–1.0		>1.0		

Source Prepared by the authors (2023)

areas have soils classified as Fluvial Neosols or Hapic Gleissols (Schaefer et al. 2017), in areas with a higher or lower level of annual sedimentation as opposed to the advance of the gleaning process, respectively. In both cases, the predominant material of Andean sedimentary origin gives rise to soils with high levels of cation exchange capacity. This property allows for high levels of exchangeable aluminium without it reaching high percentages in the sorption complex.

The sedimentary nature of floodplain soils is considered to be predominantly silt, followed by clay and sand (Pinto 2014). This information is highlighted in the studies by Pinto (2014) and Amorim et al. (2009) which showed a predominance of fine sediments (silt and clay), as well as the results of the research by Silva et al. (2018) and Abreu et al. (2007) in which the floodplain soil in these studies had a greater presence of silt and clay and a lower contribution from the sand fraction.

This sedimentary nature is explained by the fact that the finer material (silt) is carried by the Amazon River, which originates in the Andean and sub-Andean regions, but the floodplain soils also receive sedimentary influences from the tributaries that are located along the river basin. As a result, the water from the Amazon River that floods the floodplain areas tends to be shallower and carries finer sediments such as clay and silt (Amorim et al. 2009). Another reason for the predominance of silt and clay is related to the constant flooding of the area, which causes alterations or limitations in the process of pedogenesis (Lima et al. 2001).

The only plot where the predominant sedimentary nature of the floodplain soil was observed was plot 2, where silt predominated, followed by clay and sand. In plot 4, even though clay was more prevalent than silt, sand remained in a lower percentage.

However, it can be seen that the soil in plot 1 does not follow the sedimentary nature of floodplain soils, where the presence of sand (39.3%) was greater than that of silt (36.8%) and clay (23.8%), which can be explained by the fact that this plot is suffering greater anthropogenic interference from cattle breeding and human management with the digging of tanks. Plot 3 showed a very similar percentage between silt, clay and sand, which is possibly an area that is already starting to have its soil modified.

Regarding the results of the soil's chemical values, the pH result corroborates the results of exchangeable aluminium, which according to Embrapa (2017), when the pH is between 4 and 5, it indicates the presence of exchangeable aluminium. This presence of aluminium in the soil can inhibit root growth and influence the availability of other nutrients and processes such as the mineralisation of organic matter (Embrapa 2017).

The pH values in this study corroborate the study by Moraes (2016) where the pH values found in floodplain soils were between 4.5 and 5.9. In the work by Melo et al. (2013), the chemical characterisation of floodplain soils in different vegetation covers determined a pH equivalent to 5.38, indicating reduced natural acidity of the soils in these areas. In another study by Silva et al. (2018), the soil pH ranged from 4.8 to 5.4, and the Aluminium content was high, corroborating the results of the present study.

It is important to note that chemical characteristics such as high acidity and high levels of exchangeable aluminium are undesirable for plant species, causing harmful effects on plant growth and development (Sousa et al. 2007; Souza et al. 2008; Silva et al. 2018).

Considering that anthropogenic actions such as the removal of native vegetation and agricultural activities give rise to altered soils, and when this change in the soil causes a loss of functional and productive capacity, the areas are said to be degraded. The most significant losses are related to the evasion of organic matter, due to the amount of nutrients retained in them, and are largely caused by this inappropriate use of soils (Franco et al. 2003; Silva et al. 2011).

The results highlight the importance of Fabaceae species in these areas, especially *C. leiandra*, *C. laurifolia* and *B. Sericea*, which are cited in the literature for recovering degraded soils, as they are hardy and can be used as green manure in areas that are degraded or tend to be degraded by inappropriate land use, considering that they are undemanding plants in terms of soil fertility and cultivation, as well as being adapted to different soil and climate conditions (Tiecher 2016). They also have several ecological attributes, since they allow for an increase in biodiversity via natural regeneration; rapid soil cover; high production of plant biomass; possible incorporation as soil organic matter; as well as supplying other plants with N from a non-polluting source (Franco et al. 2003).

Another great benefit of using species from the Fabaceae family, such as those in this study, is the production of organic matter which, through its incorporation, is capable of stimulating various chemical and biological processes, improving soil fertility, as they are plants adapted to nutrient-poor soils, as well as having a deep and branched root system in the soil layers, encouraging the decompression of soils through their abundant roots (Embrapa 1986; Bertoni and Lombardi Neto 2008).

4 Final Remarks

This study recorded important phytosociological values, demonstrating the diversity of Fabaceae in the community of Nossa Senhora da Conceição, Oriximiná-PA. The survey found nine species of Fabaceae, but in the literature, only *C. leiandra*, *C. laurifolia* and *B. sericea* are cited for use in recovering degraded soils. However, the species were predominant *Z. latifolia* var. *communis* and *A. subdimidiata* in the area studied, emphasising the importance of future studies on the importance of these species in soil recovery.

In the particle size analysis, the only plots where the predominant sedimentary nature of floodplain soil was observed were plots 2 and 4, while plots 1 and 3 did not follow the sedimentary nature pattern, which can be explained by anthropogenic action. In the analysis of the pH and exchangeable aluminium values of the 4 study plots, it was possible to identify that in all of them, the pH result stood out as a soil that has high acidity and exchangeable aluminium is at a high level (acid), considering that these parameters are limiting to plant growth, it can be considered that this soil is undergoing alteration and possibly degradation, but to confirm this it is necessary to analyse the physical quality of the soil.

This study served to fill the gap in the scarcity of studies on floristic surveys of Fabaceae and soil analysis in the Oriximiná-PA region, and it can be emphasised that studies of this nature have the potential for recovering degraded areas because they are important for further studies aimed at surveying species to reforest degraded areas, as well as for drawing up a list of regional species in a given area that can contribute to restoration strategies.

Finally, we found limitations in carrying out the study, precisely because of the lack of studies in the Oriximiná-PA region, which prevented us from better discussing the data collected, as well as the lack of studies in the literature on the species that were found in the survey, but this study serves as an initial step towards carrying out more studies to reduce this degree of data scarcity.

References

Abreu E, Fernandes A, Ruivo M (2007) Variação temporal e vertical de atributos químicos de um gleissolo do Rio Guamá cultivado com canaranas. Rev Bras Ciênc Solo 31:277–285. https://doi.org/10.1590/S0100-06832007000200010

Albert J, Carnaval A, Flantua S, Lohmann L, Ribas C, Riff D, Carrillo J, Fan Y, Figueiredo J, Guayasamin J, Hoorn C, de Melo G, Nascimento N, Quesada C, Ulloa C, Val P, Arieira J, Encalada A, Nobre CA (2023) Human impacts outpace natural processes in the Amazon. Science 379(6630). https://doi.org/10.1126/science.abo5003

de Almeida LSD, Gama JVR, Oliveira FA, de Carvalho JOP, Gonçalves DCM, Araújo GC (2012) Fitossociologia e uso múltiplo de espécies arbóreas em floresta manejada, Comunidade Santo Antônio, município de Santarém, Estado Do Pará. Acta Amazon 42(2):185–194. https://doi.org/10.1590/S0044-59672012000200002

Angiosperm Phylogeny Group—APG (2016) An update of the Angiosperm Phylogeny Group classification for the orders and families of flowering plants: APG IV. Bot J Linn Soc 81(1):01–20

Alvarez VVH, Novais RF, Barros NF de, Cantarutti RB, Lopes AS. (1999) Interpretação dos resultados das análises de solos. Em: Ribeiro AC, Guimaraes PTG, Alvarez VVH (eds) Recomendações para o uso de corretivos e fertilizantes em Minas Gerais: 5a aproximação. Comissão de Fertilidade do Solo do Estado de Minas Gerais, Viçosa, pp 25–32

Amorim MA, Moreira-Turcq PF, Turcq BJ, Cordeiro RC (2009) Origem e dinâmica da deposição dos sedimentos superficiais na várzea do lago Grande de Curuai, Pará Brasil. Acta Amazônica 39(1):165–172. https://doi.org/10.1590/S0044-59672009000100016

Andrade DFC de, Gama JCV, Melo LO, Ruschel AR (2015) Inventário florestal de grandes áreas na floresta nacional do Tapajós, Pará, Amazônia, Brasil. Biota Amazônia 5(1):109–115. https://doi.org/10.18561/2179-5746/biotaamazonia.v5n1p109-115

Angiosperm Phylogeny Group—APG (2023) Fabaceae. http://www.mobot.org/MOBOT/research/APweb/welcome.html. Accessed 30 Jan 2023

Assis GB, Suganuma MS, Melo ACG, Durigan G (2013) Uso de espécies nativas e exóticas na restauração de matas ciliares no estado de São Paulo (1957–2008). Revista Árvore 37(4):599–609. https://doi.org/10.1590/S0100-67622013000400003

Bertoni J, Lombardi Neto F (2008) Conservação do Solo, Editora Ícone, São Paulo, p 355

Borges FS (2019) Lista de espécies com potencial para a restauração florestal na microrregião do Guamá, Pará, Amazônia, Brasil, 60 f. Trabalho de Conclusão de Curso (Graduação)—Curso de Engenharia Florestal, Campus Universitário de Belém, Universidade Federal Rural da Amazônia, Belém. http://bdta.ufra.edu.br/jspui/handle/123456789/1328. Accessed 19 Dec 2022

Brienza Jr S (2012) Enriquecimento de florestas secundárias como tecnologia de produção sustentável para a agricultura familiar. Bol Mus Para Emílio Goeldi Cienc Nat 7(3):331–337

Cardoso DBOS, Gregório BS, Carvalho CS, Lima HC (2022) Ormosia in Flora e Funga do Brasil. Jardim Botânico do Rio de Janeiro. Available at: https://floradobrasil.jbrj.gov.br/FB23103. Accessed 20 Dec 2022

Carim S, Schwartz G, Silva MFF (2007) Riqueza de espécies, estrutura e composição florística de uma floresta secundária de 40 anos no leste da Amazônia. Acta Bot Bras 21(2):293–308. https://doi.org/10.1590/S0102-33062007000200005

Chagas AP, Dutra VF (2022) Albizia in Flora e Funga do Brasil. Jardim Botânico do Rio de Janeiro. https://floradobrasil.jbrj.gov.br/FB82619. Accessed 20 Dec 2022

Chavan VS, Dattaji TR (2013) Floristic inventory: a tool for biodiversity conservation and management. Res J Bot 8(3):87–93

Chaves ADCG, Santos RM, Santos JO, Fernandes AA, Maracajá PB (2013) A importância dos levantamentos florístico e fitossociológico para a conservação e preservação das florestas. ACSA—Agropecuária Científica no Semiárido 9(2):43–48. https://doi.org/10.30969/acsa.v9i2.449

Costa JNMN, Durigan G (2010) Leucaena leucocephala (Lam.) de Wit (Fabaceae): invasora ou ruderal? Revista Árvore 34(5):825–833. https://doi.org/10.1590/S0100-67622010000500008

Cota MMT (2022) Campsiandra in Flora e Funga do Brasil. Jardim Botânico do Rio de Janeiro. https://floradobrasil.jbrj.gov.br/FB82740. Accessed 20 Dec 2022

Cravo MS, Xavier JJBN, Dias MC, Barreto JF (2002) Características, uso agrícola atual e potencial das várzeas no estado do Amazonas, Brasil. Acta Amazon 32(3):351–365. https://doi.org/10.1590/1809-43922002323365

CRIA—Centro de Referência em Informação Ambiental (2021) Specieslink—dados e ferramentas–busca centralizada. http://splink.cria.org.br/. Accessed 20 Aug 2022

Dufrêne M, Legendre P (1997) Species assemblages and indicator species: the need for a flexible asymmetrical approach. Ecol Monogr 67(3):345–366

EMBRAPA—Empresa Brasileira de Pesquisa Agropecuária (1986) Aspectos práticos ligados à formação de pastagens. Campo Grande, MS: Embrapa Gado de Corte. Circular Técnica nº 12, p 10

EMBRAPA—Empresa Brasileira de Pesquisa Agropecuária (2017) Manual de métodos de análise de solo. Embrapa, Brasília, DF, p 574

EPAMIG—Empresa de Pesquisa Agropecuária de Minas Gerais (2012) Análise do Solo: Determinações, cálculos e interpretação. Agricultura, Pecuária e Abastecimento. Minas Gerais, p 16

ESALQ/USP (2023) Coleta de amostras. Available at: https://www.esalqsolos.com.br/coleta-amostras. Accessed on 20 Jan 2023

Espirito-Santo FDB, Shimabukuro YE, de Aragão LEOC, Machado ELM (2005) Análise da composição florística e fitossociológica da floresta nacional do Tapajós com o apoio geográfico de imagens de satélites. Acta Amazon 35(2):155–173. https://doi.org/10.1590/S0044-59672005000200006

Fajardo JDV, Souza LAGS, Alfaia SS (2009) Características químicas de solos de várzeas sob diferentes sistemas de uso da terra, na calha dos rios baixo Solimões e médio Amazonas. Acta Amazônica 39(4):10. https://doi.org/10.1590/S0044-59672009000400001

FAO—Food and Agriculture Organization (1986) Food and fruit-bearing forest species 3: examples from Latin America. FAO, Roma, p 327. In: Silva R, Mary N da, Pastore JR, Floriano (orgs) Plantas da Amazônia: 450 espécies de uso geral. Universidade de Brasília, Biblioteca Central, Brasília, p 3140. http://leunb.bce.unb.br/. Accessed 29 Apr 2023

Flora do Brasil (2020) JBRJ, Rio de Janeiro. http://floradobrasil.jbrj.gov.br/. Accessed 22 Jan 2023
Franco AA, Resende AS de, Campello EFC (2003) Importância das leguminosas arbóreas na recuperação de áreas degradadas e na sustentabilidade de sistemas agroflorestais. Em: Laura VA, Alves FV, Almeida RG (eds) Sistemas Agroflorestais e Desenvolvimento Sustentável, Brasília-Embrapa, pp 1–208
Garcia FCP, Oliveira MCR, Bonadeu F (2022) Zygia in Flora e Funga do Brasil. Jardim Botânico do Rio de Janeiro. https://floradobrasil.jbrj.gov.br/FB23225. Accessed 21 Dec 2022
Kageyama P, Gandara FB (2000) Recuperação de áreas ciliares. In: Rodrigues FF, Leitão Filho HF (eds) Matas Ciliares: conservação e recuperação. Edusp/Fapesp, São Paulo, pp 249–269
Kuntz J, Vaz AMSF (2022) Schnella in Flora e Funga do Brasil. Jardim Botânico do Rio de Janeiro. https://floradobrasil.jbrj.gov.br/FB133434. Accessed 20 Dec 2022
Lima RR, Tourinho MM, Costa JPC (2001) Várzeas flúvio-marinhas da Amazônia Brasileira: características e possibilidades agropecuárias. FCAP, Belém, p 342
Junior JBT, Santos TMMS, Souza EGA, Meneses CHSGM, Soares C S (2015) Produção de fabaceas para adubação verde no agreste Paraibano. Biofarm 11(1) (ISSN 1983-4209)
Magalhães RC, Gomes RCM (2013) Mineralogia e química de solo de várzea e suas susceptibilidades no processo de terras caídas na comunidade do Divino Espírito Santo-AM. Soc Nat 25(3):609–621. https://doi.org/10.1590/S1982-45132013000300013
Melo NC, Souza LC, Costa CAS, Coelho R, Covre WP, Costa DLP (2013) Caracterização química de diferentes sistemas de uso do solo. Revista Agroecossistemas 5(2):8–13. https://core.ac.uk/download/pdf/267989232.pdf. Accessed 22 Mar 23
Moraes LLCM (2016) Fitossociologia de uma floresta de várzea e de igapó na planície de inundação do Rio Maracá, Amapá, Amazônia oriental, 46 f. Dissertação (mestrado) – Fundação Universidade Federal do Amapá, Programa de Pós-Graduação em Biodiversidade Tropical. Macapá. http://repositorio.unifap.br/jspui/bitstream/123456789/488/1/Dissertacao_FitossociologiaFlorestaVarzea.pdf. Accessed 20 Jan 2023
Mori SA (2011) From the field. In: Mori SA, Berkov A, Gracie CA, Hecklau EF (eds) Tropical plant collecting. TECC, Florianópolis, SC, pp 130–188
Niero LAC, Dechen SCF, Coelho RM, Maria IC (2010) Avaliações visuais como índice de qualidade do solo e sua validação por análises físicas e químicas em um Latossolo Vermelho distroférrico com usos e manejos distintos. Rev Bras Ciênc Solo 34(4):1271–1282. https://doi.org/10.1590/S0100-06832010000400025
Nobre CA, Borma LS, Castilla-Rubio JC, da Costa ACL, Malhi Y, Phillips OL, Ramos FM (2016) Land-use and climate change risks in the Amazon and the need of a novel sustainable development paradigm. Proc Natl Acad Sci 113(39):10759–10768
Oliveira OA (2011) Florística e Fitossociologia de fragmentos em área ecotonal Cerrado-Pantanal no município de Santo Antônio do Leverger-Mato Grosso. Dissertação. http://www.bibliotecaflorestal.ufv.br/handle/123456789/9661. Accessed 23 Mar 2023
Paisagem Nativa (2023) Viveiro de mudas. https://www.paisagemnativa.com.br/_files/ugd/2a47a1_b6f4f50467af4ab0b26686db3911bfb2.pdf?index=true. Accessed 29 Apr 2023
Pott A, Pott VJ (1994) Plantas do Pantanal. Empresa Brasileira de Pesquisa Agropecuária, Centro de Pesquisa Agropecuária do Pantanal-Corumbá, MS, EMBRAPA-SPI, p 320
Pinto ER (2014) Solos, hidrologia e estrutura populacional de pracuúbeiras em florestas de várzea do estuário amazônico. 69p. Dissertação. Programa de Pós-Graduação em Biodiversidade Tropical, Universidade Federal do Amapá. https://www2.unifap.br/ppgbio/files/2010/05/Disserta%C3%A7%C3%A3o_Emanuelle.pdf. Accessed 20 Feb 2023
Rioterra (2021) Centro de estudos Rioterra. Catálogo de espécies do Projeto Plantar–RO. https://rioterra.org.br/pt/wp-content/uploads/2021/12/Catalogo-de-Especies-2021_H.pdf. Accessed 29 Apr 2023
Rodrigues RR, Gandolfi S, Nave AG, Attanasio CM (2007) Atividades de adequação ambiental e restauração florestal do LERF/ESALQ/USP. Pesquisa Florestal Brasileira (55):7–21. https://pfb.cnpf.embrapa.br/pfb/index.php/pfb/article/view/113. Accessed 29 Apr 2023

Rodrigues RR, Martins FR, Gandolfi S (2009) Restoration of Brazilian Atlantic forest: restoring ecological corridors in Brazil. For Ecol Manage 258(10):1059–1068

dos Santos RO, Abreu JC, Lima RB, Aparício P, Sotta ED, Lima RC (2016) Distribuição diamétrica de uma comunidade arbórea na Floresta Estadual do Amapá, Brasil. Biota Amazônia 6(2):24–31

Schaefer CEGR, Lima HN de, Teixeira WG, Vale JF do, Souza KW de, Corrêia GR, Mendonça BA de, Amaral EF, Campos MCC, Ruivo MLP (2017) Solos da região amazônica. In: Curi N, Ker JC, Novais RF, Vidal-Torrado P, Schaefer CEG R (orgs) Pedologia—Solos dos Biomas Brasileiros. SBCS, pp 75–111

Scheidegger NMB, Rando JG Cassia in Flora e Funga do Brasil. Jardim Botânico do Rio de Janeiro. https://floradobrasil.jbrj.gov.br/FB22863. Accessed 20 Dec 2022

Silva RCS, Almeida JCR, Batista GT, Fortes Neto P (2011) Os indicadores físicos, químicos e biológicos da qualidade do solo e da sustentabilidade dos ambientes naturais. Repositório Eletrônico Ciências Agrárias, Coleção Ciências Ambientais, p 13

Silva SB (2018a) Análise de Solos Para Ciências Agrárias, 2nd edn. Universidade Federal Rural da Amazônia, Belém, p 167

Silva SB (2018b) Análise de Solos Para Ciências Agrárias. Universidade Federal Rural da Amazônia, p 167

Silva SB, Galvão JR, Pastana JC, Silva DR da, Almeida KC de, Souza FJL de, Nascimento ISB do (2018c) Influência das águas do estuário do Rio Pará na fertilidade do solo das ilhas de várzea em Belém, Pará. Biotemas 31(4):15–21. https://doi.org/10.5007/2175-7925.2018v31n4p15

Shephard GJ (2010) FITOPAC. Versão 2.1. Departamento de Botânica, Universidade Estadual de Campinas—UNICAMP, Campinas, SP

Sousa DMG, Miranda LN, Oliveira SA (2007) Acidez do solo e sua correção. In: Novais RF, Alvarez VVH, Barros NF, Fontes RLF, Cantarutti RB, Neves JCL (eds) Fertilidade do solo. Viçosa, MG, Sociedade Brasileira de Ciência do Solo, pp 205–274

Souza RF, Faquini V, Carvalho R, Torres PRF, Pozza AAA (2008) Atributos químicos de solos influenciados pela substituição do carbonato por silicato de cálcio. Rev Bras Ciênc Solo 32(4):1536–1572. https://doi.org/10.1590/S0100-06832008000400020

Sprent JI, Parsons R (2008) Nitrogen fixation in legume and non-legume trees. Field Crop Res 65(2–3):183–196

Suganuma MS, Assis GB, Melo ACG, Durigan G (2013) Ecossistemas de referência para restauração de matas ciliares: existem padrões de biodiversidade, estrutura florestal e atributos funcionais?. Revista Árvore 37(5):835–847. https://doi.org/10.1590/S0100-67622013000500006

Terra V, Morim MP (2022) Senegalia in Flora e Funga do Brasil. Jardim Botânico do Rio de Janeiro. https://floradobrasil.jbrj.gov.br/FB105182. Accessed 20 Dec 2022

Tiecher T (2016) Manejo e conservação do solo e da água em pequenas propriedades rurais no sul do Brasil: práticas alternativas de manejo visando a conservação do solo e da água. UFRGS—Universidade Federal do Rio Grande do Sul, 186p. ISBN 978-85-9489-010-8

Valente AA (2019) Composição florística e estrutura de 1,5 ha de floresta de terra firme na comunidade de Suruacá, Amazônia, Brasil, 30 f. Trabalho de Conclusão de Curso. Universidade Federal do Oeste do Pará (não publicado)

Vibrans AC, Sevegnani L, Lingner DV, Gasper AL, Sabbagh S (2010) Inventário florístico florestal de Santa Catarina (IFFSC): aspectos metodológicos e operacionais. Pesquisa Florestal Brasileira 30(60):291–302

Vieira DS, Oliveira MLR, Gama JRV, Machado ELM, Görgens EB, Lafetá BO, Garcia JS (2017) Phytosociology of a natural fragment of the floodplain forest in the lower Tapajós River, Brazil. Bosque 38(2):357–369

Assessment of Vegetation's Response Time to Water Availability in the Soil in the Jaru Biological Reserve, Rondônia

Gutieres Camatta Barbino, Nara Luisa Reis de Andrade, Hildevan Teixeira Oliveira Ferreira, and Alberto Dresch Webler

Abstract Vegetation plays a fundamental role in the biosphere–atmosphere system, while vegetation is extremely influenced by water availability in the atmosphere and soil. Therefore, this paper aims to evaluate the response time of vegetation to the availability of water in the soil in the Jaru Biological Reserve. The variables used to determine the response time were soil moisture and leaf area index. The response time was obtained for a time interval every four days and was investigated using the cross-correlation method to obtain the maximum correlation coefficient between the variables. From the tests carried out, it was observed that precipitation is an important environmental factor and is highly correlated with other hydrological and vegetational parameters. The response time between vegetation and soil moisture was 56 days. However, it showed a weak but significant correlation. Because of that, it was found that the availability of water in the soil is not the most relevant parameter in the development of vegetation, showing that this issue is complex and requires further studies to determine which variables influence the dynamics of vegetation in Rebio Jaru.

Keywords Leaf area index · Remote sensing · Water availability · Soil moisture

G. C. Barbino (✉)
UNIR, Campus Ji-Paraná, Ji-Paraná, Brazil
e-mail: gutieres.barbino@gmail.com

Large Scale Biosphere-Atmosphere Research Program in the Amazon, Ji-Paraná, Brazil

N. L. R. de Andrade · A. D. Webler
Department of Environmental Engineering, UNIR, Campus Ji-Paraná, Ji-Paraná, Brazil

H. T. O. Ferreira
Environmental and Sanitary Engineer, UNIR, Campus Ji-Paraná, Ji-Paraná, Brazil

W. Leal Filho et al. (eds.), *Amazon 2030—Sustainability Issues in the World's Largest Rainforest Region*, World Sustainability Series,
https://doi.org/10.1007/978-3-031-81465-5_32

1 Introduction

The Earth's surface plays fundamental roles in the water, carbon, and energy cycles in the Earth's climate system at a variety of time and spatial scales, receiving impacts from and regulating feedback to the atmosphere. Such interactions are especially strong in areas covered by dense vegetation, such as the Amazon rainforest (Li and Min 2013). This type of vegetation is among the most important habitats on Earth and plays an irreplaceable role in reducing carbon dioxide, supporting biodiversity, regulating hydrology, and consolidating soil (D'Amato et al. 2017; Oki et al. 2020).

The Amazon rainforest of South America represents 50% of the global rainforest area and is essential for absorbing carbon emissions and suppressing global warming (Amigo 2020; Fontes et al. 2018; Lyra et al. 2016). The Amazon rainforest, known as the "Lung of the Earth", is an important source of O_2 and an important sink of CH_4 and CO_2, storing approximately 120 billion tons of carbon (Atkinson et al. 2011; Phillips et al. 2009).

Vegetation indices (VI), such as leaf area index (LAI) and normalized difference vegetation index (NDVI) are important tools for studying plant dynamics, especially their photosynthetic capacity (Braz et al. 2015). Another microclimatic factor influenced by IV is evapotranspiration (Soares et al. 2017; Vourlitis et al. 2008). In accordance with Hao e Xiao-Dong (2015), VI variability can strongly impact the variability of soil evapotranspiration and evaporation. And so, resulting from the impacts on evapotranspiration, there are impacts on local and global precipitation, bringing consequences to forest ecosystems.

Therefore, it is noted that VI are an important physiological variable of the plant, which is capable of controlling climate variability, whether smaller or larger, depending on the LAI. In addition to playing a fundamental role in controlling the microclimate within forest ecosystems, (Hardwick et al. 2015). Thus, it is essential to know its dynamics in the face of limiting factors, such as water stress.

Water is a basic natural factor for the functioning of ecosystems, having importance in thermal regulation and being a limiting or boosting agent in the metabolic functions of plants (Linares and Camarero 2012). Water availability influences vegetation in a way that limits its spatial distribution and productivity (Chen et al. 2012), growth and mortality (Choat et al. 2018; Huang et al. 2016; Jolly et al. 2015; Reichstein et al. 2013), among others.

The main input of water into terrestrial systems is through precipitation. Studies have evaluated the relationship between precipitation and vegetation at different time and space scales (Camberlin et al. 2007; Hawinkel et al. 2016; Kreyling 2010). However, Kreyling (2010) in their study reported that low changes in precipitation did not cause a significant effect on vegetation growth, and this result is due to the depth and expansion of the root system of tree roots to increase the absorption of water available in the soil.

Therefore, it is essential to know the response of vegetation to water availability, since modeling studies point to a reduction in precipitation in the Amazon region (Li et al. 2008) thus intensified drought events (Lewis et al. 2011; Marengo and Espinoza

2016). Studies indicate that the climate in the Amazon will undergo a major warming (Cox et al. 2004; Li et al. 2008), it is necessary to understand how the forest will be able to respond to such changes (Souza et al. 2022). In addition, it is important to seek to understand which characteristics can influence such responses, whether they are vegetation resistance to lack of water, soil properties or adaptation to reduced rainfall.

Concerning the Amazon region, the influence of soil on vegetation may be more difficult to establish, especially in preserved regions, of which there is a gap in studies of soil water availability, mineral composition and soil texture (Rossi et al. 2004).

In this sense, understanding the interactions between vegetation and water availability in the soil is of great importance to evaluate the effects of climate on vegetation and provide a greater understanding of plant and water interaction. Therefore, this paper aims to evaluate the response time of vegetation to the availability of water in the soil in the Jaru Biological Reserve (Rebio Jaru).

2 Material and Methods

2.1 Study Area

The study area is located in the Jaru Biological Reserve—Rebio Jaru, an integral protection reserve area under the supervision of the Chico Mendes Institute for Biodiversity Conservation—ICMBio.

A micrometeorological tower managed by the Amazon Biosphere–Atmosphere Large-Scale Program—LBA Program is installed at Rebio Jaru, the location of which can be seen in Fig. 1.

According to the classification of the Brazilian Institute of Geography and Statistics – IBGE (2003), the area has vegetation classified as Submontane Open Ombrophylous Forest. Furthermore, studies such as the one carried out by Costa et al. (2010) also classify it as a semi-deciduous forest. It has an average canopy height of 30 m (Rummel et al. 2002) however, it has trees that exceed 50 m in height.

The region has a distinct climate characterization, presenting four periods, dry periods (Jul-Sep), dry-humid (Oct-Dec), wet (Jan-Mar) and wet-dry (Apr-Jun) (Franca 2015). The local microclimate can be characterized by an average annual temperature of 24.42 °C, with average monthly temperatures ranging between 23.52 °C (Feb) and 26.75 °C (Sep) (Barbino et al. 2023).

Other important variables for regulating the local microclimate, such as relative humidity and evapotranspiration, follow the same behavior as air temperature, in relation to seasonality, with average monthly values varying between 87.48% (Feb) and 64.54% (Aug), for relative air humidity, 4.47 mm.d^{-1} in (Jan) and 2.9 mm.d^{-1}in (Sep) for evapotranspiration and average annual values of 81.77% and 3.53 mm.d^{-1}, respectively (Barbino et al. 2023).

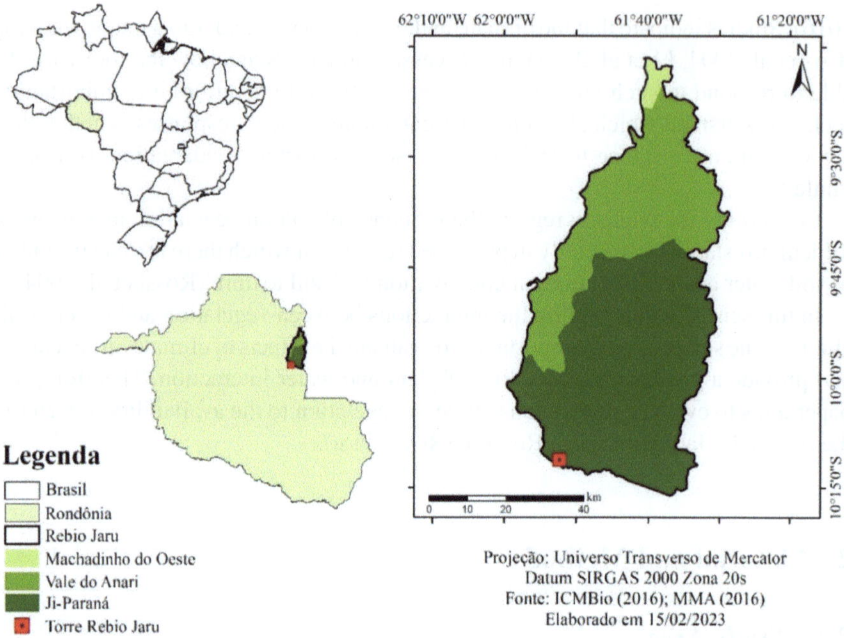

Fig. 1 Location of the Jaru biological reserve. *Source* Prepared by the authors

2.2 Data Acquisition

To carry out the research, it was necessary to acquire data on soil moisture (depth of 1 m) and precipitation, from the micrometeorological tower of the LBA Program, depth of the water table, from a piezometer well installed close to the tower and monitored monthly by the Program. LBA and IAF data, coming from the Moderate Resolution Imaging Spectroradiometer—MODIS sensor and made available, free of charge, by NASA. The data analyzed corresponds to the year 2017.

Soil moisture data was collected by the Hukseflux sensor (STP01) and the HyQuest Solutions precipitation sensor (TB4). Both variables are measured every 1 min and stored every 30 min. Storage was performed by Datalogger, model CR1000, Campbell Scientific Instrument, Utah, USA.

The IAF data comes from the MCD15A3H product and is obtained using the MODIS sensor, coupled to the Terra and Aqua satellites. The product data has a spatial resolution of 500 m for each pixel and a temporal resolution of 4 days for transmission. Along with the data, quality index values are available that attribute quality to the values, where excellent quality data infers little interference in obtaining the values. For the research, data of only excellent quality was used, corresponding to 1 pixel of the product, this pixel where the micrometeorological tower is located.

2.3 Statistical Analysis

The statistical treatment of the data consisted of descriptive statistics to characterize the studied data and inferential statistics to investigate the correlations between them.

First, the normality of the data was checked using the Shapiro–Wilk test at a significance of 5%, where it was found that the data did not adhere to this assumption. Therefore, the analyses were continued using Spearman's non-parametric correlation test.

A correlation test was carried out between the precipitation variables and the soil water level, soil moisture and LAI variables. To verify the response time of vegetation to water availability in the soil, the soil moisture and LAI variables were used. To this end, the response time was evaluated with a time interval of every 4 days. The response time for each pairing was analyzed to obtain the maximum and most significant correlation coefficient. The analysis was carried out using the cross-correlation method described by Phillips et al. (1999).

3 Results and Discussion

3.1 Characterizing Study Variables

To characterize the hydrological year and observe the dynamics of precipitation during the study period, precipitation data for the year 2017 was used, as can be seen in Fig. 2.

In the period analyzed, the highest rainfall occurred in December, with 406.1 mm of rainfall recorded. Studies carried out by Franca (2015); Gomes et al. (2015);

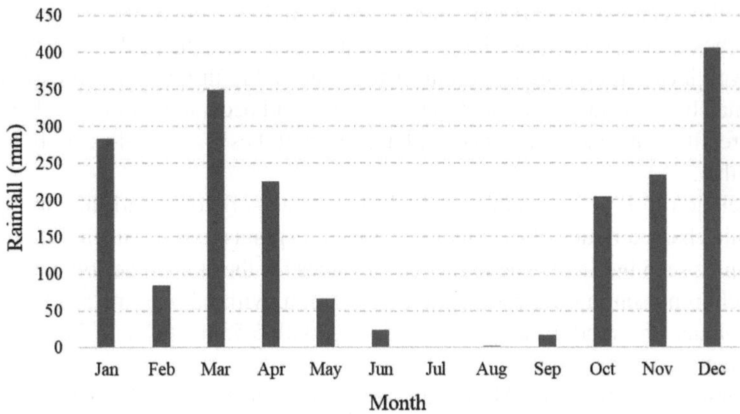

Fig. 2 Monthly precipitation in Rebio Jaru in 2017. *Source* Prepared by the authors based on the study's data

Oliveira (2014) report the highest rainfall observed in January, followed by December.

For the year in question, a high discrepancy is observed between January and December, resulting in an absolute difference of 123.2 mm less rain. This difference can be attributed to the period of the data series, since all the aforementioned studies use a data series of at least 10 years, resulting in a minimization of events that influence precipitation, such as *el niño* and *la niña.* Therefore, when analyzing a single year, we are subject to the effects of a single event, given that in 2017 there was a weak occurrence of a *la niña* event (Gweather 2021).

Another important point to be highlighted is the precipitation in February, with a rainfall of 84.83 mm being recorded. However, it is worth mentioning that this month there was a major data failure, representing 57% of data loss, making the use of these values in statistical analyzes unfeasible.

Regarding the level of the water table and soil humidity, the average monthly variation can be observed in Fig. 3.

It is observed that both variables present consistent behaviors, where it is observed that in months with higher soil humidity the water table level is closer to the surface, as observed, for example, in February and March. However, it is noted that April, the month with the groundwater level closest to the surface, does not coincide with the month with the highest soil humidity, while December, together with February and March, which present the higher soil moisture values, the observed groundwater level is one of the deepest recorded for the period studied.

This behavior is mainly related to the rainfall pattern of the study area. It is observed that in October, rains begin to occur more frequently, following the behavior in November and culminating in the highest rainfall observed in December. This results in a greater increase in soil moisture, since, through the infiltration process, this water being available in the most superficial layers of the soil causes moisture values to increase (Marinho 2022; Metcalf et al. 2018).

Nevertheless, this behavior is not repeated concerning the water table level, since the monthly recharge time occurs more slowly. Studies such as the one carried out by Sanches et al. (2009) show that there is a delay between the peak of precipitation in a given location and the elevation of the water table in that same location, this delay may be associated with the type of soil, use and occupation of the soil, canopy structure, litter and energy balance, and may each of these factors affect individually or together.

In relation to LAI, a variable studied to evaluate soil cover and the phytophysiognomic characteristics of the plant and how it responds to water availability in the soil, can be seen in Fig. 4. The highest LAI values for the studied period correspond to March, representing the end of the rainy season, with 6.35 m^2 m^{-2}. In turn, the lowest values observed occur in May and June, with 5.55 m^2 $m^{-2}$and 5.54 m^2 m^{-2}, respectively.

It is noteworthy that after March, the period with the highest LAI, there is a sharp drop in it, and it is observed that this reduction occurs for two consecutive months, where subsequently the LAI gradually increases as the months go by. However, it is

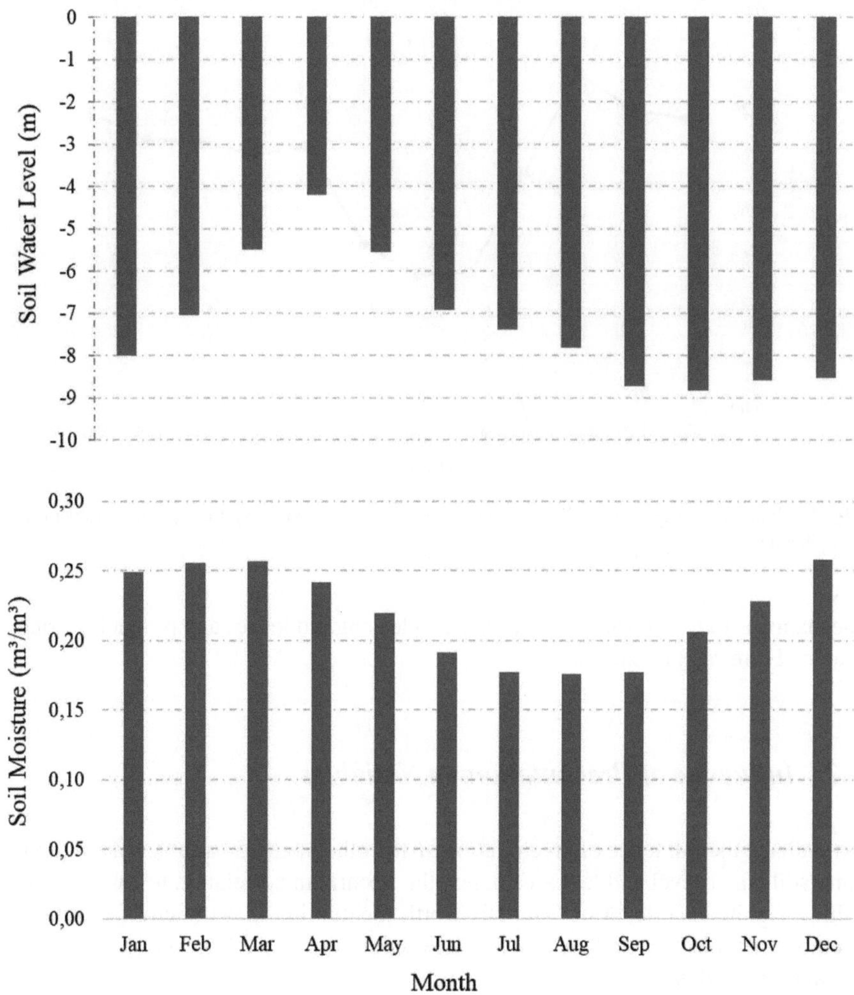

Fig. 3 Average monthly variation in soil water level and soil moisture in 2017. *Source* Prepared by the authors based on the study's data

worth highlighting that the variation that occurred for the LAI in 2017 did not differ significantly between the months (p-value > 0.05).

Similar studies were carried out by Barbino et al. (2021); Barbino et al. (2023); Danelichen et al. (2016). In these studies, the seasonal pattern of the LAI is consistent with that observed in the present study, reporting higher values in the wet period and lower values in the dry period. Also according to Barbino et al. (2023), this variation during the year is due to the great availability of water in the months with the highest rainfall, creating a favorable environment for plant development. In turn, in subsequent months, with the abrupt reduction of rainfall, plants look for

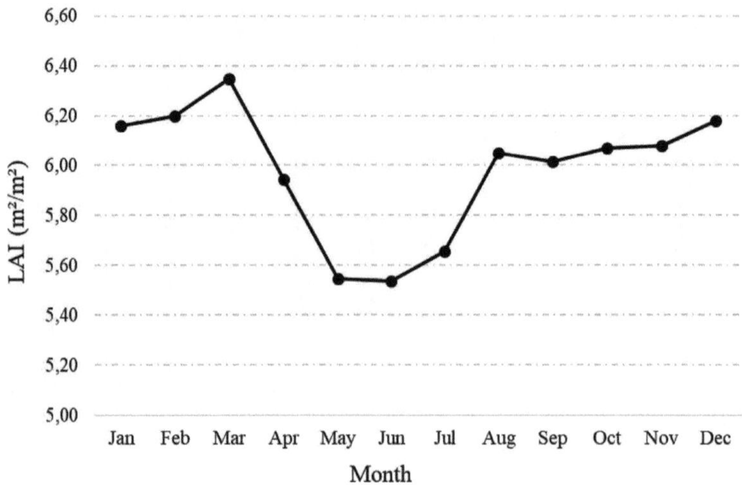

Fig. 4 Monthly average leaf area index in 2017. *Source* Prepared by the authors based on the study's data

mechanisms to avoid water stress, such as releasing old leaves and producing young leaves (Ferreira et al. 2015).

3.2 *Influence of Precipitation on Variables*

To evaluate the influence of precipitation on the other studied variables of soil moisture, soil water level and leaf area index, the Spearman correlation test was applied with a significance level of 5%. It is worth mentioning that the cross-correlation method was also carried out to check whether the variable responses occurred instantly or with some delay.

The value of the correlation coefficients, as well as their respective p-values, are shown in Table 1. Based on the literature, it is highlighted that the relationship between precipitation and variables is linear, that is, they are directly proportional (Becerra et al. 2009; Galvincío et al. 2020; Gomes et al. 2019). With the exception of the water level in the soil, which presents an inversely proportional behavior, the greater the precipitation, the lower the depth level of the water table (Sanches et al. 2009). Therefore, only the coefficients highlighted in the body of the table that represent these relationships between the variables will be considered for discussion.

It can be observed that all variables presented a strong and significant correlation, highlighting the influence that precipitation exerts on hydrological and vegetational parameters.

The variables of soil water levels and leaf area index showed a delay in response to precipitation, lasting four and two months, respectively. This behavior may be

Table 1 Cross-correlation coefficient to determine the response time of soil water level, soil moisture and leaf area index variables to precipitation in 2017

Month	Rainfall (mm)	Response time (month)	Soil water level (m)		Soil moisture (m^3 m^{-3})		LAI (m^2 m^{-2})	
			ro	p-value	ro	p-value	ro	p-value
Jan	283.0	0	0.18	> 0.05	0.97	<0.05*	0.75	<0.05*
Feb	–	-1	–	–	–	–	–	–
Mar	349.2	-2	-0.5	>0.05	0.63	<0.05*	0.8	<0.05*
Apr	225.3	-3	-0.8	<0.05*	0.21	>0.05	0.7	<0.05*
May	66.60	-4	-0.9	<0.05*	-0.35	>0.05	0.34	>0.05
Jun	24.10	-5	-0.7	<0.05*	-0.77	<0.05	– 0.2	>0.05
Jul	0.00	-6	-0.3	>0.05*	-0.93	<0.05	– 0.6	<0.05
Aug	1.30	-7	0.20	>0.05	-0.83	<0.05	– 0.9	<0.05
Sep	16.3	-8	0.70	<0.05	-0.44	>0.05	– 0.8	<0.05
Oct	204.2	-9	0.90	<0.05	0.06	>0.05	– 0.5	>0.05
Nov	234.2	-10	0.90	<0.05	0.6	<0.05*	0.02	>0.05
Dec	406.1	-11	0.50	>0.05	0.86	<0.05*	0.33	>0.05

Note p–values indicated with (*) represent significant correlations between variables
Source Prepared by the authors based on the study's data

associated with different factors, such as soil type, soil porosity, and soil composition (Sanches et al. 2009).

Furthermore, with regard to vegetation, as already reported, it is related to precipitation, but the degree of relationship, as well as the response time between the variables, is also dependent on the structure of the vegetation (Becerra et al. 2009). The authors also report that due to differences in soil cover, more established vegetation areas, such as primary forests, suffer smaller impacts due to water deficit, with no differences in photosynthetic activity and biomass production between the wet and dry periods.

Finally, soil humidity was the only variable where the response to precipitation occurred immediately, this being linked to the fact that rain increases the water content in the soil first in the most superficial layers, and only later does this water supply the water table and in fact be better used by large trees.

3.3 *Vegetation Response Time to Water in the Soil*

To respond to the main objective of the study, which is to determine the response time of vegetation to soil availability, a correlation test was carried out between leaf area index and soil moisture at a depth of 1 m. To obtain a more precise response

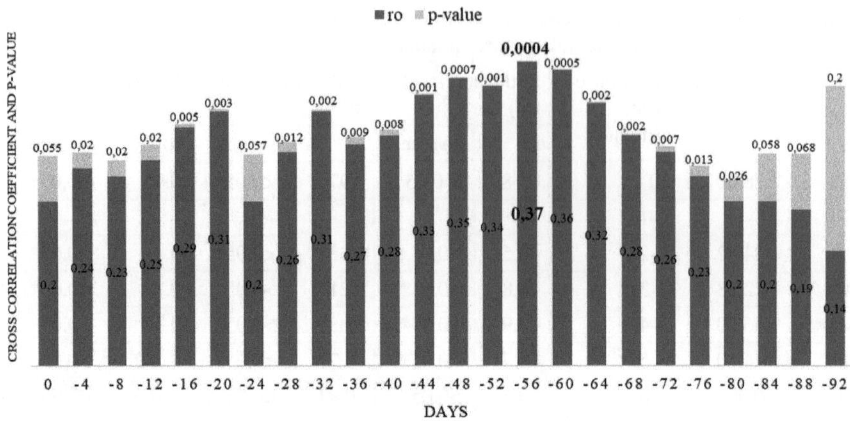

Fig. 5 Cross-correlation coefficient to determine vegetation response time to soil water availability. *Source* Prepared by the authors based on the study's data

and with a more refined response time, the test was carried out with a time interval corresponding to 04 days, referring to the temporal resolution of the MODIS product.

As observed in the previous topic, the vegetation response to precipitation occurred with a delay of 60 days. Based on this principle, it is inferred that the response time of vegetation to soil moisture occurs over a similar period, therefore, the cross-correlation coefficients, as well as their respective p-values, can be observed in Fig. 5.

As can be seen in the figure, the result was shown for an analysis carried out for a period of 92 days, and as an assumption, the values of the coefficients gradually increased until reaching the maximum coefficient 56 days after the start of the rains and increase in humidity and then decreasing again. Coincidentally, the lowest p-value observed refers to the highest coefficient obtained, corroborating the significance of the test applied.

Based on the correlation obtained, it is observed that in Rebio Jaru, the response of vegetation to the availability of water in the soil presented a weak relationship, that is, the development of vegetation may be related to other environmental factors that contribute to the increase in vegetative indices and photosynthetic production.

4 Final Remarks

From the analyses carried out, it was possible to determine the response time of the variables of soil water level, soil humidity and LAI in relation to precipitation. It was found that all variables have a strong relationship with the seasonality of rainfall in Rebio Jaru, indicating that the studied variables are strongly impacted by the occurrence of rain in the region.

Furthermore, the response time of vegetation to water availability in the soil was 56 days, showing a weak but significant correlation, showing that water availability is an important factor in vegetation development. However, this is not a parameter with a high degree of influence, and the development of vegetation may also be related to other environmental factors.

It is worth mentioning that studies on this subject are extremely necessary, since it is essential to understand the functioning of an ecosystem as important as the Amazon rainforest in the face of critical scenarios of rainfall reduction and mainly due to the scarcity of studies that investigate such relationships, given the high complexity of quality data acquisition and great complexity of soil–plant interactions. In addition, this opens up the possibility of new studies to determine which other factors contribute to the growth of vegetation, such as whether tree species are resistant to water stress or whether they are adapting to reduced rainfall.

Acknowledgements This work was carried out with the support of the Coordination for the Improvement of Higher Education Personnel—Brazil (CAPES)—Financing Code 001. We also thank the Professional Master's Program in the National Network in Management and Regulation of Water Resources—*ProfÁgua*, CAPES/ANA Project AUXPE Nº. 2717/2015, for the scientific technical support provided to date. Finally, we thank the Large Scale Biosphere-Atmosphere Research Program in the Amazon for the research grant given (Grant Term no. 0027/2023) and for all the support during the research.

References

Amigo I (2020) The amazon's fragile future. Nature 578:505–507. https://doi.org/10.1038/d41586-020-00508-4

Atkinson PM, Dash J, Jeganathan C (2011) Amazon vegetation greenness as measured by satellite sensors over the last decade. Geophys Res Lett, 38. https://doi.org/10.1029/2011GL049118

Barbino GC, Antonucci B, Ventura DJ, Andrade NLR (2021) Sensoriamento Remoto aplicado ao Índice de Área Foliar e fração da radiação fotossinteticamente ativa em áreas de floresta e pastagem na Amazônia Ocidental. Revista Brasileira De Meio Ambiente 9(2):076–090

Barbino GC, Andrade NLR, Webler AD, Sanches L, Aguiar RG, Antonucci B (2023) Índice de área foliar e sua relação com o microclima em floresta e pastagem na Amazônia Ocidental. Rev Bras Climatol, 32:311–335. https://doi.org/10.55761/abclima.v32i19.16296

Becerra JAB, Shimabukuro YE, Alvalá RCS (2009) Relação do padrão sazonal da vegetação com a precipitação na região de Cerrado da Amazônia Legal, usando índices espectrais de vegetação. Rev Bras Meteorol, 24(02):125–134. https://doi.org/10.1590/S0102-77862009000200002

Braz AM, Águas TA, Garcia PHM (2015) Análise de índices de vegetação NDVI e SAVI e Índice de Área Foliar (IAF) para a comparação da cobertura vegetal na bacia hidrográfica do córrego Ribeirãozinho, município de Selvíria—MS. Rev Percurso-NEMO, 7(2):05–22. https://doi.org/10.4025/revpercurso.v7i2.28758

Camberlin P, Martiny N, Philippon N, Richard Y (2007) Determinants of the interannual relationships between remote sensed photosynthetic activity and rainfall in tropical Africa. Remote Sens Environ, 106(02):199–216. https://doi.org/10.1016/j.rse.2006.08.009

Chen, T, Dolman HAJ, Van Der Werf GR, Jeu R (2012) Annual coupled drought and net primary productivity over global. Egu General Assembly Conf

Choat B, Brodribb TJ, Brodersen CR, Duursma RA, López R, Medlyn BE (2018)Triggers of tree mortality under drought. Nature 558(7711):531–539. https://doi.org/10.1038/s41586-018-0240-x

Costa MH, Biajoli MC, Sanches L, Malhado ACM, Hutyra LR, Rocha HR, Aguiar RG, Araújo AC (2010) Atmospheric versus vegetation controls of Amazonian tropical rain forest evapotranspiration: Are the wet and seasonally dry rain forests any different? J Geophys Res 115:9p. https://doi.org/10.1029/2009JG001179

Cox PM, Betts RA, Collins M, Harris PP, Huntingford C, Jones CD (2004) Amazonian forest dieback under climate-carbon cycle projections for the 21st century. Theor Appl Climatol 78:137–156. https://doi.org/10.1007/s00704-004-0049-4

D'Amato G, Vitale C, Rosario N, Chong Neto HJ, Chong-Silva DC, Mendonça F, Perini J, Landgraf L, Solé D, Sanchez-Borges M, Ansotegui I D'Amato M (2017) Climate change, allergy and asthma, and the role of tropical forests. World Allergy Organ J, 10(01). https://doi.org/10.1186/s40413-017-0142-7

Danelichen VHM, Biudes MS, Machado NG, Velasque MCS, Silva BB, Nogueira JS (2016) Uso do sensoriamento remoto na estimativa do índice de área foliar em floresta tropical. Ciência e Natura 38(3):1352–1360. https://doi.org/10.5902/2179460X19917

Ferreira LS, Cattânio JH, Jardim MAG (2015) Efeito da topografia e da precipitação na florística e na produção de liteira em Caxiuanã, Pará. Revista Árvore 39(6):995–1005. https://doi.org/10.1590/0100-67622015000600002

Fontes CG, Dawnson TE, Jardine K, McDowell N, Gimenez BO, Anderegg L, Negron-Juarez R, Higuchi N, Fine PVA, Araújo AC, Chambers JQ (2018) Dry and hot: the hydraulic consequences of a climate change–type drought for Amazonian trees. Philos Trans R Soc, 373.https://doi.org/10.1098/rstb.2018.0209

Franca RR (2015) Climatologia das chuvas em Rondônia—período 1981–2011. Geogr: Artig Científicos, 11(01):44–58. https://doi.org/10.35699/2237-549X..13392

Galvincío JD, Mendes SM, Souza WM, Morais YCB, Miranda RQ, Moura MSB, Santos W (2020) Correlação linear entre a precipitação e o Índice de Área Foliar do bioma Caatinga. Revista Brasileira De Geografia Física 13(07):3304–3313

Ggweather (2021) El Niño and La Niña years and intensities. Disponível em: https://ggweather.com/enso/oni.htm. Acesso em: 29 de março de 2023

Gomes ARS (2019) Estudo da Relação entre a Variabilidade dos Índices de Vegetação e Temperatura da Região Nordeste do Brasil. Revista Brasileira De Meteorologia 34(03):359–368. https://doi.org/10.1590/0102-778634305

Gomes JB, Webler AD, Aguiar RG, Aguiar LJG, Nuñes MLA (2015) Conversão de florestas tropicais em sistemas pecuários na Amazônia: quais as implicações no microclima da região? Revista Brasileira De Climatologia 17:67–81

Hao C, Xiao-Dong Z (2015) The impacts of the interannual variability of vegetation on the interannual variability of global evapotranspiration: A modeling study. Atmos Ocean Sci Lett, 05(03):225–230. https://doi.org/10.1080/16742834.2012.11446993

Hardwick SR, Toumi R, Pfeifer M, Turner EC, Nilus R, Ewers RM (2015) The relationship between leaf area index and microclimate in tropical forest and oil palm plantation: Forest disturbance drives changes in microclimate. Agric for Meteorol 201:187–195. https://doi.org/10.1016/j.agrformet.2014.11.010

Hawinkel P, Thiery W, Lhermitte A, Swinnen E, Verbist B, Van Orshoven J, Muys B (2016) Vegetation response to precipitation variability in East Africa controlled by biogeographical factors: East African Climate-vegetation response. J Geophysical Res: Biogeosci, 121(09):2422–2444. https://doi.org/10.1002/2016JG003436

Huang L, He B, Chen A, Wang H, Liu J, Lu A, Chen Z (2016) Drought dominates the interannual variability in global terrestrial net primary production by controlling semi-arid ecosystems. Scientific Reports, 06(01). https://doi.org/10.1038/srep24639

Instituto Brasileiro de Geografia e Estatística—IBGE. Mapa de Vegetação (2003) Disponível em: https://ww2.ibge.gov.br/home/presidencia/noticias/21052004biomashtml.shtm. Acesso em 15 de abril de 2019

Instituto Chico Mendes de Conservação da Biodiversidade—ICMBio (2016) Unidades de Conservação—Amazônia. Disponível em: http://www.icmbio.gov.br/portal/unidadesdeconservacao/biomas-brasileiros/amazonia/unidades-de-conservacao-amazonia/1999-rebio-do-jaru. Acesso em: 29 de setembro de 2017

Jolly WM, Cochrane MA, Freeborn PH, Holden ZA, Brown TJ, Williamson GJ, Bowman DMJS (2015) Climate-induced variations in global wildfire danger from 1979 to 2013. Nature Commun 06(01).https://doi.org/10.1038/ncomms8537

Kreyling J (2010) Winter climate change: a critical factor for temperate vegetation performance. Ecology, 91(07):1939–1948. https://doi.org/10.1890/09-1160.1

Lewis SL, Brando PM, Phillips OL, Van Der Heijden GMF, Nepstad D (2011) The 2010 Amazon drought. Science 331:554. https://doi.org/10.1126/science.1200807

Li R, Min Q (2013) Dynamic response of microwave land surface properties to precipitation in Amazon Rainforest. Remote Sens Environ 133:183–192. https://doi.org/10.1016/j.rse.2013.02.001

Li W, Fu R, Négron-Juárez RI, Fernandes K (2008) Observed change of the standardized precipitation index, its potential cause and implications to future climate change in the Amazon region. Biol Sci 363:1767–1772. https://doi.org/10.1098/rstb.2007.0022

Linares JC, Camarero JJ (2012) From pattern to process: linking intrinsic water-use efficiency to drought-induced forest decline. Glob Chang Biol 18(03):1000–1015. https://doi.org/10.1111/j.1365-2486.2011.02566.x

Lyra AA, Chou SC, Sampaio GO (2016) Sensibilidade do bioma Amazonia a projeção de mudanças climáticas em alta resolução. Acta Amazonia 46:175–188. https://doi.org/10.1590/1809-4392201502225

Marengo JA, Espinoza JC (2016) Extreme seasonal droughts and floods in Amazonia: Causes, trends and impacts. Int J Climatol 36:1033–1050. https://doi.org/10.1002/joc.4420

Marinho GO (2022) Variabilidade sazonal do efluxo de CO_2 do solo em floresta primária de terra firme na Amazônia Ocidental. Ji-Paraná: UNIR. Monografia (Bacharel em Engenharia Ambiental e Sanitária), Departamento de Engenharia Ambiental e Sanitária, Universidade Federal de Rondônia

Metcalf DB, Rocha W, Balch JK, Brando PM, Doughty CE, Malhi Y (2018) Impacts of fire on sources of soil CO_2 efflux in a dry Amazon rain forest. Glob Change Biol 24:3629–3641. https://doi.org/10.1111/gcb.14305

Ministério do Meio Ambiente—MMA. Amazônia. Disponível em:<http://www.mma.gov.br/biomas/amaz%C3%B4nia>. Acesso em: 07 out. 2016

Oki Y, Arantes-Garcia L, Costa MBM, Nunes BC, Silveira BR, Gélvez-Zúñiga I, Franco A, Fernandes GW (2020) CO_2 Fertilizer Effect on Growth, Polyphenols, and Endophytes in Two Baccharis Species. Braz Arch Biol Technol, 63. https://doi.org/10.1590/1678-4324-2020190302

Oliveira MA (2014) Caracterização da Precipitação em Área de Floresta e Pastagem no Sudoeste da Amazônia. Ji-Paraná: UNIR. Monografia (Bacharel em Estatística), Departamento de Matemática e Estatística, Universidade Federal de Rondônia

Phillips N, Oren R, Zimmermann R, Wright SJ (1999) Temporal patterns of water flux in trees and lianas in a Panamanian moist forest. Trees 14:116–123. https://doi.org/10.1007/s004680050216

Phillips O et al (2009) Drought sensitivity of the amazon rainforest. Science 323:1344–1347. https://doi.org/10.1126/science.1164033

Reichstein M, Bahn M, Ciais P, Frank D, Mahecha MD, Seneviratne SI, Zscheischler J, Beer C, Buchmann N, Frank DC, Papale D, Rammiig, Smith P, Thonicke K, van der Velde M, Vicca S, Walz A, Wattenbach M (2013) Climate extremes and the carbon cycle. Nature 500(7462):287–295. https://doi.org/10.1038/nature12350

Rossi M, Mattos IFA, Coelho RM, Menk JRF, Rocha FT, Pfeifer RM, DeMaria IC (2004) Relação solos/vegetação em área natural no Parque Estadual de Porto Ferreira, São Paulo. Rev Inst Flor 17(01):45–61

Rummel U, Ammann C, Gut A, Meixner FX, Andreae MO (2002) Eddy covariance measurements of nitric oxide flux within an Amazonian rain forest. J Geophys Res 107:17–1–17–9. https://doi.org/10.1029/2001JD000520

Sanches L, Andrade NLR, Pinto-Júnior OB, Alves MC, Nogueira JS (2009) Time lags between rainfall and groundwater levels in a forest and cattle pasture of an amazon watershed. J Environ Hydrol, 17(34)

Soares WA, Antonino ACD, Lima JRS, Hammecker C (2017) Modeling of evapotranspiration and water flow in a cultivated soil with cowpea bean. J Environ Anal Prog 02(02):115–124

Souza RA, Moura V, Paloschi RA, Aguiar RG, Webler AD, Borma LS (2022) Assessing drought response in the southwestern amazon forest by remote sensing and in situ measurements. Remote Sensing 14:1733. https://doi.org/10.3390/rs14071733

Vourlitis GL, Nogueira JS, Lobo FA, Sendall KM, Paulo SR, Dias CAA, Pinto Júnior OB, Andrade NLR (2008) Energy balance and canopy conductance of a tropical semi-deciduous forest of the Southern Amazon Basin. Water Resour Res 44:14p. https://doi.org/10.1029/2006WR005526

CO_2 Flow Dynamics in a Primary Forest Area in Western Amazon in the Dry Period of 2022

Ediane Carvalho Ferreira, Alberto Dresch Webler, Gutieres Camatta Barbino, Nara Luísa Reis de Andrade, and Barbara Antonucci

Abstract The Amazon region has been undergoing intense changes in land use, impacting the dynamics of forests. Therefore, the study aimed to analyze the hourly behavior of the microclimate and CO_2 flow in the dry period of 2022 in a forest area in the Jaru Biological Reserve. The variables of relative and specific air humidity, air temperature, photosynthetically active radiation (PAR), shortwave radiation, wind speed and CO_2 flux were used. The CO_2 flow was measured by the Eddy Covariance system. A peak in CO_2 emission was observed at dawn due to low nighttime turbulence. Relative air humidity was inversely proportional to air temperature. In turn, incident shortwave radiation and PAR reached their highest at 12:30 pm. When analyzing the days with the highest CO_2 absorption and emission, it was found that wind speed, temperature and specific air humidity presented higher values corresponding to the days with the highest CO_2 emission. Based on the above, it is noted that CO_2 has a complex relationship with the microclimate.

Keywords Eddy covariance · Microclimate · Changes in land use · Carbon

---

E. C. Ferreira (✉) · G. C. Barbino
Water Resources Management and Regulation (ProfÁgua in the Portuguese Acronym), Research Fellow in the Large-Scale Biosphere-Atmosphere Program in the Amazon, Federal University of Rondônia Foundation (UNIR in the Portuguese Acronym), Ji-Paraná Campus, Ji-Paraná, Brazil
e-mail: edianecarvalhoferreira@gmail.com

A. D. Webler · N. L. R. de Andrade
Department of Environmental Engineering, Federal University of Rondônia Foundation (UNIR in the Portuguese Acronym), Ji-Paraná Campus, Ji-Paraná, Brazil
e-mail: alberto.webler@unir.br

B. Antonucci
National Amazon Research Institute (INPA in the Portuguese Acronym), Manaus, Brazil

W. Leal Filho et al. (eds.), *Amazon 2030—Sustainability Issues in the World's Largest Rainforest Region*, World Sustainability Series,
https://doi.org/10.1007/978-3-031-81465-5_33

1 Introduction

The forests act as a net carbon dioxide sink that helps minimize impacts from human activities that have occurred since the pre-industrial period, such as the burning of fossil fuels, deforestation and other land use change activities (Friedlingstein et al. 2022; Hubau et al. 2020; Pugh et al. 2019; Quéré et al. 2018). Studies show that in recent decades the Amazon Forest has acted as a carbon sink as well as providing an ecosystem service for climate regulation (Barbino et al. 2023; Ellwanger et al. 2020; Fearnside, 2021; Marengo et al. 2018). However, with climate change, the sink function of the Amazon Rainforest could be altered, intensifying global warming (Denning 2021; Gatti et al. 2021; Hubau et al. 2020; Mitchard 2018; Rammig and Lapola 2021).

According to Gatti et al. (2021), the southeastern Amazon region acts as a carbon source and the highest emissions are found in the Eastern Amazon due to increased deforestation and intensification of the dry season. Given this scenario of imbalance in the balance of carbon absorption and emissions, which has a direct impact on global warming, a study carried out by Kim et al. (2022) referring to reversibility (capacity to restore the climate system to its initial state) concluded that in most regions affected by human activities, climate change may be irreversible, even after mitigation measures, especially in developing countries.

To perform carbon dioxide (CO_2) measurements in the atmosphere there are several methodologies. Currently, one of the most used methods is the measurement by Turbulent Vortex Covariance (Eddy Covariance), which consists of measuring the gas exchange of CO_2, water vapor and heat in the ecosystem (Vitale 2021). This technique allows long-term measurements as it does not generate changes or disturbances in the study area. However, the measurements carried out by Eddy Covariance indicate uncertainties in CO_2 measurements during the night due to the stability of the atmosphere that promotes the accumulation of air below the canopy, which affects the quantification of the energy balance in some experimental sites (Aguiar et al. 2019; Saleska et al. 2009). During the night, the speed of turbulence decreases, causing CO_2 to flow to lower places, making accurate measurements difficult (Araújo et al. 2008; Hayek et al. 2018).

Understanding the behavior of climate variables in response to the flow of CO_2 is extremely important to understand how the forest responds to changes in the ecosystem, especially during dry periods. This season results in a decrease in water availability in the soil, which has a direct impact on trees. This leads to the restriction of CO_2 assimilation due to reduced leaf transpiration, decreased stomatal conductance and closure of stomata, resulting in reduced photosynthesis processes (Mendes et al. 2020). Especially when studies indicate that the resilience of the Amazon Forest is in decline due to impacts related to climate change and land use (Boulton et al. 2022).

Despite several studies in the Amazon that attempt to understand the role of forests in the carbon cycle, their ecosystem service as a source and sink is unclear, even more so in the current scenarios of land use change (Bullock and Woodcock 2021; Marani et al. 2020). This uncertainty becomes more evident when considering the complexity of the interaction of the microclimate with the CO_2 flux and the heterogeneity of the Amazon rainforest (Green et al. 2020; Ometto et al. 2005; Heinrich et al. 2021).

In view of the above-mentioned uncertainties, this work investigates the behavior of CO_2 flux and microclimate in an area of primary forest in the dry season to better understand the dynamics of the Amazon Forest, since this biome has diverse vegetation, making it complex and constraining models that seek to infer patterns for the entire extension of the Amazon Forest. Therefore, the objective of this study was to characterize and quantify the CO_2 flow and microclimate in the Jaru Biological Reserve, located in the southwest of the Amazon in August 2022.

2 Methodology

2.1 Description of the Study Area

The area of this study is located in the Southwest of the Amazon in the Jaru Biological Reserve (Rebio Jaru) which is part of the municipality of Ji-Paraná, Machadinho d'oeste and Vale do Anari in the state of Rondônia (Fig. 1) with geographic coordinates 10°11′11.4″ south; 61°52′29.9″ west, 120 m above sea level.

The REBIO Jaru micrometeorological tower is one of the experimental humid tropical forest sites belonging to the Large-Scale Biosphere–Atmosphere research program in the Amazon—LBA, which aims to understand the behavior of the Amazon through its current change and how the flora, fauna, atmosphere, hydrosphere and humans are affected. The experimental site has a micrometeorological tower approximately 62 m high and was initially installed in 1999.

The area consists of an Open Ombrophilous Forest with approximately 346,864.20 ha (ICMBio 2023). The climate is classified according to Köppen as Aw (rainy tropical) with the average annual temperature being 25 °C, and the average annual precipitation being approximately 2,000 mm (Araújo 2022; Valadão et al. 2019). In terms of precipitation, the area has a wet period from January to March, a wet-dry transition period from April to June, a dry period from July to September and a dry-humid transition period from October to December (Gomes et al. 2015).

2.2 Analysis of Data and Micrometeorological Variables

The data in this study correspond to the dry period of 2022 (August), as they are the most recent measurements available in the database. Furthermore, the sensor that quantifies CO_2 was deactivated from 2020 until July 2022. Measurements of CO_2 flow, specific air humidity, relative air humidity, incident shortwave radiation (SWR_

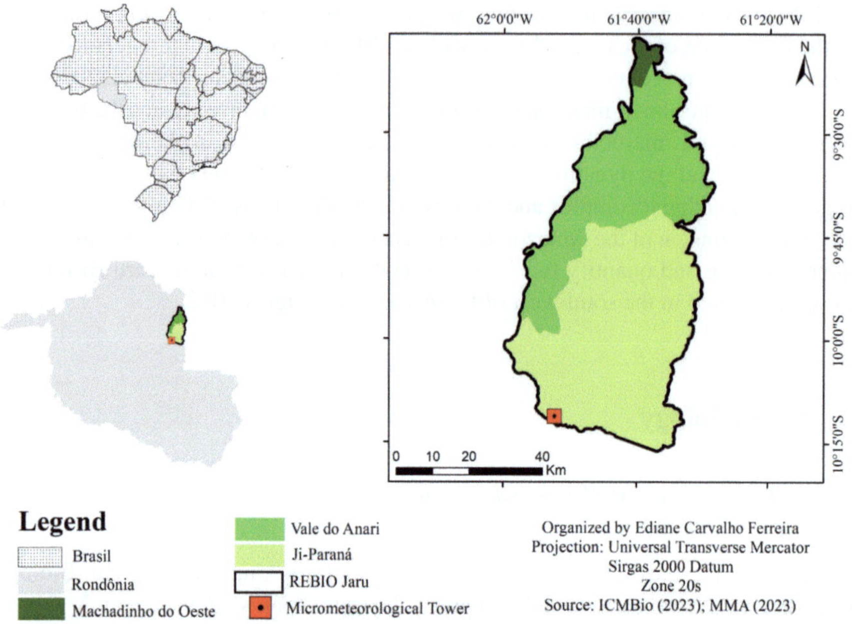

Fig. 1 Location of the Jaru biological reserve and the micrometeorological tower of the Large-Scale Biosphere–Atmosphere program in the amazon—LBA. *Source* Prepared by the authors

in) and reflected short wave (SWR_out), incident photosynthetically active radiation (PAR_in) and reflected photosynthetically active radiation (PAR_out) and wind speed. The micrometeorological variables analyzed together with their respective sensors and height are described in Table 1.

The specific air humidity was obtained through the calculation shown in Eq. 1 and the specific saturation humidity was calculated using Eq. 2.

$$q = \frac{0{,}622e}{P - 0{,}378e} \tag{1}$$

$$q = \frac{0{,}622e_s}{P - 0{,}378e_s} x1000 \tag{2}$$

where q is the specific humidity (g kg^{-1}), e is the current water vapor pressure (hPa), P is the atmospheric pressure (hPa), q_s is the saturation specific humidity (g kg^{-1}) and e_s is the saturation of pressure water vapor (hPa).

Raw data is collected biweekly via a micro-SD card and Ethernet cable and processed by the computer system EDDYPRO® LI-COR ENVIRONMENTAL (Datalogger, mod. CR1000, Campbell Scientific Instrument, Utah, USA). The high-frequency reading is taken at a frequency of 10 Hz and stored in a datalogger (CR1000x, Campbell Scientific Instrument, Utah, USA). After processing, the data

Table 1 Description of the micrometeorological elements and sensors along with their respective installation heights in the Jaru biological reserve tower

Micrometeorological variables	Sensor	Sensor height (m)
Wind direction and speed	Two-dimensional-2D sonic anemometers (A100R, Campbell)	46, 37 e 30
CO_2 flow (eddy covariance system)	Three-dimensional sonic anemometer (LI7550A, LI-COR) and LICOR (7500, LICOR Inc., Lincoln, USA)	63,5
Atmospheric pressure	Three-dimensional sonic anemometer (LI7550A, LI-COR)	63,5
Photosynthetically active radiation	Quantum licor (SQ110, Apogee, Lincoln, USA)	56
Incident and reflected shortwave radiation	Saldoradiometro (SN500, Apogee, Lincoln, USA)	56
Air temperature and relative humidity	Thermohygrometers rotronic HC2S3 (Campbell)	62

Source Prepared by the authors based on the study's data

is condensed into 30-min averages. The data is filtered and certified manually using standard maximum and minimum values in order to remove spurious and/or inconsistent values as proposed by Aguiar (2005). For data analysis, descriptive statistics tests were carried out using the programs Excel 2013 (Microsoft), Minitab 17 and SigmaPlot 11.

3 Results and Discussion

3.1 Temperature and Relative Humidity

The month of August's average air temperature profile (T_air) is 26.3 $\pm$ 4.1 °C (every value after the sign $\pm$ corresponds to the standard deviation) and the relative air humidity (RH) profile was 83.5 $\pm$ 16.9% as demonstrated in Fig. 2, with the two variables being inversely proportional.

The RH remained saturated between 3 am and 7:30 am, with the lowest value occurring at 3 pm with 54.8%. The behavior mentioned is due to the relationship that exists with solar radiation, which has a direct impact on humidity and air temperature (Shrestha; Thapa; Gautam 2019). The T_air presented the highest averages in the interval from 2 pm to 4:30 pm, with the highest value occurring at 3 pm at 32.9 °C. This behavior is due to the lag that occurs in solar radiation in which the earth takes time to heat (Paca et al. 2022). And the lowest T_air found occurred at 6:30 am at 21.2 °C.

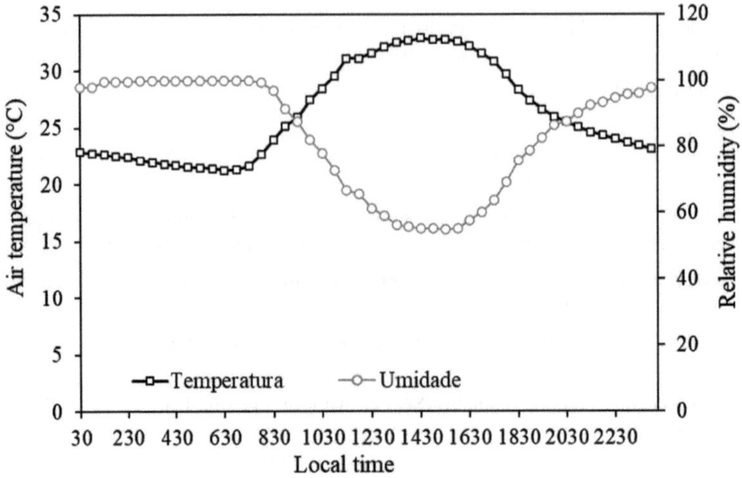

Fig. 2 Average diurnal cycle of the air temperature profile (°C) and relative air humidity (%) in the Jaru biological reserve in August 2022. *Source* Prepared by the authors based on the study's data

Barbino et al. (2023) when studying Rebio Jaru, found maximum values with monthly averages of T_air in September with 26.75 ºC and the minimum in February with an average of 23.52 ºC. The highest RH occurred in February with an average of 87.48% and the lowest in August with an average of 64.54%. Ataíde et al. (2020) when analyzing the Caxiuanã National Forest in the dry period, found a minimum average temperature of 21.7 °C and a maximum average temperature of 33.6 °C, with an average relative humidity of 83.6%. During August 2018, Souza et al. (2021) when studying the air samples collected in the Amazon Tall Tower Observatory (ATTO), verified an average air temperature of 27.2 °C and relative air humidity of 83%.

Pinto-Júnior et al. (2011) when observing an Amazon transition forest 50 km from the municipality of Sinop, Mato Grosso, observed that during August the average monthly air temperature was 24 °C while the monthly average air humidity was 59.1%.

Little is known about the effects of seasonal climate change in the Amazon Forest, in the context of which there is a mosaic of vegetation, making it heterogeneous and diverse. However, new studies like this one are extremely important to fill the remaining gaps to better understand the forest dynamics of the southwestern Amazon.

3.2 *Shortwave Radiation*

The SWR began to increase from 6:30 am, reaching peak time at 12:30 pm, where it decreased until reaching negative values from 6:30 pm onwards. The incident shortwave radiation (SWR_in) had a monthly average of 203 W m^{-2} and the shortwave radiation reflected (SWR_out) with a monthly average of 25.4 W m^{-2} (Fig. 3).

Similar findings of monthly average SWR_in during the dry period at REBIO Jaru were detailed by Aguiar (2005) with an average of 207 W m^{-2} obtained, Andrade et al. (2009) with a value of 211 W m^{-2} and Randow et al. (2004) with 206 W m^{-2}.

One of the main factors that directly impact the variability of solar radiation is cloudiness, which affects even more than the angle of the sun (Junior et al. 2023; Malhi et al. 2002; Zuluaga et al. 2021).

The hourly average of SWR_in was highest at 12:30 pm with 859 W m^{-2} and lower at 7:30 pm with –2.6 W m^{-2}, which corroborates the study carried out by Andrade et al. (2009), Oliveira and Moraes (2013) and Randow et al. (2004). The intense solar radiation during the midday period demonstrates the strong exposure to sunlight that the forest suffers, being one of the moments in which plants efficiently convert solar energy into organic compounds, boosting the production of biomass (Assis et al. 2022; Hilker et al. 2008; Ribeiro 2023).

Although one of the factors that limit the behavior of SWR_in within the forest structure is the density of the canopy, which affects the distribution of radiation, causing shading of the trees, forests that are in a state of climax are already adapted to this distribution (Albrigo 2021; Wojnowski et al. 2021).

Furthermore, SWR_in also plays an important role in regulating canopy temperature as shown in Fig. 2. Solar radiation absorbed by leaves heats the environment, creating a favorable microclimate for the biological activities of plants and other organisms. This is particularly relevant in tropical forests, where temperature can be a limiting factor for plant growth (Doughty et al. 2023; Zhong et al. 2021).

In relation to SWR_out, the maximum value was 102.3 W m^{-2}, congruent with the SWR_in time, indicating that most of the incident radiation was absorbed for heat

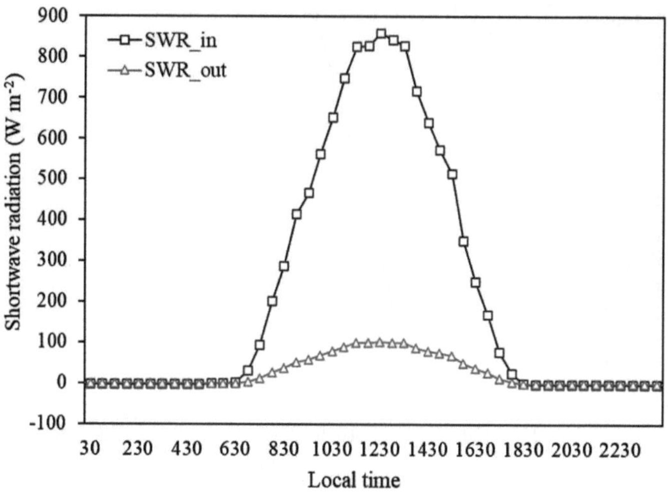

Fig. 3 Variation of the average diurnal cycle of incident and reflected shortwave radiation (W m^{-2}) in the Jaru biological reserve in August 2022. *Source* Prepared by the authors based on the study's data

conversion processes or photosynthetic processes. The greater absorption of SWR_in is of paramount importance for efficiencies in the capture and use of solar energy. The minimum value of SWR_out observed was -0.381 W m^{-2} at 8:30 pm. Understanding the diurnal pattern variation of SWR and how it consistently regulates and influences the biological processes of biomass and the forest, in general, is extremely relevant, even more so in a scenario of changes in land use.

3.3 Photosynthetically Active Radiation and CO_2 Flow

The monthly average of Incident Photosynthetically Active Radiation (PAR_in) and Reflected (PAR_out) measured was 407.3 ± 602.5 μmol m^{-2} s^{-1} and 19.4 ± 24.4 μmol m^{-2} s^{-1}, respectively. The behavior of these variables (Fig. 4) followed the same hourly pattern found for the maximum value of SWR_in, when the sun's rays fall perpendicularly, causing greater penetration within the canopy, being of 1.705 μmol m^{-2} s^{-1} and 68 μmol m^{-2} s^{-1}, in due order. However, the minimum values were at 4 am with 0.05 μmol m^{-2} s^{-1} and 1.3 μmol m^{-2} s^{-1}.

Figure 5 displays the hourly average behavior of the CO_2 flow in which negative values indicate the absorption of atmospheric CO_2 through the photosynthesis process and positive CO_2 emissions into the atmosphere through vegetation respiration within the canopy.

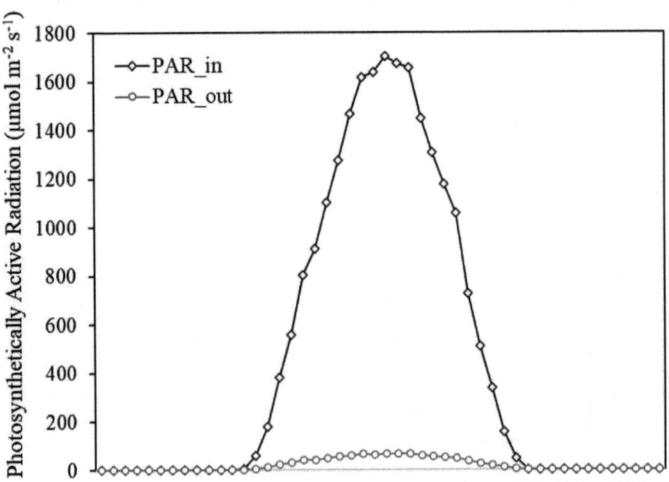

Fig. 4 Average hourly variation of incident and reflected photosynthetically active radiation (μmol m^{-2} s^{-1}) in the Jaru biological reserve in August 2022. *Source* Prepared by the authors based on the study's data

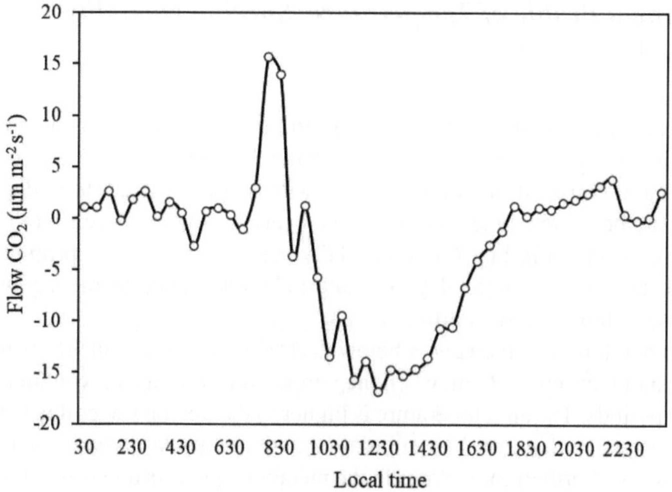

Fig. 5 Average hourly behavior of the flow of CO_2 (μmol m^{-2} s^{-1}) in the Jaru biological reserve in August 2022. *Source* Prepared by the authors based on the study's data

As seen, a peak occurs shortly after sunrise at 8 am with 15.6 μmol m^{-2} s^{-1} of CO_2 emission. This result is derived from the accumulation of nocturnal CO_2 within the canopy where, on calm nights, CO_2 drains to lower places and is released after the first solar incidence (Saleska et al. 2009; Hayek et al. 2018).

The CO_2 absorption peak occurred at 12:30 pm–16.8 μm m^{-2} s^{-1}, corroborating the period of maximum PAR in which plants are consuming CO_2 in their photosynthesis process. During this study period, the Jaru Biological Reserve behaved as a carbon dioxide sink. Furthermore, studies such as those of Antonucci et al. (2018) and Aguiar (2013) when studying REBIO Jaru, found a similar pattern in the behavior of CO_2 flow in the dry season. Marani et al. (2020) when carrying out measurements of vertical atmospheric CO_2 profiles with small aircraft in the region of the Caxiuanã National Forest, observed changes in the behavior of the CO_2 flow in recent years, where the region acted as a source of CO_2, which suggests an increase in agreement with anthropogenic activities.

Mendes et al. (2020) when studying the *caatinga* biome during the period of extreme drought (2014 and 2015) observed a reduction in the values of net CO_2 exchange (NEE) and gross primary production (GPP) in the dry season due to the plants gradually facing a decrease in the amount of water in the soil, which results in the closure of stomata and a decrease in opening and loss of water through the leaves. This, in turn, makes it difficult for CO_2 to enter, which further limits the ability to carry out net photosynthesis.

It is interesting to highlight the importance of forests as sinks of greenhouse gases, especially in the current process of changes in land use, since studies with terrestrial modeling report a decline in the southeastern Amazon Forest as a carbon sink (Gatti et al. 2021; Habau et al. 2020; Ramming and Lapola 2021).

3.4 Vertical Profile of Temperature, Specific Humidity and Wind Speed

The vertical profile of air temperature, specific air humidity and wind speed were analyzed in the daily period (2 am, 6 am, 10 am, 2 pm, 6 pm, 10 pm) of greatest absorption and emission of carbon dioxide within August 2022, to understand the characteristic behavior of the aforementioned variables with inference to CO_2.

It can be observed in Fig. 6 that the daily vertical temperature profile followed the same behavior both on the day of greatest absorption and on the day of greatest CO_2 emission, however, with different values.

The temperature remained lower below the treetops, where in the study region the average is 34 m, except at 2 pm, when the temperature was at a maximum of 22 m in both daily periods. The air temperature is higher in the treetop stages due to the dense cover characteristic that limits air circulation and direct solar radiation, causing less heat dissipation. Furthermore, plants in the metabolic process release energy through photosynthesis and water vapor through leaf transpiration, which results in greater air humidity, retaining more heat than dry air.

On the day of greatest absorption, the maximum temperature was 33 °C, while on the day of greatest emission, it was 34.6 °C. The minimum temperature occurred at 06:00 in the 12 m profile in both periods of analysis with an average of 22.6 °C for the day with the highest absorption and 21.4 °C for the day with the highest emission.

Climatic variables, such as temperature and precipitation, influenced by land use change and seasonality, are important factors that can influence the Amazon to move

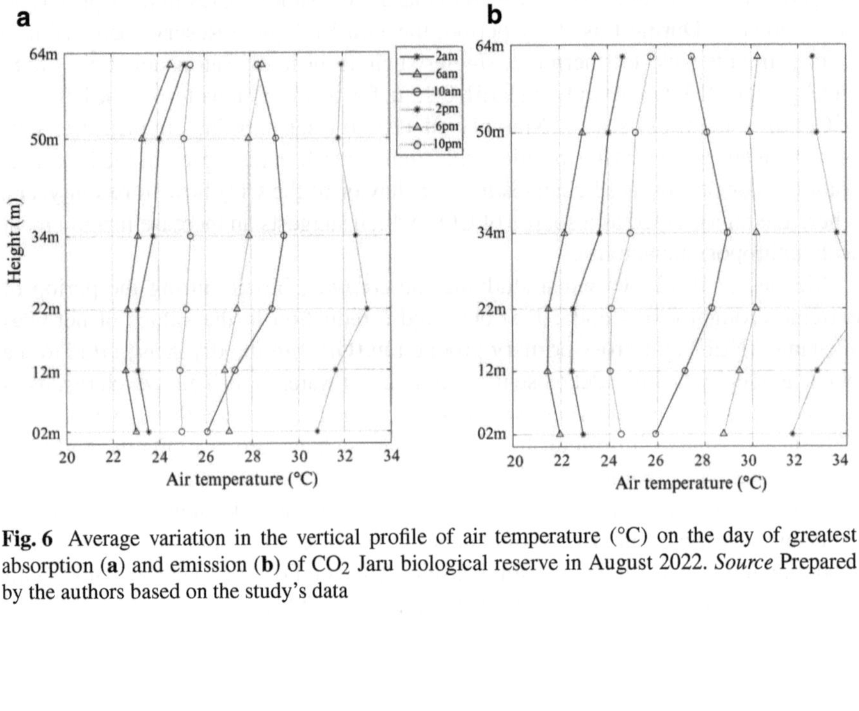

Fig. 6 Average variation in the vertical profile of air temperature (°C) on the day of greatest absorption (**a**) and emission (**b**) of CO_2 Jaru biological reserve in August 2022. *Source* Prepared by the authors based on the study's data

from a sink to an emitter of CO_2 into the atmosphere (Marani et al. 2020). Furthermore, studies such as Maia et al. (2020) report a higher incidence of CO_2 emissions and a decline in sinks in hot and dry areas due to vegetation water stress that can cause tree mortality.

The vertical profile of specific humidity (Fig. 7) followed the same hourly behavior of the air temperature in the two periods analyzed, with the maximum value of 30.3 g kg^{-1} for the day that absorbed the most and 30.9 g kg^{-1} for the day that emitted the most. The minimum value for the day of greatest carbon dioxide absorption was 15.2 g kg^{-1} and for the day with the highest emission, it was 13.3 g kg^{-1}. It is important to highlight that during the study period, there was no rain event.

It can be seen in the wind speed profile (Fig. 8) that the speed in the two study periods increases with distance from the canopy. This behavior was also observed by Santana et al. (2017) that when studying the wind speed below the canopy in the K34 tower and above by the ATTO tower in Manaus, it was verified that the wind speed suffers interference due to the structure of the vegetation, where the values were low close to the treetops and increase according to the height. This result is consistent with the structure of the forest with heterogeneous characteristics and emerging trees, in addition to interference from the roughness sublayer (Cava et al. 2022; Kaimal and Finnigan 1994; Santana et al. 2018; Poggi et al. 2004).

On the day that emitted the most, the value of wind speed at 6 pm was exacerbated compared to the day that absorbed the most CO_2. It is important to understand the behavior of wind speed below and above the treetops since this variable has a great influence on the flow of CO_2 and NEE (Komatsu et al. 2005). Furthermore, an excessive increase in wind speed causes greater plant transpiration, resulting in the

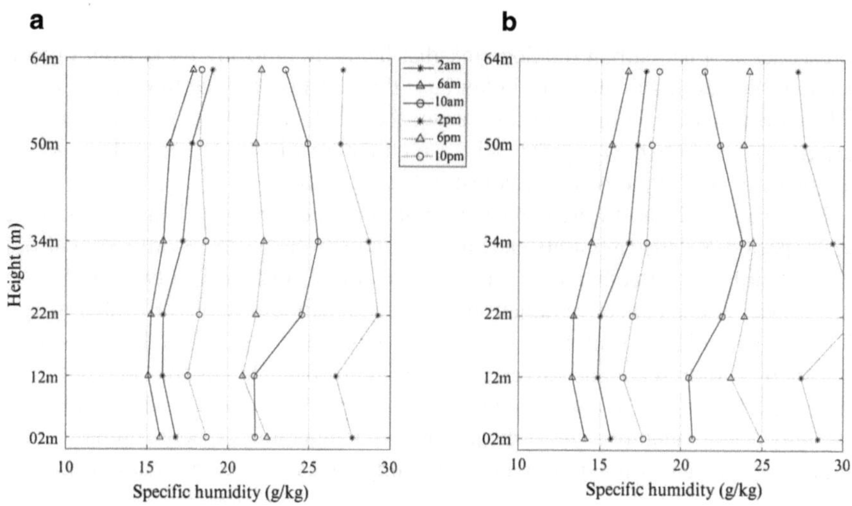

Fig. 7 Average variation of the vertical profile of specific air humidity (g kg^{-1}) of the day of greatest absorption (**a**) and emission (**b**) of CO_2 in August 2022 in the Jaru biological reserve. *Source* Prepared by the authors based on the study's data

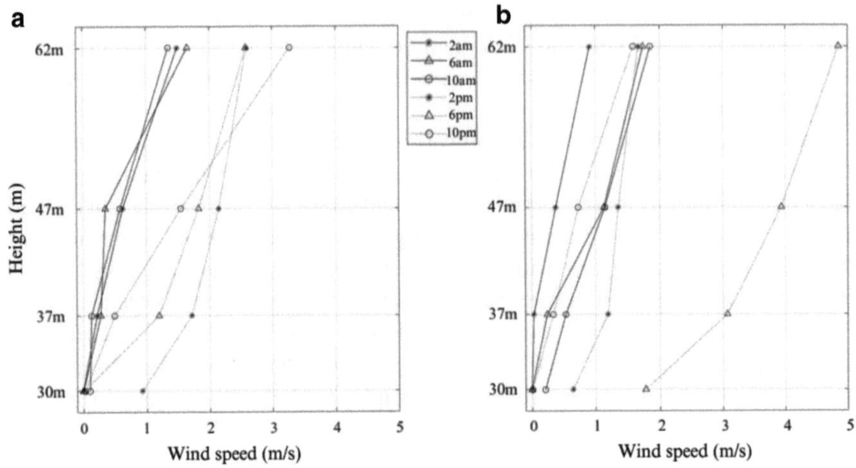

Fig. 8 Average behavior of the vertical wind speed profile (m s^{-1}) on the day that absorbed (**a**) and emitted (**b**) the most CO_2, in the Jaru biological reserve in August 2022. *Source* Prepared by the authors based on the study's data

closure of stomata and loss of leaf area. All of these factors drastically reduce the absorption of CO_2 through the photosynthesis process (Silva et al., 2021).

4 Final Remarks

Forests act as a net carbon dioxide sink that helps minimize impacts from human activities, however, with changes in land use this function may be threatened, especially in dry season regimes. In this sense, the results addressed were the diurnal variation of the CO_2 flow together with the climatic variables below and above the canopy in an area of the Amazon Forest in the dry season.

The results addressed were the behavior of micrometeorological variables during the dry period. Relative humidity and air temperature were inversely proportional and are in accordance with other studies carried out at REBIO Jaru. Shortwave radiation and photosynthetically active radiation presented maximum values at 12:30 pm The CO_2 flow peaked after sunrise related to the accumulation of carbon dioxide within the canopy and presented negative values during the day and positive values during the night. The maximum absorption of CO_2 occurred at the peak time of photosynthetically active radiation. The vertical profile of temperature, specific humidity and wind speed presented maximum and minimum values on the day that emitted the most CO_2.

These results demonstrate that the behavior of the CO_2 flow, despite being complex, is directly related to the dynamics of the climatological variables addressed. In addition, it highlights the importance of the standing Amazon rainforest as a mitigating means for climate change, which helps in the implementation of forest conservation policy programs.

For future research, it is necessary to verify the entire seasonal period of precipitation and soil moisture, in order to observe the response of vegetation to water availability and how it affects the metabolic process of plants within the Amazonian primary forest. It is important to perform the net carbon balance over a longer period, seeking to understand its variation, even though the flow database of the REBIO Jaru suffers limitations, with significant gaps in the time series since 1999. Data losses are attributed to the instability of the tower's storage, power, and maintenance system. In addition, it is necessary to evaluate the energy and carbon exchanges between the primary forest of the REBIO Jaru and a different landscape, but with similar environmental conditions.

Acknowledgements This work was carried out with the support of the research group of the Large-Scale Biosphere-Atmosphere Research Program in the Amazon (LBA) in Ji-Paraná and I thank them for the grant.

References

Aguiar RG (2013) Balanço de energia em ecossistema Amazônico por modelo de regressão robusta com bootstrap e validação cruzada Tese (Doutorado em Física Ambiental). Universidade Federal do Mato Grosso, Cuiabá, Instituto de Física

Aguiar RG, De Musis CR, Aguiar LJG, Martínez-Espinosa M, Fischer GR (2019) Energy balance closure in the Southwest Amazon Forest site—a statistical approach. Theoret Appl Climatol 136:1209–1219. https://doi.org/10.1007/s00704-018-2548-8

Aguiar RG (2005) Fluxos de massa e energia em uma floresta tropical no sudoeste da Amazônia. Dissertação (Mestrado em Física e Meio ambiente), Instituto de Ciências Exatas e da Terra, Universidade Federal de Mato Grosso, Cuiabá

Albrigo NDS (2021) Análise da produtividade de raízes finas para estimativas do estoque de carbono em um trecho de floresta primária de terra firme na Amazônia Ocidental. Monografia (Graduação em Engenharia Ambiental e Sanitária), Universidade Federal de Rondônia, Ji-Paraná

Andrade NLRD, Aguiar RG, Sanches L, Alves ÉCRDF, Nogueira JDS (2009) Partição do saldo de radiação em áreas de floresta amazônica e floresta de transição Amazônia-Cerrado. Rev Bras Meteorol 24:346–355. https://doi.org/10.1590/S0102-77862009000300008

Andreae MO, Artaxo P, Brandao C, Carswell FE, Ciccioli P, Da Costa AL, Waterloo MJ (2002) Biogeochemical cycling of carbon, water, energy, trace gases, and aerosols in Amazonia: The LBA-EUSTACH experiments. J Geophys Res: Atmos 107(D20):25–33. https://doi.org/10.1029/2001JD000524

Antonucci B, Aguiar RG, Aguiar LJG, de Andrade NLR (2018) Fluxos de Co_2 em uma área de floresta tropical úmida na Amazônia ocidental em um ano de el niño. Rev Ciência E Natura 40:119–125. https://doi.org/10.5902/2179460X30718

Araújo AC, Kruijt B, Nobre AD, Dolman AJ, Waterloo MJ, Moors EJ, de Souza JS (2008) Nocturnal accumulation of CO_2 underneath a tropical forest canopy along a topographical gradient. Ecol Appl 18(6):1406–1419. https://doi.org/10.1890/06-0982.1

Araujo SLS (2022) Padrões de chuva em uma área de floresta amazônica em Rondônia: Série temporal de 1999 a 2020. Monografia (Graduação em Engenharia Ambiental e Sanitária), Departamento de Engenharia Ambiental, Ji-Paraná

Assis RMA, Chagas JH, Leite JJF, Santos JPD, Bertolucci SKV, Pinto JEBP (2022) Effect of irradiance levels on biomass production, essential oil content and anatomy of Artemisia alba Turra. Res, Soc Dev 11(13):e377111335565. https://doi.org/10.33448/rsd-v11i13.35565

Ataide WLS, de Oliveira FDA, Pinto CAD (2020) Balanço de radiação, energia e fechamento do balanço em uma floresta prístina na Amazônia oriental. Rev Bras Geogr Física, 13(06):2603–2627. https://doi.org/10.26848/rbgf.v13.6.p2603-2627

Barbino GC, de Andrade NLR, Webler AD, Sanches L, Aguiar R, Antonucci B (2023) Índice de área foliar e sua relação com o microclima em floresta e pastagem na Amazônia Ocidental. Rev Bras Climatol, 32:311–335. https://doi.org/10.55761/abclima.v32i19.16296

Boulton CA, Lenton TM, Boers N (2022) Pronounced loss of Amazon rainforest resilience since the early 2000s. Nat Clim Chang 12(3):271–278. https://doi.org/10.1038/s41558-022-01287-8

Bullock EL, Woodcock CE (2021) Carbon loss and removal due to forest disturbance and regeneration in the Amazon. Sci Total Environ 764:142839. https://doi.org/10.1016/j.scitotenv.2020.142839

Cava D, Dias-Júnior CQ, Acevedo O, Oliveira PE, Tsokankunku A, Sörgel M, Mortarini L (2022) Vertical propagation of submeso and coherent structure in a tall and dense Amazon forest in different stability conditions Part I: Flow structure within and above the roughness sublayer. Agric for Meteorol 322:108983. https://doi.org/10.1016/j.agrformet.2022.108983

da Silva TRG, Costa MLA, Farias LRA, Santos MA, Rocha JJL, Silva JV (2021) Fatores abióticos no crescimento e floração das plantas. Pesqui, Soc E Desenvolv, 10 (4):e19710413817. https://doi.org/10.33448/rsd-v10i4.13817

Denning S (2021) Southeast Amazonia is no longer a carbon sink. Nature 595:354–355. https://doi.org/10.1038/d41586-021-01871-6

Doughty CE, Keany JM, Wiebe BC, Rey-Sanchez C, Carter KR, Middleby KB, Fisher JB (2023) Tropical forests are approaching critical temperature thresholds. Nature 621:105–111. https://doi.org/10.1038/s41586-023-06391-z

Ellwanger JH, Kulmann-Leal B, Kaminski VL, Valverde-Villegas J, Veiga ABG, Spilki FR, Chies JAB (2020) Beyond diversity loss and climate change: Impacts of Amazon deforestation on infectious diseases and public health. An Da Acad Bras Ciências, 92. https://doi.org/10.1590/0001-3765202020191375

Fearnside PM (2021) The intrinsic value of Amazon biodiversity. Biodivers Conserv 30(4):1199–1202. https://doi.org/10.1007/s10531-021-02133-7

Friedlingstein P, O'Sullivan M, Jones MW, Andrew RM, Gregor L, Hauck J, Zheng B (2022) Global carbon budget 2022. Earth System Science Data 14(11):4811–4900. https://doi.org/10.5194/essd-14-4811-2022

Gatti LV, Basso LS, Miller JB, Gloor M, Gatti LD, Cassol HL, Neves RA (2021) Amazonia as a carbon source linked to deforestation and climate change. Nature 595(7867):388–393. https://doi.org/10.1038/s41586-021-03629-6

Gomez JB, Webler AD, Aguiar RG, Aguiar LJG, Nuñes MLA (2015) Conversão de florestas tropicais em sistemas pecuários na Amazônia: Quais as implicações no microclima da região? Rev Bras Climatol 17:67–81

Green JK, Berry J, Ciais P, Zhang Y, Gentine, P (2020) Amazon rainforest photosynthesis increases in response to atmospheric dryness. Sci Adv, 6(47):eabb7232. https://doi.org/10.1126/sciadv.abb7232

Hayek MN, Wehr R, Longo M, Hutyra LR, Wiedemann K, Munger JW, Wofsy SC (2018) A novel correction for biases in forest eddy covariance carbon balance. Agric for Meteorol 250:90–101. https://doi.org/10.1016/j.agrformet.2017.12.186

Heinrich VH, Dalagnol R, Cassol HL, Rosan TM, de Almeida CT, Silva Junior CH, Aragão LE (2021) Large carbon sink potential of secondary forests in the Brazilian Amazon to mitigate climate change. Nat Commun 12(1):1785. https://doi.org/10.1038/s41467-021-22050-1

Hilker T, Coops NC, Schwalm CR, Jassal RPS, Black TA, Krishnan P (2008) Effects of mutual shading of tree crowns on prediction of photosynthetic light-use efficiency in a coastal Douglas-fir forest. Tree Physiol 28(6):825–834. https://doi.org/10.1093/treephys/28.6.825

Hubau W, Lewis SL, Phillips OL, Affum-Baffoe K, Beeckman H, Cuní-Sanchez A, Zemagho L (2020) Asynchronous carbon sink saturation in African and Amazonian tropical forests. Nature 579(7797):80–87. https://doi.org/10.1038/s41586-020-2035-0

Hutyra, LR, Munger JW, Saleska SR, Gottlieb E, Daube BC, Dunn AL, Wofsy SC (2007) Seasonal controls on the exchange of carbon and water in an Amazonian rain forest. J Geophys Res: Biogeosciences, 112(G3). https://doi.org/10.1029/2006JG000365

Instituto Chico Mendes de Conservação da Biodiversidade—ICMBio (2023) Unidades de Conservação—Amazônia. Disponível em http://www.icmbio.gov.br/portal/unidadesdeconservacao/biomas-brasileiros/amazonia/unidades-de-conservacaoamazonia/1999-rebio-do-jaru. Acesso em: 01 de março de 2023

Júnior JML, Querino CAS., de Andrade AMD, Carneiro RG, Moura MAL (2023) Balanço de radiação da cana-de-açúcar na região de tabuleiros costeiros de Alagoas, Brasil. Rev Bras Climatol, 32:83–101. https://doi.org/10.55761/abclima.v32i19.15785

Kaimal JC, Finnigan JJ (1994) Atmospheric boundary layer flows: Their structure and measurement. Oxford University Press

Kim SK, Shin J, An SI, Kim HJ, Im N, Xie SP, Yeh SW (2022) Widespread irreversible changes in surface temperature and precipitation in response to CO_2 forcing. Nat Clim Chang 12(9):834–840. https://doi.org/10.1038/s41558-022-01452-z

Komatsu H, Hotta N, Kuraji K, Suzuki M, Oki T (2005) Classification of vertical wind speed profiles observed above a sloping forest at nighttime using the bulk Richardson number. Bound-Layer Meteorol 115:205–221. https://doi.org/10.1007/s10546-004-3408-x

Le Quéré C, Andrew RM, Friedlingstein P, Sitch S, Pongratz J, Manning AC, Zhu D (2018) Global carbon budget 2017. Earth Syst Sci Data 10(1):405–448. https://doi.org/10.5194/essd-10-405-2018

Maia VA, Santos ABM, de Aguiar-Campos N, de Souza CR, de Oliveira MCF, Coelho PA, Dos Santos RM (2020) The carbon sink of tropical seasonal forests in southeastern Brazil can be under threat. Sci Adv, 6(51): eabd4548. https://doi.org/10.1126/sciadv.abd4548

Malhi Y, Nobre AD, Grace J, Kruijt B, Pereira MG, Culf A, Scott S (1998) Carbon dioxide transfer over a Central Amazonian rain forest. J Geophys Res: Atmos 103(D24):31593–31612. https://doi.org/10.1029/98JD02647

Malhi Y, Pegoraro E, Nobre AD, Pereira MG, Grace J, Culf AD, Clement R (2002) Energy and water dynamics of a central Amazonian rain forest. J Geophys Res: Atmos 107(D20):17–45. https://doi.org/10.1029/2001JD000623

Marani L, Gatti LV, Miller JB, Domingues LG, Correia CC, Gloor M, Neves RA (2020) Métodos de Estimativas de Fluxo de Gases de Efeito Estufa ea Influência da Ação Humana na Redução da Capacidade de Remoção de CO_2 na Floresta Amazônica. Rev Virtual Quim 12(5):1129–1144. https://doi.org/10.21577/1984-6835.20200091

Marengo JA, JrCM S, Thonicke K, Burton C, Halladay K, Betts RA, Soares WR (2018) Changes in climate and land use over the Amazon region: current and future variability and trends. Front Earth Sci 6:228. https://doi.org/10.3389/feart.2018.00228

Mendes KR, Campos S, da Silva LL, Mutti PR, Ferreira RR, Medeiros SS, Santos e Silva CM (2020) Seasonal variation in net ecosystem CO2 exchange of a Brazilian seasonally dry tropical forest. Sci Rep, 10(1):9454. https://doi.org/10.1038/s41598-020-66415-w

Mitchard ET (2018) The tropical forest carbon cycle and climate change. Nature 559(7715):527–534. https://doi.org/10.1038/s41586-018-0300-2

Oliveira GD, Moraes EC (2013) Validação do balanço de radiação obtido a partir de dados MODIS/TERRA na Amazônia com medidas de superfície do LBA. Acta Amazon 43:353–363. https://doi.org/10.1590/S0044-59672013000300011

Ometto JPH, Nobre AD, Rocha HR, Artaxo P, Martinelli LA (2005) Amazonia and the modern carbon cycle: Lessons learned. Oecologia 143:483–500. https://doi.org/10.1007/s00442-005-0034-3

Paca VHDM, Espinoza-Dávalos GE, da Silva R, Tapajós R, dos Santos ABG (2022) Remote sensing products validated by flux tower data in Amazon rain forest. Remote Sens 14(5):1259. https://doi.org/10.3390/rs14051259

Pinto-Júnior OB, Sanches L, de Almeida FL, Brandão AA, de Souza JN (2011) Leaf area index of a tropical semi-deciduous forest of the southern Amazon basin. Int J Biometeorol 55:109–118. https://doi.org/10.1007/s00484-010-0317-1

Poggi D, Porporato A, Ridolfi L, Albertson JD, Katul GG (2004) The effect of vegetation density on canopy sub-layer turbulence. Bound-Layer Meteorol 111:565–587. https://doi.org/10.1023/B:BOUN.0000016576.05621.73

Priante-Filho N, Vourlitis GL, Hayashi MMS, Nogueira JDS, Campelo JH Jr, Nunes PC, Silveira M (2004) Comparison of the mass and energy exchange of a pasture and a mature transitional tropical forest of the southern Amazon Basin during a seasonal transition. Glob Change Biol 10(5):863–876. https://doi.org/10.1111/j.1529-8817.2003.00775.x

Pugh TA, Lindeskog M, Smith B, Poulter B, Arneth A, Haverd V, Calle L (2019) Role of forest regrowth in global carbon sink dynamics. Proc Natl Acad Sci 116(10):4382–4387. https://doi.org/10.1073/pnas.1810512116

Rammig A, Lapola DM (2021) The declining tropical carbon sink. Nat Clim Chang 11(9):727–728. https://doi.org/10.1038/s41558-021-01135-1

Ribeiro GDS (2023) A estrutura do dossel como fator determinante da produtividade florestal: o caso do sistema agroflorestal em ondas. Dissertação (mestrado em ciências), Universidade de São Paulo

Saleska S, Da Rocha H, Kruijt B, Nobre A (2009) Ecosystem carbon fluxes and Amazonian forest metabolism. Amaz Glob Chang 186:389–407. https://doi.org/10.1029/2008GM000739

Santana, RASD, Dias-Júnior CQ, Vale RSD, Tóta J, Fitzjarrald DR (2017) Observing and modeling the vertical wind profile at multiple sites in and above the Amazon rain forest canopy. Adv Meteorol, 2017. https://doi.org/10.1155/2017/5436157

Santana RA, Dias-Júnior CQ, da Silva JT, Fuentes JD, do Vale RS, Alves EG, Manzi AO (2018) Air turbulence characteristics at multiple sites in and above the Amazon rainforest canopy. Agric For Meteorol, 260:41–54. https://doi.org/10.1016/j.agrformet.2018.05.027

Shrestha AK, Thapa A, Gautam H (2019) Solar radiation, air temperature, relative humidity, and dew point study: Damak, Jhapa, Nepal. Int J Photoenergy 2019:1–7. https://doi.org/10.1155/2019/8369231

Souza FF, Mathai PP, Pauliquevis T, Balsanelli E, Pedrosa FO, Souza EM, Huergo LF (2021) Influence of seasonality on the aerosol microbiome of the Amazon rainforest. Sci Total Environ 760:144092. https://doi.org/10.1016/j.scitotenv.2020.144092

Valadão CC, Aguiar RG, Webler AD (2019) Padrão da chuva em áreas de floresta e pastagem no Sudoeste da Amazônia. S Amn j, Rio Branco 6(7):244–260

Valadão CC, Aguiar RG, Webler AD (2019) Padrão da chuva em áreas de floresta e pastagem no Sudoeste da Amazônia. S Amn J, Rio Branco, 6(7):244–260, out

Vitale D (2021) A performance evaluation of despiking algorithms for eddy covariance data. Sci Rep 11(1):1–15. https://doi.org/10.1038/s41598-021-91002-y

Von Randow C, Manzi AO, Kruijt B, De Oliveira PJ, Zanchi FB, Silva RD, Kabat P (2004) Comparative measurements and seasonal variations in energy and carbon exchange over forest and pasture in South West Amazonia. Theoret Appl Climatol 78:5–26. https://doi.org/10.1007/s00704-004-0041-z

Wojnowski W, Wei S, Li W, Yin T, Li XX, Ow GLF, Whittle AJ (2021) Comparison of absorbed and intercepted fractions of PAR for individual trees based on radiative transfer model simulations. Remote Sens 13(6):1069. https://doi.org/10.3390/rs13061069

Zhong R, Wang P, Mao G, Chen A, Liu J (2021) Spatiotemporal variation of enhanced vegetation index in the Amazon Basin and its response to climate change. Phys Chem Earth, Parts A/B/C 123:103024. https://doi.org/10.1016/j.pce.2021.103024

Zuluaga CF, Avila-Diaz A, Justino FB, Wilson AB (2021) Climatology and trends of downward shortwave radiation over Brazil. Atmos Res 250:105347. https://doi.org/10.1016/j.atmosres.2020.105347

Yanomami People and Land in Roraima: Reviewing the Complex Nexus of Illegal Mining, Indigenous Vulnerability and Government Interventions

Ismar Borges de Lima, Raíza Meira Vieira, and Wilson de Moraes Sousa

Abstract Illegal mining has caused significant environmental and social devastation upon the Yanomami Indigenous Land (YIL) in the Brazilian Amazon, threatening the Yanomami people's traditional way of life, natural resources and well-being. The "Yanomami under Attack" report by the Socioenvironmental Institute (ISA) outlines 30 years of environmental damage and sociocultural erosion since the demarcation of YILs in 1992, with deforestation, polluted water, mercury contamination, and the spreading of diseases as consequences. The MapBiomas report shows a 3,350% increase in illegal mining activities between 2016 and 2020, leading to 3,272 hectares of devastation in 2021. The COVID-19 pandemic exacerbated the health crisis in the communities. In 2023, the Brazilian government took vigorous actions by deploying military air force and police, IBAMA, and the Ministry of Health to address illegal mining and health issues effectively. This chapter analyzes the government's interventions, including the National Indigenous People Foundation (FUNAI) role, and stresses the need for comprehensive solutions like ethnodevelopment initiatives and policies to create a more sustainable and equitable future for Yanomami. The study used diverse sources and methods, such as participant observation, literature reviews,

I. B. de Lima (✉) · W. de Moraes Sousa
Universidade Estadual de Roraima, UERR, Rua 7 de Setembro, 231, Canarinho. Boa Vista/RR. Zip Code: 69306-530, Boa Vista, Roraima, Brazil
e-mail: ismarlima@yahoo.com.br
URL: https://www.uerr.edu.br; https://www.recinatur.org/

W. de Moraes Sousa
URL: https://www.uerr.edu.br

I. B. de Lima
RECINATUR Foundation for Nature, Tourism and Science, Valdivia, Chile

R. M. Vieira
Yanomami DSEI and Technical Support Consultant - Ministry of Health (OPAS - HWO-FINATEC) & PhD candidate in Global Health and Sustainability - School of Public Health, University of São Paulo (USP), Av. Dr. Arnaldo, 715, Zip Code 01246-904, Sao Paulo, Brazil
e-mail: raiza.vieira@usp.br
URL: https://sites.usp.br/fsp/en/saude-global-e-sustentabilidade/

W. Leal Filho et al. (eds.), *Amazon 2030—Sustainability Issues in the World's Largest Rainforest Region*, World Sustainability Series,
https://doi.org/10.1007/978-3-031-81465-5_34

and reports from NGOs, government, and news articles, to provide fresh perspectives, updates, and insights on the topic.

Keywords Illegal mining · Environmental degradation · Health crisis · Yanomami vulnerability · Ethnodevelopment · Roraima · Amazon

1 Indigenous People in the Amazon: Figures, Context and Realities

In Brazil's northernmost region lies the extensive Yanomami Indigenous Land (YIL), spanning over 9 million hectares and demarcated by a 1992 decree (Brasil 1992). According to a technical report by the Amazon Environmental Research Institute (IPAM), illegal mining activities within the Yanomami Indigenous Land (YIL) expanded by nearly 20,000-fold between 1985 and 2022, leading to significant loss of Indigenous lives. This alarming growth led to devastating consequences, including significant loss of Indigenous lives, and has persisted across multiple administrations, from José Sarney to Fernando Collor de Mello, Luiz Inacio Lula da Silva, Dilma Rousseff, Michel Temer, and Jair Bolsonaro. Despite varying priorities and policies, a consistent failure to effectively enforce protections has allowed this crisis to escalate over decades. To illustrate the historical severity of the situation, it is estimated that between 1986 and 1990, approximately 20% of the Yanomami population—equivalent to 1,800 individuals—lost their lives due to disease and violence. These tragic outcomes were directly linked to the invasion of their territory by around 45,000 miners during that period (Instituto Socioambiental 2016). The gravity of the situation was highlighted in 2018 by a Yanomami shaman's statement, supported by the Brazilian army, capturing 800 prospectors in one mining hotspot (Ramos et al. 2020). Indigenous rights and land violations in Brazil have constituted a longstanding and regrettable reality dating back to colonial times. This historical narrative reflects a persistent challenge wherein the rights of Indigenous communities and the integrity of their lands have consistently been disregarded or violated.

The Indigenous population in Brazil has a history marked by genocide, forced displacement, and exploitation, tracing back to the initial encounters with the Portuguese colonizers in 1500 (Le Tourneau 2015). Approximately 5 million Indigenous individuals were believed to inhabit the region around 1500, but estimates suggest a significant decline, possibly reaching 90%, within the first century of colonization (Ribeiro 1996). In the last decades, The Brazilian Institute of Geography and Statistics (IBGE) has provided more accurate numbers and data related to Indigenous people and demographic aspects. IBGE Census of 2022 revealed that Brazil has a rich diversity of Indigenous presence, representing 305 distinct ethnic groups, with a population of approximately 1,693,535 people, constituting 0.83% of the Brazilian population. The Northern and Northeastern Regions host the majority of Indigenous people in Brazil, with the majority residing in Indigenous Lands (ILs) covering over 1 million square kilometers and 13.8% of the nation's territory. Given the historical context, the Indigenous Lands (ILs) in Brazil are critical in safeguarding and preserving traditional people, their resources, forested areas and environmental services.

Indigenous Lands primarily within the Brazilian Legal Amazon, covers over 12% of the country's landmass and significantly contributes to critical ecosystem services (Macedo et al. 2012). In terms of climate regulation, Indigenous Lands (ILs) act as natural carbon sinks, storing an estimated 15 billion tons of carbon, equivalent to 70 years of Brazil's greenhouse gas emissions (Silva-Junior et al. 2013; Nepstad et al. 2009). Notably, research by Silva-Junior et al. (2013) found that deforestation within ILs leads to a significant increase in carbon emissions, underscoring their crucial role in climate change mitigation.

In the realm of Biodiversity Conservation, the Indigenous Lands (ILs) host over 10% of the world's known plant and animal species, including numerous endangered and endemic ones (Davidson et al. 2012; Shimabakuro et al. 2012). Additionally, the research carried out by Silva-Junior et al. (2013) emphasizes the crucial role of ILs in protecting vulnerable and unique species, showcasing their irreplaceable value for biodiversity conservation. Considering the pivotal role in water security, studies have shown that indigenous land management practices significantly contribute to maintaining healthy watersheds and forested areas in the Amazon, with alliances with Indigenous people playing a critical role in conservation (Schwartzman and Zimmerman 2005); in essence, they are active guardians of the forest. As evidence accumulates, conservation scientists are reaching a strong consensus: The future of the Amazon ecosystem depends on protecting and expanding Indigenous Lands. These lands are essential due to their large size and dedicated conservation efforts. In simple terms, ensuring the well-being of the Amazon ecosystem relies on continuing to safeguard and grow Indigenous Lands. (Quintanilla et al. 2022). It is important to highlight that, despite attempts to balance economic activities with nature conservation, the constant growth of the development frontier in the Amazon poses a threat to Protected Areas (PAs).

At first, considered obstacles to exploiting resources, Protected Areas (PAs) now confront a significant challenge from a globalized economy that values land and resources as financial assets (de Lima and d'Hauteserre 2011). This situation weakens the effectiveness of PAs and hampers regional development plans (da Silva and da Silva 2022).

The active correlation between Indigenous Lands (ILs) and forest preservation is evident, as Indigenous land management practices have demonstrated the potential to significantly decrease deforestation rates by up to 66%, compared to areas beyond their jurisdiction (Stickler et al. 2013). This results in a significant reduction in greenhouse gas emissions, making ILs a cost-effective solution for climate change mitigation (Rajão et al. 2012; Silva-Junior et al. 2013; Nepstad et al. 2009). Additionally, protected areas like ILs can act as buffers against illegal logging and resource extraction, further safeguarding vital ecosystems (Schwartzman et al. 2010; Ministerio do Meio Ambiente 2008).

The global importance of ILs extends far beyond Brazil's borders. The carbon stored in these territories contributes significantly to global carbon sequestration, providing benefits for the entire planet (Nepstad et al. 2009). Their biodiversity provides essential genetic resources for future food security and medical advancements (Davidson et al. 2012). Furthermore, the wisdom and traditional knowledge possessed by indigenous communities offers valuable guidance for managing resources sustainably and adapting to climate change (Nassar et al. 2014). Recognizing and protecting Indigenous Lands is not only a matter of human rights and cultural preservation but also a strategic investment in environmental sustainability and global well-being.

Indigenous lands, including the Yanomami one, in the Amazon region, play a critical role in maintaining healthy ecosystems. Rather than solely focusing on the challenges, this chapter explores the Yanomami's resilience in the context of Amazon 2030 and the intricate challenges they face as custodians of their ancestral land. Research by Silva-Junior et al. (2013) shows that deforestation within Indigenous lands leads to significant increases in carbon emissions, emphasizing the crucial role of Indigenous Lands in mitigating climate change. These territories also host over 10% of the world's known plant and animal species, including various endangered and endemic life forms. The practices of Indigenous land management play a crucial role in sustaining healthy forests and watersheds in the Amazon. Collaborations with indigenous communities are vital for the conservation of these ecosystems.

The Brazilian government's role over the past two decades is also evaluated critically, revealing the limitations of top-down approaches that often neglect meaningful engagement with Yanomami knowledge systems and socio-cultural realities. Indigenous self-determination is emphasized as a crucial aspect of a decolonial perspective that dismantles power imbalances inherent in traditional development paradigms. Ethnodevelopment (Bonfil-Batalla, 1982) emerges as a crucial theoretical foundation for a sustainable Amazon in 2030, advocating for a future where Yanomami aspirations and ecological knowledge guide sustainable resource management practices. The chapter explores the complexities of crafting culturally appropriate ethnodevelopment initiatives, including participatory processes, knowledge exchange, and the revitalization of traditional governance structures, and Indigenous protagonism involves the active participation and leadership of Indigenous communities in the stewardship of lands and natural resources (de Lima 2014; FUNAI 2022).

Drawing upon a diverse array of secondary data sources such as academic literature, NGO and government reports, news articles, and participant observation insights, this chapter offers a fresh perspective and nuanced understanding of the Yanomami experience. It aims to demonstrate the impacts of illegal mining in the ILs as well as to indorse the relevance of the Yanomami people to becoming active stewards of their territory and heritage through a transformative vision that intertwines ethnic identity, social equity, economic prosperity, and environmental responsibility. The ultimate goal is to contribute to a vibrant future for both the Yanomami people and the Amazon ecosystem while exploring the complexities of Indigenous resilience in the face of persistent threats.

This chapter addresses a crucial question: How can the Yanomami be empowered and their land safeguarded through the integration of ethnic identity, social equity, economic prosperity, and environmental responsibility?

2 The Yanomami Crisis in Numbers: Unveiling the Scope of an Amazon Tragedy

The Yanomami Indigenous Land is currently experiencing a humanitarian crisis due to the collapse of indigenous healthcare infrastructure and illegal mining activities.

The Socio-environmental Institute (ISA), along with national media outlets, has documented this situation, underscoring the urgency for immediate action.

At the heart of this crisis lies the depletion of natural resources, contamination of water streams, and the spread of diseases (da Silva et al. 2023). Additionally, there is a preventable high mortality rate among Yanomami children, which increased by 29% from 2019 to 2022 compared to the preceding four years (Ramalho 2022; Scherf and Viana da Silva 2023). Because of the systematic dismantling of healthcare services within the territory, Yanomami communities are unable to combat endemic diseases like malaria and infectious illnesses.

Additionally, illegal mining activities in the area have surged significantly. The Hutukara Yanomami Association (HYA) document entitled, 'Yanomami Under Attack' highlights a 46% increase in illegal mining since 2020, with dire consequences for both the environment and the health of Yanomami communities. The report highlights over 16,000 individuals directly involved in illegal mining activities, with a 3,350% increase in mining activity within the territory between 2016 and 2020. Illegal gold miners in the Yanomami area have reached an estimated 20,000 according to the BBC (2021).

Factors contributing to the rise of illegal mining activity in Yanomami lands include the gold rush prompted by soaring global gold prices, loopholes in the gold supply chain resulting in dirtied "conflict gold," weakened environmental and indigenous rights laws alongside lax enforcement, economic difficulties leading individuals towards illegal mining practices, impetuous tech-informed invasions, and inconsistent policies generating a "permissive" mining atmosphere that fosters hope for future legalization.

These activities not only disrupt traditional ways of life but also worsen environmental degradation and pose a severe health threat, such as the contamination of waterways with mercury and the disruption of access to essential healthcare services. There is an undeniable close relationship between Malaria, illegal mining, and deforestation in the Yanomami land, as observed in Fig. 1.

The alarming HYA's report, 'Yanomami Under Attack', also revealed a staggering 309% increase in deforestation linked to illegal mining within the Yanomami territory from 2018 to 2022. This corresponds to a shocking loss of 5,053 hectares of rainforest, with 1,782 hectares succumbing to rampant exploitation in 2022 alone.

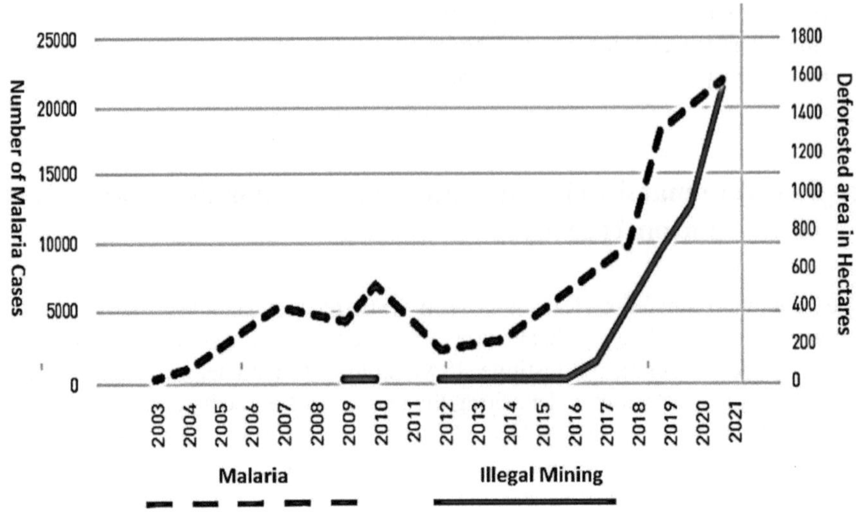

Fig. 1 Malaria spread versus illegal mining increasing from 2003 to 2021 in the Yanomami land

The extent of environmental devastation caused by illegal mining is notable. According to data from MapBiomas, there has been a 98% increase in the area affected by illegal mining activities in Brazil between 2010 and 2021, disproportionately impacting protected areas, including Indigenous Lands. The Yanomami land itself has witnessed a significant rise in deforestation and land degradation due to illegal mining, affecting over 3,272 hectares by the end of 2021. This environmental destruction poses a severe threat to the health and well-being of Yanomami communities, disrupting their traditional subsistence practices and exposing them to increased health risks (Ramalho 2022) (Figs. 2 and 3).

To fully understand the current crisis in Yanomami Indigenous Land, its historical context must be considered. Organizations like the Hutukara Yanomami Association (HYA), ISA, and various NGOs have accused the Bolsonaro administration of promoting increased mining activities in the Amazon and neglecting Yanomami communities, worsening a longstanding issue. The 2020 *Pró-Mapa* decree exemplifies the administration's approach to encourage mining in the Amazon Legal Region, leading to community destruction and institutional neglect.

The COVID-19 pandemic is making the existing crisis worse. NGOs and the media have accused the Ministry of Health of mishandling the COVID-19 response, stating that the use of chloroquine has led to a shortage of supplies meant for treating Indigenous communities. According to them, this has left Yanomami communities even more at risk of the illness. The Yanomami crisis highlights the intersection of environmental destruction, institutional neglect, and exploitation.

Addressing this situation requires a comprehensive approach that prioritizes the immediate provision of healthcare services, the dismantling of illegal mining operations, and the restoration of trust through meaningful engagement with Yanomami

Fig. 2 In the heart of the Yanomami people's ancestral land, a disconcerting scene unfolds—a mining camp operating in blatant disregard of territorial rights. The Yanomami draw attention to a troubling connection between the present environment and the heightened presence of illegal miners, sparking worries about both their safety and the preservation of their rich cultural heritage. *Source* SocioEnvironmental Institute, ISA (2022)

communities (Ramalho 2022; Scherf and Viana da Silva 2023). The international community must hold institutions accountable and advocate for indigenous rights and environmental sustainability. Only through coordinated efforts can we alleviate the suffering of the Yanomami people and ensure a future where their cultural heritage and traditional way of life can thrive.

3 The Yanomami Humanitarian Crisis, Illegal Mining and Government Interventions

The relentless pursuit of gold and minerals in Brazil's Indigenous Lands serves as a stark reminder of the monumental task facing authorities dedicated to stopping its destructive path. This illegal activity leaves behind irreparable damage, causing rampant deforestation (Kayet et al. 2021; Siqueira-Gay et al. 2020; dos Santos et al. 2021), polluted waters (Pestana et al. 2022), air pollution (Rentier and Cammeraat 2022), eroded soils (Awotwi et al. 2021), and disrupted ecosystems (Mallett et al. 2021). Beyond the environmental toll, illegal mining casts a dark shadow on society, fostering exploitative labor practices (Murta et al. 2021), fueling illicit arms and drug trades (Sonter et al. 2017), and sparking conflicts between local communities and ruthless invaders (Mataveli et al. 2022).

Fig. 3 Mining activity along the Uraricoera River in Yanomami territory, Roraima, captured in January 2022. *Source* SocioEnvironmental Institute, ISA; and Hutukara Associação Yanomami and Associação Wanasseduume Ye'kwana Report, January 2022

Rampant deforestation and illegal mining in the Legal Amazon, especially within protected areas, contradict various international agreements and commitments. This undermines the UN's Sustainable Development Goals and Brazil's pledges at COP26 to tackle climate change. It also violates ILO Convention 169, which guarantees indigenous rights and participation in managing their lands. These violations pose irreversible risks to the environment, indigenous communities, and even the reputation of legal mining in the Amazon, a vital source of income for many municipalities.

Mining in the Amazon is an unavoidable catalyst for environmental and social upheaval (Sonter et al. 2017). It disrupts landscapes, alters natural processes, and triggers a cascade of social conflicts. Indigenous communities bear the brunt of these conflicts, facing land invasions and the devastating consequences of environmental damage. Deforestation, polluted waterways, and mercury contamination plague the region, poisoning wildlife, humans, and entire ecosystems (Nepstad et al. 2014).

The "gold rush" in the mid-1980s left a harmful legacy, exposing Indigenous populations to unprecedented levels of mercury (Berzas Nevado et al. 2010). Artisanal mining alone contributes a significant 37% of global mercury emissions (Gibb and O'Leary 2014). While exact figures for mercury pollution in the Amazon are debated (Sonter et al. 2017), estimates suggest an alarming 1:1 ratio of gold and mercury used. Unofficial data from the 1980s indicates gold extraction between 1,000 and 2,000 tons, implying a matching mercury footprint (Brabo et al. 1999). For indigenous Amazonian communities near gold mines, the risk of mercury poisoning is the highest globally, especially as their primary protein source, fish, becomes contaminated by the toxic metal (Barbosa et al. 1995; Wheatley and Wheatley, 2000).

The Yanomami's ancestral home lies deep in the heart of Brazil's Amazon, stretching across Roraima and Amazonas states, and Venezuela, sheltering an Indigenous population of 97,320 people in 2022, with 16,872 Yanomami individuals in Roraima and 27,152 individuals in Amazonas (Agencia IBGE 2023). For decades, the Yanomami land has faced continuous threats from illegal resource extraction, with Ajarani I standing as an example of resilience. The village was established in the 1960s and became a sanctuary for those displaced by the agricultural frontier's relentless expansion. Today, the village grapples with the detrimental consequences of illegal mining, a pervasive threat to the Yanomami way of life (Funai's Indigenous System for Information, Accessed on December 18, 2023).

An example of the devastating threat of illegal mining in the Yanomami Indigenous Land is the Uraricoera River area, a natural gold concentrator with alarming levels of mercury (Fiocruz 2016; Vega et al. 2018). Despite its protected status since 1992, mercury contamination poses severe and lasting health risks, particularly its methylmercury form (Sonter et al. 2017; Gibb and O'Leary 2014).

Efforts to combat these problems require a comprehensive approach that respects Indigenous people's cultural values, promotes sustainable development, and enforces environmental laws. International collaboration and support from policymakers, government authorities, NGOs, and the broader community are essential to ensure the Yanomami's survival, preserve their culture, and secure their future (Fig. 4).

The Yanomami community has faced a severe threat from illegal mining, leading to heightened violence, pollution of rivers, contamination of crops, and the recruitment of young individuals. A 46% increase in illegal mining activity from 2020 to 2021 serves as a clear indicator of the escalating danger, representing a direct impact on 273 communities, 56% of the Yanomami population comprising over 16,000 individuals (Hutukara Yanomami Association 2022).

The Yanomami's initial contact with non-indigenous groups intensified with the advent of government initiatives such as the North Perimeter Highway (BR-210) in the 1970s. This highway, carved through Yanomami territory, had a chilling impact on their health, and the Radam project uncovered the mineral-rich potential of Yanomami land, leading to environmental desecration and a surge in illnesses. Recent increases in malaria cases in Palimiu community, with over 1,800 cases in 2020 alone, serve as a reminder of the challenges (Hutukara Yanomami Association 2022).

Illegal mining activities surged in the 1980s, exacerbating environmental damage and disease spread, with miners persistently returning despite removal efforts. Non-Governmental Organizations (NGOs) and the media have lodged accusations asserting that the policies implemented by the Bolsonaro government have exacerbated the prevailing crisis. This exacerbation is notably associated with a marked increase in child mortality rates attributed to preventable causes such as mercury poisoning, malnutrition, and hunger (Nicas 2023; Bridi 2023; Phillips 2023). An

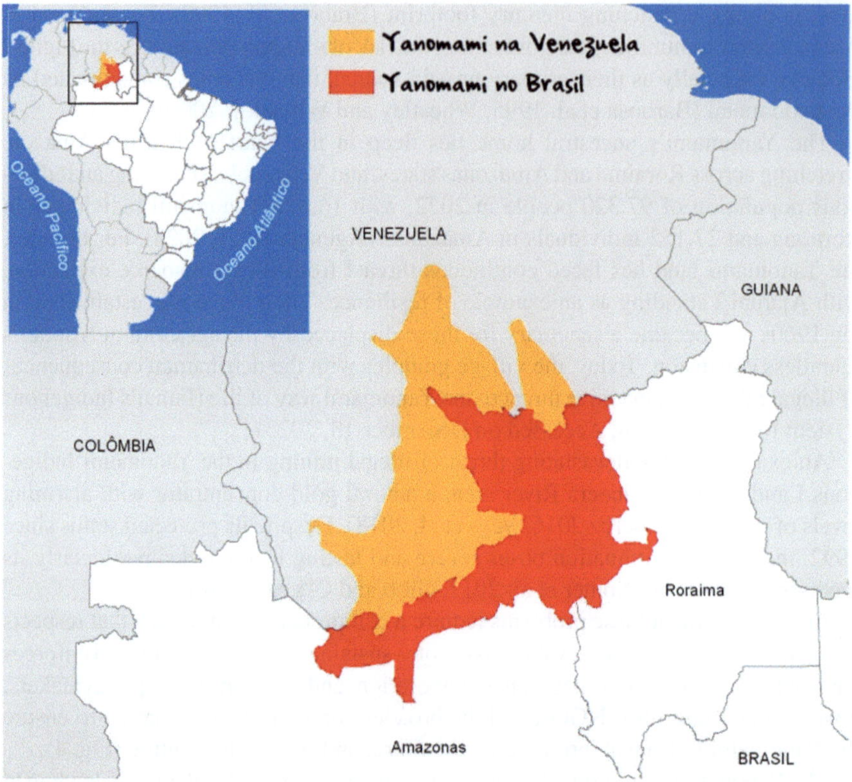

Fig. 4 Map of Yanomami Land divided between Roraima and Amazonas states in Brazil (in red), and its extension into Venezuela (in orange). *Source* SocioEnvironmental Institute, ISA, 2009, and EcoSig of the RAISG

investigation into potential genocide against the Yanomami underscores the gravity of the situation. Contrarily, the Bolsonaro administration vehemently rejected these allegations, dismissing them as 'utterly baseless.

In February 2023, during Lula's government administration, the Brazilian government collaborated with the Armed Forces to launch a joint operation to eliminate illegal miners and achieve measurable outcomes. However, sustainable solutions require an unwavering commitment to upholding indigenous rights, enforcing environmental laws, and addressing poverty and exploitation.

Mercury, a powerful neurotoxin, infiltrates human bodies, disrupting their delicate balance. Used in gold extraction for its effectiveness, it poses a threat to both the environment and the health of prospectors. Indigenous communities unknowingly become victims, exposing them to mercury absorption through inhalation of vapors released during gold manipulation (Larini 1997; Gibb and O'Leary 2014). Inhaling vapors causes neurological disorders, tremors, and digestive distress, with polluted rivers and crops contributing to mercury absorption. The Minamata Convention on Mercury (Brasil 2018) and international human rights laws, specifically those protecting indigenous minorities and cultural heritage, should be considered in protecting the Yanomami community.

Article 20, IX of the Brazilian Constitution grants ownership of extracted gold to the federal government, with gold extraction illegal on all indigenous lands, including Yanomami territory. The lack of official mining activity in Roraima, where most of the extraction occurs, further underscores its illegality. The state's mineral production was minimal, collecting only US$245 in taxes. Research on the direct clinical consequences of mercury exposure on the Yanomami community's health and well-being remains limited, despite substantial evidence pointing to elevated mercury levels in Yanomami communities in Roraima State (Vega et al 2018).

Figure 5 shows the location of the Indigenous Lands (ILs) with the most illegal mining activity in 2020 in the Legal Amazon. The Kayapó land is marked in red, Munduruku land in yellow and Yanomami land in blue. Figure 6 illustrates the locations of primary illegal mining sites, unauthorized airstrips for small aeroplanes, and areas within the Yanomami Land that are under threat of health issues.

Information from the Special Secretariat of Indigenous Health (SESAI) in 2023 reveals the presence of 366 Indigenous communities (settlements) in the Yanomami Special Indigenous Health District (DSEI), with a spatial distribution closely resembling the patterns observed in a visual survey using satellite images

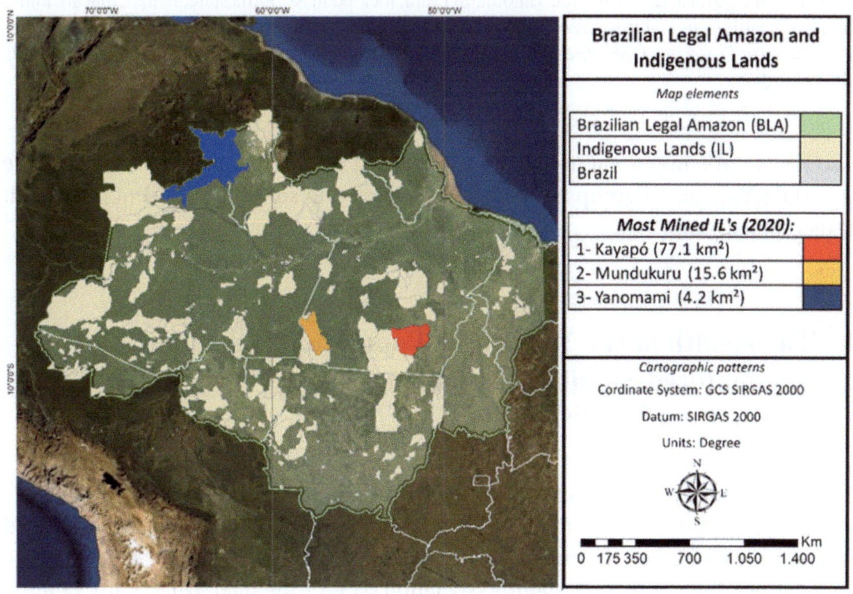

Fig. 5 This map covering the Brazilian Legal Amazon shows the location of the Indigenous Lands (ILs) with the most illegal mining activity in 2020. The Kayapó land is marked in red, Munduruku land in yellow and Yanomami land in blue. *Source* Mataveli et al. (2023), p. 3

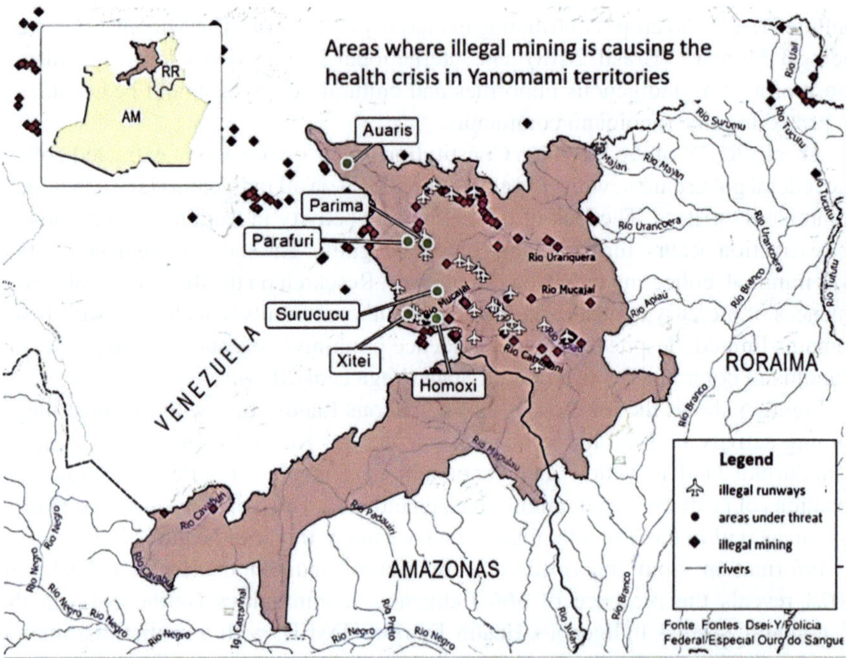

Fig. 6 The map illustrates the locations of primary illegal mining sites, unauthorized airstrips for small airplanes, and areas within the Yanomami Land, located in Roraima state, Brazil, that are under threat of health issues. https://Amazon.real.com.br/criancas-yanomami-2/ *Source* Special Indigenous Sanitary District—DSEI /Federal Police, adapted from Amazon Real by Scherf and Viana da Silva (2023)

(see Fig. 7). It's important to note, as highlighted by DSEI, that the nomadic nature of certain Yanomami groups and challenges related to survey equipment may result in incomplete data.

4 The Health of the Yanomami Population Has Encountered Significant Challenges and Threats Throughout Its History

This section provides an overview of the Yanomami population and the challenges they have faced throughout the last few decades. The information has been sourced from various internet resources, including FUNAI (National Indigenous People Foundation), Hutukara Yanomami Association (HYA), the Inter-American Commission on Human Rights (IACHR), Human Rights Watch, academic journals, and reports.

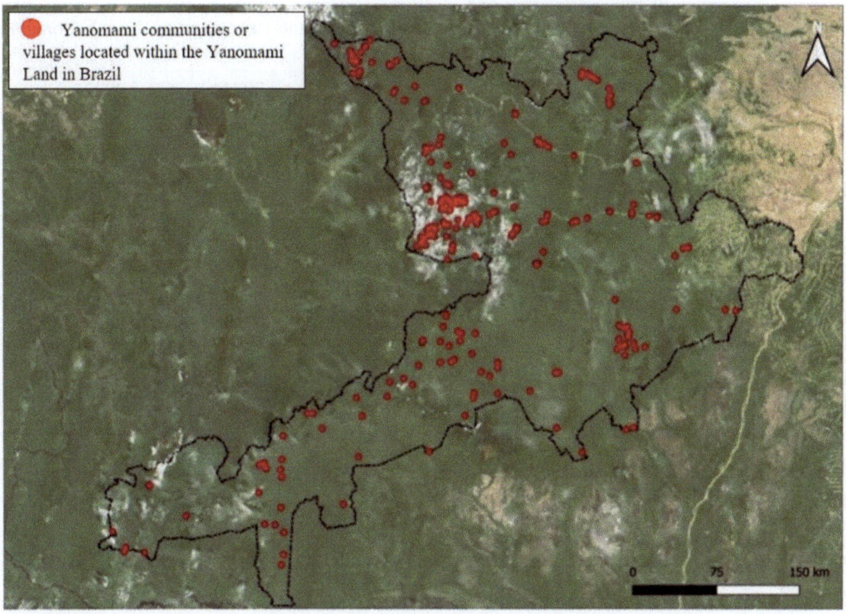

Fig. 7 Map of the Yanomami settlements' locations. *Source* SESAI/DSEI (2023)

The data reveals significant mortality rates among the Yanomami population, with estimations suggesting that thousands of Yanomami people have died since 1940. The highest mortality rates occurred during the 1960s and 1970s. Children under five years old and young adults have been the most vulnerable to death due to infectious diseases and malnutrition (refer to Table 1). A range of factors have contributed to the high mortality rates, including infectious diseases, violence, environmental degradation, and the limited access to healthcare. Despite legal protections and government interventions, the Yanomami population continues to face significant threats to their health and survival (Grupo Geo-Yanomami 2023). Illegal mining and activities that cause harm to the environment continue to pose risks to the Yanomami community (Vegas et al. 2018).

A recent article by Estadão, published on 19 June 2023, discusses how the Yanomami population has been negatively impacted by the invasion of their land, particularly by illegal miners, over the course of several decades. This has resulted in a variety of negative consequences, such as increased violence within the territory, recruitment of young individuals, contamination of rivers as well as poisoning of humans, animals, and crops due to waste from illegal mining. Additionally, the number of instances of diseases such as malaria, sexually transmitted infections, and respiratory infections has been significantly elevated. After examining related literature and institutional reports (Andrade et al. 2022), it is evident that the gravity of this situation can be further understood by examining the following impacts:

Table 1 Number of deaths and mortality rate (per 1,000 inhabitants) by age group, Dsei Yanomami, 2018 to 2022

Age Group	Population	2018	2019	2020	2021*	2022*
Less than 1 year	922	118 (12.80%)	125 (13.56%)	126 (13.67%)	69 (7.48%)	67 (7.27%)
1 to 4 years	4917	27 (0.55%)	32 (0.65%)	36 (0.73%)	51 (1.04%)	32 (0.65%)
5 to 9 years	5058	8 (0.16%)	9 (0.18%)	21 (0.42%)	8 (0.16%)	11 (0.22%)
10 to 14 years	4391	5 (0.11%)	6 (0.14%)	8 (0.18%)	6 (0.14%)	8 (0.18%)
15 to 19 years	3551	7 (0.20%)	16 (0.45%)	20 (0.56%)	14 (0.39%)	15 (0.42%)
20 to 39 years	7538	24 (0.32%)	20 (0.27%)	43 (0.57%)	31 (0.41%)	24 (0.32%)
40 to 59 years	3302	16 (0.48%)	16 (0.48%)	26 (0.79%)	26 (0.79%)	13 (0.39%)
60 to 79 years	1175	26 (22.10%)	29 (24.70%)	39 (33.20%)	32 (27.20%)	24 (20.40%)
80 years or older	163	5 (30.70%)	6 (36.80%)	13 (79.80%)	12 (73.60%)	15 (92.00%)
Total	31,017	236 (0.76%)	259 (0.84%)	332 (1.07%)	249 (0.80%)	209 (0.67%)

Note * Data for 2021 and 2022 were preliminary
Source Yanomami Expedition Report, Ministry for the Health, Brasil 2023

- Violence: The invasion of Yanomami territory by illegal miners has resulted in escalated violence, including rape, murder, and kidnapping.
- Recruitment of young people: Miners are known to recruit Yanomami youth to work in the mines under hazardous and exploitative conditions.
- Contamination of rivers: Illicit mining operations employ mercury to extract gold which can result in the contamination of rivers and streams. As a result, humans and animals can ingest this mercury and suffer from neurological damage and other health issues.
- Poisoning of crops: In addition to the contamination of rivers and streams, illegal mining operations produce waste that can be detrimental to crops, leading to food shortages and malnutrition.
- Increase in diseases: The invasion of Yanomami territory by illegal miners has also resulted in a significant increase in the prevalence of diseases such as malaria, sexually transmitted infections, and respiratory infections.

The devastating impact of the invasion of Yanomami territory by illegal miners cannot be overstated, resulting in loss of life, environmental destruction, and the spread of disease. Ongoing debates and controversies persist regarding Bolsonaro's responsibility for genocide or, as alleged by the media and, particularly by political opponents, 'ethno-genocide' of Indigenous populations. Considering the recurring Yanomami death tolls over the decades, such claims appear to be highly misleading and require careful examination of international legal definitions, historical context, and specific intent. Labeling Bolsonaro's actions as such has significant legal implications and demands substantial evidence demonstrating a willful intent to harm or exterminate an identifiable ethnic group. According to reports issued by CNN Brasil and *Revista Oeste* in February 2024, Brazil's Ministry of Health recorded a 6% increase in the number of Yanomami deaths during the first year of President Lula's administration, led by the Leftist Workers' Party, PT, compared to the previous year under President Jair Bolsonaro. By the end of July 2024, both Brazilian and international media outlets reported that President Lula had suspended the Ministry of Health from publishing new reports on Yanomami deaths, raising serious concerns about data concealment, and government transparency and accountability.

While some argue that Bolsonaro's policies have disproportionately affected Indigenous communities, it can be challenging to prove deliberate intent beyond a reasonable doubt. It is essential to recognize that the struggles faced by Indigenous Brazilians encompass historical injustices, systemic inequalities, and various ecological and economic pressures that go beyond the actions of one political figure (da Silva et al. 2023). During the Military Dictatorship era, various diseases such as malaria, measles, sexually transmitted diseases (STDs), whooping cough, and malnutrition decimated the Yanomami people in several villages between 1970 and 1980, primarily (ENS/Fiocruz 2023). The report from the National Truth Commission (CNV), released on December 10, 2014, also exposes how the Brazilian government was responsible for violence in indigenous territories.

Rather than engaging in a polarized debate focused solely on classifying a situation as 'genocide' or 'rhetoric,' a more constructive approach would involve in-depth, evidence-based analysis. To accurately assess the complexities involved, it is essential to employ an impartial, scholarly perspective informed by established academic standards. This helps prevent the dissemination of potentially misleading or manipulative narratives. This rigorous approach is indispensable for a precise comprehension of the intricacies involved, mitigating the risk of perpetuating 'sophistic narratives'.

According to the Federal Deputy Ricardo Salles (PL-SP), former Minister of the Environment, denied that the government of Jair Bolsonaro (PL) has committed genocide against the Yanomamis. He claimed that the difficulties faced by the indigenous population have a long history, constituting a historical problem spanning many decades (Carta Capital 2023), "In the Yanomami [land], there was a peak in mining activities in the 1980s, during the first gold rush in the Amazon. After being launched by the Federal Government in 1990, the Free Jungle Operation removed tens of thousands of illegal miners from these territories" (Basta 2023, p. 2).

The Haximu massacre in 1993, where 16 Yanomami Indians and one gold miner were killed, was the first and only genocide case tried in Brazil. It marked a turning point in the conflict between invaders and the Yanomami. The Federal Public Ministry filed charges against 24 gold miners for genocide, conspiracy to commit genocide, illegal mining, smuggling, concealment of a corpse, property damage, and criminal association. Only five of the miners had been fully identified. Three years later, in

December 1996, the Federal Court convicted the group of attempting to exterminate the ethnic group, not just homicide (Andrade 2023).

The Table 1 reveals the number of death and mortality rate among Yanomami by age group from 2018 to 2022 with data collection done by the DSEI Yanomami, belonging to the Brazilian Ministry for the Health.

The percentages in the table represent the mortality rate for each age group in each year. The mortality rate is calculated by dividing the number of deaths by the population and multiplying by 1,000. For example, the mortality rate for children under 1 year old in 2018 was 12.80%. This means that there were 12.80 deaths for every 1,000 children under 1 year old in 2018. Key findings: The highest number of deaths occurred in 2020 (n = 332); the highest mortality rate was among children under 1 year old in 2018 (12.80%); the mortality rate for all age groups decreased from 2018 to 2022.

5 The Yanomami and COVID-19: A Multifaceted Tragedy

Indigenous communities worldwide, including the Yanomami people of Brazil, have been disproportionately affected by the global coronavirus pandemic. Organizations such as the Socioenvironmental Institute, ISA; the Survival International; the Indigenous Peoples' Articulation of Brazil, APIB; and FIOCRUZ highlight the pre-existing vulnerabilities that make the Yanomami particularly at risk to COVID-19. These vulnerabilities include decades of marginalization resulting in malnutrition and other illnesses, inadequate healthcare infrastructure, and the encroachment of mining and logging activities that exacerbate the risk of disease transmission.

The Yanomami communities in Brazil have been hit hard by the COVID-19 pandemic, particularly in villages near illegal mining activities, with at least 1,241 confirmed cases and a 23% mortality rate among those cases (Basta 2023; Fellows et al. 2021). The impact goes beyond the tragic loss of life, disrupting traditional ways of life and adding to tensions within and between communities. The pandemic's impact is compounded by pre-existing vulnerabilities, healthcare disparities, governmental failure, and cultural complexities (Fellows et al. 2021). Only 70% of eligible Yanomami have been fully vaccinated, and only 30% of communities have access to basic healthcare, with a severe shortage of medical personnel. Environmental degradation from illegal mining activities exacerbates the risk of disease transmission and further threatens the health and well-being of the Yanomami people (Grupo Geo-Yanomami 2023). Addressing this crisis requires a multifaceted approach prioritizing culturally sensitive interventions, improving access to healthcare, and protecting the Yanomami's rights to health and ancestral territory.

The Brazilian government's response to the COVID-19 crisis in Indigenous lands has been criticized for being insufficient and poorly implemented, despite some measures such as increasing the availability of vaccines and medical personnel. Yanomami organizations have worked to raise awareness and implement initiatives to prevent the spread of the virus, and NGOs and civil society groups have provided essential supplies, medical care, and educational resources to support the Yanomami during the pandemic.

Raíza Vieira, a co-author of this chapter, spent nearly a year and a half providing dental assistance within the healthcare domain to the Yanomami people while serving as a member of the Special Indigenous Sanitary District (DSEI) for the Yanomami. With her firsthand exposure to the complex vulnerabilities faced by the Yanomami due to a land besieged by invaders, illegal mining, and deforestation, Vieira offers a participant observer's perspective in this section, fortified by her own brush with death after contracting COVID-19 while serving the Yanomami. Vieira's account includes personal anecdotes and insightful analysis, which lends unparalleled depth and authenticity to her testimony.

The DSEI Yanomami is a decentralized hub of action within the Indigenous Health Care Subsystem (SasiSUS), which is a unique service model designed for vibrant ethno-cultural spaces that transcend state boundaries. The DSEI reorganizes the existing network to deliver rationalized and high-quality healthcare measures, constructing vital administrative and managerial infrastructure for providing quality assistance. Social Control acts as a vigilant chorus, ensuring accountability and community engagement (Brasil 2021).

Brazil's innovative approach to indigenous healthcare involves 34 strategically-placed Special Indigenous Health Districts (DSEIs) within Indigenous Lands. These districts transcend rigid state borders, reflecting the dynamic movement of indigenous communities. A dedicated infrastructure of indigenous health units, base poles, and Indigenous Health Houses (CASAI) ensures crucial healthcare access throughout these districts. The DSEI model is a paradigm shift, empowering indigenous communities, promoting culturally-sensitive care, and guaranteeing the right to quality healthcare. This transformative approach holds immense promise for a healthier future throughout Brazil's indigenous populations (Brasil 2021). The included pictures offer a glimpse into the context where Vieira worked in Yanomami territory in 2021 (Figs. 8, 9, 10, 11, 12 and 13).

The following picture shows Yanomami children playing in the village of Parafuri in the Yanomami Territory. The children are wearing traditional Yanomami clothing, The children are smiling and laughing, and they seem to be enjoying themselves. The photo of Yanomami children is a reminder of the importance of preserving indigenous cultures and traditions. The Yanomami are one of the most isolated and vulnerable indigenous groups in the world, and they are under threat from illegal mining and other harmful activities. The children in this photo are the future of the Yanomami people. It is important to ensure that they have the opportunity to grow up healthy and happy and that they can preserve their culture and traditions (Fig. 14).

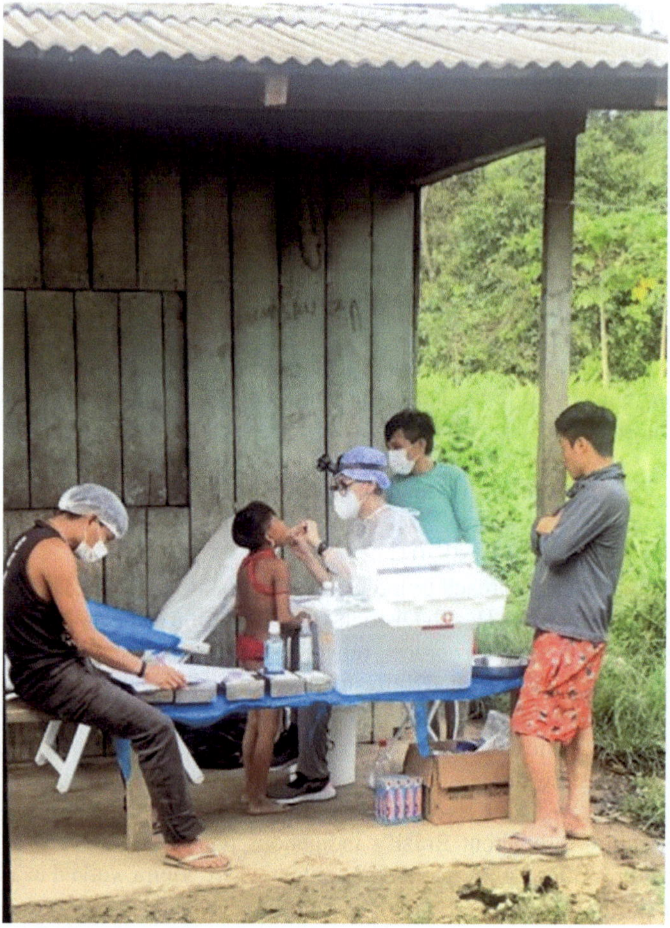

Fig. 8 Health care service at the HAXIU community Indigenous health post, Yanomami Territory, 2021– personal collection (@RaizaVieira, 2021)

6 Brazilian Policy for Territorial and Environmental Management of Indigenous Lands (PNGATI)

The Brazilian government prioritizes Indigenous healthcare, investing substantial funding in the Secretariat for Indigenous Health (SESAI), the dedicated public health system serving indigenous populations. SESAI has over 22 thousand healthcare professionals, with 52% of them being indigenous, and promotes primary healthcare and sanitation actions in a participatory and differentiated manner, respecting the epidemiological and sociocultural specificities of these communities (Gouveia 2024). Established in 2010, the Indigenous Health Secretariat (SESAI) under the Ministry

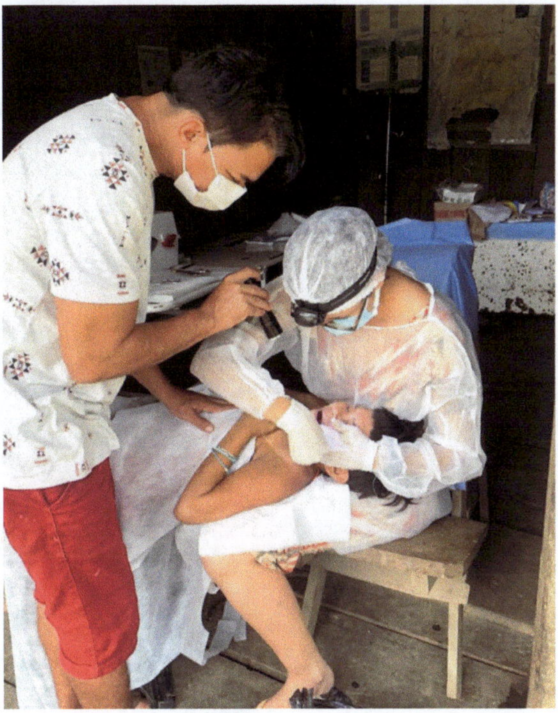

Fig. 9 Health care service at the HAXIU Indigenous community health post, Yanomami Territory, 2021– personal collection (@RaizaVieira, 2021)

of Health serves more than 762 thousand Indigenous people residing in villages throughout the country (Brasil 2024).

The government has actively worked to empower Indigenous economies, recognizing the importance of economic autonomy, through initiatives such as the National Indigenous People Foundation (FUNAI), which provides a range of services and seeks to promote ethnodevelopment initiatives and policies as well as sustainable economic utilization of indigenous lands. Additionally, the Brazilian Policy for Territorial and Environmental Management of. Indigenous Lands (PNGATI)[1] recognizes ecotourism and ethnotourism as potential tools for Indigenous self-development; the government has allocated approximately USD 6 million to activities such as rural production, fishing, and agricultural machinery acquisition, and established internet points in 197 Indigenous communities (Brasil 2022). PNGATI seeks to improve the living conditions of indigenous populations and foster their full participation in the decision-making process regarding the sustainable development of natural resources (Bavaresco and Menezes 2014; FUNAI 2013). The policy establishes guidelines to address the challenges faced by indigenous communities in the context of development, such as the creation of a communication channel for public consultation,

[1] PNGATI—Política Nacional de Gestão Ambiental e Territorial de Terras Indígenas [National Policy on Indigenous Territorial and Environmental Management—PNGATI]. Decreto Presidencial nº 7.747 de 05 de junho de 2012.

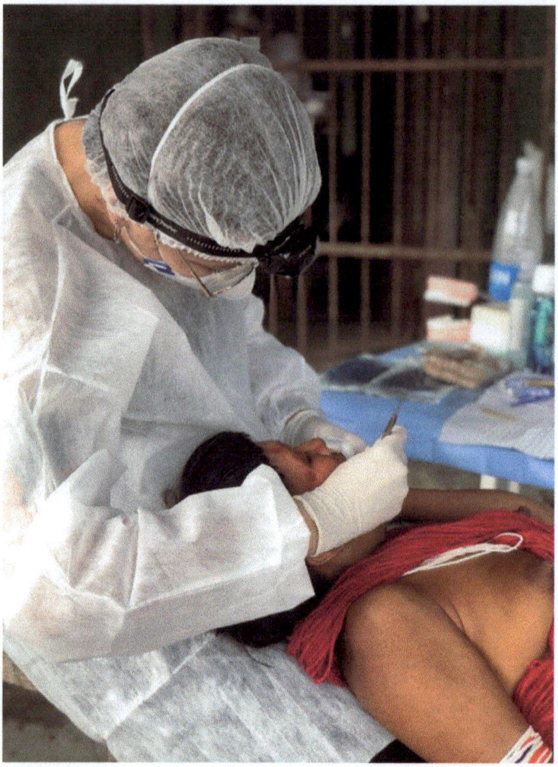

Fig. 10 Health care service at the SURUCUCU Indigenous community health post, Yanomami Territory, 2021. Personal collection (@RaizaVieira, 2021)

environmental assessments, and the elaboration of an environmental balance sheet (Le Tourneau 2015).

The Brazilian government acknowledges collaboration and respect for Indigenous autonomy, reinforcing its commitment through the 2012 National Policy on Indigenous Territorial and Environmental Management (PNGATI). PNGATI aims to promote sustainable management of their lands and territories through Indigenous community partnerships, upholding their cultural autonomy and legal framework. The government reinforces its commitment to the collaborative approach within PNGATI by empowering Indigenous communities to generate income and fostering self-sufficiency, security, dignity, and freedom (Brasil 2012; FUNAI 2013; Le Tourneau 2015). Additionally, strengthened security and legal oversight for Indigenous reservations have led to a significant decrease in violent crimes and agrarian conflicts in rural and Indigenous areas. Overall, the government continues to prioritize and invest in Indigenous communities, emphasizing their economic, social, and environmental needs, and collaborating with them to improve their quality of life.

Fig. 11 Health professionals on their way to the HATHOU indigenous community to carry out nutritional surveillance and vaccination, SURUCUCU, Yanomami Territory, 2020– personal collection (@RaizaVieira, 2021)

Fig. 12 Healthcare professionals from the EMSI, the Indigenous Multi-Disciplinary Health Team, visiting the PALIMIU community in the Yanomami Territory. Personal collection (@RaizaVieira, 2021)

Fig. 13 Healthcare professionals from the FUNAI, the Brazilian Indigenous Affairs Agency, and members of the Yanomami community, attempting to access the XEXENA community by land–personal collection (@RaizaVieira, 2021)

Fig. 14 Yanomami children playing in the village of PARAFURI—Personal collection (@RaizaVieira, 2021)

7 Ethnodevelopment: A Conceptual Framework for Empowering and Sustaining Ethnic Communities

Ethnodevelopment is a conceptual framework that prioritizes inclusion, empowerment, self-expression, and self-determination for ethnic minority communities (de Lima 2014). By focusing on these key principles, ethnodevelopment aims to assess the impact of economic initiatives on the cultural resilience, livelihoods, and economic opportunities of these communities (de Lima et al. 2020). This framework explores various themes such as gender, authenticity, indigenous knowledge, tradition, and the commodification of culture (de Lima et al. 2016).

The ethnodevelopment origins and core principles can be outlined as follows. Historically, the concept of ethnodevelopment emerged in 20th-century Mexico in response to ecological distributive conflicts and cultural destruction caused by development policies (Rodrigues 2022). Mexican anthropologist Guillermo Bonfil Batalla played a pivotal role in initial discussions on the subject, advocating for an authentic model based on cultural respect and self-determination (Rodrigues 2022). He emphasized the need for "consolidation and expansion of the scope of their own culture by strengthening the autonomous decision-making capacity of a culturally differentiated society to guide its development" (Bonfil-Batalla 1982, p. 33).

Ethnodevelopment was developed by Bonfil-Batalla and Stavenhagen, as it prioritizes respect for local ethnic identities as a fundamental pillar for the development of traditional societies and their territories. It emphasizes the unique perspectives and mindsets of these communities regarding the world, nature, and natural resources, particularly in the context of territorial management. In Latin America, especially Brazil, ethnodevelopment offers a potential transformation in public governance and territorial management, as indigenous and traditional communities often face historical marginalization and lack access to equitable opportunities and decision-making power.

As for ethnodevelopment framework applied to the context of community economy, it has to do with alignment with the social economy model, distinct from the wage-based market economy. It operates with its own legal and economic instruments, often resembling fair trade principles (Tarrega 2022). However, ethnodevelopment stands out due to its focus on cultural identity and self-managed enterprises, fostering autonomy and escaping heteronomous conditions (de Lima et al. 2016). This model integrates the concept of social economy and its legal frameworks with the additional requirement of a cultural foundation, building upon existing fair trade standards (Tarrega 2022).

Ethnodevolpment as presented in the literature seeks to integrate 'world views', 'cultural assets', and 'community values' with 'fair trade'. This has to do with the integration of ethnodevelopment projects into the fair trade movement, by offering a unique opportunity to build a network of local structures and experiences (Stavenhagen 1985; Tarrega 2022). This network can ensure diversity, cultural preservation, and empowerment of traditional communities, particularly indigenous and Quilombola[2] groups. From a legal perspective, exploring the role of fair trade and solidarity as market integrators and facilitators for ethnodevelopment proposals holds significant potential for the future of these communities (de Lima et al. 2016).

8 Applying Ethnodevelopment to the Yanomami: Challenges and Dilemmas

The Yanomami people face a multitude of threats including illegal mining, land invasion, violence, diseases, and deforestation (da Silva et al. 2023). In this context, applying ethnodevelopment, a concept emphasizing self-determined development based on indigenous knowledge and values presents both opportunities and challenges.

Challenges:

- Power imbalances: The Yanomami often lack political and economic power compared to miners, ranchers, and government actors. This can make it difficult for them to control their development and resist external pressures.
- Internal conflicts: Yanomami communities may have diverse interests and priorities, leading to internal disagreements about development projects and resource use.
- Cultural preservation: Balancing economic development with cultural preservation is a delicate act. Some development activities may unintentionally disrupt traditional practices and social structures.
- Sustainability: Ensuring that development projects are environmentally sustainable and don't further degrade the Yanomami's fragile ecosystem is crucial.
- Corruption: Corruption within government agencies, agents, stakeholders, etc. can undermine ethnodevelopment efforts and divert resources away from the Yanomami.

Dilemmas:

- Trade-offs between immediate needs and long-term sustainability: Addressing urgent needs like healthcare and food security may require compromises with long-term environmental goals.
- Balancing self-determination with external support: While self-determination is central to ethnodevelopment, some communities may need external support for certain projects or to address immediate crises.

[2] Quilombolas are Brazilian communities formed by the descendants of escaped African slaves. These communities, historically called "quilombos," were places of freedom and resistance during the era of slavery. Today, quilombolas fight for their land rights, social equality, and the preservation of their unique cultural heritage.

- Accommodation versus resistance: Deciding how much to accommodate external pressures like mining concessions or infrastructure projects can be a difficult choice.

Overall, applying ethnodevelopment to the Yanomami context is a complex and ongoing process. It requires careful consideration of the challenges and dilemmas involved, as well as ongoing dialogue and collaboration between the Yanomami people, FUNAI, and other stakeholders (Oliveira et al. 2020).

However, since the implementation of PNGATI and PGTAs (Territorial and Environmental Management Plan) within the framework of this policy as grassroots territorial management tools and guidelines in Indigenous lands in Brazil, an effective ethnodevelopment model has been established. PGTAs employ participatory approaches like 'ethno-mapping' and 'ethno-zoning'. Ethno-mapping involves indigenous communities charting areas of environmental, socio-cultural, and economic significance using their deep knowledge of the land. Ethno-zoning builds upon this, allowing communities to plan the use of these areas according to their needs and priorities (Porto Júnior et al. 2021). By integrating indigenous expertise and fostering collaborative planning, PGTAs lay the groundwork for stronger indigenous voices, self-determination, and improved well-being in the Amazon.

In the legal Amazon, there are 385 indigenous territorial planning instruments in place for the implementation of PNGATI, and Table 2 shows them. The most common instrument is the "Without instrument" category, followed by the "PGTA" instrument. It is important to note that this table only shows the number of indigenous territorial planning instruments and not the area of land they cover (FUNAI 2021–2023). It is also important to remember that this data is from 2019 and may not be current.

By prioritizing Yanomami self-determination, cultural preservation, and environmental sustainability, ethnodevelopment can offer a promising approach to addressing the many threats facing this indigenous group. It is important to note that there is no one-size-fits-all solution to the Yanomami's problems. The most effective approach will depend on the specific circumstances of each community and the ongoing dialogue between the Yanomami people and external actors.

Table 2 Number and area of indigenous territorial planning instruments in the legal Amazon (2019)

Planning tool for the implementation of PNGATI	Number of indigenous lands
PGTA	111
Ethno-mapping or Ethnozoning	39
Life plans	13
Others	19
Without instrument	203
Total	**385**

Source FUNAI 2021–2023 and TNC complement—preliminary results. Porto Junior et al. (2021)

Examples of Applying Ethnodevelopment:

- Community-based conservation projects: Yanomami communities can manage their protected areas, monitor wildlife populations, and develop sustainable resource use practices.
- Traditional medicine and healthcare: Integrating traditional Yanomami healing practices with Western medicine can improve healthcare access and outcomes.
- Education and cultural revitalization: Supporting Yanomami control over education and cultural programs can strengthen their identity and language.
- Economic development initiatives: Yanomami communities can develop income-generating activities like ecotourism or sustainable agriculture that respect their culture and environment.
- Indigenous entrepreneurship: Fostering small-scale, culturally appropriate economic activities like *artesanato* (handicrafts) or ecotourism can provide the Yanomami with income while respecting their way of life.

9 Final Considerations

This chapter presented the main problems and threats faced by the Yanomami as well as the challenges faced by the Brazilian authorities and non-governmental organizations in order to deal with the complexities of the case (Campos 2011; Brando et al. 2013). It is undeniable that Indigenous lands and people play a crucial role in nature conservation. They have faced a disastrous impact by capitalist drivers, leading to the irretrievable loss of biodiversity (Cohn et al. 2014; Chua and Fair 2019). Protecting Indigenous lands reduces atmospheric particulates, avoids associated health impacts and costs, and secures the rights of those communities. To deal with the challenges of promoting empowerment, well-being, quality of life, environmental and ecological conservation, and forest stewardship, specific actions are vital. However, a comprehensive and long-term approach that addresses the root causes of the problem, empowers the Yanomami people and fosters international collaboration is essential to ensure their survival and well-being. The initiative's success will depend on the ongoing commitment and collaboration between various stakeholders, including the Brazilian government, indigenous communities, NGOs, and the international community (Lima and d'Hauteserre 2011; ENAP 2021; Scherf and Viana da Silva 2023).

Empowering the Yanomami people through economic activities managed or controlled by them could significantly improve their lives, offer a sustainable path forward, and preserve their culture. However, developing and implementing successful economic activities requires careful planning, feasibility studies, ensuring sustainability, and providing training and support to community members while respecting their traditional values and decision-making processes. By supporting Yanomami-led economic activities, these communities can be empowered to improve their lives, preserve their culture, and secure their future (Scherf and Viana da Silva 2023).

Yanomami people can become protectors of their territories by integrating carbon markets, Indigenous land ownership, and economic benefits into their ethnodevelopment (de Lima et al. 2020; Prist et al. 2023). In 2020, FUNAI launched its Ethnodevelopment Policy to promote food and nutrition security and income generation for indigenous peoples, respecting their forms of decision-making, work organization, and traditional technologies and knowledge associated with sustainable production. To achieve these goals, FUNAI's strategic planning schedules for 2020–2023 were grounded in addressing food insecurity, unequal and exploitative insertion of indigenous peoples in the contemporary market economy, and unsustainable use of lands and natural resources. This policy reflects a promising path towards a more sustainable and equitable future for the Yanomami people and the Amazon rainforest ecosystem (FUNAI 2022).

References

Agência IBGE Notícias (2023) Brasil tem 1,7 milhão de indígenas e mais da metade deles vive na Amazônia Legal, published on 07 August 2023. https://agenciadenoticias.ibge.gov.br/agencia-noticias/2012-agencia-de-noticias/noticias/37565-brasil-tem-1-7-milhao-de-indigenas-e-mais-da-metade-deles-vive-na-Amazon-legal

Andrade R, Moret A, Moreira R, Lopes R (2022) Structures and precautions of the Fundação Nacional do Índio: analysis of national management reports 2008–2018. Int J Innov Educ Res 10:79–94. https://doi.org/10.31686/ijier.vol10.iss9.3888

Andrade C (2023) Massacre de Haximu foi o único caso julgado como genocídio. Especial Yanomami. O Tempo. https://www.otempo.com.br/especiais/yanomami/contexto/massacre-de-haximu-foi-o-unico-caso-julgado-como-genocidio-1.2813911

APIB—Articulação dos Povos Indígenas do Brasil. https://apiboficial.org/apib/?lang=en

Awotwi A, Anornu GK, Quaye-Ballard JA, Annor T, Nti IK, Odai SN, Arhin E, Gyamfi C (2021) Impact of post-reclamation of soil by large-scale, small-scale and illegal mining on water balance components and sediment yield: Pra river basin case study. Soil Tillage Res 211:105026. https://doi.org/10.1016/j.still.2021.105026

Barbosa AC et al (1995) Mercury contamination in the Brazilian Amazon: environmental and occupational aspects. Water Air Soil Pollut 80(1–4):109–121

Basta PC (2023) Gold mining in the Amazon: the origin of the Yanomami health crisis. Amazon. CSP Reports in Public Health, Thematic Section. https://doi.org/10.1590/0102-311XEN111823

Bavaresco A, Menezes M (2014) Entendendo a PNGATI: Política Nacional de Gestão Territorial e Ambiental Indígenas. GIZ/Projeto GATI/FUNAI, Brasília, DF

BBC (2021) Brazil Amazon: illegal miners fire on indigenous group. https://www.bbc.com/news/world-latin-america-57069504. Accessed 19 Dec 2023

Berzas Nevado JJ, Rodríguez Martín-Doimeadios RC, Guzmán Bernardo FJ, Jiménez Moreno M, Herculano AM, do Nascimento JL, Crespo-López ME (2010) Mercury in the Tapajós River basin, Brazilian Amazon: a review. Environ Int 36(6):593–608. https://doi.org/10.1016/j.envint.2010.03.011. Epub 2010 May 16. PMID: 20483161

Bonfil Batalla G (1982) El etnodesarrollo: sus premisas jurídicas, políticas y de organización. In VV. AA. América Latina: Etnocidio y etnodesarrollo. FLACSO

Brabo ES, Santos EO, Jesus IM, Mascarenhas AF, Faial KF (1999) Níveis de mercúrio em peixes consumidos pela comunidade indígena de Sai Cinza na Reserva Munduruku, Município de Jacareacanga, Estado do Pará. Brasil. Cadernos De Saúde Pública 15(2):325–331

Brando PM, Coe MT, DeFries R, Azevedo AA (2013) Mapping the fate of Amazon in forests in the twenty-first century. Philos Trans R Soc Lond B Biological Sci 368(1628):20120152
Brasil (1992) Decreto de 25 de maio de 1992. Homologa a demarcação administrativa da Terra Indígena Yanomami, nos Estados de Roraima e Amazonas (Decree approves the administrative demarcation of the Yanomami indigenous land, in the states of Roraima and Amazonas). Diário Oficial da República Federativa do Brasil, Brasília, Seção 1, p 6457
Brasil (2012) Política Nacional de Gestão Ambiental e Territorial de Terras Indígenas—PNGATI. Decreto Presidencial nº 7.747 de 05 de junho de 2012
Brasil (2018) Decreto nº 9.470, de 14 de agosto de 2018. Promulgates the Minamata Convention on Mercury. Diário Oficial da União, Brasília, Seção 1, p 65
Brasil (2021) Ministry for the health, Gov.Br. Distrito Sanitário Especial Indígena—DSEI. Acesso à Informação. https://www.gov.br/saude/pt-br/composicao/sesai/estrutura/dsei
Brazil (2022) Brazil guarantees protection to the indigenous population. Serviços e Informações do Brasil. Latest News. https://www.gov.br/en/government-of-brazil/latest-news/brazil-guarantees-protection-to-the-indigenous-population#:~:text=Brazil%20has%20already%20invested%20around,also%20installed%20in%20indigenous%20communities
Brazil (2023) Relatório Missão Yanomami [Yanomami expedition report]. Ministério da Saúde, 127 p. https://www.gov.br/saude/pt-br/assuntos/noticias/2023/fevereiro/arquivos/RelatorioYanomamiversao_FINAL_07_02.p
Brasil (2024) Ministério da Saúde. Notícias. Sobre a SESAI. https://www.gov.br/saude/pt-br/composicao/sesai
Bridi C (2023) Brazil declares public health emergency for Yanomami people. Associated Press. https://apnews.com/article/
Campos C (2011) Diversidade socioambiental de Roraima: subsídios para debater o futuro sustentável da região (2nd edn). Instituto Socioambiental
Carta Capital (2023) Salles nega culpa de Bolsonaro por drama yanomami. Carta Capital. https://www.cartacapital.com.br/politica/salles-nega-culpa-de-bolsonaro-por-drama-yanomami-a-responsabilidade-e-da-sociedade/
Chua L, Fair H (2019) Anthropocene. In: Stein F (ed) The open encyclopedia of anthropology (Facsimile of the first edition in The Cambridge Encyclopedia of Anthropology). https://doi.org/10.29164/19anthro
Cohn AS et al (2014) Anthropogenic disturbance in the Amazon basin: assessing land-use changes, fire, and climate change initiatives over the past five decades. Proc Natl Acad Sci 111(40):14343–14348
Da Silva CFA, de Andrade M, Oliveira F, Falcão VA, Martins SFS (2023) The drivers of illegal mining on indigenous lands in the Brazilian Amazon. Extr Ind Soc 16:1–5. https://doi.org/10.1016/j.exis.2023.101354
Davidson EA et al (2012) The Amazon basin in transition. Nature 481:321–328
de Lima IB, d'Hauteserre AM (2011) Community capitals and ecotourism for enhancing Amazon forest livelihoods. Anatolia Int J Tour Hosp Res 22(2):184–203. https://doi.org/10.1080/13032917.2011.597933
de Lima IB (ed) (2014) Etnodesenvolvimento e gestão territorial: comunidades indígenas e quilombolas. CRV, Curitiba
de Lima IB, Kumble PA, de Almeida MG et al (2016) Ecotourism community enterprises and ethnodevelopment: modelling the Kalunga empowerment possibilities in the Brazilian savannah. Braz J Sci Technol 3:1. https://doi.org/10.1186/s40552-016-0013-8
de Lima IB, King VT, Leal Filho W (2020) Amazon in transformation: the need for indigenous empowerment, innovative ethnodevelopment models and ethnic-ecological accountability. In: Leal Filho W et al (eds) Indigenous Amazon, regional development and territorial dynamics. Springer Nature, Switzerland AG, pp 421–432. https://doi.org/10.1007/978-3-030-29153-2_18
ENAP (2021) O papel dos órgãos de assistência aos povos indígenas. Acesso aos direitos fundamentais: uma abordagem da pauta indígena. Enap Escola Nacional de Administração Pública. Módulo 4, Enap, Brasília, DF

ENSP/Fiocruz (2023) O garimpo ilegal e o genocídio yanomami [The illegal mining and the Yanomami genocide]. Mapa de Conflitos. https://mapadeconflitos.ensp.fiocruz.br/conflito/rr-invasao-de-posseiros-e-garimpeiros-em-terra-yanomami/

Fellows M, Paye V, Alencar A, Nicácio M, Castro I, Coelho ME et al (2021) Under-reporting of COVID-19 cases among indigenous peoples in Brazil: a new expression of old inequalities. Front Psych 12:638359

FIOCRUZ. https://portal.fiocruz.br/

Fiocruz (2016) Avaliação da exposição ambiental ao mercúrio proveniente da atividade garimpeira de ouro na terra indígena Yanomami, Roraima, Amazônia, Brasil. Fiocruz, Rio de Janeiro. https://www.socioambiental.org/sites/blog.socioambiental.org/files/diagnostico_contaminacao_mercurio_terra_indigena_yanomami.pdf

FUNAI (2013) Planos de Gestão Territorial e Ambiental de Terras Indígenas: Orientações para Elaboração. FUNAI, Brasília

FUNAI (2021–2023). Planos de Dados Abertos. Relatórios Nacionais de Gestão—Fundação Nacional do Índio. Brasília, FUNAI. Disponível em https://www.gov.br/funai/pt-br/acesso-a-informacao/dados-abertos-1

FUNAI (2022) Política de Etnodesenvolvimento. Carteira de Políticas Públicas do Ministério da Justiça e Segurança Pública—MJSP. Disponível em, https://www.gov.br/funai/pt-br/acesso-a-informacao/gestao-estrategica/instrumentos-de-gestao-estrategica/carteira-de-politicas-publicas/etnodesenvolvimento.pdf

Gibb H, O'Leary KG (2014) Mercury exposure and health impacts among individuals in the artisanal and small-scale gold mining community: a comprehensive review. Environ Health Perspect 122(7):667–672

Gouveia A (2024) Ministério da Saúde planeja incluir medicina indígena no SUS. Questão Indígena. Correio Braziliense. https://www.correiobraziliense.com.br/brasil/2024/01/6791731-ministerio-da-saude-planeja-incluir-medicina-indigena-no-sus.html

Gov.Br—Serviços e Informações do Brasil (2022) Brazil guarantees protection to the indigenous population, published on September 12, 2022. https://www.gov.br/en/government-of-brazil/latest-news/brazil-guarantees-protection-to-the-indigenous-population

Grupo Geo-Yanomami (2023) Cenários de impacto para direcionamento de intervenções na crise da Terra Indígena Yanomami: Uma abordagem das geotecnologias. Nota Técnica do Grupo Geo-Yanomami. Retrieved from Arca Fiocruz. https://www.arca.fiocruz.br/bitstream/handle/icict/59153/Yanomami%20Resultados%20an%C3%A1lise%20preliminar_12_06_AR.pdf?sequence=2&isAllowed=y

Hutukara Yanomami Association (2022) Yanomami under attack: illegal mining on Yanomami indigenous land proposals to combat it. https://acervo.socioambiental.org/sites/default/files/documents/yad00613_en.pdf

Indigenous System of Information (2023) Ministry of Justiça. FUNAI. Brazilian government. http://sii.funai.gov.br/funai_sii/informacoes_indigenas/visao/visao_aldeia_indigenas.wsp?tmp.v=0

ISA—Instituto Socioambiental. https://www.socioambiental.org/

Instituto Socioambiental (2016, May 17) The Yanomami people are contaminated with the mercury from gold-digging. Social environmental stories. Retrieved from https://medium.com/social-environmental-stories/the-yanomami-peopleis-contaminated-with-the-mercury-from-gold-digging-6ef8740bd4d0

Kayet N, Pathak K, Kumar S, Singh CP, Chowdary VM, Chakrabarty A, Sinha N, Shaik I (2021) Deforestation susceptibility assessment and prediction in hilltop mining-affected forest region. J Environ Manage 289:112504. https://doi.org/10.1016/j.jenvman.2021.112504

Le Tourneau F-M (2015) The sustainability challenges of indigenous territories in Brazil's Amazonia. Curr Opin Environ Sustain 14:213–220. https://doi.org/10.1016/j.cosust.2015.07.017

Macedo M, Defries R, Morton D, Stickler C, Galford G, Shimabukuro Y (2012) Decoupling of deforestation and soy production in the southern Amazon during the late 2000s. Proc Natl Acad Sci 109:1341–1346

Mallett A, França ELB, Alves Í, Mills L (2021) Environmental impacts of mining in Brazil and the environmental licensing process: changes needed for changing times? Extr Ind Soc 8(3):100952. https://doi.org/10.1016/j.exis.2021.100952

Mataveli G, Chaves M, Guerrero J, Escobar-Silva EV, Conceição K, de Oliveira G (2022) Mining is a growing threat within indigenous lands of the Brazilian Amazon. Remote Sens 14:1–13. https://doi.org/10.3390/rs14164092

Mataveli G, Chaves M, Guerrero J, Escobar-Silva EV, Conceição K, de Oliveira G (2023) Mining is a growing threat within Indigenous lands of the Brazilian Amazon. Remote Sens 15(11):2809. https://doi.org/10.3390/rs15112809

Ministério do Meio Ambiente (2008) Plano Nacional de Mudança do Clima. Ministério do Meio Ambiente, Brasília

Murta FLG, Marques LLG, Santos APC, Batista TSB, Mendes MO, Silva ED, Neto AVS, Fabiano M, Rodovalho SR, Monteiro WM, Lacerda MVG (2021) Perceptions about malaria among Brazilian gold miners in an Amazon n border area: perspectives for malaria elimination strategies. Malaria J 20(286). https://doi.org/10.1186/s12936-021-03817-9

Nassar A et al (2014) Brazil's pathway to low-emission rural development. AgroIcone and Earth Innovation Institute

Nepstad D et al (2009) The end of deforestation in the Brazilian Amazon. Science 326(5958):1350–1351

Nepstad D et al (2014) Slowing Amazon deforestation through public policy and interventions in beef and soy supply chains. Science 344(6188)

Nicas J (2023) The Amazon's largest isolated tribe is dying. NY Times. https://www.nytimes.com/2023/03/25/world/americas/brazil-amazon-indigenous-tribe.html

Oliveira CA, AYRCA, AMYK (2020) Ecotourism in the Yanomami Land: a proposal for territorial management and indigenous ethnodevelopment. In: Filho WL, King VT, de Lima IB (eds) Indigenous Amazon, regional development and territorial dynamics. Springer, pp 159–179

Pestana IA, de Rezende CE, Almeida R, de Lacerda LD, Bastos WR (2022) Let's talk about mercury contamination in the Amazon (again): the case of the floating gold miners' village on the Madeira River. Extr Ind Soc 11:101122. https://doi.org/10.1016/j.exis.2022.101122

Phillips T (2023) Indigenous children suffer most from illegal miners' Amazon invasion. The Guardian. https://www.theguardian.com/world/2023/mar/17/amazon-mining-indigenous-children-health-lula

Porto Júnior FGR, Alves VB, Oliveira ISC, Santana da Costa J (2021) A view about the importance of the legal Amazon indigenous peoples environmental management. Palmas 7(1):1–17. https://doi.org/10.20873/uft.2447-4266.2021v7n1a3en

Prist PR, Sangermano F, Bailey A, Bugni V, Villalobos-Segura MDC, Pimiento-Quiroga N, Daszak P, Zambrana-Torrelio C (2023) Protecting Brazilian Amazon indigenous lands reduces atmospheric particulates and avoids associated health impacts and costs. Commun Earth Environ 4 (Article 34). https://doi.org/10.1038/s43247-023-00704-w

Quintanilla M, Guzmán León A, Josse C (2022) The Amazon against the clock: a regional assessment on where and how to protect 80% by 2025. AVAAZ, Wild Heritage, One Earth, and Amazon Watch. https://Amazon80x2025.earth/

Rajão R, Azevedo A, Stabile MCC (2012) The impact of government policy on deforestation in the Brazilian Amazon: a counterfactual analysis of the soy moratorium. Public Adm Dev 32(3):229–244

Ramalho Y (2022) Terra Yanomami: números mostram maior devastação causada pelo garimpo em 30 anos [Yanomami land: figures show the greatest devastation in 30 years caused by mining]. G1 Roraima. https://g1.globo.com/rr/roraima/noticia/2022/04/25/terra-yanomami-numeros-mostram-maior-devastacao-causada-pelo-garimpo-em-30-anos.ghtml

Ramos ARA, de Oliveira KA, dos Rodrigues F, S (2020) Mercury-based mining in Yanomami indigenous lands and accountabilities. Revista Ambiente e Sociedade 23:1–20. https://doi.org/10.1590/1809-4422asoc20180326r2vu2020L5AO

Rentier ES, Cammeraat LH (2022) The environmental impacts of river sand mining. Sci Total Environ 838(Part 1):155877. https://doi.org/10.1016/j.scitotenv.2022.155877

Ribeiro D (1996) Os Indígenas e a civilização: a integração das populações indígenas no Brasil moderno [The indigenous and the civilization: the integration of indigenous populations in modern Brazil]. Companhia das Letras (Original work published in 1970), São Paulo

Rodrigues KP (2022) O etnodesenvolvimento como diretriz de governança para povos e comunidades tradicionais: uma revisão normativa. In: Castro AC, Carvalho AC, de Carvalho AV (eds) Meio ambiente e a outra economia dos povos e comunidades tradicionais (Capítulo 6). Científica Digital. https://doi.org/10.37885/22020787

dos Santos AM, Assunção da Silva CF, de Almeida M, Junior P, Rudke AP, Nogueira de Melo S (2021) Deforestation drivers in the Brazilian Amazon: assessing new spatial predictors. J Environ Manage 294:113020. https://doi.org/10.1016/j.jenvman.2021.113020

Scherf EDL, Viana da Silva MV (2023) Brazil's Yanomami health disaster: addressing the public health emergency requires advancing criminal accountability. Front Public Health 11:1166167. https://doi.org/10.3389/fpubh.2023.1166167

Schwartzman S, Alencar A, Zarin H, Santos Souza AP (2010) The carbon budget of the Brazilian Amazon: a reanalysis. J Environ Dev 19(2):274–282

Schwartzman S, Zimmerman B (2005) Conservation alliances with indigenous peoples of the Amazon. Conserv Biol 19(3):721–727. https://doi.org/10.1111/j.1523-1739.2005.00695

Siqueira-Gay J, Sonter LJ, Sánchez LE (2020) Exploring potential impacts of mining on forest loss and fragmentation within a biodiverse region of Brazil's northeastern Amazon. Resour Policy 67:101662. https://doi.org/10.1016/j.resourpol.2020.101662

Shimabakuro YE, dos Santos JR, Formaggio AR, Duarte V, Rudorff BFT (2012) Global forest monitoring from earth observation, pp 167–183. https://doi.org/10.1201/b13040-10

Silva-Junior CHL et al (2013) Persistent collapse of biomass in Amazon n forest edges following deforestation leads to unaccounted carbon losses. Sci Rep

da Silva VV, da Silva RGC (2022) Amazon, frontier and protected areas: dialectic between economic expansion and nature conservation. Ambiente Sociedade 25(1):1–21. https://doi.org/10.1590/1809-4422asoc20200224r1vu2022L3OA

Sonter LJ, Herrera D, Barrett DJ, Galford GL, Moran CJ, Soares-Filho BS (2017) Mining drives extensive deforestation in the Brazilian Amazon. Nat Commun 8:1013. https://doi.org/10.1038/s41467-017-00557-w

Stavenhagen R (1985) Etnodesenvolvimento: uma dimensão ignorada no pensamento desenvolvimentista. Anuário Antropológico/84. Rio de Janeiro

Stickler CM, Nepstad DC, Azevedo AA, McGrath DG (2013) Defending public interests in private lands: compliance, costs and potential environmental consequences of the Brazilian forest code in Mato Grosso. Philos Trans 368:20120160. https://doi.org/10.1098/rstb.2012.0160

Survival International. https://www.survivalinternational.org/tribes

Tarrega MCVB (2022) Fair trade—possibilidades para o Etno-Desenvolvimento Quilombola. Consinter. https://doi.org/10.19135/revista.consinter.00015.22

Vega CM, Orellana JDY, Oliveira MW, Hacon SS, Basta PC (2018) Human mercury exposure in yanomami indigenous villages from the Brazilian Amazon. Int J Environ Res Public Health 15(6):1051. https://doi.org/10.3390/ijerph15061051

Wheatley B, Wheatley MA (2000) Methylmercury and the health of indigenous peoples: a risk management challenge for physical and social sciences and for public health policy. Sci Total Environ 259(1–3):23–29

Correction to: Integrated Diagnosis of Biodiversity, Water Quality, and Aquatic Resources as an Insight for Conservation Strategies and Sustainability of Artisanal Fishing on the Tapajós River

Edenise Garcia, André L. C. Canto, Charles H. Faria Junior, Dávia M. Talgatti, Diego M. Zacardi, Frank R. V. Ribeiro, Heloísa N. M. Meneses, Iracenir A. Santos, Priscila S. Miorando, and Tony M. P. Braga

Correction to:
Chapter 21: W. Leal Filho et al. (eds.), *Amazon 2030—Sustainability Issues in the World's Largest Rainforest Region*, World Sustainability Series, https://doi.org/10.1007/978-3-031-81465-5_21

In the original version of the book, the following belated corrections have been incorporated: The author name "Diogo M. Zacardi" has been changed to "Diego M. Zacardi" in the Frontmatter and in Chapter 21. The book has been updated with the changes.

The updated version of this chapter can be found at https://doi.org/10.1007/978-3-031-81465-5_21

W. Leal Filho et al. (eds.), *Amazon 2030—Sustainability Issues in the World's Largest Rainforest Region*, World Sustainability Series,
https://doi.org/10.1007/978-3-031-81465-5_35

MIX
Papier aus verantwortungsvollen Quellen
Paper from responsible sources
FSC® C105338
FSC
www.fsc.org

If you have any concerns about our products,
you can contact us on
ProductSafety@springernature.com

In case Publisher is established outside the EU,
the EU authorized representative is:
Springer Nature Customer Service Center GmbH
Europaplatz 3, 69115 Heidelberg, Germany

Printed by Libri Plureos GmbH
in Hamburg, Germany